Handbook of Spectral Lines in Diamond

Bernhard Dischler

Handbook of Spectral Lines in Diamond

Volume 1: Tables and Interpretations

With 203 Tables

Springer

Bernhard Dischler
Sonnhalde 3
79104 Freiburg
Germany
Retired from: Fraunhofer Institut (IAF) Freiburg, Germany
bernhard.dischler@t-online.de

ISBN 978-3-642-22214-6 e-ISBN 978-3-642-22215-3
DOI 10.1007/978-3-642-22215-3
Springer Heidelberg Dordrecht London New York

Library of Congress Control Number: 2012934366

Printed on acid-free paper

Springer is part of Springer Science+Business Media (www.springer.com)

Preface

The number of observed spectral lines in diamond is steadily increasing. For example, the remarkable review by J. Walker in 1979 described 30 centers with some 100 lines [Wal79]. The compilation by J. E. Field in 1992 had 118 entries [Fie92]. A. M. Zaitsev in his valuable data collections, had 465 entries in his book chapter of 1998 [Zai98] and 838 entries in his book of 2001 [Zai01]. This book has more than 2,000 entries from some 300 centers. The main entries are presented in 23 tables (see Chaps. 2–7), according to the type of sample, the method of observation, etc. (see Sect. 1.4). Specific tables follow in Chaps. 8–13.

Of special interest for the present author is (1) the interpretation of spectra and (2) the assignment of structures to the optical centers.

In the field of **interpretation**, it was found that in 113 cases line groups arise from resolved donor–acceptor pair (DAP) transitions (see Chap. 10). Only two of them have been published before [Dis94a, Dis94b, Ste99a, Ste99b], i.e., 111 DAPs of this book are unpublished [Dis03]. Unresolved DAP transitions were proposed in 1965 to explain the band-A luminescence [Dea65a]. With the present knowledge, a final proof for this hypothesis is provided (see Table 10.2.2).

The **structure assignment** of optical centers is established for some 15 cases (e.g., $A = N_2^{\circ}, B = V_1N_4^{\circ}, C = N_1^{\circ}, NV = V_1N_1, N = V_1N_3^{\circ}, GR1 = V_1^{\circ}$). However, for the large number of some 200 proposed structures (preceded by *), confirmation is desirable (e.g., $H3 = *V_1N_2^{\circ}, TH5 = *V_2^{\circ}, 5RL = *(C_2)_iN_1^{\circ}$).

In contrast to previous reviews, where up to three possible structures are cited, only the most probable structure is presented here. This selection is based on the available information, like (1) isotope shifts, (2) frequencies of local vibrational modes (LVM), or (3) of quasilocal vibrational modes (QLVM), (4) involvement of intrinsic defects (vacancies or self-interstitials) as evidenced by the connection to a special type of DAP transitions (the sideband DAP, see Chap. 10), and (5) relationships of the value of the DAP dielectric factor with the number of electrons involved (see Sect. 10.4) or with the charge state of the defect (see Sect. 10.5).

Certainly, the large amount of data (especially the new structure proposals) require a critical examination. Each reader is urgently invited to inform the author about errors, improvements, and new results.

The author regrets that no figures are included in this book. It was decided to collect the figures and diagrams in a separate book. Fortunately, the experienced diamond researcher Prof. Alexander M. Zaitsev has agreed to write this book (together with the present author):

Handbook of Spectral Lines in Diamond, Volume 2, Figures and Diagrams by A. M. Zaitsev and B. Dischler, Springer-Verlag, to be published.

This second book provides a good opportunity to update the present data on spectral lines in diamond.

In the meantime, the reader will find figures for most of the lines and bands in two previous books [Zai98, Zai01]. The respective figure numbers are indicated in an extra column in the tables.

Acknowledgments

First of all, I am grateful to Prof. Peter Koidl, who established and headed the diamond research group at the Fraunhofer Institut für Angewandte Festkörperphysik (IAF, Institute for Applied Solid State Physics) in Freiburg. The present author has been concerned with the spectroscopy of diamond like carbon (DLC) since 1982 and with the optical spectra of diamond since 1992.

I want to thank the members of this diamond group for the fruitful cooperation: Gernot Brandt, Achim Bubenzer, Roland Locher, Karin Maier, Wolfgang Müller-Sebert, Manfred Ramsteiner, Wolfram Rothemund, Ram Ekwal Sah, Harald Schneider, Joachim Wagner, and Christoph Wild.

Special thanks are expressed to Dr. habil. Claus Ascheron, Executive Editor Physics at Springer-Verlag Heidelberg, for continued advice and support, while this book was written.

Invaluable support came from my wife Heidrun, who made a substantial contribution to this book by her persistent encouragement, a critical reading of the manuscript, valuable discussions, and new ideas.

This book is dedicated to the many diamond researchers worldwide, who contributed to the impressive amount of data, which is now available. Instrumental for the present book were the exhaustive data collections by Alexander M. Zaitsev, both in a review article [Zai98] and in a book [Zai01]. Of great help were the reviews of John Walker [Wal79] and of Alan T. Collins [Col97].

Freiburg, Germany *Bernhard Dischler*

Contents

Chapter 1
Introduction

1.1 Diamond in History and Research

The word "diamond" (or "adamas," "adamant") is of Greek origin with the meaning "indomitable" or "invincible." Diamond is a material with many outstanding properties (a) extreme hardness (ten on the Mohs scale), (b) chemical inertness, (c) high refractive index and dispersion ($n = 2.463$ at 400 nm and $n = 2.405$ at 700 nm), (d) large optical transparency range, and (e) high thermal conductivity (better than copper).

A large number of optical centers can be observed in diamond, due to a combination of (d) the extended optical transparency and (a) the extreme hardness. Diamond is transparent from the bandgap absorption edge in the ultraviolet (at 5.49 eV) through the visible and infrared spectral range to the microwave region. Because of the hardness, defect migration (e.g., vacancies to the surface) requires high activation energies, and most defects remain immobile up to temperatures of 420–2, 500 °C (see Sects. 8.3.1, 9.1.6, and 9.5.6).

The number of lines (ca. 2,000) is much larger than the corresponding number of centers (ca. 300), because most zero phonon lines (ZPL) have typically 1–10 phonon or vibrational sidebands (see Chap. 11). The centers with resolved donor–acceptor pair transitions form groups of 3–23 lines (see Chap. 10).

To give a first impression, 45 typical defects are listed in Tables 1.1.1 and 1.1.2. These Tables are given in matrix form, because the majority of centers consist of associates. For some ten centers in Tables 1.1.1 and 1.1.2 the structure is generally accepted, while for the remaining 35 centers (marked by a preceding *) the structure is proposed in the literature or in this book.

In five sections, the principal types of diamonds are described: Natural diamond (Sect. 1.2.1, including Table 1.2), high pressure synthetic (HPHT) diamond (Sect. 1.2.2), low pressure synthetic (CVD) diamond (Sect. 1.2.3), modified diamond (Sect. 1.2.4), and diamond-related materials (Sect. 1.2.5).

The introduction ends with a short description of the optical techniques (Sect. 1.3) and explanations on the organization of the tables (Sect. 1.4).

B. Dischler, *Handbook of Spectral Lines in Diamond*,

DOI 10.1007/978-3-642-22215-3_1,

Table 1.1.1 Fundamental defects in diamond[a,b,c] (mostly with vacancies, see Chap. 8)

Defect-composition	–	+1N	+2N	+3N	+4N	+1B	+1H
–	–	**N_1° = C center** 0.105–0.167, 4.059	**N_2° = A center** 0.060–0.159, 3.928, 4.567	–	–	**B_1° = boron acc.** 0.304–0.372	–
V_1^-	*V_1^- = R10, ND1 3.150, DAP93	***$V_1N_1^-$ = 638 nm** 1.943, 4.328	***$V_1N_2^-$ = H2** 1.256, DAP77/78	***$V_1N_3^-$** 2.526, DAP76	***$V_1N_4^-$ = N9** 5.252, DAP63/64	***$V_1B_1^-$** 3.684, DAP67	–
V_1°	*V_1 = GR1* 1.673, DAP33/34	**$V_1N_1^\circ$ = NV** 2.154, DAP46/47	***$V_1N_2^\circ$ = H3** 2.464, 3.361 DAP 50/51,52	**$V_1N_3^\circ$ = N3** 2.985 DAP 23/24	**$V_1N_4^\circ$ = B cent.** 0.093–0.165 4.191, DAP82	***$V_1B_1^\circ$** 2.395, DAP74 3.120, DAP73	***$(V_1H_1)^\circ$** 0.1742, 0.3852, DAP41
V_1^+	*V_1^+ = GR 2–8 2.781, DAP32	–	*$V_1N_2^+$ = S1(b) 2.463, DAP43	–	–	–	–

V_2°	* *V_2° = TH5* 2.543, DAP15	* **$V_2N_1^{\circ}$** 2.329	–	* *$V_2N_3^{\circ}$* 2.916	* *$N_3V_2N_1$ = H4(b)* 2.498, DAP37/38	–	–
$V_1Ni_1^{+}$	* **$V_1Ni_1^{+}$ = 1.40 eV** 1.404, DAP25/26	–	–	–	–	–	–
(V_2Ni_1)	–	–	+ **N_{2+0+0} = S3** 2.496–2.800 sta. DAP58	+ **N_{2+0+1} = S2** 2.515–3.341 sta. DAP57	–	–	–
* $(V_3Ni_2)^{\circ}$	* **$(V_3Ni_2)^{\circ}$** 2.562, DAP106	–	–	–	–	–	–
* $(V_3Si_2)^{\circ}$	* **$(V_3Si_2)^{\circ}$** 2.052, DAP65	–	–	–	–	–	–

[a]Energies in eV
[b]**Bold = natural**, *italics = irradiated*
[c]DAP donor–acceptor pair transitions (see Chap. 10)

Table 1.1.2 Fundamental defects in diamond[a,b,c] (with split interstitials, see Sect. 8.2)

Defect-composition	–	+1N	+2N	+1–3B
$(C_2)_i^-$	$^*(C_2)_i^- = TR12$ ***2.638, DAP35/36***	$^*(C_2)N_1^- = 3.19$ eV, **3.188, DAP72**	$^*(C_2)_iN_2^-$ ***2.367, DAP88/89***	–
$(C_2)_i^\circ$	$^*(C_2)_i^\circ = R2$ ***1.685, DAP71***	$^*(C_2)_iN_1^\circ = 5RL$ ***4.582, DAP79/80***	$^*(C_2)_iN_2^\circ$ ***1.979***	$^*(C_2)_iB_1^\circ =$ *2BD(F)*, 4.777, DAP97 $^*(C_2)_iB_2^\circ =$ *2BD(G)*, 4.698, $^*(C_2)_iB_3^\circ =$ *2BD(C)*, 4.803
$(C_2)_i^+$	$^*(C_2)_i^+ = 3H$ ***2.462, DAP86***	$^*(C_2)_iN_1^+$, *4.676*	$^*(C_2)_iN_2^+$ ***2.535, DAP90***	–
$(C_2)_i^{2+}$	–	–	$^*(C_2)_iN_2^{2+}$ ***2.322***	–
$(C_{2n})_i^\circ$	**Platelets = D center** ***0.041–0.177 1.528, DAP96***	–	–	–
$(N_1C_1)_i^-$	$^*(N_1C_1)_i^- = R11$ ***3.988, DAP98***	–	–	$^*(N_1C_1)_iB_1^-$ ***2.992, DAP101***
$(N_1C_1)_i^\circ$	–	–	–	$^*(N_1C_1)_iB_1^\circ$ ***2.792, DAP99***
$(N_1C_1)_i^+$	$^*(N_1C_1)_i^+$ *0.212*, *2.807, DAP98*	–	–	$^*(N_1C_1)_iB_1^+$ ***3.092, DAP100***

[a]Energies in eV
[b]**Bold = natural**, *italics = irradiated*,
unmarked* = CVD diamond, as-grown or irradiated. (* = normal type)
[c]DAP donor–acceptor pair transitions (see Chap. 10)

Table 1.2 Classification scheme for diamond types (see Sect. 1.2.1)

Defect/ property	Types				
	IaA	IaB	Ib	IIa	IIb
A-center $=N_2{}^\circ$	**Dominant**	Weak	Weak	Absent	Absent or weak
B-center $=V_1N_4{}^\circ$	Weak	**Dominant**	Weak	Absent	Absent or weak
C-center $=N_1{}^\circ$	Weak	Weak	**Dominant**	Absent	Absent or weak
Defining defect	A-center	B-center	C-center	**Absent**	Boron-acceptor
Boron content	Low	Low	Low	Very low	**Dominant**
Typical natural hydrogen content	500 ppm	200 ppm	70 ppm	$<10^{18}$ cm^{-3}	$<10^{16}$ cm^{-3}
Electrical resistance (Ω cm)	$>10^{15}$	$>10^{15}$	$>10^{15}$	$>10^{16}$	Semiconducting

1.2 Types of Diamond Samples

1.2.1 Natural Diamond

In its long history (from 400 BC), natural diamond gem stones (clear or colored) came first from India and later from South America, South Africa, and Russia [Mey98]. In a recent study [Shi02], it was shown that natural diamonds are very old. There are three groups with an age of 3.3, 2.9, and 1.9 billion (10^9) years. Originally, they grew at a depth of 150–300 km under the surface from carbon-containing compounds like CO_2 or $CaCO_3$. From there the diamonds have been transported to regions near the surface by volcanic activity. They are either still enclosed in the volcanic rocks kimberlite or lamproite (primary sources in diamond mines), or have been liberated by erosion and transported by rivers toward the sea (secondary sources, e.g., in Namibia).

The value of natural diamond gemstones is closely related to the four "C"s: (1) carat, (2) cut, (3) color, and (4) clarity.

1. The weight is given in the traditional unit "carat." One carat = 100 points = 200 mg, (the weight of a dried carobtree seed).
2. A good cut (e.g., "brilliant") produces the "fire" from reflected light, making use of the high refractive index and dispersion: $n = 2.463$ at 400 nm (3.10 eV) and $n = 2.405$ at 700 nm (1.77 eV) [Zai01].
3. The diamonds can be absolutely colorless ("River"), white ("Wesselton"), or colored. Common colors are yellow, brown, or gray. More seldom colors are red, blue, green, pink, or violet (see Chap. 6 in [Zai01]).
4. Most natural diamonds contain inclusions. Their number and dimension determines the value, i.e., their absence in clear diamonds makes a gemstone very valuable.

In 1934 Robertson et al. introduced a classification scheme for natural diamonds [Rob34]: Type I diamonds (98% abundance) are transparent at > 300 nm(<4.13 eV), and type II diamonds (2% abundance) are transparent at > 230 nm (< 5.39 eV). After the detection of boron, type II diamonds were subdivided into IIa (< 10 ppm boron) and IIb (dominant boron, semiconducting) [Cus52]. Finally, characteristic infrared absorption allowed discrimination between three different nitrogen centers, and type I was subdivided: A centers (type IaA), B centers (type IaB), and C centers (type Ib) [Dye65b] (see Table 1.2).

Many lines are observed in natural diamonds, see Tables 2.1.1.1–2.4.3.

1.2.2 High Pressure Synthetic Diamond (HPHT)

The successful synthesis of diamond in a high pressure and high temperature apparatus was reported in 1955 [Bun55]. Graphite is transformed into diamond in the presence of transition metals (Fe, Co, Ni, Mn, or Pt), which act as a solvent–catalyst. Typical conditions are a pressure of 6–7 GPa and a temperature of 1,500–1,700 °C [Set94]. The standard HPHT diamonds are of type Ib, and the isolated substitutional nitrogen (C center) causes a yellow color. By the use of a nitrogen getter (e.g., Al or Ti), the concentration of C centers can be reduced [Kan98]. Lines from HPHT samples are given in Tables 3.1.1.1–3.4.3.

1.2.3 Low Pressure Synthetic Diamond (CVD)

In 1981 and 1982 it was shown that low temperature synthetic (CVD) diamonds can be deposited with practical growth rates (up to 1 μm/h) [Mat81, Kam82]. With the standard CVD (chemical vapor deposition) method diamond films are grown on a heated substrate [Spe94, Dis98, Pre98]. In a typical MW-CVD apparatus the feed gas (3% CH_4, 97% H_2) is decomposed in a 2.45-GHz *microwave* plasma (200 mbar pressure), and the (polycrystalline) diamond film is deposited on a silicon substrate (at 850 °C). Instead of the MW plasma a *hot filament* (2,000 °C) can be used for the decomposition of the feed gas (HF-CVD). Alternate methods use a *combustion flame* or a *plasma jet* (see e.g., [Dis98]).

Typical for CVD diamond films is the incorporation of silicon and sometimes boron (from the substrate), nickel (from the steel vessel), and also Ta or W (from the hot filament), see Tables 4.1.1.1–4.4.3.

It has been shown that the prominent hydrogen lines of CVD diamond films arise from hydrogen in an amorphous carbon surrounding (grain boundaries), and are essentially identical to the hydrogen lines in diamond-like carbon (DLC), see Tables 9.3.2–9.3.5 [Dis93]. Therefore, these CVD hydrogen lines are classified as diamond related and listed in Tables 6.1.1.1–6.3.2.

1.2.4 Modified Diamond

When diamond is irradiated with high energy electrons, fast neutrons, or X-rays, the typical intrinsic defects (vacancies and $\langle 1\ 0\ 0 \rangle$ split self-interstitials) are created. At anneal temperatures of $T > 420\,°C$ for interstitials and $T > 500\,°C$ for vacancies, migration of the intrinsic defects starts and the typical associates with impurities are formed (see Tables 5.1.1.1–5.4.4 and Chap. 8).

1.2.5 Diamond-Related Materials

There are several carbon compounds, which differ in their structure from diamond, but have similar properties (see Tables 6.1.1.1–6.3.2). Five of these are natural:

1. *Lonsdalite* (LON), a carbon mineral with a hexagonal structure
2. *Carbonado* (CAR)
3. *Ballas* (BAL), a naturally occurring polycrystalline diamond
4. *Natural brown/gray diamond* (NBD/NGD), often with high H content
5. *Natural diamond coat* (NDC)

The remaining two materials are synthetic films:

6. *Diamond-like carbon (DLC)*
7. *Polymer-like amorphous carbon (PAC)*

Typical films of DLC are obtained by high frequency (2.3 MHz) plasma deposition from a C_6H_6 feed gas (30-μ bar pressure). In the negative self-bias voltage of 1 kV, the ionized fragments arrive at the substrate with high kinetic energy, forming a compact and hard film. The structure is amorphous (also named a-C:H) with a high concentration of 13–38 at. % hydrogen, which is bonded to carbon. The infrared analysis yields a typical carbon hybridization of 68% sp^3, 30% sp^2, and 2% sp^1 (see Sect. 9.3).

The low cost and flexible deposition methods of DLC films opened a wide field of applications, making use of the high hardness, low friction, and exceptional smoothness [Gri99]. More recently, high power deposition methods allowed the preparation of "superhard" hydrogen-free DLC (named "tetrahedral" carbon, taC) [Gri99].

PAC films are obtained, if the kinetic energy is lowered. These films are of scientific interest (see Tables 9.3.2 and 9.3.3), but have no practical applications.

In the next section, the optical techniques (Sect. 1.3) and the organization (Sect. 1.4) of the tables in Chaps. 2–7 are explained.

In Chap. 8, the intrinsic defects (lattice vacancy or split self-interstitial) and their associates are treated.

Impurity defects (with sections on N, B, H, Si, Ni, Co, and others) are discussed in Chap. 9.

Very important new information on donor–acceptor pair (DAP) transitions is contained in Chap. 10.

The Chaps. 11–14 deal with vibrational frequencies (Chap. 11), modification of diamond by irradiation and heat (Chap. 12), isotopic line shifts (Chap. 13), and spectroscopic discrimination between natural and nonnatural diamond (Chap. 14).

The book ends with conclusions and outlook (Chap. 15) and an Appendix.

1.3 Optical Techniques

1.3.1 Absorption

The absorption coefficient α (in cm^{-1}) is defined by (1.3a)

$$I(E_1) = I_o(E_1)e^{-\alpha x}, \tag{1.3a}$$

here, $I(E_1)$ is the measured intensity of monochromatic light at energy E_1, $I_o(E_1)$ is the initial intensity (corrected for reflection loss), and x is the path length (in cm). With a scan of E_1, an absorption spectrum is obtained.

1.3.2 Photoluminescence Excitation

The definition is given by (1.3b), which is similar to (1.3a)

$$I(E_1) = I_o(E_2)e^{-\alpha x}. \tag{1.3b}$$

However, now E_1 is fixed (monitoring energy of a luminescence line), while E_2 is scanned ($E_1 < E_2$). This photo luminescence excitation (PLE) spectrum is (in general) similar to an absorption spectrum, but it can reveal relationships between different lines of the same defect center.

1.3.3 Luminescence

A luminescence spectrum is obtained, when the optical emission of a sample is monochromatically scanned. The standard luminescence is excited (a) by light (photoluminescence, PL), (b) by an electron beam (cathodoluminescence, CL), or (c) by X-rays (XL). Typical light sources for photoluminescence are mercury lamps, argon, or krypton lasers. Cathodoluminescence is a "near surface" technique. The penetration depth is 9–18 μm for a typical 50-kV electron beam [Col92b]. In X-ray luminescence, the high energy can modify the centers.

In contrast to the *quantitative* absorption, the luminescence intensity is only *qualitatively* related to the concentration of the respective defect center.

In time resolved luminescence, the detection starts at a gated time delay after the excitation pulse. This allows determination of radiative decay times. A tabulation of decay times for 21 centers (ranging from 17 ns to 18 ms) is given in [Per94c].

1.3.4 Raman Scattering

This technique is used to investigate the crystal perfection, which influences the position and linewidth of the principal Raman line at 165 meV ($1,332\,cm^{-1}$). Also, nondiamond inclusions (graphitic or grain boundaries) can be detected. A review (with a listing of 83 lines or bands) is given in Chap. 4 of [Zai01], see also [Zai00b].

1.3.5 Photoconductivity

If the photocurrent is measured as a function of monochromatic illumination, a photoconductivity spectrum is obtained. A listing of some 60 photoconductivity features is given in Chap. 10 of [Zai01]. In photo-Hall measurements, the type of released carriers (electrons for n-type, holes for p-type) can be determined [Col79a, Col97].

1.4 Organization of the Tables in Chaps. 2–7

The great French scientist René Descartes (1596–1650) proposed a four-step method for the presentation of data "(a) split up the data, (b) put the data in order, (c) establish surveys of the data, and (d) make sure the data are reliable and intelligible." After several hundred years, this is still a good advice.

A basic idea of how to split up the 2,000 spectral lines in diamond was found for the present book, using two criteria:

1. The type of sample
2. The method of observation

The lines are presented in 150 tables (see Chaps. 2–7).

The five types of samples are:

N = natural diamond (Sect. 1.2.1 and Tables 2.1.1.1–2.4.3)
H = HPHT diamond (Sect. 1.2.2 and Tables 3.1.1.1–3.4.3)

L = low pressure (CVD) synthetic diamond (Sect. 1.2.3 and Tables 4.1.1.1–4.4.3)
M = modified diamond (Sect. 1.2.4 and Tables 5.1.1.1–5.4.4)
R = diamond-related materials (Sect. 1.2.5 and Tables 6.1.1.1–6.3.2)

The three methods of observation are:

A = absorption (see Sect. 1.3.1)
E = excitation (PLE, see Sect. 1.3.2)
L = luminescence (see Sect. 1.3.3)
B = the extra criterion for broad bands

For clarity and for referencing purposes, a six-digit label for individual lines or bands is introduced, consisting of two letters and a four-digit number (e.g., NA2154 = NV):

The **first** letter identifies the type of diamond sample (**N, H, L, M, R**).
The **second** letter stands for the method of observation (**A, E, L, B**).
The four-digit number indicates the energy in meV units.

The labels are very important for the identification of line groups belonging to the same center. Group members can be easily recognized from the attached extensions:

The letters a, b, c, etc., identify different transitions of the same center. In the case of DAP transitions, the letters identify different shells (see Chap. 10). The extension $-z$ is used for the ZPL of sideband DAPs.
In the infrared spectral range, the letters a′, b′, etc., identify different vibrations. Phonon sidebands are denoted by -s1, -s2, etc., for increasing energies. Only important sidebands have their own entry.
Lines from isotopes (see Tables 7.1–7.5 and Chap. 13) are characterized by a prefix (i2 to i8- for ^{2}H, i13- for ^{13}C, i15- for ^{15}N, i57- etc. for Si_2, i60- etc. for Ni_1, and i118- etc. for Ni_2).

In order to present the relevant data in a concise form, a nine-column format was chosen for the tables. The line label in column 1 is followed by three columns (2, 3, 4) giving the energy (in eV or meV), the frequency (in cm^{-1} or 10^{3} cm^{-1}), and the wavelength (in nm or μm), with four significant digits. These three columns take into account, that in the literature, the optical spectra and line positions are given as a function of energy, frequency, or wavelength.

For the conversion of energy into frequency, the relation $1\,\text{eV} = 8,066\,\text{cm}^{-1}$ is used.

The tables in Sect. 2.1–6.3 are subdivided according to the spectral range, as given in Table 1.3. Visible colors (in luminescence) are not observed from single lines, but only from broad bands which are listed under the energy of the band maximum. The considerable spectral overlap must be taken into account, therefore some bands named "blue" have their maximum in the ultraviolet.

Many lines have a name(see Tables A.1.1–A.1.3) and this is given in column 5. From the literature, the impurity or the structure of several optical centers is

Table 1.3 Spectral regions (infrared, visible, ultraviolet) used in Tables 2.1.1.1–6.3.2

Spectral region	Energy (eV)	Wavelength (nm)
Far infrared	0.04–0.18	30,000–7,000
Mid infrared	0.18–1.24	7,000–1,000
Near infrared	1.24–1.77	1,000–700
Purple	1.77–1.88	700–660
Red	1.88–2.03	660–610
Orange	2.03–2.10	610–590
Yellow	2.10–2.18	590–570
Green	2.18–2.43	570–510
Blue	2.43–2.58	510–480
Ultramarine	2.58–2.75	480–450
Violet	2.75–3.10	450–400
Near ultraviolet (UV-A)	3.10–3.94	400–315
Mid ultraviolet (UV-B)	3.94–4.43	315–280
Far ultraviolet (UV-C)	4.43–5.90 (−12.40)	280–210 (−100)

known or suspected, as listed in column 6, including proposals by the present author. Additional information on the respective line is collected in the comment column 7. Because no spectra are shown in the present book, but valuable collections of spectra are contained in the reviews of Zaitsev [Zai98, Zai01], these figures are referenced in column 8 by 3–6.x for Chaps. 3–6 in [Zai01], and by 7.x for Chap. 7 in [Zai98]. The last column (9) contains a limited number of references, especially for the cited figures. Additional information on the respective lines can be found in the review literature (e.g., [Bie67, Wal79, Fie92, Col97, Zai98] and especially in [Zai01] with almost 1,300 references).

Some statistical numbers can be given: In Chaps. 2–7, more than 2,500 entries are presented in 151 tables. In Chaps. 2–5, ca. 2,000 entries from ca. 400 centers are given. The corresponding structures are established for 40 centers (published before), are almost certain for 120 centers (mostly proposed for the first time), and are incomplete or unknown for 240 centers. It has to be taken into account that multiple entries for a single center can often occur:

1. Centers with line groups, e.g., from DAP transitions (DAP, see Chap. 10).
2. Centers with parallel entries, e.g., the $N3 = V_1N_3{}^{\circ}(c)$ center at 2.985 eV with seven entries: NA2985, NE2985, NL2985, HA2985, HL2985, LA2985, and LL2985.

Chapter 2
Spectral Lines in Natural Diamond

In this chapter ca. 600 lines (and bands), which are observed in natural diamond (see Sect. 1.2.1), are presented in 32 tables. The entries are assigned to 120 centers. The corresponding defect structure is established for 18 centers, is almost certain for 20 centers, and is incomplete or unknown for 82 centers.

Chapter 2 is subdivided into *absorption lines* (Sect. 2.1: 230 **NA** lines in 11 tables (see Tables 2.1.1.1–2.1.8)), into *photoluminescence excitation (PLE) lines* (Sect. 2.2: 50 **NE** lines in three tables (see Tables 2.2.1–2.2.3)), into *luminescence lines* (Sect. 2.3: 280 **NL** lines in 15 tables (see Tables 2.3.1–2.3.11.3)), and into *broad bands* (Sect. 2.4: 40 **NB** bands in three tables (see Tables 2.4.1–2.4.3)).

The progress achieved within the last 30 years can be estimated from the following comparison. In 1979 the comprehensive review "Optical Absorption and Luminescence in Diamond" was published [Wal79]. For natural diamond 20 centers are described, but only four corresponding structures are given (A center = $N_2{}^{\circ}$, C center = $N_1{}^{\circ}$, boron acceptor = $B_1{}^{\circ}$, and N3 center = $V_1N_3{}^{\circ}$). The present knowledge is an increase of 500%.

B. Dischler, *Handbook of Spectral Lines in Diamond*,
DOI 10.1007/978-3-642-22215-3_2, © Springer-Verlag Berlin Heidelberg 2012

2.1 Absorption lines (NA)

Table 2.1.1.1 Natural diamond: absorption lines, far infrared, Part 1 (0.04–0.144 eV)

Line-label	Energy (meV)	Frequ. (cm^{-1})	Wavel. (μm)	Name	Impur./defect	Comment	Figs. **3–6.** [Zai01]; Fig. **7.** [Zai98]	References
NA0041a′	40.91	330.0	30.30	D-cent.(a′)	$*(C_{2n})_i{}^\circ$	Six IR lines a′–f′(177.3 meV), also named platelets or B′, luminescence line *D(a) at 1.528 eV, see Table 8.2.2	...	[Bok86, Jon92, Zai01]
NA0058	57.65	465.0	21.51	...	...	...	**3**.7	[Fer96, Zai01]
NA0059	59.02	476.1	21.01	...	...	Observed at $T > 450\,°C$	...	[Bie67, Zai01]
NA0060a′	60.00	484.0	20.66	A-cent.(a′)	$N_2{}^\circ$	Five IR lines a′–e′(158.9 meV), two UV lines A(a, b) at 3.928, 4.470 eV, see Table 9.1.1.3	**3**.29, **7**.31	[Ang65, Bok86, Zai01]
NA0094a′	93.50	754.0	13.26	B-cent.(a′)	$V_1N_4{}^\circ$	Seven IR lines a′–g′(165.1 meV); a′ var.(93.5–96.7 meV); three UV lines: *B(a1–a3) at 4.184–4.197 eV, see Table 8.2.4	**3**.7	[Fer96, Zai01]
NA0105a′	105.4	850	11.76	C-cent.(a′)	$N_1{}^\circ$	Seven IR lines a′–g′(166.6 meV), three UV lines: *C(a, b) indirect, *C(c) at 4.059 eV, see Table 9.1.1.1	...	[Zai01]
NA0118a′	117.8	950.0	10.58	E(a′) = X(a′)	$N_1{}^+$	Seven IR lines a′–g′(165.1 meV), see Table 9.1.1.2	**3**.20	[Woo83, Law98]
NA0121a′	121.0	976.0	10.25	F-cent.(a′)	$V_1N_4(C_2)_i{}^\circ$	Nine IR lines a′–i′(196.0 eV), three lines *F(a–c) at 2.145, 2.721, 4.567 eV, see Table 8.1.3.6	**3**.24; **7**.26	[Woo83, Cla84b, Dav99a]

NA0094b′	125.2	1,010	9.902	B-cent.(b′)	$V_1N_4{}^{\circ}$	…	**3**.27, **3**.29, **7**.24	[Ang65, Cla84b, Law98]
NA0121b′	126.7	1,022	9.785	F-cent.(b′)	$V_1N_4(C_2)_i{}^{\circ}$	…	**3**.14, **3**.20, **7**.26	[Cla84b, Col82b, Woo83]
NA0105b′	129.6	1,045	9.566	C-cent.(b′)	$N_1{}^{\circ}$	…	………	[Zai01]
NA0118b′	129.7	1,046	9.560	E(b′) = X(b′)	$^{*}N_1{}^{+}$	…	**3**.20, **3**.30, **7**.26	[Woo83, Cla84b, Zai01]
NA0094c′	134.5	1,085	9.309	B-cent.(c′)	$V_1N_4{}^{\circ}$	…	**3**.27; **7**.24, **7**.169	[Cla84b, Law98, Zai01]
NA0060b′	135.1	1,090	9.177	A-cent.(b′)	$N_2{}^{\circ}$	…	**3**.22, **3**.29; **7**.31	[Ang65, Bok86, Law93b]
NA0105c′	136.0	1,097	9.116	C-cent.(c′)	$N_1{}^{\circ}$	…	**3**.28; **7**.24, **7**.170	[Cla84b, Law93b, Zai01]
NA0118c′	138.2	1,115	8.969	E(c′) = X(c′)	$^{*}N_1{}^{+}$	…	**7**.26	[Cla84b, Zai01]
NA0105d′	140.1	1,130	8.849	C-cent.(d′)	$N_1{}^{\circ}$	**Dominant** C-cent. line; var.(139–141 meV)	**3**.28, **7**.24, **7**.170	[Cla84b, Law93b, Enc94]
NA0121c′	144.1	1,162	8.604	F-cent.(c′)	$V_1N_4(C_2)_i{}^{\circ}$	…	**3**.14, **3**.20, **3**.24	[Col82b, Cla84b, Woo83]

Table 2.1.1.2 Natural diamond: absorption lines, far infrared, Part 2 (0.145–0.18 eV)

Line-label	Energy (meV)	Frequ. (cm^{-1})	Wavel. (μm)	Name	Impur./defect	Comment	Figs. **3–6.** [Zai01]; Fig. **7.** [Zai98]	References
NA0094d′	145.7	1175	8.509	B-cent.(d′)	$V_1N_4{}^{\circ}$	**Dominant** B-cent. line	**3.**27, **7.**24, **7.**169	[Cla84b, Law98, Zai98]
NA0118d′	146.9	1185	8.439	E(d′) = X(d′)	$^*N_1{}^{+}$	Overlapping band	**7.**26	[Cla84b, Zai01]
NA0060c′	147.0	1186	8.412	A-cent.(c′)	$N_2{}^{\circ}$	...	**3.**29	[Ang65]
NA0105e′	149.5	1206	8.292	C-cent.(e′)	$N_1{}^{\circ}$	...	...	[Zai01]
NA0041b′	150.0	1210	8.265	D-cent.(b′)	$^*(C_{2n})_i{}^{\circ}$	...	**3.**25; **7.**24	[Cla84b, Zai01]
NA0060d′	150.6	1215	8.232	A-cent.(d′)	$N_2{}^{\circ}$	...	**3.22, 3.**29; **7.**31	[Ang65, Cla84b, Bok86]
NA0094e′	150.6	1215	8.232	B-cent.(e′)	$V_1N_4{}^{\circ}$	Overlapping band	**3.**27	[Law98]
NA0118e′	152.5	1230	8.130	E(e′) = X(e′)	$^*N_1{}^{+}$	Overlapping band	**7.**26	[Cla84b, Zai01]
NA0121d′	154.0	1242	8.050	F-cent.(d′)	$V_1N_4(C_2)_i{}^{\circ}$	**Dominant** F-cent. line...	**3.**14, **3.**24; **7.**26	[Col82d, Cla84b, Dav99a]
NA0041c′	156.1	1259	7.942	D-cent.(c′)	$^*(C_{2n})_i{}^{\circ}$	...	**3.**25; **7.**24	[Cla84b, Zai01]
NA0121e′	157.5	1270	7.872	F-cent.(e′)	$V_1N_4(C_2)_i{}^{\circ}$	...	**3.**14, **3.**24; **7.**26	[Col82d, Cla84b, Dav99a]
NA0060e′	158.9	1282	7.802	A-cent.(e′)	N_2	**Dominant,** last A-cent. line, FWHM=4%	**3.**22, **7.**24, **7.**31	[Cla84b, Bok86, Law93b]
NA0094f′	158.9	1282	7.802	B-cent.(f′)	$V_1N_4{}^{\circ}$	Overlapping band	**3.**27	[Law98]
NA0160	159.8	1289	7.756	......	B?	LVM of B?...	...	[Gos99, Zai01]
NA0118f′	159.9	1290	7.752	E(f′) = X(f′)	$^*N_1{}^{+}$	Overlapping band	**7.**26	[Cla84b, Zai01]
NA0121f′	161.0	1299	7.698	F-cent.(f′)	$V_1N_4(C_2)_i{}^{\circ}$	...	...	
NA0105f′	161.2	1300	7.692	C-cent.(f′)	$N_1{}^{\circ}$	var. 159.9–162.4 meV...	**7.**24, **7.**170	[Cla84b, Enc94]

NA0121g′	164.0	1323	7.559	F-cent.(g)	$V_1N_4(C_2)_i{}^\circ$	…	**3**.14	[Col82d]
NA0041d′	164.6	1328	7.532	D-cent.(d′)	$^*(C_{2n})_i{}^\circ$	**Dominant,** FWHM=1%, asymm. shape	**3**.25	[Cla84b, Zai01]
NA0094g′	165.1	1332	7.509	B-cent.(g′)	$V_1N_4{}^\circ$	Last B-cent. line, FWHM=0.6%	**3**.27, **7**.24, **7**.169	[Cla84b, Law98, Zai98]
NA0118g′	165.1	1332	7.509	E(g′) = X(g′)	$^*N_1{}^+$	Sharp, last E-cent. line; see Table 9.1.1.2	**7**.26	[Cla84b, Zai01]
NA0105g′	166.6	1344	7.442	C-cent.(g′)	$N_1{}^\circ$	Sharp, last C-cent. line; FWHM=0.2%	**7**.24, **7**.170	[Cla84b, Enc94]
NA0041d′	167.6	1352	7.396	D-cent.(d′)	$^*(C_{2n})_i{}^\circ$		**3**.20, **3**.26	[Woo83, Cla84b]
NA0041e′1	168.0	1355	7.380	D-cent.(e′1)	$^*(C_{2n})_i{}^\circ$	var. 168 (large) to 172 meV (small platelets)	**3**.7, **3**.20, **7**.26	[Woo83, Cla84b, Fer96]
NA0041e′2	172.0	1387	7.210	D-cent.(e′2)	$^*(C_{2n})_i{}^\circ$	Small platelets	**3**.20, **3**.26	[Woo83, Cla84b]
NA0174a	174.2	1405	7.117	CH bend	$^*(V_1H_1)^\circ$	Nat. CH, two lines a, b (385.2 meV), Table 9.3.1	**3**.7, **3**.8, **3**.26	[Fri91a, Fer96, Rei98]
NA0041f′	177.3	1430	6.993	D-cent.(f′)	$^*(C_{2n})_i{}^\circ$	Last D-cent. line, weak; see Table 8.2.2	**3**.26, **7**.5	[Fri91a]

Table 2.1.2.1 Natural diamond: absorption lines, mid infrared, Part 1 (0.18–0.343 eV)

Line-label	Energy (meV)	Frequ. (cm^{-1})	Wavelength (μm)	Name	Impur./defect	Comment	Figs. **3–6.** [Zai01]; Fig. **7.** [Zai98]	References
NA0182a	182.2	1470	6.803	NH bend	$^*(V_1N_1H_1)$	Nat. NH, two lines a, b (401.2 meV), Table 9.3.1	**3.**13	[Che94b]
NA0121h	191.4	1544	6.477	F-cent.(h′)	$^*V_1N_4(C_2)_i{}^\circ$	...	**3.**14	[Col82b]
NA0192a	191.8	1547	6.464	NH bend	$^*(V_1N_4H_1)^\circ$	Nat. NH, two lines a, b (423.8 meV), Table 9.3.1	**3.**7, **3.**14, **3.**26	[Col82b, Fri91a, Fer96]
NA0121i	195.9	1580	6.329	F-cent.(i′)	$^*V_1N_4(C_2)_i{}^\circ$	Last F-cent. line	**3.**14	[Col82b]
NA0225a	225.0	1815	5.509	2LP(a)	Lattice	Two latt.-phonons, 17 lines a–q(330), shoulder	...	[Zai01] (Table 3.3)
NA0225b	232.0	1872	5.343	2LP(b)	Lattice	Shoulder	...	[Zai01] (Table 3.3)
NA0225c	244.0	1968	5.080	2LP(c)	Lattice	Strong peak, abs. coeff. 12.3–14.9 cm^{-1}	**3.**7, **3.**10, **7.**171	[Jan91, Enc94, Fer96]
NA0225d	247.1	1993	5.018	2LP(d)	Lattice	Shoulder	...	[Zai01] (Table 3.3)
NA0225e	251.1	2025	4.938	2LP(e)	Lattice	Strong peak, abs. coeff. 12.3–14.9 cm^{-1}	**3.**7, **3.**10, **7.**171	[Jan91, Enc94, Fer96]
NA0225f	253.1	2041	4.899	2LP(f)	Lattice	Shoulder	...	[Zai01] (Table 3.3)
NA0225g	258.1	2082	4.804	2LP(g)	Lattice	Weak peak	**3.**10	[Enc94, Zai01] (Table 3.3)
NA0225h	262.0	2114	4.731	2LP(h)	Lattice	Shoulder	...	[Zai01] (Table 3.3)

NA0225i	267.0	2154	4.642	2LP(i)	Lattice	Strong peak, abs. coeff. 12.3–14.9 cm^{-1}	**3.**7, **3.**10, **7.**171	[Jan91, Enc94, Fer96]
NA0225j	270.0	2178	4.591	2LP(j)	Lattice	Shoulder	...	[Zai01] (Table 3.3)
NA0225k	274.0	2210	4.524	2LP(k)	Lattice	Shoulder	...	[Zai01] (Table 3.3)
NA0225-l	281.0	2267	4.411	2LP(l)	Lattice	Shoulder	...	[Zai01] (Table 3.3)
NA0225m	289.0	2331	4.290	2LP(m)	Lattice	Shoulder	...	[Zai01] (Table 3.3)
NA0225n	292.0	2356	4.245	2LP(n)	Lattice	Shoulder	...	[Zai01] (Table 3.3)
NA0225o	302.1	2436	4.104	2LP(o)	Lattice	Medium strong peak, abs. coeff. 4.9 cm^{-1}	**3.**7, **3.**10, **7.**171	[Jan91, Enc94, Fer96]
NA0225p	315.0	2541	3.935	2LP(p)	Lattice	Medium strong peak, abs. coeff. 4.9 cm^{-1}	**3.**7, **3.**10, **7.**171	[Jan91, Enc94, Fer96]
NA0225q	330.0	2662	3.756	2LP(q)	Lattice	Shoulder	...	[Zai01] (Table 3.3)
Na0304a	304.2	2454	4.076	B-acc.(a)	B_1	Boron-acceptor, 18 lines a–r(0.3720 eV); 1a, b → 2	**5.**2, **7.**172	[Smi62, Enc94, Zai01]
Na0304b	335.6	2707	3.694	B-acc.(b)	B_1	1b → 3	**5.**2	[Smi62, Zai01] (Table 5.1)
Na0304c	337.3	2721	3.676	B-acc.(c)	B_1	1a → 3	...	[Zai01] (Table 5.1)
Na0304d	340.4	2746	3.642	B-acc.(d)	B_1	see Table 9.2.1	...	[Zai01] (Table 5.1)
Na0304e	341.5	2754	3.631	B-acc.(e)	B_1	1b → 4	**5.**2	[Smi62, Zai01] (Table 5.1)
Na0304f	342.1	2759	3.624	B-acc.(f)	B_1	see Table 9.2.1	...	[Zai01] (Table 5.1)

Table 2.1.2.2 Natural diamond: absorption lines, mid infrared, Part 2 (0.343–0.47 eV)

Line-label	Energy (meV)	Frequ. (cm^{-1})	Wavel. (μm)	Name	Impur./defect	Comment	Figs. **3–6.** [Zai01]; Fig. **7.** [Zai98]	References
NA0304g	343.5	2771	3.609	B-acc.(g)	$B_1{}^\circ$	1a → 4	...	[Zai01] (Table 5.1)
NA0304h	345.1	2784	3.593	B-acc.(h)	$B_1{}^\circ$	see Table 9.2.1	...	[Zai01] (Table 5.1)
NA0174a–s1	345.4	2786	3.589	CH bend	$^*(V_1H_1)^\circ$	2 × nat. CH-bend, see Table 9.3.1	**3.**7; **3.**26, **7.**10	[Dav84b, Fri91a, Fer96]
NA0304i	346.4	2794	3.579	B-acc.(i)	$B_1{}^\circ$	see Table 9.2.1	...	[Zai01] (Table 5.1)
NA0304j	347.1	2800	3.572	B-acc.(j)	$B_1{}^\circ$	Dominant line; 1b → 5	**5.**2, **7.**172	[Smi62, Enc94, Zai01]
NA0304k	349.1	2816	3.551	B-acc.(k)	$B_1{}^\circ$	1a → 5	**5.**2	[Smi62, Zai01]
NA0304-l	352.4	2842	3.518	B-acc.(l)	$B_1{}^\circ$	1b → 6	...	[Zai01] (Table 5.1)
NA0304m	354.6	2860	3.497	B-acc.(m)	$B_1{}^\circ$	1a → 6	...	[Zai01] (Table 5.1)
NA0304n	355.8	2870	3.485	B-acc.(n)	$B_1{}^\circ$	1b → 7	...	[Zai01] (Table 5.1)
NA0304-o	357.9	2887	3.464	B-acc.(o)	B_1	1a → 7	...	[Zai01] (Table 5.1)
NA0192a–s1	362.0	2920	3.425	NH bend	$^*(V_1N_1H_1)^\circ$	2 × nat. NH-bend, see Table 9.3.1	**3.**26, **7.**5	[Fri91a]
NA0304p	362.7	2926	3.418	B-acc.(p)	$B_1{}^\circ$	1b → 8, medium strong line	**5.**2, **7.**172	[Smi62, Enc94, Zai01]
NA0304q	364.6	2941	3.400	B-acc.(q)	$B_1{}^\circ$	1a → 8	...	[Zai01] (Table 5.1)

NA0304r	372.0	3001	3.333	B-acc.(r)	$B_1{}^\circ$	Bound to free?	...	[Zai01]
NA0174b	385.2	3107	3.219	CH stretch	$*(V_1H_1)^\circ b$	**Nat. CH stretch**, see Table 9.3.1; ZPL of DAP41 = NA0884a–e	**3**.7, **3**.26, **7**.10	[Dav84b, Fri91a, Fer96]
NA0386a	386.0	3113	3.212	DAP3a	$*(N_1C_1)_i{}^\circ$	Seven lines a–g(420.8 meV); shell = 38; calc. 385.7 meV; ZPL at 211.5 meV	**3**.9, **7**.8	[Woo83]
NA0386b	389.9	3145	3.180	DAP3b	$*(N_1C_1)_i{}^\circ$	$s = 36$; calc. 390.4 meV	**3**.9, **7**.8	[Woo83]
NA0386c	394.4	3181	3.144	DAP3c	$*(N_1C_1)_i{}^\circ$	$s = 34$; calc. 395.6 meV	**3**.9	[Woo83]
NA0192b	401.2	3236	3.090	NH stretch	$*(V_1N_1H_1)^\circ$	Nat. NH stretch, see Table 9.3.1	**3**.7, **3**.26, **7**.10	[Dav84b, Fri91a, Fer96]
NA0386d	406.6	3280	3.049	DAP3d	$*(N_1C_1)_i{}^\circ$	$s = 30$; calc. 407.5 meV	**3**.9, **7**.8	[Woo83]
NA0386e	410.4	3310	3.021	DAP3e	$*(N_1C_1)_i{}^\circ$	$s = 29$; calc. 411.8 meV	**3**.9, **7**.8	[Woo83]
NA0087b	414.5	3343	2.991	CH strech	$*(V_1H_1)^\circ$	Acetylenic sp_1CH_1 strech after natural irradiation, see Table 9.3.2.2	**3**.9, **7**.8	[Woo83]
NA0386f	418.1	3372	2.966	DAP3f	$*(N_1C_1)_i{}^\circ$	$s = 27$; calc. 419.1 meV	**3**.9, **7**.8	[Woo83]
NA0386g	420.8	3394	2.946	DAP3g	$*(N_1C_1)_i{}^\circ$	$s = 26$; calc. 422.1 meV	**3**.9, **7**.8	[Woo83]
NA0192b	423.8	3418	2.926	NH stretch	$*(V_1N_4H_1)^\circ$	Nat. NH stretch, see Table 9.3.1	**3**.7	[Fer96]
NA0304a–s1	466.0	3759	2.660	B-acc.(a)-s1	$B_1{}^\circ$	B-acceptor (a) + phonon(162 meV)	**5**.2	[Smi62]

Table 2.1.2.3 Natural diamond: absorption lines, mid infrared, Part 3 (0.50–1.24 eV)

Line-label	Energy (meV)	Frequ. (10^3 cm^{-1})	Wavel. (μm)	Name	Impur./ defect	Comment	Figs. **3–6.** [Zai01]; Fig. **7.** [Zai98]	References
NA0304j–s1	508.0	4.098	2.440	B-acc.(j)-s1	$B_1{}^{\circ}$	B-acc.(j) + phonon(161 meV)	**5.**2	[Smi62]
NA0174a–s2	516.7	4.168	2.399	CH bend	$^*(V_1H_1)^{\circ}a$	3 × nat. CH-bend; see Table 9.3.1	**3.**5, **3.**7, **7.**10	[Fri91a, Fer96, Dav84b]
NA0520a	520.0	4.194	2.384	amber(a)	...	Two lines a, b (0.7700), two bands(2.2 and 3.3 eV)	...	[Wal79, Zai01]
NA0304p–s1	527.0	4.251	2.352	B-acc.(p)-s1	$B_1{}^{\circ}$	B-acc.(p) + phonon (165 meV)	**5.**2	[Smi62]
NA0174b–s1	557.8	4.499	2.223	CH combin.	$^*(V_1H_1)$	Nat. CH stretch + bend; see Table 9.3.1	**3.**5, **3.**7, **3.**26	[Fri91a, Fer96]
NA0192b–s1	583.2	4.704	2.126	NH combin.	$^*(V_1N_1H_1)^{\circ}$	Nat. NH stretch + bend; see Table 9.3.1	**3.**5, **7.**5	[Fri91a]
NA0304a–s2	625.0	5.041	1.984	B-acc.(a)-s2	$B_1{}^{\circ}$	B-acceptor(a) +2× phonon (161 meV)	**5.**2	[Smi62]
NA0304j–s2	670.0	5.404	1.850	B-acc.(j)-s2	$B_1{}^{\circ}$	B-acceptor(j) +2× phonon (162 meV)	**5.**2	[Smi62]
NA0174a–s3	688.7	5.555	1.800	CH bend	$^*(V_1H_1)^{\circ}a$	4 × Nat. CH-bend; see Table 9.3.1	**3.**5, **7.**5	[Fri91a]

NA0174b–s2	729.0	5.880	1.701	CH combin.	$*(V_1H_1)^\circ$	Nat. CH stretch +2× bend; see Table 9.3.1	**3**.5, **7**.5	[Fri91a]
NA0174b–s3	752.5	6.070	1.647	CH stretch	$*(V_1H_1)^\circ$b	2 × Nat. CH stretch (Table 9.3.1)	**3**.5, **7**.5	[Fri91a]
NA0520b	770.0	6.211	1.610	amber(b)	...	SBs at 83 and 89 meV	...	[Wal79, Zai01]
NA0304j–s3	830.0	6.695	1.494	B-acc.(j)-s3	B_1°	B-acceptor(j) +3× phonon (161 meV)	...	[Zai01]
NA0884a	884.0	7.130	1.493	(V_1H_1)-DAP41a	$(V_1H_1)^\circ$b	Five lines a–e(1.068 eV), $s = 7$; calc. 0.898 eV	...	[Fri91a]
NA0884b	929.8	7.500	1.333	(V_1H_1)-DAP41b	$(V_1H_1)^\circ$b	$s = 6$; calc. 0.929 eV	**3**.5, **7**.5	[Fri91a]
NA0884c	973.2	7.850	1.274	(V_1H_1)-DAP41c	$(V_1H_1)^\circ$b	$s = 5$; calc. 0.978 eV	**3**.5, **7**.5	[Fri91a]
NA0884d	1018	8.215	1.217	(V_1H_1)-DAP41d	$(V_1H_1)^\circ$b	$s = 4$; calc. 1.018 eV	**3**.5, **7**.5	[Fri91a]
NA0884e	1068	8.615	1.161	(V_1H_1)-DAP41e	$(V_1H_1)^\circ$b	$s = 3$; calc. 1.109 eV	**3**.5, **7**.5	[Fri91a]

Table 2.1.3 Natural diamond: absorption lines, near infrared (1.24–1.77 eV)

Line-label	Energy (eV)	Frequ. (10^3 cm^{-1})	Wavel. (nm)	Name	Impur./defect	Comment	Figs. **3–6.** [Zai01]; Fig. **7.** [Zai98]	References
NA1401a	1.401	11.30	884.9	1Ni1.40a	$^*V_1Ni_1{}^+$	Two lines a, b (2.7 meV splitting); ^{58}Ni, other isotopes; see Table 7.5	...	[Zai01]
NA1401b	1.404	11.32	883.2	1Ni1.40b	$^*V_1Ni_1{}^+$	^{58}Ni; ZPL of DAP25, see HA1485a–k	...	[Zai01]
NA1500	1.500	12.10	826.5	**N1**-line	...	SBs at +60 and +180 meV	**5**.14, **7**.60	[Cla56a, Wal79]
NA1563a	1.563	12.60	793.4	*S5-DAP60a	$(V_2Ni_1)N_x$	Seven lines a–g(2.320 eV), lines b–g; see ME, $s = 162$, calc. 1.564 eV; ESR-cent. NE 8?	...	[Zai01]
NA1673	1.673	13.49	741.0	...	...	ZPL of C-band, 2SBs with 23 meV	**5**.29, **5**.137	[Wig71, Rei98]
NA1682a	1.681	13.56	737.5	*2Si1.68 eV	$(V_3Si_2)^+$	ZPL of DAP53a–m, see LA, LE, Table 8.1.6	...	[Sol72, Zai01]
NA1743a	1.743	14.06	711.2	DAP9a	$^*N_1{}^+$ + $B_1{}^\circ$	Twenty-three lines a–w(2.270 eV); $s = 162$; calc. 1.745; lines c, e–i, p, s–w, see NL; $L = 1.641$ eV	**5**.137	[Rei98]
NA1743b	1.766	14.24	702.2	DAP9b	$^*N_1{}^+$ + $B_1{}^\circ$	$s = 106$; calc. 1.770 eV; line DAP9c; see NL	**5**.137	[Rei98]

Table 2.1.4 Natural diamond: absorption lines, visible: purple-to-ultramarine (1.77–2.75 eV)

Line-label	Energy (eV)	Frequ. ($10^3\,cm^{-1}$)	Wavel. (nm)	Name	Impur./ defect	Comment	Figs. **3–6.** [Zai01]; Fig. **7.** [Zai98]	References
NA1743d	1.795	14.48	690.6	DAP9d	$*N_1{}^+ + B_1{}^\circ$	$s = 74$; calc. 1.795 eV; lines DAP9e–i, NL	**5.**137	[Rei98]
NA1743j	1.913	15.43	648.1	DAP9j	$*N_1{}^+ + B_1{}^\circ$	$s = 24$; calc. 1.912 eV	**5.**137	[Rei98]
NA1743k	1.929	15.56	642.7	DAP9k	$*N_1{}^+ + B_1{}^\circ$	$s = 22$; calc. 1.923 eV	**5.**137	[Rei98]
NA1743-l	1.934	15.60	641.0	DAP9-l	$*N_1{}^+ + B_1{}^\circ$	$s = 20$; calc. 1.938 eV	**5.**137	[Rei98]
NA1743m	1.952	15.74	635.3	DAP9m	$*N_1{}^+ + B_1{}^\circ$	$s = 18$; calc. 1.953 eV	**5.**137	[Rei98]
NA1743n	1.967	15.87	630.1	DAP9n	$*N_1{}^+ + B_1{}^\circ$	$s = 16$; calc. 1.972 eV	**5.**137	[Rei98]
NA1743-o	2.003	16.16	618.8	DAP9-o	$*N_1{}^+ + B_1{}^\circ$	$s = 14$; calc. 1.996 eV; line DAP9p see NL	**5.**137	[Rei98]
NA1743q	2.029	16.37	610.9	DAP9q	$*N_1{}^+ + B_1{}^\circ$	$s = 12$; calc. 2.023 eV	**5.**137	[Rei98]
NA1743r	2.057	16.59	602.8	DAP9r	$*N_1{}^+ + B_1{}^\circ$	$s = 10$; calc. 2.060 eV; lines DAP9s–w, see NL	**5.**137	[Rei98]
NA2145	2.145	17.30	578.0	...	...	ZPL of red luminescence band	...	[Zai01]
NA2152	2.152	17.36	576.0	...	...	NBD	...	[Col02]
NA2154	2.154	17.38	575.5	**NV**	$V_1N_1{}^\circ$	Named 575 nm center	...	[Zai01]
NA2202	2.202	17.76	563.0	...	H?	NBD; QLVM SBs at +32 and +72 meV	**7.**74	[Fri91a]
NA2370	2.370	19.12	523.2	**"A"** line	$*V_1Ni_3{}^\circ$	ZPL of DAP91a–g(2.641), see MA	...	[Nad93, Zai01]
NA2424	2.424	19.55	511.5	....	....	NBD	...	[Zai01]

(continued)

Table 2.1.4 (continued)

Line-label	Energy (eV)	Frequ. (10^3 cm^{-1})	Wavel. (nm)	Name	Impur./defect	Comment	Figs. **3–6.** [Zai01]; Fig. **7.** [Zai98]	References
NA2464a	2.464	19.88	503.2	**H3**a	$^*V_1N_2^\circ$	Two lines a, b (3.361 eV); ZPL of DAP50, see MA	...	[Zai01]
NA2490	2.467	19.90	502.5	...	...	NBD; var. 2.460–2.490 eV	**5.**87, **6.**2	[Wel94, Col97]
NA2496a	2.496	20.13	496.7	**S3**-DAP58a	$^*(V_2Ni_1)N_2$	Eight lines a–h(2.652); $s = 162$, calc. 2.494 eV	...	[Zai01]
NA2526	2.526	20.38	490.8	...	N?	SBs at 80, 106, 143 meV	...	[Zai01]
NA2496b	2.592	20.91	478.3	**S3**-DAP58b	$^*(V_2Ni_1)N_2$	$s = 50$, calc. 2.582 eV	...	[Zai01]
NA2596	2.596	20.94	477.6	**N2**-line	...	Three SBs at +89, +154, and +241 meV	**5.**14, **7.**60, **5.**126	[Cla56a, Col97]
NA2515e	2.597	20.95	477.4	**S2**-DAP57e	$^*(V_2Ni_1)N_3$	Fifteen lines a–o, see H/MA, $s = 64$, calc. 2.598 eV, **C**-line	**7.**104	[Yel92a, Yel92b, Nad93]
NA2496c	2.599	20.97	477.0	**S3**-DAP58c	$^*(V_2Ni_1)N_2$	$s = 42$, calc. 2.601 eV	...	[Zai01]
NA2615	2.615	21.10	474.0	...	H?	NBD	**7.**74	[Fri91a]
NA2496d	2.622	21.15	472.8	**S3**-DAP58d	$^*(V_2Ni_1)N_2$	$s = 34$, calc. 26244 eV	...	[Zai01]
NA2515f	2.623	21.16	472.7	**S2**-DAP57f	$^*(V_2Ni_1)N_3$	$s = 50$, calc. 2.621 eV, **D**-line, ESR: NE3	**7.**104	[Yel92a, Yel92b, Nad93]
NA2515h	2.634	21.25	470.7	**S2**-DAP57h	$^*(V_2Ni_1)N_3$	$s = 42$, calc. 2.640 eV, named **E**-line	**7.**104	[Yel92a, Yel92b, Nad93]
NA2496e	2.658	21.44	466.4	**S3**-DAP58e	$^*(V_2Ni_1)N_2$	$s = 26$, calc. 2.663 eV	...	[Zai01]

Table 2.1.5 Natural diamond: absorption lines, visible: violet (2.75–3.10 eV)

Line-label	Energy (eV)	Frequ. (10^3 cm^{-1})	Wavel. (nm)	Name	Impur./defect	Comment	Figs. **3–6** [Zai01]; Fig. **7** [Zai98]	References
NA2767	2.767	22.32	448.1	...	H?	NGD	...	[Fri91a]
NA2792	2.792	22.52	444.0	...	H?	NGD	...	[Fri91a]
NA2818	2.818	22.73	439.9	...	Si?	NBD	**7**.74, **7**.89	[Fri91a]
NA2870	2.870	23.15	432.0	...	H?	NBD	**7**.74	[Fri91a]
NA2910	2.910	23.47	426.0	...	$*N_2{}^+$?	NGD	**5**.137	[Rei98]
NA2916	2.916	23.52	425.2	...	H?	NGD	**7**.89	[Fri91a]
NA2649e	2.960	23.88	418.8	DAP18e	$*N_2{}^+ + B_1{}^\circ$	DAP18a–h, $s = 4$, calc. 2.967 eV, lines a–d, see NL	**3**.137	[Rei98, Col82b]
NA2985	2.985	24.08	415.3	**N3**(c)	$V_1N_3{}^\circ$(c)	Lines **N3** (a, b) see NL; ZPL of DAP23 a–p, see NA	**7**.89, **7**.141	[Cla65, Fri91a]
NA2649f	3.060	24.68	405.2	DAP18f	$*N_2{}^+ + B_1{}^\circ$	$s = 3$, calc. 3.058 eV, lines g, h, see NL	**3**.137	[Rei98, Col82b]
NA3070	3.070	24.76	403.8	...	H?	NGD	**7**.89	[San94]
NA3075a	3.075	24.80	403.2	V_1N_3(c)-DAP23a	$V_1N_3{}^\circ$(c)	Fifteen lines a–o(3.762); $s = 162$, calc. 3.075 eV	**7**.141	[Cla65, Fuc96]

Table 2.1.6 Natural diamond: absorption lines, near ultraviolet (3.10–3.94 eV)

Line-label	Energy (eV)	Frequ. ($10^3\,cm^{-1}$)	Wavel. (nm)	Name	Impur./defect	Comment	Figs. **3–6** [Zai01]; Fig. **7** [Zai98]	References
NA3075b	3.131	25.26	396.0	V_1N_3(c)-DAP23b	$V_1N_3^\circ$	$s = 64$, calc. 3.129 eV	**7**.141	[Cla65, Fuc96]
NA3131	3.131	25.26	396.0	...		NGD (pink)	**6**.2	[Wel94]
NA3075c	3.136	25.29	395.4	V_1N_3(c)-DAP23c	$V_1N_3^\circ$	$s = 58$, calc. 3.137 eV	**7**.141	[Cla65, Fuc96]
NA3075d	3.141	25.34	394.7	V_1N_3(c)-DAP23d	$V_1N_3^\circ$	$s = 54$, calc. 3.142 eV	**7**.141	[Cla65, Fuc96]
NA3075e	3.146	25.38	394.0	V_1N_3(c)-DAP23e	$V_1N_3^\circ$	$s = 50$, calc. 3.147 eV	**7**.141	[Cla65, Fuc96]
NA3075f	3.167	25.54	391.5	V_1N_3(c)-DAP23f	$V_1N_3^\circ$	$s = 42$, calc. 3.163 eV	**7**.141	[Cla65, Fuc96]
NA3179	3.179	25.64	390.0	...		NGD (pink)	...	[Zai01]
NA3075g	3.181	25.66	389.7	V_1N_3(c)-DAP23g	$V_1N_3^\circ$	$s = 34$, calc. 3.183 eV	**7**.141	[Cla65, Fuc96]
NA3075h	3.229	26.04	383.9	V_1N_3(c)-DAP23h	$V_1N_3^\circ$	$s = 22$, calc. 3.231 eV	**7**.141	[Cla65, Fuc96]
NA3260	3.260	26.30	380.3	...	H?	var. 3.229–3.262 eV	...	[Zai01]
NA3075i	3.292	26.55	376.6	V_1N_3(c)-DAP23i	$V_1N_3^\circ$	$s = 14$, calc. 3.292 eV	**7**.141	[Cla65, Fuc96]
NA3075j	3.317	26.75	373.8	V_1N_3(c)-DAP23j	$V_1N_3^\circ$	$s = 12$, calc. 3.317 eV	**7**.141	[Cla65, Fuc96]

NA3075k	3.331	26.87	372.1	V_1N_3(c)-DAP23k	$V_1N_3{}^{\circ}$	$s = 11$, calc. 3.336 eV	**7**.141	[Cla65, Fuc96]
NA3333b	3.378	27.25	367.0	*S4-DAP59b	*$(V_2Ni_1)N_3$	$s = 50$, calc. 3.378 eV; other lines see MA	. . .	[Zai01]
NA3075-l	3.391	27.35	365.6	V_1N_3(c)-DAP23-l	$V_1N_3{}^{\circ}$	$s = 8$, calc. 3.392 eV	**7**.141	[Cla65, Fuc96]
NA3075m	3.542	28.57	350.0	V_1N_3(c)-DAP23m	$V_1N_3{}^{\circ}$	$s = 4$, calc. 3.532 eV	**7**.141	[Cla65, Fuc96]
NA3593	3.593	28.98	345.1	. . .		. . .	. . .	[Zai01]
NA3075n	3.603	29.06	344.1	V_1N_3(c)-DAP23n	$V_1N_3{}^{\circ}$	$s = 3$, calc. 3.610 eV; named **N4**-line	. . .	[Wal79, Fuc96, Zai01]
NA3075-o	3.682	29.70	336.7	V_1N_3(c)-DAP23-o	$V_1N_3{}^{\circ}$	$s = 2$, calc. 3.653 eV	. . .	[Wal79, Fuc96, Zai01]
. . .. . .NA3075p	3.762	30.34	329.6	V_1N_3(c)-DAP23p	$V_1N_3{}^{\circ}$	$s = 1$, calc. 3.730 eV; named **N5**-line	. . .	[Wal79, Fuc96, Zai01]
NA3877	3.877	31.27	319.8	. . .		. . .	. . .	[Rob34, Cla56a, Ily71b]
NA3901	3.901	31.46	317.8	DAP52g	*$V_1N_2{}^{\circ}$	$s = 4$, calc. 3.911 eV; lines (a–f), see MA	. . .	[Wal79, Zai01]
NA3928	3.928	31.68	315.6	**N6**-line= *A(a)	$N_2{}^{\circ}$	Two lines a, b (4.470 eV), see Table 9.1.1.3	. . .	[Wal79, Zai01]

Table 2.1.7 Natural diamond: absorption lines, mid ultraviolet (3.94–4.43 eV)

Line-label	Energy (eV)	Frequ. (10^3 cm^{-1})	Wavel. (nm)	Name	Impur./defect	Comment	Figs. **3–6** [Zai01]; Fig. **7** [Zai98]	References
NA4022	4.022	32.44	308.2	...		...	...	[Zai01]
NA3928-s1	4.041	32.60	306.8	**N7** = DAP52h	$^*V_1N_2{}^\circ$	$s = 2$, calc. 4.034 eV	...	[Wal79, Zai01]
NA4059	4.059	32.74	305.4	**C**-cent.(c)	$N_1{}^\circ$	One UV line; Seven IR lines a′–g′(0.1666 eV); see Table 9.1.1.1; 2 QLVM at +61 meV (3C)	**5**.164	[Naz87, Col97]
NA4088	4.088	32.97	303.3	...	?	var. 4.084–4.088, N6+160 meV?	...	[Wal79, Zai01]
NA4137a	4.137	33.37	299.7	...	N?	Four lines a–d(4.355)	...	[Dav77c]
NA4184f	4.184	33.75	296.3	**B**-cent.(a1)	$V_1N_4{}^\circ$	Three UV lines a1–a3(4.197 eV), seven IR lines a′–g′(0.165 eV); see Table 8.1.3.5; SB +36 meV	...	[Naz87, Zai01]
NA4190	4.190	33.80	295.9	**N8**-line	N?	...	...	[Wal79, Zai01]
NA4184g	4.191	33.81	295.8	**B**-cent.(a2)	$V_1N_4{}^\circ$	ZPL of DAP82a–e; SB at +36 meV (QLVM)	...	[Naz87, Zai01]
NA4184h	4.197	33.85	295.4	**B**-cent.(a3).	$V_1N_4{}^\circ$	SBs at +36, +72 meV	...	[Naz87, Zai01]
NA4137b	4.270	34.44	290.3	...	N?	...	...	[Dav77c]
NA4307a	4.307	34.74	287.9	B(a)-DAP82a	$V_1N_4{}^\circ$	Five lines a–e(4.438 eV); $s = 130$, calc. 4.310 eV	...	[Naz87]
NA4137c	4.316	34.81	287.3	...	N?	...	...	[Dav77c]
NA4307b	4.350	35.09	285.0	B(a)-DAP82b	$V_1N_4{}^\circ$	$s = 72$, calc. 4.350 eV	...	[Naz87]
NA4137d	4.355	35.13	284.7	...	N?	...	...	[Dav77c]
NA4374	4.374	35.28	283.4	...		Weak line	...	[Zai01]
NA4307c	4.400	35.49	281.8	B(a)-DAP82c	$V_1N_4{}^\circ$	$s = 42$, calc. 4.400 eV	...	[Naz87]
NA4307d	4.423	35.68	280.3	B(a)-DAP82d	$V_1N_4{}^\circ$	$s = 34$, calc. 4.423 eV	...	[Naz87]

Table 2.1.8 Natural diamond: absorption lines, far ultraviolet (4.43–5.90 eV)

Line-label	Energy (eV)	Frequ. ($10^3\,cm^{-1}$)	Wavel. (nm)	Name	Impur./ defect	Comment	Figs. **3–6** [Zai01]; Fig. **7** [Zai98]	References
NA4307e	4.438	35.80	297.4	B(a)-DAP82e	$V_1N_4{}^{\circ}$	$s = 30$, calc. 4.437 eV	...	[Naz87]
NA4468	4.470	36.06	277.4	*A(b)	$N_2{}^{\circ}$	See Table 9.1.1.3	...	[Naz87]
NA4567	4.567	36.84	271.5	**F**-cent.(c)	$V_1N_4(C_2)i^{\circ}$	2 vis. +1 UV line; Nine IR lines a–i; see Table 8.1.3.6 ZPL of DAP81a–g; SBs +55, 106, 136, 154 meV	**5**.164	[Naz87, Col97]
NA4646	4.646	37.48	266.8	...		...	...	[Zai01]
NA4721a	4.721	38.08	262.6	F(c)-DAP81a	$V_1N_4(C_2)i^{\circ}$	Seven lines a–g(4.906 eV); $s = 86$, calc. 4.722 eV	...	[Naz87]
NA4721b	4.748	38.30	261.1	F(c)-DAP81b	$V_1N_4(C_2)i^{\circ}$	$s = 64$, calc. 4.747 eV	...	[Naz87]
NA4721c	4.775	38.52	259.6	F(c)-DAP81c	$V_1N_4(C_2)i^{\circ}$	$s = 50$, calc. 4.770 eV	...	[Naz87]
NA4721d	4.814	38.83	257.5	F(c)-DAP81d	$V_1N_4(C_2)i^{\circ}$	$s = 34$, calc. 4.814 eV	...	[Naz87]
NA4721e	4.835	39.00	256.4	F(c)-DAP81e	$V_1N_4(C_2)i^{\circ}$	$s = 30$, calc. 4.831 eV	...	[Naz87]
NA4721f	4.880	39.36	254.1	F(c)-DAP81f	$V_1N_4(C_2)i^{\circ}$	$s = 22$, calc. 4.876 eV	...	[Naz87]
NA4721g	4.906	39.57	252.7	F(c)-DAP81g	$V_1N_4(C_2)i^{\circ}$	$s = 18$, calc. 4.908 eV	...	[Naz87]
NA4990a	4.990	40.25	248.5	**N10**a–line		Two lines a, b (5.165 eV), closely related to N9	...	[Bok86, Zai01]
NA4990b	5.165	41.66	240.0	**N10b**-line		...	...	[Bok86, Zai01]
NA5252a	5.252	42.36	236.1	**N9a**	$*V_1N_4{}^{-}$	Three lines a–c(5.277 eV), ZPL of N9a–DAP63a–g	**5**.158, **7**.169a	[Bok86, Col97]

(continued)

Table 2.1.8 (continued)

Line-label	Energy (eV)	Frequ. (10^3 cm^{-1})	Wavel. (nm)	Name	Impur./ defect	Comment	Figs. **3–6** [Zai01]; Fig. **7** [Zai98]	References
NA5252b	5.262	42.44	235.6	**N9b**	$^*V_1N_4^-$	ZPL of N9b (similar DAP as-N9a–DAP63)	**5**.158, **7**.169a	[Bok86, Col97]
NA5277	5.277	42.56	234.9	**N9c**	$^*V_1N_4^-$	Very weak line (2% of lines a, b)	**5**.158	[Col97]
NA5340a	5.340	43.07	232.2	N9a–DAP63a	$^*V_1N_4^-$	Four lines (+3 PLE lines, see NE), $s = 198$, calc. 5.340 eV; also photocond. Spectrum [Den67]	**5**.158	[Col97]
NA5350a	5.350	43.15	231.7	N9b-DAP-a	$^*V_1N_4^-$	Three lines a–d(5.410 eV); $s = 162$, calc. 5.350 eV	**5**.158	[Col97]
NA5340b	5.370	43.31	230.9	N9a–DAP63b	$^*V_1N_4^-$	Three lines b–d(5.420 eV); $s = 106$, calc. 5.372 eV	**5**.158	[Col97]
NA5340c	5.388	43.46	230.1	N9a–DAP63c	$^*V_1N_4^-$	$s = 78$, calc. 5.392 eV	**5**.158	[Col97]
NA5350c	5.398	43.54	229.6	N9b-DAP-b	$^*V_1N_4^-$	$s = 78$, calc. 5.402 eV	**5**.158	[Col97]
NA5340d	5.410	43.64	229.1	N9a-DAP63d	$^*V_1N_4^-$	$s = 64$, calc. 5.406 eV; N9a-DAP63e–g, see NE	**5**.158	[Col97]
NA5350d	5.420	43.72	228.7	N9b-DAP-c	$^*V_1N_4^-$	$s = 50$, calc. 5.416 eV	**5**.158, **7**.169a	[Bok86, Col97]

2.2 Excitation of Photoluminescence Lines (NE)

Table 2.2.1 Natural diamond: excitation of photoluminescence lines (PLE), visible: blue to violet (2.43–3.10 eV)

Line-label	Energy (eV)	Frequ. (10^3 cm^{-1})	Wavel. (nm)	Name	Impur./ defect	Comment	Figs. **3–6** [Zai01]; Fig. **7** [Zai98]	References
NE2496a	2.496	20.13	496.7	*Ni-S3-DAP58a	$(V_2Ni_1)N_{2+0+0}$	Eight lines a–h(2.652); $s = 162$, calc. 2.494 eV, named **S3**a	. . .	[Zai01]
NE2537b	2.597	20.95	477.4	*Ni-S2-DAP57b	$(V_2Ni_1)N_{2+0+1}$	Seven lines a–g(3.341 eV), lines (a–e) see NL; $s = 64$, calc. 2.598 eV, named **S2**b, **C**-line; ESR: NE3	. . .	[Zai01]
NE2496c	2.599	20.97	477.0	*Ni-S3-DAP58c	$(V_2Ni_1)N_{2+0+0}$	$s = 42$, calc. 2.601 eV, named **S3**c; lines a, b; see NA	. . .	[Zai01]
NE2496d	2.622	21.15	472.8	*Ni-S3-DAP58d	$(V_2Ni_1)N_{2+0+0}$	$s = 34$, calc. 26244 eV, named **S3**d	. . .	[Zai01]
NE2496e	2.652	21.39	467.5	*Ni-S3-DAP58e	$(V_2Ni_1)N_{2+0+0}$	$s = 26$, calc. 2.660 eV, named **S3**e	. . .	[Zai01]
NE2721	2.721	21.95	455.6	PLE of Yellow 2band	$*N_2^{\circ} + B_1^{-}$	NBD, sharp feature on the 2.91-eV band (SBs at 38.5 meV); DAP18c, $s = 12$, calc. 2.719 eV; see NA, NL	. . .	[Zai01]
NE2985	2.985	24.08	415.3	**N3**(c) = V_1N_3(c)	$V_1N_3^{\circ}$	ZPL of DAP23; PLE of lines N3 (a, b)	5.93 = 7.99, 5.140	[Sob76, Iak00a]
NE3075a	3.075	24.80	403.2	V_1N_3(c)-DAP23a	$V_1N_3^{\circ}$	PLE of N3(a–c); 15 lines a–o(3.762 eV); $s = 162$, calc. 3.075 eV	5.93, 5.140, 7.151	[Sob76, Iak00a, Cla70]

Table 2.2.2 Natural diamond: excitation of photoluminescence lines (PLE), near ultraviolet (3.10–3.94 eV)

Line-label	Energy (eV)	Frequ. ($10^3\,cm^{-1}$)	Wavel. (nm)	Name	Impur./defect	Comment	Figs. **3–6** [Zai01]; Fig. **7** [Zai98]	References
NE2496f	3.234	26.09	383.4	*Ni-S3-DAP58f	$(V_2Ni_1)N_{2+0+0}$	$s = 2$, calc. 3.240 eV, named **S3**f	...	[Zai01]
NE3075b	3.131	25.26	396.0	V_1N_3(c)-DAP23b	$V_1N_3{}^\circ$	$s = 64$, calc. 3.129 eV	7.151	[Cla70]
NE3075c	3.141	25.34	394.7	V_1N_3(c)-DAP23c	$V_1N_3{}^\circ$	$s = 54$, calc. 3.142 eV	5.140, 7.151	[Iak00a, Cla70]
NE3075d	3.146	25.38	394.1	V_1N_3(c)-DAP23d	$V_1N_3{}^\circ$	$s = 50$, calc. 3.147 eV	7.151	[Cla70]
NE3075e	3.167	25.54	391.5	V_1N_3(c)-DAP23e	$V_1N_3{}^\circ$	$s = 42$, calc. 3.163 eV	7.151	[Cla70]
NE3075f	3.181	25.66	389.7	V_1N_3(c)-DAP23f	$V_1N_3{}^\circ$	$s = 34$, calc. 3.183 eV	5.93 = 7.99	[Sob76]
NE3075g	3.229	26.04	383.9	V_1N_3(c)-DAP23g	$V_1N_3{}^\circ$	$s = 22$, calc. 3.231 eV	5.140, 7.151	[Iak00a, Cla70]
NE2537f	3.234	26.09	383.4	*Ni-S2-DAP57f	$(V_2Ni_1)N_{2+0+1}$	$s = 64$, calc. 3.240 eV, named **S2**f	...	[Zai01]
NE3075h	3.292	26.55	376.6	V_1N_3(c)-DAP23h	$V_1N_3{}^\circ$	$s = 14$, calc. 3.292 eV	5.93, 5.140, 7.151	[Sob76, Iak00a, Cla70]

NE3075i	3.317	26.75	373.8	V_1N_3(c)-DAP23i	$V_1N_3{}^\circ$	$s = 12$, calc. 3.317 eV	5.140, 7.151	[Iak00a, Cla70]
NE2537g	3.341	26.95	371.1	*Ni-S2-DAP57g	$(V_2Ni_1)N_{2+0+1}$	$s = 1$, calc. 3.335 eV, named **S3**g	...	[Zai01]
NE3333b	3.380	27.26	366.8	*Ni-S4-DAP59b	$(V_2Ni_1)N_{1+2+0}$	$s = 50$, calc. 3.378; other lines see MA	...	[Zai01]
NE3075j	3.331	26.87	372.2	V_1N_3(c)-DAP23j	$V_1N_3{}^\circ$	$s = 11$, calc. 3.336 eV	**7**.151	[Cla70]
NE3075k	3.391	27.35	365.6	V_1N_3(c)-DAP23k	$V_1N_3{}^\circ$	$s = 8$, calc. 3.392 eV	**5**.140, **7**.151	[Iak00a, Cla70]
NE3075-l	3.542	28.57	350.0	V_1N_3(c)-DAP23-l	$V_1N_3{}^\circ$	$s = 4$, calc. 3.532 eV	**5**.93 = **7**.99	[Sob76]
NE3075m	3.603	29.06	344.1	V_1N_3(c)-DAP23m	$V_1N_3{}^\circ$	$s = 3$, calc. 3.610 eV; named **N4**	**7**.151	[Cla70]
NE3075n	3.682	29.70	336.7	V_1N_3(c)-DAP23n	$V_1N_3{}^\circ$	$s = 2$, calc. 3.653 eV; line (o) see NA	**7**.151	[Cla70]

Table 2.2.3 Natural diamond: excitation of photoluminescence lines (PLE), far ultraviolet (4.43–5.90 eV)

Line-label	Energy (eV)	Frequ. ($10^3\,cm^{-1}$)	Wavel. (nm)	Name	Impur./defect	Comment	Figs. **3–6** [Zai01]; Fig. **7** [Zai98]	References
NE4646	4.646	37.48	266.8	...	...	PLE of green–yellow band	...	[Zai01]
NE4990	4.990	40.25	248.5	**N10a**-line	...	PLE of A bands, closely related to N9	...	[Zai01]
NE5165a	5.156	41.66	240.0	**N10b**-line	...	PLE of N10a and A bands; 166 meV SB of N10a?	**7**.160	[Dea64c]
NE5165b	5.248	42.33	236.2	...	...	Splitting 5 meV	...	[Zai01]
NE5252a	5.252	42.36	236.1	**N9a** = DAP63z	$^*V_1N_4^-$	ZPL of N9a-DAP63a–g	**5**.135, **5**.140, **7**.160	[Iak00a, Dea64c]
NE5252b	5.262	42.44	235.6	**N9b**	$^*V_1N_4^-$	ZPL of N9b, (similar DAP as N9a)	**5**.135, **5**.140, **7**.160	[Iak00a, Dea64c]
NE5340a	5.340	43.07	232.2	N9a-DAP63a	$^*V_1N_4^-$	Seven lines a–g(5.527 eV), shell = 198, calc. 5.340 eV; PLE of A-bands, N10	**5**.135, **5**.140, **7**.160	[Iak00a, Dea64c, Den67]
NE5340b	5.370	43.31	230.9	N9a-DAP63b	$^*V_1N_4^-$	s = 106, calc. 5.372 eV	**5**.140, **7**.160	[Iak00a, Dea64c]
NE5340c	5.388	43.46	230.1	N9a-DAP63c	$^*V_1N_4^-$	s = 78, calc. 5.392 eV	**5**.135, **5**.140, **7**.160	[Iak00a, Dea64c]
NE5340d	5.410	43.64	229.2	N9a-DAP63d	$^*V_1N_4^-$	s = 64, calc. 5.406 eV	**5**.135, **5**.140, **7**.160	[Iak00a, Dea64c, Den67]
NE5340e	5.425	43.76	228.5	N9a-DAP63e	$^*V_1N_4^-$	s = 50, calc. 5.426 eV	**7**.160	[Dea64c]

NE5340f	5.497	44.34	225.5	N9a-DAP63f	$^*V_1N_4^-$	$s = 26$, calc. 5.495 eV	**7**.160	[Dea64c]
NE5340g	5.527	44.58	224.3	N9a-DAP63g	$^*V_1N_4^-$	$s = 20$, calc. 5.529 eV	**5**.135, **5**.140, **7**.160	[Iak00a, Dea64c]
NE5521a	5.521	44.53	224.6	DAP21a	...	Twelve lines a–l(5.889 eV); $s = 98$, calc. 5.526 eV, PLE of A bands	**7**.160	[Dea64c]
NE5521b	5.571	44.94	222.5	DAP21b	...	$s = 50$, calc. 5.572 eV, PLE of A bands	**7**.160	[Dea64c]
NE5521c	5.619	45.32	220.6	DAP21c	...	$s = 30$, calc. 5.620 eV, PLE of A bands	**5**.135, **7**.160	[Iak00a, Dea64c]
NE5521d	5.630	45.41	220.2	DAP21d	...	$s = 26$, calc. 5.635 eV, PLE of A bands	**7**.160	[Dea64c]
NE5521e	5.650	45.57	219.4	DAP21e	...	$s = 24$, calc. 5.645 eV, PLE of A bands	...	[Den67]
NE5521f	5.669	45.73	218.7	DAP21f	...	$s = 20$, calc. 5.667 eV, PLE of A bands	**7**.160	[Dea64c]
NE5521g	5.704	46.01	217.4	DAP21g	...	$s = 16$, calc. 5.697 eV, PLE of A bands	...	[Den67]
NE5521h	5.719	46.13	216.8	DAP21h	...	$s = 14$, calc. 5.716 eV, PLE of A bands	**7**.160	[Dea64c]
NE5521i	5.739	46.29	216.0	DAP21i	...	$s = 12$, calc. 5.741 eV, PLE of A bands	**5**.135, **7**.160	[Iak00a, Dea64c]
NE5521j	5.775	46.58	214.7	DAP21j	...	$s = 10$, calc. 5.773 eV, PLE of A bands	**7**.160	[Dea64c]
NE5521k	5.809	46.86	213.4	DAP21k	...	$s = 8$, calc. 5.815 eV, PLE of A bands	...	[Dea64c]
NE5521-l	5.889	47.50	210.5	DAP21-l	...	$s = 6$, calc. 5.877 eV, PLE of A bands	...	[Dea64c]

2.3 Luminescence Lines (NL)

Table 2.3.1 Natural diamond: luminescence lines, mid infrared (0.18–1.24 eV)

Line-label	Energy (eV)	Frequ. (10^3 cm^{-1})	Wavel. (nm)	Name	Impur./defect	Comment	Figs. **3–6** [Zai01]; Fig. **7.** [Zai98]	References
NL0995a	0.995	8.026	1246	D(a)-DAP96a	$^*(C_{2n})_i{}^\circ$	Six lines a–f(1.305 eV), $s = 4$, calc. 0.992 eV	**5.9**	[Ruo91a]
NL1053	1.053	8.493	1177	…	…	…	…	[Ruo91a]
NL0995b	1.125	9.074	1102	D(a)-DAP96b	$^*(C_{2n})_i{}^\circ$	$s = 8$, calc. 1.128 eV	**5.9**	[Ruo91a]
NL0995c	1.170	9.437	1060	D(a)-DAP96c	$^*(C_{2n})_i{}^\circ$	$s = 10$, calc. 1.170 eV	**5.9**	[Ruo91a]
NL0995d	1.240	10.00	1000	D(a)-DAP96d	$^*(C_{2n})_i{}^\circ$	$s = 16$, calc. 1.245 eV	**5.9**	[Ruo91a]

Table 2.3.2 Natural diamond: luminescence lines, near infrared (1.24–1.77 eV)

Line-label	Energy (eV)	Frequ. (10^3 cm^{-1})	Wavel. (nm)	Name	Impur./defect	Comment	Figs. **3–6** [Zai01]; Fig. **7.** [Zai98]	References
NL0995e	1.262	10.18	982.4	D(a)-DAP96e	$^*(C_{2n})_i{}^\circ$	$s = 18$, calc. 1.261 eV	**5.9**	[Ruo91a]
NL1264	1.264	10.20	980.8	…	…	SBs −60 meV	…	[Dav77c]
NL0995f	1.305	10.53	950.0	D(a)-DAP96f	$^*(C_{2n})_i{}^\circ$	$s = 64$, calc. 1.386 eV	**5.9**	[Ruo91a]
NL1328	1.328	10.71	933.6	…	…	…	**5.11** = **7.56**	[Ruo91a]
NL1360	1.360	10.97	911.6	…	…	…	**5.11** = **7.56**	[Ruo91a]
NL1401a	1.401	11.30	884.9	1Ni1.40a	$^*V_1Ni_1{}^+$	^{58}Ni, ZPL of Ni1.4-DAP26a–k, see HL	…	[Zai01]
NL1401b	1.404	11.32	883.2	1Ni1.40b	$^*V_1Ni_1{}^+$	^{58}Ni, see Table 7.4b (intensities a : b = 40 : 60%)	…	[Zai01]

NL0995z	1.526	12.31	812.4	D(a)-DAP96z	$^*(C_{2n})_i{}^\circ$	ZPL of D(a)-DAP96	**5.9**	[Ruo91a]
NL1559	1.559	12.58	795.2	...	...	...	...	[Bok86]
NL1563	1.563	12.60	793.4	*NiS5-DAP60a	$(V_2Ni_1)N_{1+x+y}$	Seven lines a–g(2.320 eV), lines b–g see ME, $s = 162$, calc. 1.564 eV; ESR-cent. NE8?	**7.63**a	[Bok86, Zai01]
NL1573	1.573	12.69	788.2		N?		**5.11** = **7.56**, **7.63**a	[Ruo91a, Bok86]
NL1606	1.606	12.96	771.8	DAP116a	...	Six lines a–f(1.701 eV), $s = 22$, calc. 1.606 eV	...	[Sol72]
NL1619	1.619	13.06	765.6	DAP116b	...	$s = 24$, calc. 1.618 eV	...	[Sol72]
NL1629	1.629	13.14	760.9	DAP116c	...	$s = 26$, calc. 1.628 eV	...	[Sol72]
NL1652	1.652	13.33	750.3	DAP116d	...	$s = 32$, calc. 1.652 eV	...	[Sol72]
NL1658	1.658	13.38	747.6	DAP116e	...	$s = 34$, calc. 1.659 eV	...	[Sol72]
NL1673	1.673	13.49	741.0	(C-band)	...	ZPL of C-band(1.33 eV), 11 SBs at 53–40 meV	**5.29** = **7.55**a	[Wig71]
NL1633b	1.689	13.62	734.0	DAP4b	$^*N_2{}^+ + ?$	$s = 64$, calc. 1.689 eV, other DAP4 lines see ML	...	[Sol72]
NL1701	1.701	13.72	729.0	DAP116f	...	$s = 54$, calc. 1.702 eV	...	[Sol72]
NL1733f	1.708	13.77	726.2	DAP4f	$^*N_2{}^+ + ?$	$s = 52$, calc. 1.708 eV, other DAP4 lines see ML	...	[Sol72]
NL1743a	1.743	14.06	711.3	DAP9a	$^*N_1{}^+ + B_1{}^\circ$	Twenty-three lines a–w(2.270 eV), lines b, j, l, m, o, see NA; $s = 162$; calc.1.745 eV	**5.29**	[Wig71]
NL1762	1.762	14.21	703.6	...	...	SBs at −33 meV...	...	[Dav77c]
NL1770	1.770	14.28	700.4	...	...	SBs at −40 and −70 meV	**5.11** = 7.56, 7.63	[Ruo91a, Bok86]

Table 2.3.3 Natural diamond: luminescence lines, visible: purple + red (1.77–2.03 eV)

Line-label	Energy (eV)	Frequ. (10^3 cm^{-1})	Wavel. (nm)	Name	Impur./defect	Comment	Figs. **3–6**. [Zai01]; Fig. **7.** [Zai98]	References
NL1743c	1.775	14.32	698.5	DAP9c	$^*N_1^+ + B_1^\circ$	$s = 98$; calc. 1.775 eV	**5.**29...	[Wig71]
NL1785	1.785	14.40	694.4	...	...	...	...	[Sol72]
NL1792a	1.792	14.45	692.1	DAP20a	$^*N_2^\circ + B_1^\circ$	Seven lines a–g(2.105 eV), $s = 10$, calc. 1.785	**7.**132a	[Moh82b]
NL1743d	1.795	14.48	690.7	DAP9d	$^*N_1^+ + B_1^\circ$	$s = 74$; calc. 1.795 eV	**5.**29...	[Wig71]
NL1743e	1.818	14.66	681.9	DAP9e	$^*N_1^+ + B_1^\circ$	$s = 58$; calc. 1.815 eV	**5.**29...	[Wig71]
NL1819	1.819	14.67	681.6	...	...	NBD	...	[Sol72, Zai01]
NL1743f	1.830	14.76	677.5	DAP9f	$^*N_1^+ + B_1^\circ$	$s = 50$; calc. 1.828 eV	**5.**29...	[Wig71, Col82d]
NL1792b	1.831	14.77	677.1	DAP20b	$^*N_2^\circ + B_1^\circ$	$s = 8$, calc. 1.829 eV	**7.**132a	[Moh82b]
NL1844a	1.844	14.87	672.3	DAP65a	$^*(V_3Si_2)^\circ$	Five lines a–e(1.976 eV), $s = 34$, calc. 1.846 eV	**7.**85a	[San94]
NL1743g	1.862	15.02	665.8	DAP9g	$^*N_1^+ + B_1^\circ$	$s = 36$; calc. 1.862 eV	**5.**29...	[Wig71]
NL1792c	1.867	15.06	664.0	DAP20c	$^*N_2^\circ + B_1^\circ$	$s = 7$, calc. 1.867 eV	**7.**132a	[Moh82b]
NL1743h	1.869	15.08	663.3	DAP9h	$^*N_1^+ + B_1^\circ$	$s = 34$; calc. 1.868 eV	**5.**29...	[Wig71]
NL1792d	1.890	15.24	656.0	DAP20d	$^*N_2 + B_1$	$s = 6$, calc. 1.894 eV	**7.**132a	[Moh82b]
NL1743i	1.899	15.32	652.9	DAP9i	$^*N_1^+ + B_1^\circ$	$s = 26$; calc. 1.901 eV	**5.**29...	[Wig71, Col82d]
NL1844b	1.902	15.34	651.8	DAP65b	$^*(V_3Si_2)^\circ$	$s = 64$, calc. 1.902 eV	**7.**85a, **7.**87a, **7.**94a	[San94]

NL1844c	1.915	15.45	647.4	DAP65c	$^*(V_3Si_2)^\circ$	$s = 78$, calc. 1.916 eV	**7.**85a, **7.**87a, **7.**94a	[San94]
NL1743k	1.929	15.56	642.7	DAP9k	$^*N_1{}^+ + B_1{}^\circ$	$s = 22$; calc. 1.923 eV	**5.**29...	[Wig71, Col82d]
NL1792e	1.937	15.62	640.0	DAP20e	$^*N_2{}^\circ + B_1{}^\circ$	$s = 5$, calc. 1.938 eV	**7.**132a	[Moh82b]
NL1946	1.946	15.70	637.1	...	...	NBD, SB at −42 meV	...	[Naz85a, Fie92]
NL1844d	1.951	15.74	635.5	DAP65d	$^*(V_3Si_2)^\circ$	$s = 130$, calc. 1.946 eV	**7.**85a, **7.**87a, **7.**94a	[San94]
NL1743n	1.967	15.87	630.3	DAP9n	$^*N_1{}^+ + B_1{}^\circ$	$s = 16$; calc. 1.972 eV	**5.**29...	[Wig71, Col82d]
NL1844e	1.976	15.94	627.4	DAP65e	$^*(V_3Si_2)^\circ$	$s = 228$, calc. 1.973 eV	**7.**85a, **7.**87a, **7.**94a	[San94]
NL1792f	1.980	15.97	626.1	DAP20f	$^*N_2{}^\circ + B_1{}^\circ$	$s = 4$, calc. 1.974 eV	**7.**132a	[Moh82b]
NL1750c	2.001	16.14	619.4	DAP73c	$^*V_1B_1{}^\circ$(a)	$s = 8$, calc. 1.997 eV; a–f(2.240 eV) see ML	...	[Dea65a, Zai01]
NL2006a	2.006	16.18	618.0	N3(a)-DAP55a	$V_1N_3{}^\circ$(a)	5 lines a–e (2.207 eV), $s = 16$, calc. 2.010 eV	**5.**93 = **7.**99	[Sob76]
NL1743p	2.007	16.19	617.7	DAP9p	$^*N_1{}^+ + B_1{}^\circ$	$s = 13$; calc. 2.011 eV	**5.**29...	[Wig71]
NL1743q	2.023	16.32	612.8	DAP9q	$^*N_1{}^+ + B_1{}^\circ$	$s = 12$; calc. 2.023 eV	...	[Col82d]
NL2024	2.024	16.33	612.4	...	...	...	**5.**64	[Zai01]

Table 2.3.4 Natural diamond: luminescence lines, visible: orange (2.03–2.10 eV)

Line-label	Energy (eV)	Frequ. (10^3 cm^{-1})	Wavel. (nm)	Name	Impur./ defect	Comment	Figs. **3–6** [Zai01]; Fig. **7.** [Zai98]	References
NL2034	2.034	16.41	609.5	…	…	…	…	[Gom60a, Zai01]
NL2041a	2.041	16.46	607.4	DAP75a	$*(V_3Si_2)^-$	Ten lines a–j(2.450 eV), $s = 6$, calc. 2.042	**7.**118	[Gra94]
NL2006b	2.049	16.53	605.1	DAP55b	$V_1N_3{}^\circ$(a)	$s = 22$, calc. 2.051	**5.**93 = **7.**99	[Sob76]
NL2051	2.051	16.54	604.5	…	…	NBD, 8 SBs with 28–36 meV	**7.**132	[Moh82b]
NL1844z	2.053	16.56	603.9	DAP65z	$*(V_3Si_2)^\circ$	ZPL of DAP65a–e(1.976 eV), QLVM(2Si) 45 meV	**7.**85a, **7.**87, **7.**94	[San94]
NL1743r	2.057	16.59	602.7	DAP9r	$*N_1{}^+ + B_1{}^\circ$	$s = 10$; calc. 2.060 eV	**5.**29…	[Wig71, Col82d]
NL1750d	2.062	16.63	601.2	DAP73d	$*V_1B_1{}^\circ$(a)	$s = 12$, calc. 2.070	…	[Dea65a]
NL2070a	2.070	16.70	598.9	DAP1a	$*N_1{}^\circ + B_1{}^\circ$	Twelve lines a–m(2.668 eV), $s = 50$, calc. 2.067, obs. 2.10		[Dea64c]
NL2082a	2.082	16.79	595.5	…	…	NBD…	**7.**87a, **7.**94a	[San94]
NL2088	2.088	16.84	593.8	…	…	NBD…	**7.**87a, **7.**94a	[San94]
NL1743s	2.089	16.85	593.5	DAP9s	$*N_1{}^+ + B_1{}^\circ$	$s = 9$; calc. 2.089 eV	**5.**29…	[Wig71]
NL2095a	2.095	16.90	591.8	DAP19a	$*N_2{}^\circ + B_1{}^-$	NBD; ten lines a–j(2.645 eV), $s = 32$, calc. 2.089 eV	**7.**132	[Moh82b]
NL2096	2.096	16.91	591.5	…	…	NBD, SB at −30 meV	…	[Jor83, Per86]
NL2099	2.099	16.93	590.6	…	…	…	…	[Bok86]
NL2100a	2.100	16.94	590.4	S1b-DAP43a	$*V_1N_1{}^\circ$	Seven lines a–g(2.372 eV), $s = 16$, calc. 2.099 eV	**7.**110	[Sob69c, Zai01]

Table 2.3.5 Natural diamond: luminescence lines, visible: yellow (2.10–2.18 eV)

Line-label	Energy (eV)	Frequ. ($10^3 cm^{-1}$)	Wavel. (nm)	Name	Impur./defect	Comment	Figs. **3–6** [Zai01]; Fig. **7.** [Zai98]	References
NL1792g	2.105	16.98	589.0	DAP20g	$^*N_2^\circ + B_1^\circ$	s = 2, calc. 2.100	**7.**132a	[Moh82b]
NL2070b	2.108	17.00	588.1	DAP1b	$^*N_1^\circ + B_1^\circ$	s = 34, calc. 2.105, obs. 2.10		[Dea64c]
NL1743t	2.112	17.06	586.3	DAP9t	$^*N_1^+ + B_1^\circ$	s = 8; calc. 2.109	...	[Col82d]
NL2006c	2.116	17.07	585.9	DAP55c	$V_1N_3^\circ$(a)	s = 42, calc. 2.119	**5.**93 = **7.**99	[Sob76]
NL2131	2.131	17.18	581.9	...	...	NBD, SB at −30 meV	...	[Per86]
NL2041b	2.132	17.20	581.5	DAP75b	$^*(V_3Si_2)^-$	s = 8, calc. 2.141	**7.**118	[Gra94]
NL2134	2.134	17.21	581.0	...	...	SBs at −30 meV	...	[Zai01]
NL2137	2.137	17.24	580.1	...	...	NBD, SBs at −30 meV	...	[Per86]
NL2100b	2.137	17.24	580.1	S1b-DAP43b	$^*V_1N_1^\circ$	s = 18, calc. 2.131 eV	**7.**110	[Sob69c, Zai01]
NL1743u	2.138	17.25	579.9	DAP9u	$^*N_1^+ + B_1^\circ$	s = 7; calc. 2.150 eV	**5.**29...	[Wig71]
NL2095b	2.140	17.26	579.3	DAP19b	$^*N_2^\circ + B_1^-$	NBD; s = 20 calc. 2.142 eV	**5.**130, **7.**132	[Per88, Moh82b]
NL2143	2.143	17.28	578.6	...	...	...	...	[Zai01]
NL2145	2.145	17.30	578.0	...	...	ZPL of red band (NB1805); 10 SBs with −30 meV	**7.**92a	[San90]
NL2151	2.151	17.35	576.4	...	...	NBD, SB at −30 meV	...	[Per86]
NL2154	2.154	17.38	575.5	**NV**	$V_1N_1^\circ$	var. 2.153–2.156 eV; ZPL of DAP47a–h, see ML	**5.**64, **7.**64, **7.**94b	[Bok86, San94]

(continued)

Table 2.3.5 (continued)

Line-label	Energy (eV)	Frequ. ($10^3 cm^{-1}$)	Wavel. (nm)	Name	Impur./defect	Comment	Figs. **3–6** [Zai01]; Fig. **7.** [Zai98]	References
NL2157	2.157	17.40	574.8	...	...	NBD; related to NL2082?	**7.**87b	[Zai01]
NL2052b	2.163	17.45	573.2	...	...	Related to NL2053 ($*(V_3Si_2)^\circ$ – DAP65)?	...	[San94]
NL2164a	2.164	17.45	572.9	DAP76a	$*V_1N_3^-$	Thirteen lines a–g(2.457 eV), $s = 8$, calc. 2.162 eV	**5.**120b	[Naz92]
NL2095c	2.165	17.46	572.6	DAP19c	$*N_2^\circ + B_1^-$	NBD; $s = 16$ calc. 2.171 eV	**5.**130, **7.**132	[Per88, Moh82b]
NL2167	2.167	17.48	572.1	...	...	SBs at −30 meV	...	[Zai01]
NL2006d	2.175	17.54	570.0	DAP55d	$V_1N_3^\circ$(a)	$s = 86$, calc. 2.173	**5.**93 = **7.**99	[Sob76]
NL2082b	2.175	17.54	570.0	...	...	NBD, related to 2.082 line?	**7.**87b, **7.**94b	[San94]
NL1743v	2.180	17.58	568.7	DAP9v	$*N_1^+ + B_1^\circ$	$s = 6$; calc. 2.183 eV	...	[Col82d]

Table 2.3.6.1 Natural diamond: luminescence lines, visible: green, Part 1 (2.18–2.36 eV)

Line-label	Energy (eV)	Frequ. ($10^3\,cm^{-1}$)	Wavel. (nm)	Name	Impur./ defect	Comment	Figs. **3–6** [Zai01]; Fig. **7.** [Zai98]	References
NL2164b	2.189	17.66	566.4	DAP76b	$*V_1N_3^-$	$s = 10$, calc. 2.200 eV	**5.**120b	[Naz92]
NL2070c	2.202	17.76	563.0	DAP1c	$*N_1^\circ + B_1^\circ$	$s = 16$, calc. 2.2202, obs. 2.20 eV		[Dea64c]
NL2205	2.205	17.79	562.3	...	...	NBD, var 2.205–2.218 eV	**7.**87b, **7.**94b	[San94]
NL2100c	2.206	17.79	562.0	S1b-DAP43c	$*V_1N_1^\circ$	$s = 30$, calc. 2.205 eV	**7.**110	[Sob69c, Zai01]
NL2006e	2.207	17.80	561.7	DAP55e	$V_1N_3^\circ$(a)	$s = 162$, calc. 2.207 eV	**5.**93 = **7.**99	[Sob76]
NL2041c	2.215	17.87	559.7	DAP75c	$*(V_3Si_2)^-$	$s = 12$, calc. 2.211	**7.**118	[Gra94]
NL2095d	2.217	17.88	559.2	DAP19d	$*N_2^\circ + B_1^-$	NBD; $s = 12$ calc. 2.215	**5.**130, **7.**132	[Per88, Moh82b]
NL2164c	2.227	17.96	556.7	DAP76c	$*V_1N_3^-$	$s = 12$, calc. 2.228 eV	**5.**120b	[Naz92]
NL2070d	2.240	18.07	553.5	DAP1d	$*N_1^\circ + B_1^\circ$	$s = 13$, calc. 2.239, obs. 2.24 eV		[Dea64c]
NL2164d	2.244	18.10	552.5	DAP76d	$*V_1N_3^-$	$s = 14$, calc. 2.250 eV	**5.**120b	[Naz92]
NL2095e	2.254	18.18	550.0	DAP19e	$*N_2^\circ + B_1^-$	NBD; $s = 10$ calc. 2.246 eV	**5.**130, **7.**132	[Per94b, Moh82b]
NL2100d	2.263	18.25	547.8	S1b-DAP43d	$*V_1N_1^\circ$	$s = 50$, calc. 2.264 eV	**7.**110	[Sob69c, Zai01]
NL1743w	2.270	18.31	546.2	DAP9w	$*N_1^+ + B_1^\circ$	$s = 4$; calc. 2.270 eV	...	[Col82d]
NL2041d	2.280	18.39	543.8	DAP75d	$*(V_3Si_2)^-$	$s = 20$, calc. 2.281	**7.**118	[Gra94]

(continued)

Table 2.3.6.1 (continued)

Line-label	Energy (eV)	Frequ. (10^3 cm^{-1})	Wavel. (nm)	Name	Impur./defect	Comment	Figs. **3–6** [Zai01]; Fig. **7.** [Zai98]	References
NL2070e	2.284	18.42	542.8	DAP1e	$^*N_1{}^\circ + B_1{}^\circ$	$s = 10$, calc. 2.285 eV, obs. 2.26 eV	…	[Dea64c]
NL2297a	2.297	18.53	539.7	N3(a)	$V_1N_3{}^\circ$(a)	ZPL of DAP55(= NL2006z) and of N3(a) band (2.12eV), delayed luminescence, see Table 8.1.3.2	**5**.93 = **7**.99	[Sob76]
NL2164e	2.306	18.60	537.6	DAP76e	$^*V_1N_3{}^-$	$s = 22$, calc. 2.306 eV	**5**.120b	[Naz92]
NL2100e	2.312	18.65	536.2	S1b-DAP43e	$^*V_1N_1{}^\circ$	$s = 86$, calc. 2.311 eV	**7**.110	[Sob69c, Zai01]
NL2070f	2.331	18.80	531.9	DAP1f	$^*N_1{}^\circ + B_1{}^\circ$	$s = 8$, calc. 2.331, obs. 2.30	…	[Dea64c]
NL2041e	2.335	18.83	531.0	DAP75e	$^*(V_3Si_2)^-$	$s = 34$, calc. 2.338	**7**.118	[Gra94]
NL2100f	2.337	18.85	530.5	S1b-DAP43f	$^*V_1N_1{}^\circ$	$s = 130$, calc. 2.339 eV	**7**.110	[Sob69c, Zai01]
NL2164f	2.342	18.89	529.4	DAP76f	$^*V_1N_3{}^-$	$s = 32$, calc. 2.344 eV	**5**.120b	[Naz92]
NL2095f	2.347	18.93	528.2	DAP19f	$^*N_2{}^\circ + B_1{}^-$	NBD; $s = 6$ calc. 2.349	**5**.130, **7**.132	[Per88, Moh82b]
NL2041f	2.357	19.01	526.0	DAP75f	$^*(V_3Si_2)^-$	$s = 42$, calc. 2.355	**7**.118	[Gra94]
NL2164g	2.357	19.01	526.0	DAP76g	$^*V_1N_3{}^-$	$s = 36$, calc. 2.355 eV	**5**.120b	[Naz92]

Table 2.3.6.2 Natural diamond: luminescence lines, visible: green, Part 2 (2.36–2.43 eV)

Line-label	Energy (eV)	Frequ. (10^3 cm^{-1})	Wavel. (nm)	Name	Impur./ defect	Comment	Figs. **3–6** [Zai01]; Fig. **7.** [Zai98]	References
NL2363	2.363	19.06	524.7	...	...	NBD, delayed (1 ms) luminescence	...	[Fie92]
NL2041g	2.368	19.10	523.6	DAP75f	$^*(V_3Si_2)^-$	$s = 50$, calc. 2.371	**7.**118	[Gra94]
NL2370	2.370	19.11	523.3	**Ni**-2.37-**A**	$^*V_1Ni_3{}^+$	ZPL of DAP92a–f, NL three lines b, c, e; others ML	**7.**103	[Ily71a]
NL2100g	2.372	19.13	522.7	S1b-DAP43g	$^*V_1N_1{}^\circ$	$s = 228$, calc. 2.373 eV	**7.**110	[Sob69c, Zai01]
NL2070g	2.373	19.14	522.4	DAP1g	$^*N_1{}^\circ + B_1{}^\circ$	$s = 7$, calc. 2.369 eV, obs. 2.37 eV	...	[Dea64c]
NL2164h	2.377	19.17	521.6	DAP76h	$^*V_1N_3{}^-$	$s = 48$, calc. 2.377 eV	**5.**120b	[Naz92]
NL2164i	2.382	19.21	520.5	DAP76i	$^*V_1N_3{}^-$	$s = 50$, calc. 2.381 eV	**5.**120b	[Naz92]
NL2391	2.391	19.29	518.4	...		NBD, SBs at −29 and −47 meV	...	[Naz85a, Fie92]
NL2393a	2.393	19.30	518.1	**N3b**-DAP56a	$V_1N_3{}^\circ$(b)	Four lines a–d(2.562 eV), delayed, $s = 16$, calc. 2.402	**5.**93 = **7.**99	[Sob76]
NL2164j	2.395	19.32	517.7	DAP76j	$^*V_1N_3{}^-$	$s = 64$, calc. 2.397 eV	**5.**120b	[Naz92]
NL2095g	2.403	19.38	515.9	DAP19g	$^*N_2{}^\circ + B_1{}^-$	NBD; $s = 4$, calc. 2.424, named yellow A line	**5.**130, **7.**132	[Per88, Moh82b]
NL2041h	2.410	19.44	514.4	DAP75h	$^*(V_3Si_2)^-$	$s = 98$, calc. 2.414	**7.**118	[Gra94]
NL2070h	2.413	19.46	513.8	DAP1h	$^*N_1{}^\circ + B_1{}^\circ$	$s = 6$, calc. 2.398 eV, obs.: 2.37 eV	...	[Dea64c]
NL2164k	2.420	19.52	512.3	DAP76k	$^*V_1N_3{}^-$	$S = 98$, calc. 2.422 eV	**5.**120b	[Naz92]
NL2424	2.424	19.55	511.5	...	...	NBD...	...	[Zai01]
NL2429a	2.429	19.59	510.4	**S1**a	$^*V_1N_1{}^\circ$	NBD, two lines a, b(2.463 eV), exc. st. spl., a = weak	**7.**110	[Sob69c, Zai01]
NL2393b	2.430	19.60	510.2	N3b-DAP56b	$V_1N_3{}^\circ$	$s = 22$, calc. 2.443	**5.**93 = **7.**99	[Sob76]

Table 2.3.7 Natural diamond: luminescence lines, visible: blue (2.43–2.58 eV)

Line-label	Energy (eV)	Frequ. ($10^3\,cm^{-1}$)	Wavel. (nm)	Name	Impur./defect	Comment	Figs. **3–6** [Zai01]; Fig. **7.** [Zai98]	References
NL2041i	2.437	19.66	508.7	DAP75i	$*(V_3Si_2)^-$	$s = 162$, calc. 2.438	**7.**118	[Gra94]
NL2440a	2.440	19.68	508.1	N3c-DAP24a	$V_1N_3{}^{\circ}$	Seventeen lines a–q(2.897 eV); $s = 4$, calc. 2.438 eV	…	[Dea65a]
NL2164-l	2.442	19.70	507.7	DAP76-l	$*V_1N_3{}^-$	$s = 162$, calc. 2.445 eV	**5.**120b	[Naz92]
NL2041j	2.450	19.76	506.0	DAP75j	$*(V_3Si_2)^-$	$s = 228$, calc. 2.452	**7.**118	[Gra94]
NL2070i	2.456	19.81	504.8	DAP1i	$*N_1{}^{\circ} + B_1{}^{\circ}$	$s = 5$, calc. 2.444 eV, obs.: 2.45 eV	…	[Dea64c]
NL2164m	2.457	19.82	504.6	DAP76m	$*V_1N_3{}^-$	$s = 228$, calc. 2.458 eV	**5.**120b	[Naz92]
NL2100z	2.463	19.87	503.4	**S1**b-DAP43z	$*V_1N_1{}^{\circ}$	NBD, ZPL of DAP43a–g, see Table 8.1.3.1	**7.**110	[Sob69c, Zai01]
NL2464	2.464	19.88	503.2	**H3**a	$V_1N_2{}^{\circ}$	var. 2.462–2.465; ZPL of DAP51a–f, see ML	…	[Zai01]
NL2476a	2.476	19.97	500.7	DAP74a	$*V_1B_1$(b)	Six lines a–f(2.966 eV), $s = 2$, calc. 2.467 eV	…	[Dea65a]
NL2095h	2.478	19.99	500.3	DAP19h	$*N_2{}^{\circ} + B_1{}^-$	NBD; $s = 3$, calc. 2.501, named yellow B line	**5.**130, **7.**132	[Per88, Moh82b]
NL2490	2.490	20.08	497.9	…	…	NBD, 1 ms delay	**7.**134b	[Zai01]
NL2496a	2.496	20.13	496.7	**S3**-DAP58a	$*(V_2Ni_1)N_{2+0+0}$	Five lines a–e(2.652 eV, see MA), $s = 162$, calc. 2.494	**7.**63a	[Bok86]
NL2393c	2.500	20.17	495.9	N3b-DAP56c	$V_1N_3{}^{\circ}$	$s = 42$, calc. 2.502	**5.**93, **7.**99	[Sob76]
NL2515	2.515	20.29	492.9	…	Si?	NBD …	…	[Per94b]

NL2520	2.520	20.33	492.0		B?	In IIb diamonds	**7**.117	[Wig71]
NL2523	2.523	20.35	491.4	DAP75z	$^{*}(V_3Si_2)^-$	NBD, ZPL of DAP75a–j; QLVM(2Si): −44 meV	**7**.118	[Gra94]
NL2164z	2.526	20.38	490.8	DAP76z	$^{*}V_1N_3{}^-$	ZPL of DAP76a–m; QLVM(3N + 1C): −46 meV	**5**.120b, **7**.119	[Naz92, Col82e]
NL2537a	2.537	20.46	488.7	**S2**-DAP57a	$(V_2Ni_1)\ N_{2+0+1}$	Nine lines a–i(4.938 eV); var. 2.527–2.537 eV, $s = 162$, calc. 2.533 eV, named S2-B line; SBs at −46 meV	**7**.103	[Ily71a]
NL2095i	2.550	20.57	486.2	DAP19i	$^{*}N_2{}^{\circ} + B_1{}^-$	NBD; $s = 2$, calc. 2.543	**5**.130, **7**.132	[Per88, Moh82b]
NL2556	2.556	20.62	485.0	...	...	NBD, 1 ms delay	...	[Fie92]
NL2393d	2.562	20.67	483.9	N3b-DAP56d	$V_1N_3{}^{\circ}$	$s = 86$, calc. 2.556	**5**.93 = **7**.99	[Sob76]
NL2565	2.565	20.69	483.3	...	...	NBD, rel. to 2.818 eV center?	...	[Per94b]
NL2070k	2.568	20.71	482.8	DAP1k	$^{*}N_1{}^{\circ} + B_1{}^{\circ}$	$s = 3$; calc. 2.565 eV, obs.: 2.54 eV	...	[Dea64c]
NL2440b	2.580	20.81	480.5	N3c-DAP24b	$V_1N_3{}^{\circ}$	$s = 8$, calc. 2.578 eV; named D_8 [Dea65a, Dea65b]	**7**.117a, **7**.141	[Wig71, Cla65]

Table 2.3.8 Natural diamond: luminescence lines, visible: ultramarine (2.58–2.75 eV)

Line-label	Energy (eV)	Frequ. ($10^3\,cm^{-1}$)	Wavel. (nm)	Name	Impur./defect	Comment	Figs. **3–6.** [Zai01]; Fig. **7.** [Zai98]	References
NL2476b	2.590	20.89	478.7	DAP74b	$^*V_1B_1(b)$	$s = 4$, calc. 2.586 eV	...	[Dea65a]
NL2440c	2.591	20.90	478.5	N3c-DAP24c	$V_1N_3{}^{\circ}$	$s = 9$, calc. 2.595 eV	**7.**141	[Cla65, Dea65a]
NL2537b	2.597	20.95	477.4	S2-DAP57b	$(V_2Ni_1)\,N_{2+0+1}$	$s = 64$, calc. 2.598 eV, named S2-C line	**7.**103	[Ily71a]
NL2070–l	2.623	21.16	472.7	DAP1-l	$^*N_1{}^{\circ} + B_1{}^{\circ}$	$s = 2$, calc. 2.611 eV, obs. 2.57 eV	...	[Dea64c]
NL2095j	2.623	21.16	472.7	DAP19j	$^*N_2{}^{\circ} + B_1{}^{-}$	NBD; $s = 1$ calc. 2.618	**5.**130, **7.**132	[Per88, Moh82b]
NL2537c	2.623	21.16	472.7	**S2**-DAP57c	$(V_2Ni_1)\,N_{2+0+1}$	var. 2.618–2.627, $s = 50$, calc. 2.621 eV, S2-D line	**7.**103	[Ily71a]
NL2537d	2.634	21.25	470.7	**S2**-DAP57d	$(V_2Ni_1)\,N_{2+0+1}$	$s = 42$, calc. 2.640 eV, named S2-E line	**7.**103	[Ily71a]
NL2649a	2.649	21.37	468.0	DAP18a	$^*N_2{}^{+} + B_1{}^{\circ}$	NBD; eight lines a–h(3.204 eV), DAP18e–g see NA; $s = 18$, calc. 2.648 eV	**5.**137, **7.**132	[Moh82b]
NL2440d	2.650	21.37	467.8	**N3c**-DAP24d	$V_1N_3^{\circ}$	$s = 12$, calc. 2.653 eV; named D_7 [Dea65a, Dea65b]	**7.**141a	[Cla65, Dea65a]
NL2652a	2.652	21.39	467.5	DAP22a	$^*V_1N_3{}^{+}$	Thirteen lines a–m(3.296 eV), $s = 1$, calc. 2.652 eV	...	[Dea64c, Dea65a]
NL2070m	2.668	21.52	464.7	DAP1m	$^*N_1{}^{\circ} + B_1{}^{\circ}$	$s = 1$, calc. 2.694 eV, obs. 2.60 eV	...	[Dea64c]
NL2440e	2.678	21.60	462.9	**N3c**-DAP24e	$V_1N_3{}^{\circ}$	$s = 14$, calc. 2.678 eV	**7.**117a, **7.**141a	[Wig71], [Cla65]

NL2393z	2.680	21.62	462.6	**N3b**-DAP56z	$V_1N_3{}^{\circ}$	ZPL of N3b-DAP56a–d, delayed	**5.**93 = **7.**99	[Sob76]
NL2681a	2.681	21.62	462.4	DAP66a	$^*V_1Al_1{}^{\circ}$	Nine lines a–i(2.874 eV), $s = 22$, calc. 2.683 eV	**5.**138a, **7.**140a	[Per93]
NL2649b	2.699	21.77	459.3	DAP18b	$^*N_2{}^{+} + B_1{}^{\circ}$	NBD; $s = 14$, calc. 2.691 eV, named yellow D line	**7.**132	[Moh82b]
NL2476c	2.720	21.94	455.8	DAP74c	$^*V_1B_1$(b)	$s = 8$, calc. 2.722 eV	...	[Dea65a]
NL2649c	2.721	21.95	455.6	DAP18c	$^*N_2{}^{+} + B_1{}^{\circ}$	NBD; $s = 12$, calc. 2.719 eV, SBs with 34–37 meV	**5.**129, **7.**132	[Col82d, Moh82b]
NL2681b	2.722	21.96	455.5	DAP66b	$^*V_1Al_1{}^{\circ}$	$s = 30$, calc. 2.725 eV	**5.**138a = **7.**140a	[Per93]
NL2725	2.725	21.98	455.0	...	...	...	...	[Wig71]
NL2652b	2.731	22.03	454.0	DAP22b	$^*V_1N_3{}^{+}$	$s = 2$, calc. 2.731 eV	...	[Dea64c,Dea65a]
NL2681c	2.739	22.09	452.6	DAP66c	$^*V_1Al_1{}^{\circ}$	$s = 34$, calc. 2.739 eV	**5.**138a = **7.**140a	[Per93]
NL2440f	2.741	22.11	452.3	**N3c**-DAP24f	$V_1N_3{}^{\circ}$	$s = 22$, calc. 2.739 eV; named D_6 [Dea65a, Dea65b]	**7.**117a, **7.**141a	[Wig71], [Cla65]
NL2649d	2.748	22.16	451.2	DAP18d	...	NBD; $s = 10$, calc. 2.756 eV, named yellow E line	**7.**132	[Moh82b]

Table 2.3.9.1 Natural diamond: luminescence lines, visible: violet, Part 1 (2.75–2.94 eV)

Line-label	Energy (eV)	Frequ. ($10^3\,cm^{-1}$)	Wavel. (nm)	Name	Impur./ defect	Comment	Figs. **3–6.** [Zai01]; Fig. **7.** [Zai98]	References
NL2681d	2.758	22.25	449.5	DAP66d	$*V_1Al_1{}^{\circ}$	$s = 42$, calc. 2.760 eV	**5.**138a = **7.**140a	[Per93]
NL2681e	2.779	22.42	446.1	DAP66e	$*V_1Al_1{}^{\circ}$	$s = 50$, calc. 2.779 eV	**5.**138a = **7.**140a	[Per93]
NL2440g	2.805	22.62	442.0	**N3c**-DAP24g	$V_1N_3{}^{\circ}$	$s = 42$, calc. 2.807 eV; named D_5 [Dea65a, Dea65b]	**7.**141a	[Cla65, Dea65a]
NL2815	2.815	22.71	440.4	...	V?, N?	Paramagnetic	...	[Zai01]
NL2818	2.818	22.73	439.9	...	...	NBD	**5.**130 = **7.**134a	[Per88]
NL2681f	2.823	22.73	439.9	DAP66f	$*V_1Al_1{}^{\circ}$	$s = 86$, calc. 2.823 eV	**5.**138a = **7.**140a	[Per93]
NL2440h	2.823	22.77	439.2	N3c-DAP24h	$V_1N_3{}^{\circ}$	$s = 50$, calc. 2.823 eV; named D_4 [Dea65a, Dea65b]	**7.**141a	[Cla65, Dea65a]
NL2440i	2.827	22.80	438.5	N3c-DAP24i	$V_1N_3{}^{\circ}$	$s = 54$, calc. 2.828 eV	**7.**117a, **7.**141a	[Wig71, Cla65]
NL2833	2.833	22.85	437.6	...	...	Type II b,	...	[Dav77c]
NL2440j	2.834	22.86	437.5	N3c-DAP24j	$V_1N_3{}^{\circ}$	$s = 58$, calc. 2.833 eV	**7.**141a	[Cla65, Dea65a]
NL2440k	2.836	22.88	437.2	N3c-DAP24k	$V_1N_3{}^{\circ}$	$s = 62$, calc. 2.839 eV	Table 5.8	[Zai01] (p. 322)
NL2681g	2.836	22.88	437.2	DAP66g	$*V_1Al_1{}^{\circ}$	$s = 106$, calc. 2.836 eV	**5.**138a = **7.**140a	[Per93]

NL2440-l	2.841	22.92	436.4	N3c-DAP24-l	$V_1N_3{}^\circ$	$s = 64$, calc. 2.841 eV	Table 5.8	[Zai01] (p. 322)
NL2440m	2.853	23.01	434.6	N3c-DAP24m	$V_1N_3{}^\circ$	$s = 74$, calc. 2.851 eV	Table 5.8	[Zai01] (p. 322)
NL2652c	2.855	23.03	434.2	DAP22c	$*V_1N_3{}^+$	$s = 4$, calc. 2.855 eV	...	[Dea64c,Dea65a]
NL2440n	2.864	23.10	432.9	N3c-DAP24n	$V_1N_3{}^\circ$	$s = 86$, calc. 2.861 eV	Table 5.8	[Zai01] (p. 322)
NL2440-o	2.872	23.17	431.7	N3c-DAP24-o	$V_1N_3{}^\circ$	$s = 106$, calc. 2.873 eV; named D_3 [Dea65a, Dea65b]	**7.**141a	[Cla65, Dea65a]
NL2681i	2.874	23.18	431.4	DAP66i	$*V_1Al_1{}^\circ$	$s = 228$, calc. 2.877 eV	**5.**138a = **7.**140a	[Per93]
NL2476d	2.876	23.20	431.1	DAP74d	$*V_1B_1$(b)	$s = 22$, calc. 2.880 eV	...	[Dea65a]
NL2440p	2.889	23.30	429.1	N3c-DAP24p	$V_1N_3{}^\circ$	$s = 130$, calc. 2.884 eV	**7.**141a	[Cla65, Dea65a]
NL2440q	2.897	23.37	428.0	N3c-DAP24q	$V_1N_3{}^\circ$	$s = 162$, calc. 2.895 eV; named D_2 [Dea65a, Dea65b]	**7.**117a, **7.**141a	[Wig71, Cla65]
NL2920	2.920	23.55	424.6	...	B?	...	...	[Zai01]
NL2985-s1	2.921	23.56	424.4	N3c-s1	$V_1N_3{}^\circ$	SB, −64 meV	Table 5.8	[Zai01] (p. 322)
NL2476e	2.930	23.63	423.1	DAP74e	$*V_1B_1$(b)	$s = 34$, calc. 2.928 eV	...	[Dea65a]
NL2985-s2	2.933	23.66	422.7	N3c–s2	$V_1N_3{}^\circ$	SB, −52 meV	Table 5.8	[Zai01] (p. 322)
NL2985-s3	2.938	23.70	422.0	N3c–s3	$V_1N_3{}^\circ$	SB, −47 meV	Table 5.8	[Zai01] (p. 322)

Table 2.3.9.2 Natural diamond: luminescence lines, visible: violet, Part 2 (2.94–3.10 eV)

Line-label	Energy (eV)	Frequ. ($10^3\,cm^{-1}$)	Wavel. (nm)	Name	Impur./ defect	Comment	Figs. **3–6.** [Zai01]; Fig. **7.** [Zai98]	References
NL2652d	2.940	23.71	421.7	DAP22d	$*V_1N_3{}^+$	$s = 6$, calc. 2.934 eV	. . .	[Dea64c,Dea65a]
NL2985-s4	2.942	23.73	421.4	N3c–s4	$V_1N_3{}^\circ$	SB, −43 meV; named D_1 [Dea65a, Dea65b]	Table 5.8	[Zai01] (p. 322)
NL2985-s5	2.949	23.79	420.4	N3c–s5	$V_1N_3{}^\circ$	SB, −36 meV	Table 5.8	[Zai01] (p. 322)
NL2681z = NL2964a	2.964	23.91	418.3	DAP66z	$*V_1Al_1{}^\circ$	Two lines a, b(2.974 eV); ZPL of DAP66a–i; line (a) dominant at $T < 50$ K, (b) at $T > 50$ K	**5.**138a = **7.**140a	[Per93]
NL2476f	2.966	23.92	418.0	DAP74f	$*V_1B_1$(b)	$s = 50$, calc. 2.962 eV	. . .	[Dea65a]
NL2964b	2.974	23.99	416.9	. . .	$*V_1Al_1{}^\circ$	Slow (delay 0.1–2.0 ms), similar DAP as DAP66	**5.**138a = **7.**140a	[Per93]
NL2985 = NL2440z	2.985	24.08	415.3	**N3c**	$V_1N_3{}^\circ$	ZPL of DAP24a–q(2.897 eV); 0.59 meV excited state splitting; named D_0 [Dea65a, Dea65b]	**7.**99, **7.**140b	[Sob76, Per93]
NL2652e	2.990	24.12	414.6	DAP22e	$*V_1N_3{}^+$	$s = 8$, calc. 2.998 eV	. . .	[Dea64c,Dea65a]
NL2652f	3.041	24.53	407.7	DAP22f	$*V_1N_3{}^+$	$s = 10$, calc. 3.041 eV	. . .	[Dea64c,Dea65a]
NL2652g	3.077	24.82	402.9	DAP22g	$*V_1N_3{}^+$	$s = 12$, calc. 3.074 eV	. . .	[Dea64c,Dea65a]

Table 2.3.10 Natural diamond: luminescence lines, near ultraviolet (3.10–3.94 eV)

Line-label	Energy (eV)	Frequ. (10^3 cm^{-1})	Wavel. (nm)	Name	Impur./defect	Comment	Figs. **3–6** [Zai01]; Fig. **7.** [Zai98]	References
NL2652h	3.120	25.17	397.4	DAP22h	$^*V_1N_3{}^+$	$s = 16$, calc. 3.120 eV	...	[Dea64c,Dea65a]
NL2476z	3.120	25.17	397.4	DAP74z	$^*V_1B_1$(b)	ZPL of DAP74a–f	...	[Dea65a]
NL2652i	3.139	25.32	395.0	DAP22i	$^*V_1N_3{}^+$	$s = 18$, calc. 3.137 eV	...	[Dea64c,Dea65a]
NL2652j	3.165	25.53	391.7	DAP22j	$^*V_1N_3{}^+$	$s = 22$, calc. 3.162 eV	...	[Dea64c,Dea65a]
NL2649h	3.204	25.84	386.9	DAP18h	$^*N_2{}^+ + B_1{}^\circ$	NBD; $s = 1$ calc. 3.196, named yellow H line	**7.**132	[Moh82b]
NL2652k	3.213	25.92	385.9	DAP22k	$^*V_1N_3{}^+$	$s = 34$, calc. 3.213 eV	...	[Dea64c,Dea65a]
NL3224	3.216	25.94	385.5	**I-line**	...	NBD, closely related to $^*N_2{}^+ + B_1{}^\circ$ – DAP18	**7.**132	[Moh82b]
NL3224	3.224	26.00	384.5	**J-line**	...	NBD, closely related to $^*N_2{}^+ + B_1{}^\circ$ – DAP18	**7.**132	[Moh82b]
NL2652-l	3.246	26.18	381.9	DAP22-l	$^*V_1N_3{}^+$	$s = 50$, calc. 3.248 eV	...	[Dea64c,Dea65a]
NL2652m	3.296	26.59	376.1	DAP22m	$^*V_1N_3{}^+$	$s = 98$, calc. 3.296 eV	...	[Dea64c,Dea65a]
NL3414-s1	3.340	26.94	371.2	...	$^*V_1N_3{}^+$	SB(QLVM:2N) −74 meV, dominant line	...	[Dea64c,Dea65a]
NL3414 = NL2652z	3.414	27.54	363.1	DAP22z	$^*V_1N_3{}^+$	ZPL of DAP22a–m (2.652–3.296 eV), very weak line	...	[Dea64c,Dea65a]

Table 2.3.11.1 Natural diamond: luminescence lines, far ultraviolet, Part 1 (4.43–5.10 eV)

Line-label	Energy (eV)	Frequ. ($10^3\,cm^{-1}$)	Wavel. (nm)	Name	Impur./defect	Comment	Figs. **3–6** [Zai01]; Fig. **7.** [Zai98]	References
NL5356a–s1	4.755	38.35	260.7	BE-B(a)-s1	B	SB −601 meV (= 142 + 3 × 153?); named $\mathbf{D_4}$	**5.**169 = **7.**157	[Dea65b]
NL4810a	4.810	38.80	257.7	**N9a**-DAP64a	$^*V_1N_4^-$	Nine lines a–i(5.165 eV), $s = 8$, calc. 4.814 eV	**7.**161	[Wig67]
NL5409a–s1	4.810	38.80	257.7	FE(a)-s1	...	SB −599 meV (= 4 × 150(TO)?), named $\mathbf{B_4}$	**5.**169 = **7.**157	[Dea65b]
NL4846a	4.846	39.09	255.8	**T-line**	...	Two lines a, b = T, T′(4.855 eV) in type IaB	...	[Dav77c]
NL4846b	4.855	39.16	255.4	T′-line	...	SB (LVM) at −148 meV	...	[Dav77c]
NL5356a–s2	4.890	39.44	253.5	BE-B(a)-s2	B	SB −466 meV (= 140 + 2 × 153?); named $\mathbf{D_3}$	**5.**152 = **7.**159 **5.**169 = **7.**157	[Dea65b]
NL5356b-s1	4.903	39.55	252.9	BE-B(b)-s1	B	SB 464 meV (= 140 + 2 × 152?); named $\mathbf{D_3}'$	**5.**152 = **7.**159 **5.**169 = **7.**157	[Dea65b]
NL5409a–s2	4.952	39.94	250.4	FE(a)-s2	...	SB −457 meV (= 3 × 152(TO)?), named B_3	**5.**152 = **7.**159 **5.**169 = **7.**157	3× [Dea65b]
NL4810b	4.957	39.98	250.1	N9a-DAP64b	$^*V_1N_4^-$	$s = 18$, calc. 4.960 eV	**7.**161	[Wig67]
NL4810c	4.975	40.13	249.2	N9a-DAP64c	$^*V_1N_4^-$	$s = 20$, calc. 4.975 eV	**7.**161	[Wig67]

NL4986a	4.986	40.22	248.7	line N	...	Four lines a–d, named N, M, L, K(5.092 eV)	**5**.154	[Den92]
NL5409a–s3	4.990	40.25	248.5	FE(a)-s3	...	SB −419 meV (=87(TA)+2 × 166(LO)?), named A_3	**5**.152, **7**.157, **5**.155b	2× [Dea65b]
								[Ste96b, Ron99]
NL4986b	4.999	40.32	248.0	line M	...	Dominant line, 4 SBs (LVM) with −150 meV	**5**.154	[Den92]
NL5145-s1	5.005	40.37	247.7	*BE-Y	?	SB–140 meV (TO), named E_1	**5**.155b, 5.169	Ste96b, Dea65b
NL4810d	5.010	40.41	247.5	N9a-DAP64d	$^*V_1N_4{}^-$	$s = 26$, calc. 5.009 eV	**7**.161	[Wig67]
NL5356a–s3	5.023	40.52	246.8	BE-B(a)-s3	B	SB −333 meV (= 2 × 166?); named D_2''	**5**.152	[Dea65b]
NL4986c	5.037	40.63	246.1	Lines L, L′	...	Split into 5.032 and 5.042 eV	**5**.154	[Den92]
NL4810e	5.040	40.65	246.0	N9a-DAP64e	$^*V_1N_4{}^-$	$s = 34$, calc. 5.040 eV	**7**.161	[Wig67]
NL5356a–s4	5.048	40.72	245.6	BE-B(a)-s4	B	SB −308 meV (= 141 + 167?); named D_2	**5**.152, **7**.159	3× [Dea65b]
							5.169 = **7**.157	
NL5356b–s2	5.060	40.81	245.0	BE-B(b)-s2	B	SB −307 meV (= 141 + 166?); named D_2'	**5**.152, **7**.159	3× [Dea65b]
							5.169 = **7**.157	
NL4810f	5.089	41.06	243.6	N9a-DAP64f	$^*V_1N_4{}^-$	$s = 58$, calc. 5.089 eV	**7**.161	[Wig67]
NL4986d	5.092	41.07	243.5	Line K	...	...	**5**.154	[Den92]
NL4810g	5.098	41.12	243.2	N9a-DAP64g	$^*V_1N_4{}^-$	$s = 64$, calc. 5.098 eV	**7**.161	[Wig67]

Table 2.3.11.2 Natural diamond: luminescence lines, far ultraviolet, Part 2 (5.10–5.37 eV)

Line-label	Energy (eV)	Frequ. ($10^3\,cm^{-1}$)	Wavel. (nm)	Name	Impur./ defect	Comment	Figs. **3–6** [Zai01]; Fig. **7.** [Zai98]	References
NL4810h	5.112	41.23	242.5	N9a-DAP64h	$*V_1N_4^-$	$s = 198$, calc. 5.165 eV	**7**.161	[Wig67]
NL5409a–s4	5.115	41.26	242.4	FE(a)-s4	…	SB −294 meV (= 2 × 147(TO)?), named B_2	**5**.152, **7**.159	3× [Dea65b]
							5.169 = **7**.157	
NL5409a–s5	5.151	41.55	240.7	FE(a)-s5	…	SB −258 meV (=87(TA)+171(LO)?), named A_2	**5**.155b	[Ste96b, Ron99]3×
							5.152, **7**.159	[Dea65b]
							5.169 = **7**.157	
NL4810i	5.165	41.66	240.0	N9a-DAP64i	$*V_1N_4^-$	$s = 198$, calc. 5.165 eV	**7**.161	[Wig67]
NL5356a–s5	5.193	41.89	238.7	BE-B(a)-s5	B	SB −163 meV; named $\mathbf{D_1}''$	**5**.152, **7**.159	[Dea65b]
NL5356a–s6	5.215	42.06	237.7	BE-B(a)-s6	B	SB −141 meV; named $\mathbf{D_1}$	**5**.152, **7**.159	[Dea65b]
							5.169 = **7**.157	
NL5356b-s3	5.227	42.16	237.2	BE-B(b)-s3	B	SB −140 meV; named $\mathbf{D_1}'$	**5**.152, **7**.159	[Dea65b]
							5.169 = **7**.157	
NL5409a–s6	5.246	42.31	236.3	FE(a)-s6	…	SB −163 meV (=LO); named C_1; very weak	**5**.152, **7**.159	[Dea65b]
NL5252a	5.252	42.36	236.1	**N9(a)**	$*V_1N_4^-$	ZPL of N9a-DAP64a–h; two lines a, b(5.262 eV)	**7**.161	[Wig67]

NL5409b-s1	5.253	42.37	236.0	FE(b)-s1	. . .	SB −163 meV (=LO); named $C_1{}'$; very weak	**5.**152, **7.**159	[Dea65b]
NL5252b	5.262	42.44	235.6	**N9(b)**	$*V_1N_4{}^-$	ZPL	**7.**161	[Wig67]
NL5409a–s7	5.268	42.49	235.3	FE(a)-s7	. . .	SB −141 meV (=TO); named $\mathbf{B_1}$	**5.**152, **7.**159	[Dea65b]
							5.169 = **7.**157	
							5.155b	[Ste96b, Ron99]
NL5409b-s2	5.275	42.55	235.0	FE(b)-s2	. . .	SB −141 meV (=TO); named $\mathbf{B_1}'$	**5.**152, **7.**159	[Dea65b]
							5.169 = **7.**157	
							5.155b	[Ste96b, Ron99]
NL5409a–s8	5.322	42.93	233.0	FE(a)-s8	. . .	SB −87 meV (=TA); named $\mathbf{A_1}$	**5.**152, **7.**159	[Dea65b]
NL5409b-s3	5.329	42.98	232.6	FE(b)-s3	. . .	SB −87 meV (=TA); named $\mathbf{A_1}'$	**5.**152, **7.**159	[Dea65b]
NL5356a	5.356	43.20	231.5	BE-B(a)	B	NPL(upper VB); two lines a, b(5.367 eV), line $\mathbf{D_0}$	**7.**159	[Dea65b]
NL5356b	5.367	43.29	231.0	BE-B(b)	B	NPL(lower VB), named line $\mathbf{D_0}'$	**7.**159	[Dea65b]

Table 2.3.11.3 Natural diamond: luminescence lines, far ultraviolet, Part 3 (5.40–5.90 eV)

Line-label	Energy (eV)	Frequ. ($10^3\,cm^{-1}$)	Wavel. (nm)	Name	Impur./ defect	Comment	Figs. **3–6** [Zai01]; Fig. **7.** [Zai98]	References
NL5409a	5.409	43.63	229.2	FE(a)	...	NPL(upper VB); two energies a, b(5.416 eV), ***forbidden*** NPL of A_{1-3}(TA), B_{1-4}(TO), C_1(LO)	**7.**159	[Dea65b]
NL5409b	5.416	43.68	228.9	FE(b)	...	***Forbidden*** NPL(lower VB),	**7.**159	[Dea65b]
NL5409a–s9	5.436	43.85	228.1	FE(a)-s9	...	Antistokes SB (550 K) + 46 meV (= +87 − 41(temp. shift))	**7.**163	[Dea65b]
NL5409a–s10	5.490	44.28	225.8	FE(a)-s10	...	Antistokes SB (550 K) + 81 meV (= +141 − 60(temp. shift))	**7.**163	[Dea65b]

2.4 Broad Bands (NB)

Table 2.4.1 Natural diamond: broad bands, mid and near infrared (0.40–1.77 eV)

Line-label	Energy (eV)	Frequ. ($10^3\,cm^{-1}$)	Wavel. (nm)	Name	Impur./ defect	Comment	Figs. **3–6** [Zai01]; Fig. **7.** [Zai98]	References
NB0399a	0.3986	3.215	3110	3LP(a)	...	**A**; three lattice phonons, var. 0.396–0.399 eV; 2 bands a, b(0.4500); $w = 23\%$; abs. coeff. $= 1.7–3.0\,cm^{-1}$	**3.**7, **3.**8, **7.**10 **7.**170, **7.**171	[Zai01]
NB0399b	0.4500	3.630	2755	3LP(b)	...	**A**; three lattice phonons; var. 0.444–0.450 eV; $w = 8\%$; abs. coeff. $= 1.7–3.0\,cm^{-1}$	**3.**7, **3.**8, **7.**10 **7.**170, **7.**171	[Zai01]
NB0659	0.6590	5.315	1881	...	B	**A**; $w = 22\%$	**7.**25	[Bok86]
NB1250	1.250	10.08	992.0	...	...	**CL, PL**; related to platelets; $w = 24\%$	**5.**9, **7.**53	[Ruo91a]
NB1300	1.300	10.49	953.7	IR C band	...	**CL**; $w = 17\%$; ZPL at 1.673 eV; 11 SBs with 53–40 meV	**5.**29 = **7.**55, **7.**117	[Wig71, Rei98]
NB1719	1.719	13.87	721.2	...	...	**A**; in hydrogen rich nat. diam. $w = 14\%$	**7.**89	

Observation: ***A*** absorption, ***CL, EL, IL, PL, XL*** luminescence, ***PLE*** photoluminescence excitation

Table 2.4.2 Natural diamond: broad bands, visible: purple to violet (1.77–3.10 eV)

Line-label	Energy (eV)	Frequ. (10^3 cm^{-1})	Wavel. (nm)	Name	Impur./defect	Comment	Figs. **3–6** [Zai01]; Fig. **7.** [Zai98]	References
NB1800	1.800	14.52	688.8	B band, DAP9	$^*N_1{}^+ + B_1{}^\circ$	**A, CL, PL**; $w = 20\%$; structure of 20 lines (1.743–2.138 eV, see DAP9=NA/NL1743);	**5**.29, **5**.137	[Wig71, Rei98]
NB1805	1.805	14.56	686.9	Red band	...	**CL, PL**; $w = 17\%$; ZPL at 2.145 eV, 10 SBs with 30–35 meV	**7**.92a	[San90]
NB1900	1.900	15.33	652.5	Yellow1 band	...	**CL**; $w = 12\%$; ZPL at 2.051, eight SBs with 32 meV	**7**.132	[Moh82b]
NB1980	1.980	15.97	626.1	...	...	**PL**; $w = 9\%$	**7**.94a	
NB2000	2.000	16.13	619.9	...	...	**A**; $w = 12\%$, NBD	...	[Orl73a, Bok86]
NB2070	2.070	16.70	598.9	vis. C band	...	**A**; $w = 15\%$; ZPL at 1.673 eV; SBs 26 meV	**5**.137	[Rei98]
NB2100	2.100	16.94	590.4	A1 band	...	**XL**; $w = 30\%$; in type IIb diamonds; see A_{semic}(a) Table 10.2.2	...	[Dea65a]
NB2138	2.150	17.34	576.6	N3(a) band	$V_1N_3{}^\circ$	**CL, PL**(slow); ZPL at 2.297; $w = 17\%$; DAP55a–e; A_{insul}(a) band (Table 10.2.2)	**5**.93 = **7**.99	[Sob76, Dea65a]
NB2170	2.170	17.50	571.3	...	...	**CL, PL**; $w = 22\%$; related to H3 and large platelets	**7**.91	[Col82d, Bok86]
NB2200	2.200	17.75	563.5	DAP73 band	$V_1B_1{}^\circ$(a)	**CL, PL**(slow); $w = 23\%$, ZPL at 2.490 eV; type IIb, A_{semic}(b) band (Table 10.2.2)	**7**.134b	[Per88]
NB2230	2.230	17.99	556.0	...	...	**A**; $w = 16\%$	...	[Dav77c]

NB2250	2.250	18.15	551.0	...	...	**A**; $w = 15\%$; gray, pink, violet H rich diamonds	**5**.87; **7**.89	[Col97, Fri91a]
NB2350	2.350	18.96	527.6	"Green"-band	$*N_2{}^{\circ} + B_1{}^{-}$	**CL, PL**; NBD; $w = 20\%$; DAP19a–j	**5**.130, **7**.132	[Per88, Moh82b]
NB2440	2.440	19.68	508.1	...	...	**PL**; NBD, ZPL at 2.818	**5**.130 = **7**.134a	[Per88]
NB2450	2.450	19.76	506.0	**N3(b)** band	$V_1N_3{}^{\circ}$(b)	**CL, PL**(slow); ZPL at 2.680 eV; $w = 21\%$; DAP56a–d; A_{insul}(b) band (Table 10.2.2)	**5**.93 = **7**.99	[Sob76, Dea65a]
NB2570	2.570	20.73	482.4	...	...	**A, PLE**; $w = 11\%$; ZPL at 2.145 eV, SBs with 23 meV	**7**.92b, **7**.125	[San90, Col82d]
NB2740	2.740	22.10	452.5	DAP66 band	$*V_1Al_1{}^{\circ}$	**PL**; $w = 11\%$; NBD; ZPL at 2.964 and 2.974 eV	**5**.138a, **7**.140a	[Per93]
NB2750	2.750	22.18	450.8	N3(c) band	$V_1N_3{}^{\circ}$	**PL**; ZPL at 2.985 eV, $w = 13\%$; DAP24a–q; A_{insul}(c) band (Table 10.2.2)	**7**.117b, **7**.141	[Wig71, Cla65, Dea65a]
NB2800	2.800	22.58	442.8	DAP74 band	$V_1B_1{}^{\circ}$(b)	**CL**; $w = 12\%$; type IIb; A_{semic}(d) band (Table 10.2.2)	**7**.117a	[Dea65a]
NB2910	2.910	23.47	426.0		$*N_2{}^{+} + B_1{}^{\circ}$	**CL**; NBD; $w = 16\%$; DAP18a–h	**7**.132	[Moh82b]
NB3000	3.000	24.20	413.3	DAP22-band	$*V_1N_3{}^{+}$	**CL, PL**; $w = 20\%$; A_{insul}(d) band (Table 10.2.2)	...	[Dea64c, Dea65a]

Observation: ***A*** absorption, ***CL, EL, IL, PL, XL*** luminescence, ***PLE*** photoluminescence excitation

Table 2.4.3 Natural diamond: broad bands, near and far ultraviolet (3.10–5.90 eV)

Line-label	Energy (eV)	Frequ. (10^3 cm^{-1})	Wavel. (nm)	Name	Impur./ defect	Comment	Figs. **3–6** [Zai01]; Fig. **7.** [Zai98]	References
NB3180	3.180	25.65	390.0	...	...	**A, PLE** of yellow two band; NBD; $w = 17\%$; ZPL at 2.721 eV	**6**.2	[Wel94, Per88]
NB3263	3.263	26.32	380.0	**N3(c)** band	$V_1N_3{}^\circ$(c)	**CL, PL**(slow), **PLE**; ZPL (DAP23) at 2.985 eV	**5**.93 = **7**.99, **6**.4	[Sob76, Zai01]
NB3300	3.300	26.62	375.7	...	N?	**A**; polarized E perpendicular to $\langle 1\ 1\ 1 \rangle$	...	[Kop86]
NB3300	3.308	26.62	375.7	...	...	**CL, PL, XL;** w = 12%A_{insnl}(e) band (Table 10.2.2)	...	[Dea65a]
NB3750	3.750	30.25	330.6	...	...	**PLE** of 2.35-eV band; $w = 12\%$; ZPL at 2.818 eV	**5**.130b, **7**.134a	[Per94b, Per88]
NB3800	3.800	30.65	326.3	...	$^*V_1Al_1{}^\circ$	**PLE** of 2.74-eV band (DAP66); NBD; $w = 12\%$; var. 3.75–3.90 eV; ZPL at 2.964 and 2.974 eV	**5**.138b, **7**.140b	[Per93]
NB4430	4.430	35.73	279.9	...	...	**A**; weak, related to platelets?	...	[Zai01]
NB4590	4.590	37.02	270.1	...	...	**A**; related to platelets?	...	[Dav94a, Sob68d]
NB4600	4.600	37.10	269.5	...	N?	**A**; $w = 12\%$	**7**.154	[Dav94a, Sob68d]
NB4690	4.690	37.83	264.3	...	...	**A, PL**; weak	...	[Bok86, Sob68d]
NB5480	5.480	44.20	226.2	**Indirect bandgap**	...	**A**; begin of indirect bandgap absorption	**5**.160, **5**.161	[Zai01, Col97]
NB5370	5.370	43.31	230.9	N9-DAP63 band	$^*V_1N_4{}^-$	**A, PLE** of N9, N3(c) and other A bands; $w = 5\%$; ZPL at 5.252 eV	**5**.158	[Col97, Den67, Dea64c]
NB5500	5.500	44.36	225.4	DAP21 band	...	**PLE** of N3(c) and other A bands; $w = 8\%$	**5**.140	[Iak00a, Dea64c]
NB7120	7.120	57.43	174.1	**Direct bandgap**	...	**A**; begin of direct bandgap absorption	**5**.160	[Zai01]

Observation: ***A*** absorption, ***CL, EL, IL, PL, XL*** luminescence, ***PLE*** photoluminescence excitation

Chapter 3
Spectral Lines in High Pressure Synthetic (HPHT) Diamond

In this chapter ca. 330 lines (and bands), which are observed in high pressure synthetic (HPHT) diamond (see Sect. 1.2.2), are presented in 20 tables. The entries are assigned to 115 centers. The corresponding defect structure is established for 14 centers, is almost certain for 23 centers, and is incomplete or unknown for 78 centers.

Chapter 3 is subdivided into *absorption lines* (Sect. 3.1: 150 HA lines in seven tables (see Tables 3.1.1.1–3.1.6)), into *photoluminescence excitation* (*PLE*) *lines* (Sect. 3.2: three HE lines in one table (see Table 3.2.1)), into *luminescence lines* (Sect. 3.3: 137 HL lines in nine tables (see Tables 3.3.1–3.3.7.2)), and into *broad bands* (Sect. 3.4:40 HB bands in three tables (see Tables 3.4.1–3.4.3)).

B. Dischler, *Handbook of Spectral Lines in Diamond*,
DOI 10.1007/978-3-642-22215-3_3, © Springer-Verlag Berlin Heidelberg 2012

3.1 Absorption Lines (HA)

Table 3.1.1.1 High pressure synthetic diamond (HPHT): absorption lines, far infrared, Part 1 (0.04–0.14 eV)

Line-label	Energy (meV)	Frequ. (cm^{-1})	Wavel. (μm)	Name	Impur./defect	Comment	Figs. **3–6** [Zai01]; Fig. **7** [Zai98]	References
HA0041a′	40.91	330.0	30.30	D-center (a′) = **platelets**	$^*(C_{2n})_i{}^\circ$	Six IR lines a′–f′ (177.3 meV), also named platelets or B′, luminescence line *D(a) at 1.526 eV, see Table 8.2.2	…	[Zai01]
HA0054a	53.93	435.0	22.99	**MX**(a)	Met., N?	Ten lines a–h (119 meV); melt: Mn, Fe, Co, Ni	**7**.29	[Bok86, Zai01]
HA0060a′	60.00	484.0	20.66	**A**-center (a′)	$N_2{}^\circ$	Five IR lines a′–e′ (158.9 meV), two UV lines A(a,b) at 3.928, 4.470 eV, see Table 9.1.3	…	[Zai01]
HA0054b	60.13	485.0	20.62	MX(b)	Met., N?	Melt: Mn, Fe, Co, Ni	**7**.29	[Bok86, Zai01]
HA0054c	62.73	506.0	19.76	MX(c)	Met., N?	Melt: Mn, Fe, Co, Ni	**7**.29	[Bok86, Zai01]
HA0054d	70.42	568.0	17.61	MX(d)	Met., N?	Melt: Mn, Fe, Co, Ni	**7**.29	[Bok86, Zai01]
HA0054e	77.49	625.0	16.00	MX(e)	Met., N?	Melt: Mn, Fe, Co, Ni	**7**.29	[Bok86, Zai01]
HA0082	81.82	660.0	15.15	…	O?	Melt with TiO_2		[Zai01]
HA0054f	90.50	730.0	13.70	MX(f)	Met., N?	Melt: Mn, Fe, Co, Ni	**7**.29	[Bok86, Zai01]
HA0054g	101.9	822.0	12.17	MX(g)	Met., N?	Melt: Mn, Fe, Co, Ni	**7**.29	[Bok86, Zai01]
HA0105a′	105.4	850.0	11.76	**C**-center (a′)	$N_1{}^\circ$	Seven IR lines a′–g′ (166.6 meV), three UV lines: *C(a, b = indirect), *C(c) at 4.059 eV, see Table 9.1.1	**3**.28	[Law93b, Zai01]
HA0054h	107.0	863.0	11.59	MX(h)	Met., N?	Melt: Mn, Fe, Co, Ni	**7**.29	[Bok86, Zai01]

HA0278z	112.8	910.0	10.99	DAP115z	$V_1As_1{}^{\circ}$	As doped; ZPL of DAP115a-f (578 meV)	**7**.47	[Bok86]
HA0054i	114.1	920.0	10.87	MX(i)	Met., N?	Melt: Mn, Fe, Co, Ni	**7**.29	[Bok86, Zai01]
HA0118a′	117.8	950.0	10.52	**E**(a′) = X(a′)	$*N_1{}^{+}$	Seven IR lines a′–g′ (165.1 meV), see Table 9.1.2	**3**.21	[Law93b, Law98]
HA0054j	119.0	960.0	10.42	MX(j)	Met., N?	Melt: Mn, Fe, Co, Ni	**7**.29	[Bok86, Zai01]
HA0123a′	122.7	989.7	10.10	**D**′(a′)	B, N, Ni?	Three lines a–c (160.1 meV), not related to D center	. . .	[Zai01]
HA0118b′	129.7	1046	9.559	**E**(b′) = X(b′)	$*N_1{}^{+}$	See Table 9.1.2	**3**.21	[Law93b, Law98]
HA0105b′	129.6	1045	9.566	C-center (b′)	$N_1{}^{\circ}$	. . .	**3**.28, **7**.23a	[Law93b, Col82f]
HA0130	130.1	1050	9.527	. . .	O?	. . .	. . .	[Zai01]
HA0123b′	132.0	1065	9.392	**D**′(b′)	B, N, Ni?	. . .	. . .	[Zai01]
HA0060b′	135.1	1090	9.177	**A**-center (b′)	$N_2{}^{\circ}$	. . .	**3**.22	[Law93b, Zai01]
HA0105c′	136.0	1097	9.116	**C**-center (c′)	$N_1{}^{\circ}$	. . .	**3**.28, **7**.23a	[Law93b, Col82f]

Table 3.1.1.2 High pressure synthetic diamond (HPHT) diamond: absorption lines, far infrared, Part 2 (0.14–0.18 eV)

Line-label	Energy (meV)	Frequ. (cm^{-1})	Wavel. (μm)	Name	Impur./defect	Comment	Figs. **3–6** [Zai01]; Fig. **7** [Zai98]	References
HA0118c′	138.2	1115	8.971	E(c′) = X(c′)	$^*N_1{}^+$	See Table 9.1.2	**3**.21	[Law93b, Law98]
HA0105d′	140.1	1130	8.849	C-center (d′)	$N_1{}^\circ$	**Dominant** C-ce. line; var. (139–141 meV)	**3**.28, **7**.23a	[Law93b, Col82f]
HA0118d′	146.9	1185	8.440	E(d′) = X(d′)	$^*N_1{}^+$	See Table 9.1.2	**3**.21	[Law93b, Law98]
HA0105e′	149.5	1206	8.292	C-center (e′)	$N_1{}^\circ$	...	...	[Zai01]
HA0060d′	150.6	1215	8.232	A-center (d′)	$N_2{}^\circ$	...	**3**.22	[Law93b, Zai01]
HA0118e′	152.5	1230	8.130	E(e′) = X(e′)	$^*N_1{}^+$	See Table 9.1.2	**3**.21	[Law93b]
HA0060e′	158.9	1282	7.802	A-center (e′)	N_2	**Dominant**, last A-center line, *FWHM = 4%*	**3**.22	[Law93b, Zai01]
HA0118f′	159.9	1290	7.753	E(f′) = X(f′)	$^*N_1{}^+$	See Table 9.1.2	**3**.21	[Law93b, Law98]
HA0123c′	160.1	1291	7.744	D′(c′)	B, N, Ni?	...	...	[Zai01]
HA0105f′	161.2	1300	7.692	C-center (f′)	$N_1{}^\circ$	var. 159.9–162.4 meV...	**3**.28, **7**.23a	[Law93b, Col82f]
HA0118g′	165.1	1332	7.509	E(g′) = X(g′)	$^*N_1{}^+$	Sharp, last E-center line; see Table 9.1.2	**3**.21	[Law93b, Law98]
HA0105g′	166.6	1344	7.442	C-center (g′)	$N_1{}^\circ$	Sharp, last C-center line; FWHM = 0.2%	**3**.28	[Law93b, Zai01]
HA0041e′1	168.0	1355	7.380	D-center (e′1)	$^*(C_{2n})_i{}^\circ$	var.168 (large) to 172 meV (small platelets)	...	[Zai01]
HA0041e′2	172.0	1387	7.210	D-center (e′2)	$^*(C_{2n})_i{}^\circ$	Small platelets	...	[Zai01]
HA0041f′	177.3	1430	6.993	D-center (f′)	$^*(C_{2n})_i{}^\circ$	Last D-center line, weak; see Table 8.2.2	...	[Zai01]

Table 3.1.2 High pressure synthetic diamond (HPHT) diamond: absorption lines, mid infrared (0.18–1.24 eV)

Line-label	Energy (meV)	Frequ. (cm^{-1})	Wavel. (μm)	Name	Impur./defect	Comment	Figs. **3–6** [Zai01]; Fig. **7** [Zai98]	References
HA0207	207.0	1670	5.988	...	...	As doped (not $V_1As_1{}^{\circ}$)	...	[Kli75c]
HA0213	213.2	1720	5.814	...	...	As doped (not $V_1As_1{}^{\circ}$)	...	[Kli75c]
HA0225a	225.0	1815	5.509	**2LP**(a)	...	Two lattice phonons, 17 lines a–q (330), see NA	...	[Zai01] (Table 3.3)
HA0225c	244.0	1968	5.080	2LP(c)	...	Strong peak, abs. coeff. 12.3–14.9 cm^{-1}	**7**.25b, **7**.47	[Bok86]
HA0248	248.0	2000	5.000	...	Al?	Al doped ...	...	[Kli75c]
HA0225e	251.1	2025	4.938	2LP(e)	...	Strong peak, abs. coeff. 12.3–14.9 cm^{-1}	**7**.25b, **7**.47	[Bok86]
HA0225i	267.0	2154	4.642	2LP(i)	...	Strong peak, var. 0.2671–0.2690 eV, see NA	**7**.25, **7**.47	[Bok86]
HA0278a	278.0	2242	4.460	DAP115a	$V_1As_1{}^{\circ}$	As doped; six lines a–f (0.578 eV), $s = 50$, calc. 0.274	**5**.5, **7**.47	[Rot83, Bok86]
HA0278b	295.0	2379	4.203	DAP115b	$V_1As_1{}^{\circ}$	As doped; $s = 42$, calc. 0.290	**5**.5, **7**.47	[Rot83, Bok86]
HA0225o	302.1	2436	4.104	2LP(o)	...	Medium strong peak, var. 0.3013–0.3021 eV	**7**.25, **7**.47	[Bok86]
HA0225p	315.0	2541	3.935	2LP(p)	...	Medium strong peak, var. 0.3150–0.3174 eV	**7**.25, **7**.47	[Bok86]
HA0304j	347.1	2800	3.572	**B-acc.**(j)	B_1	Dominant line; see Table 9.2.1 and NA	Table 5.1	[Zai01, p. 127]
HA0278c	348.0	2.807	3.563	DAP115c	$V_1As_1{}^{\circ}$	As doped; $s = 22$, calc. 0.357	5.5, 7.47	[Rot83, Bok86]
HA0304p	362.7	2.926	3.418	B-acc.(p)	B_1	See Table 9.2.1 and NA	Table 5.1	[Zai01, p. 127]
HA0304r	372.0	3.001	3.333	B-acc.(r)	B_1	See Table 9.2.1 and NA	Table 5.1	[Zai01, p. 127]
HA0278d	443.0	3.573	2.800	DAP115d	$V_1As_1{}^{\circ}$	As doped; $s = 12$, calc. 0.443	5.5, 7.47	[Rot83, Bok86]
HA0278e	517.0	4.170	2.400	DAP115e	$V_1As_1{}^{\circ}$	As doped; $s = 8$, calc. 0.517	5.5, 7.47	[Rot83, Bok86]
HA0278f	578.0	4.662	2.140	DAP115f	$V_1As_1{}^{\circ}$	As doped; $s = 6$, calc. 0.579	5.5, 7.47	[Rot83, Bok86]
HA1212a	1212	9.776	1.023	***S8**-DAP16a	$^*(V_2Ni_1)^-$	12 lines a–l (1.383 eV); shell = 51, calc. 1.212 eV	**5**.7 = **7**.51b	[Law93c]
HA1212b	1213	9.784	1.022	*S8-DAP16b	$^*(V_2Ni_1)^-$	Shell = 50, calc. 1.214 eV	**5**.7 = **7**.51b	[Law93c]
HA1212c	1216	9.808	1.020	*S8-DAP16c	$^*(V_2Ni_1)^-$	Shell = 49, calc. 1.216 eV	**5**.7 = **7**.51b	[Law93c]
HA1212d	1218	9.824	1.018	*S8-DAP16d	$^*(V_2Ni_1)^-$	Shell = 48, calc. 1.217 eV	**5**.7 = **7**.51b	[Law93c]
HA1212e	1219	9.832	1.917	*S8-DAP16e	$^*(V_2Ni_1)^-$	Shell = 47, calc. 1.220 eV	**5**.7 = **7**.51b	[Law93c]
HA1212f	1226	9.889	1.011	*S8-DAP16f	$^*(V_2Ni_1)^-$	Shell = 44, calc. 1.225 eV	**5**.7 = **7**.51b	[Law93c]
HA1212g	1228	9.905	1.010	*S8-DAP16g	$^*(V_2Ni_1)^-$	Shell = 43, calc. 1.228 eV	**5**.7 = **7**.51b	[Law93c]
HA1212h	1230	9.921	1.008	*S8-DAP16h	$^*(V_2Ni_1)^-$	Shell = 42, calc. 1.230 eV	**5**.7 = **7**.51b	[Law93c]

Table 3.1.3 HPHT diamond: absorption lines, near infrared (1.24–1.77 eV)

Line-label	Energy (eV)	Frequ. ($10^3 cm^{-1}$)	Wavel. (nm)	Name	Impur./defect	Comment	Figs. **3–6** [Zai01]; Fig. **7** [Zai98]	References
HA1212i	1.283	10.35	966.3	*S8-DAP16i	$*(V_2Ni_1)^-$	Shell = 26, calc. 1.282 eV	**7**.51a	[Law93c, Zai01]
HA1212j	1.332	10.74	930.8	*S8-DAP16j	$*(V_2Ni_1)^-$	Shell = 18, calc. 1.332 eV	**7**.51a	[Law93c, Zai01]
HA1212k	1.371	11.06	904.3	*S8-DAP16k	$*(V_2Ni_1)^-$	Shell = 14, calc. 1.372 eV	**7**.51a	[Law93c, Zai01]
HA1212–l	1.383	11.16	896.4	*S8-DAP16-l	$*(V_2Ni_1)^-$	Shell = 13, calc. 1.388 eV	**7**.51a	[Law93c, Zai01]
HA1401a = HA1485z	1.401	11.30	884.9	**1Ni-1.4 eV**(a)	$*V_1Ni_1{}^+$	Two lines a, b from 2.7-meV ground state splitting; intensities a:b = 4:6; ZPL of 1Ni-DAP25a–k; four SBs with +26 meV, LVM at +89, 101, 165 meV	**5**.141 = **7**.142 **5**.165 = **7**.51a	[Col90e, Law93c, Zai01]
HA1401b	1.404	11.32	883.2	1Ni-1.4 eV(b)	$*V_1Ni_1{}^+$	^{58}Ni, 4 SBs: ^{60}Ni, ^{61}Ni, ^{62}Ni, ^{64}Ni; see Table 7.4b	Dito (a)	Dito (a)
HA1485a	1.485	11.98	834.9	1Ni-DAP25a	$*V_1Ni_1{}^+$	11 lines a–k (2.070 eV); $s = 198$, calc. 1.485 eV	**5**.165 = **7**.51a	[Law93c]
HA1485b	1.503	12.12	824.9	1Ni-DAP25b	$*V_1Ni_1{}^+$	$s = 130$, calc. 1.504 eV	**5**.165 = **7**.51a	[Law93c]
HA1485c	1.548	12.49	800.9	1Ni-DAP25c	$*V_1Ni_1{}^+$	$s = 64$, calc. 1.547 eV	**5**.165 = **7**.51a	[Law93c]

HA1485d	1.563	12.61	793.2	1Ni-DAP25d	*$V_1Ni_1^+$	$s = 50$, calc. 1.565 eV	**5**.165 = **7**.51a	[Law93c]
HA1563a	1.563	12.61	793.2	***S5**-DAP60a	*$(V_2Ni_1)N_x^\circ$	11 lines a–k (see MA/ME/ML), $s = 162$, calc. 1.564 eV	...	[Kan99, Kup99, Zai01]
HA1485e	1.643	13.25	754.6	1Ni-DAP25e	*$V_1Ni_1^+$	$s = 22$, calc. 1.649 eV	**5**.165 = **7**.51a	[Law93c]
HA1681	1.682	13.57	737.1	Si-DAP53z	*$(V_3Si_2)^+$	var. 1.667–1.691 eV; ZPL of DAP53, see LA	...	[Zai01]
HA1693a	1.693	13.66	732.3	*Ni 1.69a	*$V_1Ni_2^+$(a)	3 lines a-c; b=1.940, (c) = 1.991 eV Weaker doublet line at 1.695 eV; 3 SBs 51 meV	...	[Zai01]
HA1704a = HA1816z	1.704	13.74	727.6	*Ni 1.70a	*$V_1Ni_1^\circ$(a)	2 lines a, b(2.401 eV), ZPL of DAP83a-f	5.165 = 7.51a	[Law93c]
HA1485f	1.714	13.83	723.3	1Ni-DAP25f	*$V_1Ni_1^+$	$s = 14$, calc. 1.710 eV	**5**.165 = **7**.51a	[Law93c]
HA1722	1.722	13.89	720.0	...	Ti, In?	Ti + In doped	...	[Kli75b, But76]
HA1744	1.744	14.06	711.0	...	Ni + N?	...	...	[Yel96]
HA1746	1.746	14.08	710.1	...	Ni?	...	...	[Yel92b, Yel95a]
HA1485g	1.765	14.24	702.4	1Ni-DAP25g	*$V_1Ni_1^+$	$s = 10$, calc. 1.767 eV	**5**.165 = **7**.51a	[Law93c]

Table 3.1.4 HPHT diamond: absorption lines, visible: purple to yellow (1.77–2.18 eV)

Line-label	Energy (eV)	Frequ. ($10^3 cm^{-1}$)	Wavel. (nm)	Name	Impur./defect	Comment	Figs. **3–6** [Zai01]; Fig. **7** [Zai98]	References
HA1784	1.784	14.39	695.0	...	As?	As doped...	...	[Kli75c]
HA1796	1.796	14.49	690.3	...	Ni, B?	Stimulated by B doping	...	[Zai01]
HA1797	1.797	14.49	689.9	...	Ni, N?	Related to 1.744 eV line?	...	[Yel96]
HA1485h	1.816	14.65	682.7	1Ni-DAP25h	$*V_1Ni_1{}^+$	$s = 8$, calc. 1.810 eV	**5**.165 = **7**.51a	[Law93c]
HA1816a	1.816	14.65	682.7	Ni-DAP83a	$*V_1Ni_1{}^\circ$	$s = 106$, calc. 1.817 eV	**5**.165 = **7**.51a	[Law93c]
HA1485i	1.870	15.08	663.0	1Ni-DAP25i	$*V_1Ni_1{}^+$	$s = 6$, calc. 1.872 eV	**5**.165 = **7**.51a	[Law93c]
HA1816b	1.870	15.08	663.0	Ni-DAP83b	$*V_1Ni_1{}^\circ$	$s = 50$, calc. 1.867 eV	**5**.165 = **7**.51a	[Law93c]
HA1878	1.878	15.15	660.2	...	Ti, In?	Ti + In doped, related to 1.722 eV line?	...	[Kli75b, But76]
HA1883 = HA1958z	1.883	15.19	658.4	Ni-DAP107z	$*(V_3Ni_2)^\circ$	Three lines a, b (1.906), –c (1.914 eV), LVM + 61 meV, ZPL of DAP107a–g (2.300 eV); ^{13}C see Table 7.2	7.57, 7.77	[Col83c, Col90h, Col88c, Naz93c]
HA1920	1.920	15.49	645.7	...	...	...	...	[Kli75c]
HA1485j	1.936	15.62	640.4	1Ni-DAP25j	$*V_1Ni_1{}^+$	$s = 4$, calc. 1.950 eV	**5**.165, **7**.51a	[Law93c]
HA1816c	1.936	15.62	640.4	Ni-DAP83c	$*V_1Ni_1{}^\circ$	$s = 26$, calc. 1.932 eV	**5**.165, **7**.51a	[Law93c]
HA1693b	1.940	15.65	639.1	*Ni 1.69b	$*V_1Ni_2(b)$	SB + 59 meV	...	[Zai01]

HA1958a	1.958	15.79	633.2	Ni-DAP107a	$^*(V_3Ni_2)^\circ$	Seven lines a–g (2.300 eV), $s = 228$, calc. 1.960 eV	**7**.77	[Col90h, Col88c]
HA1816d	1.975	15.93	627.7	Ni-DAP83d	$^*V_1Ni_1{}^\circ$	$s = 18$, calc. 1.977 eV	**5**.165 = **7**.51a	[Law93c]
HA1958b	1.986	16.02	624.3	Ni-DAP107b	$^*(V_3Ni_2)^\circ$	$s = 130$, calc. 1.986 eV	**7**.77	[Col90h, Col88c]
HA1693c	1.991	16.06	622.7	*Ni 1.69c	$^*V_1Ni_2(b)$	SBs + 59 meV	...	[Zai01]
HA1816e	2.020	16.29	613.7	Ni-DAP83e	$^*V_1Ni_1{}^\circ$	$s = 14$, calc. 2.015 eV	**5**.165 = **7**.51a	[Law93c]
HA1958c	2.022	16.31	613.1	Ni-DAP107c	$^*(V_3Ni_2)^\circ$	$s = 78$, calc. 2.016 eV	**7**.77	[Col90h, Col88c]
HA1958d	2.033	16.40	609.8	Ni-DAP107d	$^*(V_3Ni_2)^\circ$	$s = 64$, calc. 2.030 eV	**7**.77	[Col90h, Col88c]
HA1485k	2.070	16.70	598.9	1Ni-DAP25k	$^*V_1Ni_1{}^+$	$s = 2$, calc. 2.070 eV	**5**.165 = **7**.51a	[Law93c]
HA2070a	2.070	16.70	598.9	DAP1a	$^*N_1{}^\circ + B_1{}^\circ$	13 lines a–m (see MA), HA: j–l, $s = 50$	...	[Law93b]
HA1816f	2.070	16.70	598.9	Ni-DAP83f	$^*V_1Ni_1{}^\circ$	$s = 10$, calc. 2.071 eV	**5**.165 = **7**.51a	[Law93c]
HA1958e	2.090	16.86	593.2	Ni-DAP107e	$^*(V_3Ni_2)^\circ$	$s = 34$, calc. 2.084 eV	**7**.77	[Col90h, Col88c]

Table 3.1.5 HPHT diamond: absorption lines, visible: green to violet (2.18–3.10 eV)

Line-label	Energy (eV)	Frequ. ($10^3 cm^{-1}$)	Wavel. (nm)	Name	Impur./defect	Comment	Figs. **3–6** [Zai01]; Fig. **7** [Zai98]	References
HA1958f	2.200	17.75	563.5	Ni-DAP107f	$^*(V_3Ni_2)^\circ$	$s = 14$, calc. 2.199 eV	**7**.77	[Col90h, Col88c]
HA2245	2.245	18.11	552.2	...	Ni?	...	...	[Yel96]
HA2267	2.267	18.29	546.9	...	Ni?	...	...	[Zai01]
HA2295	2.295	18.51	540.2	...	Ni?	var. 2.292–2.298 eV	...	[Zai01]
HA1958g	2.300	18.55	539.0	Ni-DAP107g	$^*(V_3Ni_2)^\circ$	$s = 8$, calc. 2.999 eV	**7**.77	[Col90h, Col88c]
HA1704b	2.401	19.37	516.4	*Ni1.70 b	$^*V_1Ni_1{}^\circ$	ZPL of DAP29(a-f) see ME	...	[Col83c, Zai01]
HA2464a	2.464	19.88	503.2	H3a-DAP50z	$^*V_1N_2{}^\circ$(a)	Two lines a, b (3.361 eV); ZPL of DAP50 (see MA)	...	[Zai01]
HA2070j	2.495	20.12	496.9	DAP1j	$^*N_1 + B_1{}^\circ$	$s = 4$, calc. 2.481 eV	...	[Law93b]
HA2496a	2.496	20.13	496.7	S3-DAP58a	$(V_2Ni_1)N_2$	Eight lines (see MA/ML), $s = 162$, calc. 2.494 eV	...	[Nad93, Zai01]
HA2509a	2.509	20.24	494.1	2.51 eV(a)	Ni?	...	**7**.57, **7**.77	[Col83c, Col90h]
HA2509b	2.510	20.25	493.9	2.51 eV(b)	Ni?	...	**7**.57, **7**.77	[Col83c, Col90h]
HA2509c	2.511	20.25	493.7	2.51 eV(c)	Ni?	...	**7**.57, **7**.77	[Col83c, Col90h]
HA2515a	2.515	20.59	493.0	...	$(V_2Ni_1)N_3$	13 lines (see MA/ME/ML), $s = 228$, calc. 2.516 eV	...	[Nad93, Zai01]
HA2530	2.530	20.41	490.0	...	Al?	Al doped	...	[Kli75c]

HA2553	2.553	20.59	485.6	...	Ni?	Weak line	...	[Zai01]
HA2562a	2.562	20.67	483.9	2Ni-2.56 eV	$^*(V_3Ni_2)^+$	Four lines a–d (3.076 eV), (lines a, b weak in absorption, but dominant in luminescence (HL))	**5**.165. . ..	[Law93c, Zai01]
HA2070k	2.568	20.71	482.8	DAP1k	$^*N_1 + B_1{}^\circ$	$s = 3$, calc. 2.565 eV	...	[Law93b]
HA2562b	2.588	20.87	479.0	2Ni-DAP105z	$^*(V_3Ni_2)^+$	ZPL of DAP105a–d (3.221 eV)	...	[Col83c, Zai01]
HA2070-l	2.623	21.16	472.7	DAP1-l	$^*N_1 + B_1{}^\circ$	$s = 2$, calc. 2.611 eV	...	[Law93b]
HA2637	2.637	21.27	470.1	...	P?	P doped	...	[Kli75c]
HA2780a	2.780	22.42	446.0	2Ni-DAP105a	$^*(V_3Ni_2)^+$	Four lines a–d (3.221 eV), $s = 50$, calc. 2.777 eV	**5**.165	[Law93c]
HA2780b	2.980	24.04	416.0	2Ni-DAP105b	$^*(V_3Ni_2)^+$	$s = 12$, calc. 2.976 eV; DAP105a, b weak in abs.	**5**.165	[Law93c]
HA2985	2.985	24.08	415.3	N3c	$V_1N_3{}^\circ$	ZPL of DAP23a–p (3.762 eV); 0.59 meV see NA excited state splitting	...	[Zai01]
HA2562c	3.065	24.72	404.5	2Ni-DAP105c	$^*(V_3Ni_2)^+$	$s = 8$, calc. 3.062 eV, four QLVM 26 meV (2Ni + 2C)	**5**.141 = **7**.142	[Col90e, Zai01]
HA2562c	3.076	24.81	403.0	2Ni-DAPc	$^*(V_3Ni_2)^+$	11 meV split-off line, two QLVM 26 meV	**5**.141 = **7**.142	[Col90e, Zai01]

Table 3.1.6 HPHT diamond: absorption lines, near to far ultraviolet (3.10–5.90 eV)

Line-label	Energy (eV)	Frequ. ($10^3 cm^{-1}$)	Wavel. (nm)	Name	Impur./defect	Comment	Figs. **3–6** [Zai01]; Fig. **7** [Zai98]	References
HA3100	3.100	25.00	400.0	...	...	...	...	[Zai01]
HA2780d	3.221	25.98	384.9	2Ni-DAP105d	$^*(V_3Ni_2)^+$	$s = 4$, calc. 3.225 eV	**5**.141	[Col90e]
HA3333	3.333	26.88	372.0	*__S4__-DAP59a	$V_2Ni_1N_3{}^+$	13 lines (see MA/ME), $s = 86$, calc. 3.333 eV	...	[Nad93, Zai01]
HA2464b	3.361	27.11	368.9	H3b-DAP52z	$^*V_1N_2{}^\circ$(b)	ZPL of DAP52 (see MA)	...	[Zai01]
HA3400	3.400	27.42	364.6	...	...	...	...	[Kli75c]
HA3540	3.540	28.55	350.2	...	...	...	...	[Kli75c]
HA3810	3.810	30.73	325.4	...	...	...	...	[Kli75c]
HA3862a	3.862	31.15	321.0	DAP67a	$^*N_2{}^+ + B_1{}^\circ$	Nine lines a–i (4.280 eV), $s = 50$, calc. 3.858 eV; see Table 9.1.4	5.168	[Col97]
HA3862b	3.898	31.44	318.1	DAP67b	$^*N_2{}^+ + B_1{}^\circ$	$s = 34$, calc. 3.896 eV	5.168	[Col97]
HA3862c	3.926	31.96	312.9	DAP67c	$^*N_2{}^+ + B_1{}^\circ$	$s = 26$, calc. 3.927 eV	5.168	[Col97]
HA3862d	3.990	32.18	310.7	DAP67d	$^*N_2{}^+ + B_1{}^\circ$	$s = 16$, calc. 3.993 eV	5.168	[Col97]
HA3862e	4.016	32.39	308.7	DAP67e	$^*N_2{}^+ + B_1{}^\circ$	$s = 14$, calc. 4.016 eV	5.168	[Col97]
HA3862f	4.039	32.58	307.0	DAP67f	$^*N_2{}^+ + B_1{}^\circ$	$s = 12$, calc. 4.042 eV	5.168	[Col97]
HA4059	4.059	32.74	305.4	**C(c)**	$N_1{}^\circ$	2 SBs + 61 meV, ZPL of C(γ2) band (4.55 eV)	**5**.164	[Col97, Naz87]
HA3862g	4.076	32.88	304.2	DAP67g	$^*N_2{}^+ + B_1{}^\circ$	$s = 10$, calc. 4.076 eV	5.168	[Col97]
HA3862h	4.100	33.07	302.4	DAP67h	$^*N_2{}^+ + B_1{}^\circ$	$s = 6$, calc. 4.189 eV	5.168	[Col97]
HA3862i	4.280	34.52	289.7	DAP67i	$^*N_2{}^+ + B_1{}^\circ$	$s = 4$, calc. 4.272 eV	5.168	[Col97]
HA4567	4.567	36.84	271.5	***F(c)**	$^*V_1N_4(C_2)_i$	Seven lines a–g (4.906 eV), see NA; ZPL of DAP81a–g (4.721–4.906 eV), see Table 8.1.3.6	...	[Naz87, Zai01]

3.2 Excitation of Photoluminescence Lines (HE)

Table 3.2.1 HPHT diamond: Excitation of photoluminescence lines (PLE)[a], visible, violet (2.75–3.10 eV)

Line-label	Energy (eV)	Frequ. ($10^3 cm^{-1}$)	Wavel. (nm)	Name	Impur./defect	Comment	Figs. **3–6** [Zai01]; Fig. **7** [Zai98]	References
HE3065c	3.065	24.72	404.5	2Ni-2.56c	$^*(V_3Ni_2)^+$	Four lines a–d (3.076 eV), PLE of line a (2.562 eV) = DAP106z, and line b (2.588 eV), see HL	...	[Col90e, Zai01]
HE3075a	3.075	24.80	403.2	N3c-DAP23a	$V_1N_3{}^\circ$	PLE of N3 (a–c); 15 lines a–o (3.762 eV), see NE; DAP23a, $s = 162$, calc. 3.075 eV	...	[Cla70, Sob76, Iak00a]
HE3065d	3.076	24.81	403.0	2Ni-2.56d	$^*(V_3Ni_2)^+$	PLE of line a (2.562 eV)	...	[Col90e, Zai01]

[a]Possibly, the ME lines of the centers S2-DAP57a–m (2.515–3.341 eV), S3-DAP58a–i (2.496–2.890 eV), and S4-DAP60a–l (3.333–3.820 eV) are also observed as HE lines (see Table 2, column 1, in [Nad93])

3.3 Luminescence Lines (HL)

Table 3.3.1 HPHT diamond: luminescence lines, mid infrared (0.18–1.24 eV)

Line-label	Energy (eV)	Frequ. ($10^3 cm^{-1}$)	Wavel. (nm)	Name	Impur./defect	Comment	Figs. **3–6** [Zai01]; Fig. **7** [Zai98]	References
HL1000a	1.000	8.066	1240	1Ni-DAP26a	$^*V_1Ni_1{}^+$	11 lines a–k (1.323 eV), $s = 8$, calc. 0.996 eV	**5**.12 = **7**.58	[Col83c, Dea65a, Dea65b]
HL1000b	1.071	8.639	1158	1Ni-DAP26b	$^*V_1Ni_1{}^+$	$s = 12$, calc. 1.071 eV; named Ni-D_6	**5**.12 = **7**.58	[Col83c, Dea65a, Dea65b]
HL1000c	1.111	8.961	1116	1Ni-DAP26c	$^*V_1Ni_1{}^+$	$s = 16$, calc. 1.116 eV	...	[Col89a]
HL1000d	1.133	9.139	1094	1Ni-DAP26d	$^*V_1Ni_1{}^+$	$s = 18$, calc. 1.132 eV; named Ni-D_5	**5**.12 = **7**.58	[Col83c, Dea65a, Dea65b]
HL1000e	1.177	9.494	1053	1Ni-DAP26e	$^*V_1Ni_1{}^+$	$s = 26$, calc. 1.177 eV; named Ni-D_4	**5**.12 = **7**.58	[Col83c, Dea65a, Dea65b]
HL1000f	1.190	9.599	1042	1Ni-DAP26f	$^*V_1Ni_1{}^+$	$s = 30$, calc. 1.193 eV	...	[Col89a]
HL1000g	1.238	9.986	1001	1Ni-DAP26g	$^*V_1Ni_1{}^+$	$s = 50$, calc. 1.241 eV; named Ni-D_3	**5**.12 = **7**.58	[Col83c, Dea65a, Dea65b]

Table 3.3.2.1 High pressure synthetic diamond: luminescence lines, near infrared, Part 1 (1.24–1.63 eV)

Line-label	Energy (eV)	Frequ. ($10^3 cm^{-1}$)	Wavel. (nm)	Name	Impur./defect	Comment	Figs. **3–6** [Zai01]; Fig. **7** [Zai98]	References
HL1000h	1.245	10.04	995.8	1Ni-DAP26h	$^*V_1Ni_1{}^+$	$s = 54$, calc. 1.246 eV	**5**.12 = **7**.58	[Col83c, Dea65a, Dea65b]
HL1000i	1.276	10.29	971.6	1Ni-DAP26i	$^*V_1Ni_1{}^+$	$s = 86$, calc. 1.279 eV	**5**.12 = **7**.58	[Col83c, Dea65a, Dea65b]
HL1000j	1.303	10.51	951.5	1Ni-DAP26j	$^*V_1Ni_1{}^+$	$s = 130$, calc. 1.302 eV, named Ni-D_2	**5**.12 = **7**.58	[Col83c, Dea65a, Dea65b]
HL1000k	1.323	10.67	937.1	1Ni-DAP26k	$^*V_1Ni_1{}^+$	$s = 198$, calc. 1.321 eV, together with 1.358 eV line peak at 1.340 eV, named Ni-D_1	**5**.12 = **7**.58	[Col83c, Dea65a, Dea65b]
HL1401-s1	1.358	10.95	912.9	1Ni-1.40-s1	$^*V_1Ni_1{}^+$	SB, −45 meV; 1Ni-QLVM, calc. −43.6 meV	**5**.12 = **7**.58	[Col83c, Dea65a, Dea65b]
HL1401a = DAP26z	1.401	11.30	884.9	1Ni-1.40a	$^*V_1Ni_1{}^+$	Two lines a, b from 2.7 meV ground state splitting; named 1.4 eV center ZPL of 1Ni-DAP26a–k (1.000)	**5**.12 = **7**.58	[Col83c, Dea65a, Dea65b]
HL1401b = DAP26z	1.404	11.32	883.2	1Ni-1.40b	$^*V_1Ni_1{}^+$	Main line (intensities a:b = 4:6), named Ni-D_0, Ni_1 from isotopes (see Table 7.5)	**5**.12 = **7**.58	[Col83c, Dea65a, Dea65b]
HL1443a	1.443	11.64	859.2	DAP5a	$^*(V_2Zr_1)^\circ$	Ten lines a–j (1.565 eV); $s = 228$, calc. 1.444 eV	**5**.18	[Sit95]
HL1443b	1.457	11.75	850.9	DAP5b	$^*(V_2Zr_1)^\circ$	$s = 162$, calc. 1.455 eV	**5**.18	[Sit95]
HL1443c	1.469	11.85	844.0	DAP5c	$^*(V_2Zr_1)^\circ$	$s = 130$, calc. 1.470 eV	**5**.18	[Sit95]
HL1472	1.472	11.87	842.2	...	Co?	Width 20 meV, related to Co 1.99 eV center?	**5**.13b	[Kan99, Sit96]
HL1443d	1.482	11.95	836.6	DAP5d	$^*(V_2Zr_1)^\circ$	$s = 106$, calc. 1.481 eV	**5**.18	[Sit95]

(continued)

Table 3.3.2.1 (continued)

Line-label	Energy (eV)	Frequ. ($10^3 cm^{-1}$)	Wavel. (nm)	Name	Impur./defect	Comment	Figs. **3–6** [Zai01]; Fig. **7** [Zai98]	References
HL1443e	1.496	12.07	828.7	DAP5e	$^*(V_2Zr_1)^\circ$	$s = 86$, calc. 1.494 eV	**5**.18	[Sit95]
HL1443f	1.508	12.16	822.1	DAP5f	$^*(V_2Zr_1)^\circ$	$s = 72$, calc. 1.506 eV	**5**.18	[Sit95]
HL1443g	1.522	12.28	814.6	DAP5g	$^*(V_2Zr_1)^\circ$	$s = 58$, calc. 1.522 eV	**5**.18	[Sit95]
HL1443h	1.534	12.37	808.2	DAP5h	$^*(V_2Zr_1)^\circ$	$s = 50$, calc. 1.534 eV; dominant line	**5**.18	[Sit95]
HL1541	1.541	12.43	804.5	...	Si, Ni?	...	...	[Sit95, Sit96]
HL1443i	1.549	12.49	800.4	DAP5i	$^*(V_2Zr_1)^\circ$	$s = 42$, calc. 1.550 eV	**5**.18	[Sit95]
HL1563	1.563	12.60	793.4	***S5**-DAP60a	$^*(V_2Ni_1)N_x{}^\circ$	11 lines a–k (see MA/ME/ML), $s = 162$, calc. 1.564 eV	....	[Kan99, Kup99]
HL1443j	1.565	12.62	792.2	DAP5j	$^*(V_2Zr_1)^\circ$	$s = 36$, calc. 1.565 eV	**5**.18	[Sit95]
HL1568a	1.568	12.65	790.7	Ni-DAP84a	$^*V_1Ni_1{}^\circ$	Three lines a–c (1.627 eV), $s = 72$, calc. 1.567 eV	...	[Kup99]
HL1568b	1.621	13.07	764.8	Ni-DAP84b	$^*V_1Ni_1{}^\circ$	$s = 198$, calc. 1.621 eV	...	[Kup99]
HL1568c	1.627	13.12	762.0	Ni-DAP84c	$^*V_1Ni_1{}^\circ$	$s = 228$, calc. 1.627 eV	...	[Kup99]

Table 3.3.2.2 HPHT diamond: luminescence lines, near infrared, Part 2 (1.67–1.77 eV)

Line-label	Energy (eV)	Frequ. ($10^3 cm^{-1}$)	Wavel. (nm)	Name	Impur./defect	Comment	Figs. **3–6** [Zai01]; Fig. **7** [Zai98]	References
HL1673	1.673	13.49	741.0	(C-band)	...	ZPL of C-band (1.33 eV), 11 SBs at 53–40 meV	...	[Wig71, Wal79]
HL1679 = HL1487z	1.679	13.55	738.2	2Si-DAP54z	$^*(V_3Si_2)^+$	ZPL of DAP54a–g (1.487 eV), see LL; 2Si from isotopes (Table 7.4)	**5**.47a	[Ste95, Cla95]
HL1681	1.682	13.57	737.1	Si-1.7 eV	$^*(V_3Si_2)^\circ$	...	**5**.47a	[Ste95]
HL1704	1.704	13.74	727.6	Ni-DAP84z	$^*V_1Ni_1{}^\circ$	ZPL of DAP84a–c (1.568 eV)	...	[Kup99, Zai01]
HL1716a	1.716	13.84	722.6	...	Ni?	Two lines a, b (1.720 eV) from excited state splitting	...	[Zai01]
...HL1716b	1.720	13.88	720.6	...	Ni?	...	...	[Zai01]
HL1744	1.744	14.06	711.0	...	Ni, N?	...	...	[Yel96]

Table 3.3.3 HPHT diamond: luminescence lines, visible: purple to yellow (1.77–2.18 eV)

Line-label	Energy (eV)	Frequ. ($10^3 cm^{-1}$)	Wavel. (nm)	Name	Impur./defect	Comment	Figs. **3–6** [Zai01]; Fig. **7** [Zai98]	References
HL1772a	1.772	14.29	699.8	**Co**-DAP112a	$^*(V_3Co_2)^\circ$	Six lines a–f (1.914 eV), s = 30, calc. 1.772 eV	**5**.13b	[Law96]
......HL1779a	1.779	14.35	696.9	Ni-DAP108a	$^*(V_3Ni_2)^\circ$	Two lines a, b (1.808 eV), s = 130, calc. 1.780 eV	...	[Naz93c]
HL1796	1.796	14.49	690.3	...	Ni, B?	...	...	[Zai01]
HL1779b	1.808	14.58	685.7	Ni-DAP108b	$^*(V_3Ni_2)^\circ$	s = 228, calc. 1.806 eV	...	[Naz93c]
HL1819	1.819	14.68	681.4	...	...	...	...	[Zai01]
HL1772b	1.828	14.74	678.4	Co-DAP112b	$^*(V_3Co_2)^\circ$	s = 54, calc. 1.827 eV	**5**.13b, **5**.62a	[Law96]
HL1854	1.854	14.96	668.6	...	Ni, Si?	...	...	[Sit95, Sit96]
HL1772c	1.862	15.02	665.8	Co-DAP112c	$^*(V_3Co_2)^\circ$	s = 86, calc. 1.862 eV	**5**.13b, **5**.62a	[Law96]
HL1772d	1.870	15.08	663.1	Co-DAP112d	$^*(V_3Co_2)^\circ$	s = 98, calc. 1.870 eV	**5**.13b, **5**.62a	[Law96]
HL1883 = HL1779z	1.883	15.19	658.4	Ni-1.88 eV a	$^*(V_3Ni_2)^\circ$	ZPL of DAP108<u>a</u>, b (1.779 eV); in HA 3 ZPL lines b (1.906), −c (1.914 eV); QLVM(2Ni2C) −26 meV	...	[Naz93c]
HL1888	1.888	15.23	656.5	...	Ni, Si?	...	...	[Sit95, Sit96]

HL1772e	1.890	15.24	656.2	Co-DAP112e	$^{*}(V_3Co_2)^{\circ}$	$s = 130$, calc. 1.885 eV	**5**.13b, **5**.62a	[Law96]
HL1910a	1.910	15.41	649.1	NV-DAP47a	$V_1N_1{}^{\circ}$	Eight lines a–h (2.059 eV), 8 lines DAP47(a-h) see ML	**7**.106	[Fre93b, Zai96b]
HL1772f	1.914	15.44	647.7	Co-DAP112b	$^{*}(V_3Co_2)^{\circ}$	$s = 228$, calc. 1.914 eV	**5**.13b, **5**.62a	[Law96]
HL1940a	1.940	15.65	639.1	...	$^{*}V_1Ni_1{}^{\circ}$	Two lines a, b (1.991 eV), SB (LVM) −59 meV	...	[Zai01]
HL1984b = HL1772z	1.989	16.04	623.3	Co-1.99 eV	$^{*}(V_3Co_2)^{\circ}$	Three lines a (1.984), b, c (1.991 eV) from excited state splitting; ZPL of DAP112a-f (1.772 eV)	**5**.13b, **5**.62a	[Law96]
HL1693c	1.991	15.48	646.1	*Ni 1.69c	$^{*}V_1Ni_2{}^{+}$(c)	ZPL of DAP111(a-f) SB (LVM) −59 meV	...	[Zai01]
HL2135a	2.135	17.22	580.7	Co-DAP104	$^{*}(V_2Co_1)N_x$	Eight lines a–h (2.820 eV) of Co-S5-DAP104, see ML and Table 9.6	**5**.13b	[Kan99, Law96]
HL2136	2.136	17.23	580.3	...	Ni?	...	...	[Yel99]
HL2139	2.139	17.25	579.6	...	Co, Si?	...	**5**.13b	[Law96]
HL2154 = HL1910z	2.154	17.38	575.7	DAP47z = **NV center**	$V_1N_1{}^{\circ}$	ZPL of DAP47a-h (1.910 eV), see ML; named **575-nm** center, ^{15}N see Table 7.3	**7**.106	[Fre93b, Zai96b]

Table 3.3.4 HPHT diamond: luminescence lines, visible: green (2.18–2.43 eV)

Line-label	Energy (eV)	Frequ. ($10^3 cm^{-1}$)	Wavel. (nm)	Name	Impur./defect	Comment	Figs. **3–6** [Zai01]; Fig. **7** [Zai98]	References
HL2.237	2.237	18.04	554.2	**H3a**-DAP54a	V_1N_2°	Six lines a–f (2.323 eV), see ML, $s = 42$, calc. 2.238	**7**.115	[Col92b]
HL2238	2.238	18.05	554.0	...	Co + Si?	...	...	[Sit95, Sit96]
HL2245	2.245	18.11	552.2	...	Ni?	...	...	[Zai01]
HL2370 = HL2170z	2.370	19.12	523.1	Ni-DAP92z	*V_1Ni_3°	ZPL of DAP92a–f (2.170 eV), see ML, named A, excited with *S4-DAP59 (3.333–3.820 eV)	...	[Zai01]
HL2379	2.379	19.19	521.1	...	...	...	...	
HL2383a	2.383	19.22	520.3	2Ni-DAP106a	*$(V_3Ni_2)^+$	Six lines a–f (2.466 eV), $s = 42$, calc. 2.381 eV	**5**.122, **5**.125	[Col83c, Zai01]
HL2385	2.385	19.24	519.8	...	Co?	...	...	
HL2383b	2.393	19.30	518.1	2Ni-DAP106b	*$(V_3Ni_2)^+$	$s = 50$, calc. 2.397 eV	**5**.122, **5**.125	[Col83c, Zai01]
HL2395	2.395	19.32	517.7	DAP73z	*V_1B_1°(a)	ZPL of DAP73a–f (see ML); three SBs with 49 meV	**7**.106	[Fre93b]
HL2383c	2.404	19.39	515.7	2Ni-DAP106c	*$(V_3Ni_2)^+$	$s = 54$, calc. 2.403 eV	**5**.122, **5**.125	[Col83c, Zai01]
HL2408	2.408	19.42	514.9	...	Ni?	...	...	
HL2383d	2.417	19.50	512.9	2Ni-DAP106d	*$(V_3Ni_2)^+$	$s = 64$, calc. 2.416 eV	**5**.122, **5**.125	[Col83c, Zai01]

Table 3.3.5 High pressure synthetic diamond: luminescence lines, visible: blue (2.43–2.58 eV)

Line-label	Energy (eV)	Frequ. ($10^3 cm^{-1}$)	Wavel. (nm)	Name	Impur./defect	Comment	Figs. **3–6** [Zai01]; Fig. **7** [Zai98]	References
HL2383e	2.439	19.67	508.3	2Ni-DAP106e	$*(V_3Ni_2)^+$	$s = 86$, calc. 2.436 eV	5.122, 5.125	[Col83c, Zai01]
HL2440a	2.440	19.68	508.1	**N3c**-DAP24a	$V_1N_3{}^{\circ}$	17 lines a–q (2.897 eV), see NL; $s = 4$, calc. 2.438 eV	. . .	[Zai01]
HL2464a = HL2237z	2.464	19.88	503.2	**H3a**-DAP51z	$V_1N_2{}^{\circ}$	Two lines a, b (3.361 eV = DAP52z); ZPL of DAP51a–f (2.323 eV), see ML	7.112	[Col92b]
HL2383f	2.466	19.89	502.7	2Ni-DAP106f	$*(V_3Ni_2)^+$	$s = 162$, calc. 2.470 eV	5.122, 5.125	[Col83c, Zai01]
HL2468	2.468	19.91	502.3	. . .	Ni?	SBs −43 meV	. . .	[Yel99]
HL2496a	2.496	20.13	496.7	**S3**-DAP58a	$(V_2Ni_1)N_2$	Eight lines (see MA/ME/ML), $s = 162$, calc. 2.494 eV	7.115	[Yel92b, Zai01]
HL2499	2.499	20.16	496.1	. . .	Ni?	Two lines a, b (2.512 eV)	. . .	[Vin88a, Vin88b]
HL2510	2.510	20.23	494.0	. . .	N?	2 SBs with 40 meV; related to HB2504?	5.119a	[Ger71]
HL2515a	2.515	20.59	493.0	**S2**-DAP57	$(V_2Ni_1)N_3$	13 lines (see ML), $s = 228$, calc. 2.516 eV; lines "B" (2.537), "C" (2.597), "D" (2.623), "E" (2.634)	. . .	[Nad93, Zai01]
HL2520	2.520	20.33	492.0	. . .	. . .	. . .	. . .	[Zai01, Zai96b]
HL2562-s1	2.538	20.47	488.5	2Ni-2.56a–s1	$*(V_3Ni_2)^+$	SB (−24 meV), QLVM of 2Ni2C, calc. −26.2	5.125	[Col83c]
HL2553	2.553	20.59	485.6	. . .	Ni?	Weak line, observed in ML	. . .	[Zai01]
HL2562a = HL2383z	2.562	20.67	483.9	2Ni-2.56a = 2Ni-DAP106z	$*(V_3Ni_2)^+$	Three lines, (b) 2.588, (c1,2) 3.064, 3.076 eV; ZPL of DAP106a–f (2.383 eV)[a]; named 484 nm ce., very slow luminescence (140 μs at 2.3 K)	5.124, 5.125, 5.122, 5.139b	[Col97, Col83c, Zai01, Kan98]

[a]The 2.562 eV line is part of a quintet from Ni_2 isotopes (m = 116, 118, 120, 122, 124, see Table 7.5) with 1.56 meV separation, each quintet line is further split into a triplet with 0.33 meV separation; 9 of 15 lines are observed in the range 2.558–2.565 eV [Col97]

Table 3.3.6 HPHT diamond: luminescence lines, visible: ultramarine, violet (2.58–3.10 eV)

Line-label	Energy (eV)	Frequ. ($10^3 cm^{-1}$)	Wavel. (nm)	Name	Impur./defect	Comment	Figs. **3–6** [Zai01]; Fig. **7** [Zai98]	References
HL2562b	2.588	20.87	479.0	2Ni-2.56b	$^{*}(V_3Ni_2)^+$	Weak ZPL (in absorption ZPL of DAP105)	**5**.124, **5**.125	[Col97, Col83c]
HL2694	2.694	21.73	460.2	...	Al, Ti?	...	...	[Kli75c, Dav94a]
HL2889	2.889	23.30	429.1	...	Co?	...	...	[Fie92, Zai01]
HL2950	2.950	23.80	420.3	...	Al, B, Si?	...	...	[Zai01]
HL2964a	2.964	23.91	418.3	DAP66z	$^{*}V_1Al_1{}^{\circ}$	Two lines a, b (2.974 eV); ZPL of DAP66a–i (see NL); dominant line at $T < 50$ K	...	[Zai01]
HL2972	2.972	23.97	417.2	...	Ni, Ti, Al?	...	**5**.139	[Kan98]
HL2964b	2.974	23.99	416.9	...	$^{*}V_1Al_1{}^{\circ}$	Dominant at $T > 50$ K; slow (delay 0.1–2.0 ms), similar DAP sidebands like DAP66 from line a		[Zai01]
HL2985	2.985	24.08	415.3	N3c	$V_1N_3{}^{\circ}$	ZPL of DAP24a–q (2.897 eV); 0.59 meV excited state splitting; named D_0 (see NL)	...	[Zai01]

Table 3.3.7.1 HPHT diamond: luminescence lines, far ultraviolet, Part 1 (4.43–5.07 eV)

Line-label	Energy (eV)	Frequ. ($10^3 cm^{-1}$)	Wavel. (nm)	Name	Impur./defect	Comment	Figs. **3–6** [Zai01]; Fig. **7** [Zai98]	References
HL4711a	4.711	38.00	263.2	DAP2a	$P_1 + B_1{}^\circ$	23 lines a–w (5.215 eV); shell = 50; calc. 4.711 eV	...	[Ste99a, Ste99b]
HL4711b	4.736	38.21	261.8	DAP2b	$P_1 + B_1{}^\circ$	$s = 39$, calc. 4.739 eV	...	[Ste99a, Ste99b]
HL4711c	4.745	38.27	261.3	DAP2c	$P_1 + B_1{}^\circ$	$s = 37$, calc. 4.746 eV	...	[Ste99a, Ste99b]
HL4711d	4.763	38.42	260.3	DAP2d	$P_1 + B_1{}^\circ$	$s = 32$, calc. 4.763 eV	...	[Ste99a, Ste99b]
HL4711e	4.769	38.47	260.0	DAP2e	$P_1 + B_1{}^\circ$	$s = 31$, calc. 4.769 eV	...	[Ste99a, Ste99b]
HL4711f	4.774	38.51	259.7	DAP2f	$P_1 + B_1{}^\circ$	$s = 30$, calc. 4.774 eV	...	[Ste99a, Ste99b]
HL4711g	4.796	38.68	258.5	DAP2g	$P_1 + B_1{}^\circ$	$s = 26$, calc. 4.796 eV	...	[Ste99a, Ste99b]
HL4711h	4.801	38.72	258.2	DAP2h	$P_1 + B_1{}^\circ$	$s = 25$, calc. 4.799 eV	...	[Ste99a, Ste99b]
HL4711i	4.806	38.76	258.0	DAP2i	$P_1 + B_1{}^\circ$	$s = 24$, calc. 4.804 eV	...	[Ste99a, Ste99b]
HL4711j	4.820	38.88	257.2	DAP2j	$P_1 + B_1{}^\circ$	$s = 22$, calc. 4.818 eV	...	[Ste99a, Ste99b]
HL4711k	4.834	38.99	256.5	DAP2k	$P_1 + B_1{}^\circ$	$s = 20$, calc. 4.833 eV	...	[Ste99a, Ste99b]
HL4711-l	4.846	39.09	255.8	DAP2-l	$P_1 + B_1{}^\circ$	$s = 19$, calc. 4.844 eV	...	[Ste99a, Ste99b]
HL4711m	4.853	39.14	255.5	DAP2m	$P_1 + B_1{}^\circ$	$s = 18$, calc. 4.851 eV	...	[Ste99a, Ste99b]
HL4711n	4.861	39.21	255.0	DAP2n	$P_1 + B_1{}^\circ$	$s = 17$, calc. 4.864 eV	...	[Ste99a, Ste99b]
HL4711-o	4.870	39.28	254.6	DAP2-o	$P_1 + B_1{}^\circ$	$s = 16$, calc. 4.872 eV	...	[Ste99a, Ste99b]
HL4711p	4.890	39.44	253.5	DAP2p	$P_1{}^\circ + B_1{}^\circ$	$s = 15$, calc. 4.888 eV	...	[Ste99a, Ste99b]
HL4711q	4.900	39.52	253.0	DAP2q	$P_1{}^\circ + B_1{}^\circ$	$s = 14$, calc. 4.898 eV	...	[Ste99a, Ste99b]
HL4711r	4.929	39.76	251.5	DAP2r	$P_1{}^\circ + B_1{}^\circ$	$s = 12$, calc. 4.930 eV	...	[Ste99a, Ste99b]
HL4711s	4.955	39.97	250.2	DAP2s	$P_1{}^\circ + B_1{}^\circ$	$s = 11$, calc. 4.954 eV	...	[Ste99a, Ste99b]
HL5409a–s3	4.990	40.25	248.5	A_3 = FE(a)–s3	...	SB −419 meV (=87(TA)+2 × 166(LA/O)?)	...	[Zai01]
HL4711t	4.997	40.31	248.1	DAP2t	$P_1 + B_1{}^\circ$	$s = 9$, calc. 5.004 eV	...	[Ste99a, Ste99b]
HL5170-s1	5.015	40.51	247.2	FE-eh-drop-s1	...	SB −155 meV from 5.170 line	5.156	[Tho00]
HL5356a–s4	5.048	40.72	245.6	BE-B(a)–s4	B	SB −308 meV (= 141 + 167?); named D_2	...	[Zai01]
HL5356b–s2	5.060	40.81	245.0	BE-B(b)–s2	B	SB −307 meV (= 141 + 166?); named $D_2{}'$	...	[Zai01]
HL4711u	5.067	40.87	244.7	DAP2u	$P_1 + B_1{}^\circ$	s = 7a, b, calc. 5.073 eV; 6 meV splitting	...	[Ste99a, Ste99b]

Table 3.3.7.2 HPHT diamond: luminescence lines, far ultraviolet, Part 2 (5.11–5.90 eV)

Line-label	Energy (eV)	Frequ. ($10^3 cm^{-1}$)	Wavel. (nm)	Name	Impur./defect	Comment	Figs. **3–6** [Zai01]; Fig. **7** [Zai98]	References
HL5409a–s4	5.115	41.26	242.4	$\mathbf{B_2}$ = FE(a)–s4	....	SB −294 meV (= 2 × 147(TO)?)	...	[Zai01]
HL5409a–s5	5.151	41.55	240.7	$\mathbf{A_2}$ = FE(a)–s5	....	SB −258 meV (=87(TA)+171(LO)?)	...	[Zai01]
HL4711v	5.155	41.58	240.5	DAP2v	$P_1 + B_1{}^{\circ}$	s = 5, calc. 5.162 eV	...	[Ste99a, Ste99b]
HL5170	5.170	41.70	239.8	FE-eh-drop	...	Intense 5.66 eV pulse excitation	5.156	[Tho00]
HL5356a–s5	5.193	41.89	238.7	BE-B(a)–s5	B	SB −163 meV (= LO); named $\mathbf{D_1}''$	...	[Zai01]
HL5356a–s6	5.215	42.06	237.7	BE-B(a)–s6	B	SB −141 meV; named D_1	...	[Zai01]
HL4711w	5.215	42.06	237.7	DAP2w	$P_1 + B_1{}^{\circ}$	s = 4, calc. 5.207 eV	...	[Ste99a, Ste99b]
HL5356-s3	5.227	42.16	237.2	BE-B(b)–s3	B	SB −140 meV (= TO); named D_1'	....	[Zai01]
HL5409a–s6	5.246	42.31	236.3	$\mathbf{C_1}$ = FE(a)–s6	...	SB −163 meV (= LO)	**7**.162	[Col94c]
HL5258	5.258	42.41	235.8	$\mathbf{S_0}$-center	...	...	...	[Dea65], [Dav77c]
HL5409a–s7	5.268	42.49	235.3	$\mathbf{B_1}$ = FE(a)–s7	...	SB −141 meV (=TO)	**5**.156, **7**.162	[Tho00, Col94c]
HL5409b–s2	5.275	42.55	235.0	$\mathbf{B_1}'$ = FE(b)–s2	...	SB −141 meV (=TO)	**7**.162	[Col94c]
HL5409a–s8	5.322	42.93	233.0	$\mathbf{A_1}$ = FE(a)–s8	...	SB −87 meV (=TA)	**5**.156, **7**.162	[Tho00, Col94c]
...HL5409b–s3	5.329	42.98	232.6	$\mathbf{A_1}'$ = FE(b)–s3	...	SB −87 meV (=TA)	...	[Zai01]
HL5356a	5.356	43.20	231.5	BE-B(a)	B	ZPL (upper VB); two lines a, b (5.367 eV)	...	[Zai01]
HL5356b	5.367	43.29	231.0	BE-B(b)	B	ZPL (lower VB)	...	[Zai01]
HL5409a	5.409	43.63	229.2	FE(a)	...	ZPL (upper VB); two lines a, b (5.416 eV); indirect	...	[Zai01]
HL5409b	5.416	43.68	228.9	FE(b)	...	ZPL (lower VB); indirect determined	...	[Zai01]

3.4 Broad Bands (HB)

Table 3.4.1 HPHT diamond: broad bands, infrared (0.04–1.77 eV)

Line-label	Energy (eV)	Frequ. ($10^3 cm^{-1}$)	Wavel. (nm)	Name	Impur./defect	Comment	Figs. **3–6** [Zai01]; Fig. **7** [Zai98]	References
HB0105	0.1054	0.850	11765	...	...	**A**; $w = 25\%$	3.28	[Law93b]
HB0399a	0.3986	3.215	3110	3LP(a)	...	**A**; three lattice phonons, var. 0.396–0.399 eV; two bands a, b (0.4500);$w = 23\%$; abs. coeff.=1.7–3.0 cm^{-1}	...	[Zai01]
HB0399b	0.4500	3.630	2755	3LP(b)	...	**A**; three lattice phonons; var. 0.444–0.450 eV; $w = 8\%$; abs. coeff.= $1.7 - -3.0\,cm^{-1}$	...	[Zai01]
HB0659	0.6590	5.315	1881	...	B	**A**; $w = 22\%$	7.25a	[Bok86]
HB1400	1.400	11.29	885.6	...	Ni?	**A**; $w = 36\%$	...	[Col97, Zai01]
HB1550	1.550	12.50	799.9	...	...	**EL**	...	[Lev79, Zai01]
HB1700	1.700	13.71	729.3	...	B?	**PL**; $w = 11\%$	7.76	[Fre93a]

A absorption; *CL*, *EL*, *IL*, *PL*, *XL* luminescence; *PLE* photoluminescence excitation

Table 3.4.2 HPHT diamond: broad bands, visible (1.77–3.10 eV)

Line-label	Energy (eV)	Frequ. ($10^3 cm^{-1}$)	Wavel. (nm)	Name	Impur./defect	Comment	Figs. **3–6** [Zai01]; Fig. **7** [Zai98]	References
HB1790	1.790	14.44	692.6	...	...	**CL**; $w = 28\%$	...	[Col91b, Den92]
HB1820	1.820	14.68	681.2	Ni-DAP83	$V_1Ni_1{}^{\circ}$	A; $w = 9\%$	**5**.165, **7**.51a	[Law93c]
HB1840	1.840	14.84	673.8	...	B?	**PL**; $w = 10\%$; ZPL at 2.051 eV	**7**.76	[Fre93a]
HB1870	1.870	15.08	663.0	...	N?	A; $w = 23\%$	**7**.78	[Col89c]
HB1900	1.900	15.33	652.5	...	...	A; $w = 23\%$	...	[Zai01]
HB1904	1.904	15.36	651.0	...	Ni?	**PL**, **EL**	...	[Zai01]
HB1907	1.907	15.38	650.0	1Ni-1.40 eV	$V_1Ni_1{}^{+}$	**PLE**(a) of 1Ni-1.40 eV-DAP26	p. 140	[Zai01]
HB2100	2.100	16.94	590.4	"Yellow" band	B?	**PL**; $w = 8\%$	**7**.76	[Fre93a]
HB2175	2.175	17.54	570.0	...	Dislocations?	**PL**, **XL**; var. 2.14–2.21 eV	p. 231	[Zai01]
HB2200	2.200	17.75	563.5	A_{semic}(b)	$V_1B_1{}^{\circ}$(a)	**CL**; $w = 15\%$, see Table 10.2.2	**7**.144	[Law95]
HB2320	2.320	18.71	534.4	"Green" band	B?	**PL**; $w = 17–34\%$; var. 2.10–2.73 eV	**7**.76	[Fre93a]
HB2455	2.455	19.80	505.0	...	...	**PL, XL, Gamma-ray-L**	...	[Zai01]
HB2504	2.500	20.17	495.9	A_{semic}(c)	B?	**CL**; $w = 14\%$, see Table 10.2.2	**7**.144	[Law95]
HB2530	2.530	20.41	490.0	...	B?	**PL**, **XL**; var. 2.48–2.58 eV	p. 276	[Zai01]
HB2600	2.600	20.97	476.8	...	...	**PL**, **XL**, Gamma-ray-L;	...	[Zai01]
HB2740	2.740	22.10	452.5	...	Ni, Al	**PL**; $w = 11\%$; NBD; ZPL at 2.964 eV	**5**.139	[Kan98]
HB2750	2.750	22.18	450.8	...	...	**PL**, **XL**; $w = 18\%$;excited at 4.75 eV	...	[Yel99, Zai01]
HB2800	2.800	22.58	442.8	A_{semic}(d)	$V_1B_1{}^{\circ}$(b)	**CL**; $w = 12\%$, see Table 10.2.2	**7**.144	[Law95]
HB2990	2.990	24.12	414.6	...	Al?	**A**	...	[Kli75c]

A absorption; *CL*, *EL*, *IL*, *PL*, *XL* luminescence; *PLE* photoluminescence excitation

Table 3.4.3 HPHT diamond: broad bands, ultraviolet (3.10–5.90 eV)

Line-label	Energy (eV)	Frequ. ($10^3 cm^{-1}$)	Wavel. (nm)	Name	Impur./defect	Comment	Figs. **3–6** [Zai01]; Fig. **7** [Zai98]	References
HB3120	3.120	25.17	397.4	...	B?	**CL**; $w = 21\%$	...	[Law95]
HB3300	3.300	26.62	375.7	C(α2)	$N_1{}^{\circ}$	**A**; $w = 6\%$; ZPL at 2.80 eV; see Table 9.1.1.1	**5**.164, **7**.154	[Col97, Kop86]
HB3470	3.470	27.99	357.3	...	B?	**CL**; $w = 10$–17%; var. 3.35–3.50 eV	...	[Zai01]
HB3750	3.700	29.84	335.1	H3a-2.46 eV	$V_1N_2{}^{\circ}$	**PLE**(a) of H3a-DAP51	p. 267	[Zai01]
HB3900	3.900	31.46	317.9	C(β2)	$N_1{}^{\circ}$	**A**; $w = 6\%$; ZPL at 3.40 eV; see Table 9.1.1.1	**5**.164, **7**.154	[Col97, Kop86]
HB3999	4.000	32.26	310.0	1Ni-1.40 eV	$V_1Ni_1{}^{+}$	**PLE**(b) of 1Ni-1.40 eV-DAP26	p. 140	[Zai01]
HB4500	4.500	36.30	275.5	...	B?	**CL**; $w = 10\%$; var. 4.4–4.6 eV	...	[Zai01]
HB4600a	4.600	37.10	269.5	C(γ 2)	$N_1{}^{\circ}$	**A**; $w = 12\%$; ZPL at 4.06 eV; see Table 9.1.1.1	**5**.164, **7**.154	[Col97, Kop86]
HB4600b	4.600	37.10	269.5	H3a-2.46 eV	$V_1N_2{}^{\circ}$	**PLE**(b) of H3a-DAP51	p. 267	[Zai01]
HB4700	4.700	37.91	263.8	B(α 2)	$V_1N_4{}^{\circ}$	**A**; $w = 20\%$; ZPL at 4.191 eV; see Table 8.1.3.5	**5**.164	[Col97]
HB4750	4.750	38.31	261.0	...	...	**PLE** of 2.75 eV band	p. 303	[Zai01]
HB4768	4.768	38.46	260.0	1Ni-1.40 eV	$V_1Ni_1{}^{+}$	**PLE**(c) of 1Ni-1.40 eV-DAP26	p. 140	[Zai01]
HB4800	4.800	38.72	258.3	...	B?	**PLE** of 2.53 eV band	p. 276	[Zai01]
HB5000	5.000	40.33	248.0	...	...	**PLE** of A bands; $w = 8\%$; type Ib diamond	**5**.135b	[Iak00a]
HB5300	5.300	42.75	233.9	H3a-2.46 eV	$V_1N_2{}^{\circ}$	**PLE**(c) of H3a-DAP51	p. 267	[Zai01]
HB5520	5.520	44.52	224.6	...	...	**PLE** of A bands; $w = 7\%$ type Ib diamond	**5**.135b	[Iak00a]

A absorption; *CL*, *EL*, *IL*, *PL*, *XL* luminescence; *PLE* photoluminescence excitation

Chapter 4
Spectral Lines in Low Pressure Synthetic (CVD) Diamond

In this chapter ca. 340 lines (and bands), which are observed in low pressure synthetic (CVD) diamond (see Sect. 1.2.3) are presented in 22 tables. The entries are assigned to 93 centers. The corresponding defect structure is established for 7 centers, is almost certain for 30 centers, and is incomplete or unknown for 56 centers.

Chapter 4 is subdivided into *absorption lines* (Sect. 4.1: 71 **LA** lines in 6 tables) (see Tables 4.1.1.1–4.1.4), into *photoluminescence excitation (PLE) lines* (Sect. 3.2: three **HE** lines in one table), into *photoluminescence excitation (PLE) lines* (Sect. 4.2: 144 **LE** lines in 2 tables), (see Tables 4.2.1–4.2.2), into *luminescence lines* (Sect. 4.3: 190 **LL** lines in 11 tables, (see Tables 4.3.1.1–4.3.9)), and into *broad bands* (Sect. 4.4: 33 **LB** bands in 3 tables, (see Tables 4.4.1–4.4.3).

B. Dischler, *Handbook of Spectral Lines in Diamond*,
DOI 10.1007/978-3-642-22215-3_4, © Springer-Verlag Berlin Heidelberg 2012

4.1 Absorption Lines (LA)

Table 4.1.1.1 Low pressure synthetic diamond (CVD): absorption lines; far infrared, part 1 (0.04–0.15 eV)

Line-label	Energy (meV)	Frequ. (cm^{-1})	Wavel. (μm)	Name	Impur./ defect	Comment	Figs. **3–6**. [Zai01]; Fig. **7.** [Zai98]	References
LA0057	57.03	460.0	21.740	...	Si?	...	**3.**17	[Ghe92b]
LA0069a	68.56	553.0	18.08	1LP(a)	...	One lattice phonon, 17 lines a–q (0.1667 eV)	**3.**3a, Table 3.1	[Kle92, Zai01]
LA0069b	94.72	764.0	13.09	1LP(b)	...	...	**3.**3a, Table 3.1	[Kle92, Zai01]
LA0069c	102.8	829.0	12.06	1LP(c)	...	...	**3.**3a, Table 3.1	[Kle92, Zai01]
LA0105a/	105.4	850	11.76	**C**-cent.(a/)	N_1°	7 IR lines a/–g/ (166.6 meV), 3 UV lines: *C(a, b); indirect, *C(c) at 4.059 eV, see Table 9.1.1.1	...	[Zai01]
LA0069d	113.8	918.0	10.89	1LP(d)	...	...	**3.**3a, Table 3.1	[Kle92, Zai01]
LA0069e	121.2	978.0	10.23	1LP(e)	...	...	**3.**3a, Table 3.1	[Kle92, Zai01]
LA0069f	123.0	992.0	10.08	1LP(f)	...	...	**3.**3a, Table 3.1	[Kle92, Zai01]
LA0069g	126.3	1,019	9.814	1LP(g)	...	...	**3.**3a, Table 3.1	[Kle92, Zai01]
LA0126a	126.5	1,020	9.804	...	B?	2 lines a, b (133.9 meV)	**3.**13	[Che94b]
LA0069h	128.1	1,033	9.681	1LP(h)	...	...	**3.**3a, Table 3.1	[Kle92, Zai01]
LA0069i	129.2	1,042	9.596	1LP(i)	...	...	**3.**3a, Table 3.1	[Kle92, Zai01]

LA0105b′	129.6	1,045	9.566	C-cent.(b′)	N_1°	See Table 9.1.1.1	...	[Zai01]
LA0069j	132.9	1,072	9.329	1LP(j)	...	...	**3.**3a, Table 3.1	[Kle92, Zai01]
LA0126b	133.9	1,080	9.259	...	B?	...	**3.**13	[Che94b]
LA0105c′	136.0	1,097	9.116	C-cent.(c′)	N_1°	See Table 9.1.1.1	...	[Zai01]
LA0069k	137.7	1,111	9.003	1LP(k)	...	...	**3.**3a, Table 3.1	[Kle92, Zai01]
LA0105d′	140.1	1,130	8.849	C-cent.(d′)	N_1°	**Dominant** C-ce. line; var.(139–141 meV)	...	[Zai01]
LA0069-l	142.1	1,146	8.725	1LP(l)	...	...	**3.**3a, Table 3.1	[Kle92, Zai01]
LA0069m	147.7	1,191	8.394	1LP(m)	...	...	**3.**3a, Table 3.1	[Kle92, Zai01]
LA0105e	149.5	1,206	8.292	C-cent.(e′)	N_1°	See Table 9.1.1.1	...	[Zai01]

Table 4.1.1.2 Low pressure synthetic diamond (CVD): absorption lines; far infrared, part 2 (0.15–0.18 eV)

Line-label	Energy (meV)	Frequ. (cm^{-1})	Wavel. (μm)	Name	Impur./defect	Comment	Figs. **3–6.** [Zai01]; Fig. **7.** [Zai98]	References
LA0069n	151.3	1,220	8.194	1LP(n)	...	...	**3**.3a, Table 3.1	[Kle92, Zai01]
LA0069-o	153.6	1,239	8.071	1LP(o)	...	...	**3**.3a, Table 3.1	[Kle92, Zai01]
LA0155a	155.2	1,252	7.987	C–H_1 bend	H	Lattice sp^3CH_1; Table 9.3.2.1	**3**.18...	[Jan92]
LA0069p	155.7	1,256	7.963	1LP(p)	...	...	**3**.3a, Table 3.1	[Kle92, Zai01]
LA0160	159.8	1,289	7.756	...	B?	LVM of B?; calc. 159 meV	...	[Zai01]
LA0105f′	161.2	1,300	7.692	C-cent.(f′)	N_1°	var. 159.9–162.4 meV, see Table 9.1.1.1	...	[Zai01]
LA0165	165.1	1,332	7.508	*(V_1H_4)	*(V_1H_4)	2 lines a, b (412.0 meV); ZPL of DAP39, see i13-LA	**3**.18, **3**.19a	[Jan92, Fuc95b]
LA0069q	165.2	1,333	7.505	1LP(q)	...	Raman frequency	**3**.3a, Table 3.1	[Kle92, Zai01]
LA0105g′	166.6	1,344	7.442	C-cent.(g′)	N_1°	Last C-center line; sharp, FWHM=0.2%	...	[Zai01]

Table 4.1.2.1 Low pressure synthetic diamond (CVD): absorption lines; mid infrared, part 1 (0.18–0.51 eV)

Line-label	Energy (meV)	Frequ. (cm^{-1})	Wavel. (μm)	Name	Impur./defect	Comment	Figs. **3–6.** [Zai01]; Fig. **7.** [Zai98]	References
LA0205	204.6	1,650	6.061	...	B?	...	...	[Zai01]
LA0216	215.6	1,739	5.750	...	O?	LVM?	**3.**10 = **7.**7	[Jan91]
LA0225c	244.0	1,968	5.080	2LP(c)	...	var. 244.0–245.5 meV, strong peak; abs. coefficient (12.3–14.9 cm^{-1}); other 2LP lines (a–q) see NA	**3.**4a,b, **3.**10 **7.**3, **7.**18	[Kle92, Jan91] [Kle92, Phi92]
LA0225e	251.1	2,025	4.938	2LP(e)	...	var. 251.1–251.7 meV, strong peak (12.3–14.9 cm^{-1})	**3.**4a, b, **3.**10	[Kle92, Jan91]
LA0260	259.7	2,095	4.774	2LP(g)	...	Weak line	**3.**4a, b	[Kle92]
LA0225i	267.0	2,154	4.642	2LP(i)	...	var. 267.0–269.0 meV, strong peak (12.3–14.9 cm^{-1})	**3.**4a, b, **3.**10	[Kle92, Jan91]
LA0225o	302.1	2,436	4.104	2LP(o)	...	Medium strong peak (4.6–5.2 cm^{-1})	**3.**4a, **3.**10	[Kle92, Jan91]
LA0304a	304.2	2,454	4.076	B-acc.(a)	B_1	B acceptor; other lines (a–r) see NA	**5.**2b	[Ert95]
LA0225p	315.0	2,541	3.935	2LP(p)	...	Medium strong peak (4.6–5.2 cm^{-1})	**3.**4a, **3.**10	[Kle92, Jan91]
LA0333	333.2	2,688	3.721	...	H?	Weak line in flame grown CVD film	**7.**18	[Phi92]

(continued)

Table 4.1.2.1 (continued)

Line-label	Energy (meV)	Frequ. (cm^{-1})	Wavel. (μm)	Name	Impur./defect	Comment	Figs. **3–6.** [Zai01]; Fig. **7.** [Zai98]	References
LA0344	343.8	2,773	3.606	…	H?	Weak line	7.12	[Dis93]
LA0304j	347.1	2,800	3.572	B-acc.(j)	B_1	Dominant line; other lines (a–r) see NA	**5**.2b	[Ert95]
LA0155b	349.9	2,822	3.544	C–H_1 stretch	H	Lattice sp^3CH_1; Table 9.3.3; var. 0.3495–0.3510 eV	**7**.11, **7**.12	[Dis93]
LA0304p	362.7	2,926	3.418	B-acc.(p)	B_1	Other lines (a–r) see NA	**5**.2b	[Ert95]
LA0304r	372.0	3,001	3.333	B-acc.(r)	B_1	Bound to free?	**5**.2b	[Ert95]
LA0240i	387.3	3,124	3.201	*(V_1H_4)a = –DAP39i	*(V_1H_4)	$s = 26$, calc. 387.3; other DAP39 lines see i13-LA0237 homo epit. film; ^{12}C line (a) expected at 239.8 meV	…	[Fuc95b]
LA0378	390.5	3,150	3.175	N–H_1 stretch	*(V_1H_1)N_3^0	See Table 9.3.1	…	[Mil95]
LA0165b = LA0599z	412.0	3,323	3.009	*(V_1H_4)b = DAP40z	*(V_1H_4)	ZPL of DAP40a-g(985.6 meV); homo epitaxial diamond film	…	[Fuc95b]
LA0304a-s1	466.0	3,759	2.660	B-acc.(a)-s1	B_1	B-acceptor(a) + phonon(162 meV)	**5**.2b	[Ert95]
LA0304j-s1	508.0	4,098	2.440	B-acc.(j)-s1	B_1	B-acceptor(j) + phonon(161 meV)	**5**.2b	[Ert95]

Table 4.1.2.2 Low pressure synthetic diamond (CVD): absorption lines; mid infrared, part 2 (0.59–1.24 eV)

Line-label	Energy (meV)	Frequ. (10^3 cm^{-1})	Wavel. (μm)	Name	Impur./ defect	Comment	Figs. **3–6.** [Zai01]; Fig. **7.** [Zai98]	References
LA599a	599.4	4.835	2.068	(V_1H_4)b-DAP40a	(V_1H_4)	7 lines a–g (0.9856 eV); homo epitaxial film; *shell* = 42; calc. 0.6040 eV	...	[Fuc95b]
LA599b	616.4	4.972	2.011	(V_1H_4)b-DAP40b	(V_1H_4)	$s = 36$; calc. 0.6180 eV	...	[Fuc95b]
LA599c	690.8	5.572	1.795	(V_1H_4)b-DAP40c	(V_1H_4)	$s = 20$; calc. 0.6890 eV	...	[Fuc95b]
LA599d	851.4	6.867	1.456	(V_1H_4)b-DAP40d	(V_1H_4)	$s = 8$; calc. 0.85500 eV	...	[Fuc95b]
LA599e	897.1	7.236	1.362	(V_1H_4)b-DAP40e	(V_1H_4)	$s = 7$; calc. 0.8880 eV	...	[Fuc95b]
LA599f	913.2	7.366	1.358	(V_1H_4)b-DAP40f	(V_1H_4)	$s = 6$; calc. 0.9170 eV	...	[Fuc95b]
LA599g	985.6	7.950	1.258	(V_1H_4)b-DAP40g	(V_1H_4)	$s = 4$; calc. 1.000 eV	...	[Fuc95b]

Table 4.1.3 Low pressure synthetic diamond (CVD): absorption lines; near infrared (1.24–1.77 eV)

Line-label	Energy (eV)	Frequ. (10^3 cm^{-1})	Wavel. (nm)	Name	Impur./ defect	Comment	Figs. **3–6.** [Zai01]; Fig. **7.** [Zai98]	References
LA1310	1.310	10.57	946.4	…	…	…	…	[All95, Col97]
LA1423	1.423	11.48	871.2	…	…	…	…	[All95, Col97]
LA1446	1.446	11.66	857.4	…	…	…	…	[All95, Col97]
LA1493	1.493	12.05	830.2	…	…	…	…	[All95, Col97]
LA1681	1.682	13.57	736.9	2Si-1.68 eV center	$*(V_3Si_2)^+$	var. 1.667–1.691 eV; Si_2 from QLVM; ZPL of DAP53a-m, see LE	**5.**48a	[Iak00a, Zai01]

Table 4.1.4 Low pressure synthetic diamond (CVD): absorption lines; visible: purple to violet (1.77–3.10 eV)

Line-label	Energy (eV)	Frequ. (10^3 cm^{-1})	Wavel. (nm)	Name	Impur./defect	Comment	Figs. **3–6.** [Zai01]; Fig. **7.** [Zai98]	References
LA1785a	1.785	14.40	694.6	2Si-DAP53a	$^*(V_3Si_2)^+$	13 lines a–m (2.350 eV), see LE; $s = 162$, calc. 1.791 eV	...	[Fen93, Zai01]
LA1785b	1.810	14.60	685.0	2Si-DAP53b	$^*(V_3Si_2)^+$	$s = 118$, calc. 1.810 eV	...	[Fen93, Zai01]
LA1785d	1.836	14.81	675.3	2Si-DAP53d	$^*(V_3Si_2)^+$	$s = 86$, calc. 1.831 eV	...	[Fen93, Zai01]
LA1785f	1.859	14.99	666.9	2Si-DAP53f	$^*(V_3Si_2)^+$	$s = 64$, calc. 1.855 eV	...	[Fen93, Zai01]
LA1785g	1.878	15.15	660.2	2Si-DAP53g	$^*(V_3Si_2)^+$	$s = 50$, calc. 1.878 eV	...	[Fen93, Zai01]
LA1883	1.883	15.19	658.4	Ni-DAP107z	$^*(V_3Ni_2)^+$	Split into 3 lines: 1.883 (dominant), 1.906, 1.914 eV; ZPL of DAP107a-g, see HA	...	[Zai01]
LA2154	2.154	17.38	575.7	**NV** = 575 nm-center	$V_1N_1^\circ$	var. 2.154-2.156 eV; ZPL of DAP46a-f, see MA	...	[Zai01]
LA2985	2.985	24.08	415.3	**N3c**	$V_1N_3^\circ$	ZPL of DAP23a-p (3.762 eV), see NA; 0.59 meV excited state splitting	...	[Zai01]

4.2 Excitation of Photo Luminescence Lines (LE)

Table 4.2.1 Low pressure synthetic diamond (CVD): excitation of photo luminescence lines (PLE), visible: purple to blue (1.77–2.58 eV)

Line-label	Energy (eV)	Frequ. (10^3 cm^{-1})	Wavel. (nm)	Name	Impur./ defect	Comment	Figs. **3–6.** [Zai01]; Fig. **7.** [Zai98]	References
LE1785a	1.785	14.40	694.6	2Si-DAP53a	$^*(V_3Si_2)^+$	13 lines a–m (2.350 eV), lines (a, b) see LA; $s = 162$, calc. 1.791 eV; ZPL at 1.682 eV	...	[Fen93, Zai01]
LE1808a	1.808	14.58	685.7	***S7**-DAP11a	$^*(V_2Ni_1)N_x^o$	14 lines a–n (2.620 eV); lines a–d, f see LL; sta-DAP; $s = 228$, calc. 1.815 eV; L at 1.724 eV	...	[Iak00a, Zai01]
LE1785b	1.810	14.60	685.0	2Si-DAP53b	$^*(V_3Si_2)^+$	PLE of ZPL at 1.682 eV; $s = 118$, calc. 1.810 eV	...	[Fen93, Zai01]
LE1785c	1.817	14.66	682.3	2Si-DAP53b	$^*(V_3Si_2)^+$	PLE of ZPL at 1.682 eV; $s = 106$, calc. 1.817 eV	**5.**48b	[Iak00a, Zai01]
LE1785d	1.836	14.81	675.3	2Si-DAP53d	$^*(V_3Si_2)^+$	PLE of ZPL at 1.682 eV; $s = 86$, calc. 1.831 eV	**5.**48b	[Iak00a, Fen93]
LE1785e	1.848	14.81	675.3	2Si-DAP53e	$^*(V_3Si_2)^+$	PLE of ZPL at 1.682 eV; $s = 72$, calc. 1.845 eV	**5.**48b	[Iak00a, Zai01]
LE1785f	1.859	14.99	666.9	2Si-DAP53f	$^*(V_3Si_2)^+$	PLE of ZPL at 1.682 eV; $s = 64$, calc. 1.855 eV	**5.**48b	[Iak00a, Fen93]
LE1785g	1.878	15.15	660.2	2Si-DAP53g	$^*(V_3Si_2)^+$	PLE of ZPL at 1.682 eV; $s = 50$, calc. 1.878 eV	**5.**48b	[Iak00a, Fen93]
LE1785h	1.897	15.30	653.5	2Si-DAP53h	$^*(V_3Si_2)^+$	PLE of ZPL at 1.682 eV; $s = 42$, calc. 1.897 eV	**5.**48b	[Iak00a, Zai01]
LE1808e	2.020	16.29	613.7	DAP11e	$^*(V_2Ni_1)N_x^o$	PLE of DAP11a-c; $s = 26$, calc. 1.811 eV	**5.**84b	[Iak00a, Zai01]

LE1785i	2.070	16.70	598.9	2Si-DAP53i	$^*(V_3Si_2)^+$	PLE of ZPL at 1.682 eV; $s = 12$, calc. 2.082 eV	**5**.48b	[Iak00a, Zai01]
LE1808g	2.099	16.93	590.6	DAP11g	$^*(V_2Ni_1)N_x^{\circ}$	PLE of DAP11a-c; $s = 14$, calc. 2.095 eV	**5**.84b	[Iak00a, Zai01]
LE1785j	2.125	17.14	583.4	2Si-DAP53j	$^*(V_3Si_2)^+$	PLE of ZPL at 1.682 eV; $s = 10$, calc. 2.120 eV	**5**.48b	[Iak00a, Zai01]
LE1785k	2.150	17.34	576.6	2Si-DAP53k	$^*(V_3Si_2)^+$	PLE of ZPL at 1.682 eV; $s = 8$, calc. 2.172 eV	**5**.48b	[Iak00a, Zai01]
LE1808h	2.190	17.66	566.1	DAP11h	$^*(V_2Ni_1)N_x^{\circ}$	PLE of DAP11a-g; $s = 9$, calc. 2.193 eV	**5**.84b	[Iak00a, Zai01]
LE1785-l	2.260	18.23	548.6	2Si-DAP53-l	$^*(V_3Si_2)^+$	PLE of ZPL at 1.682 eV; $s = 6$, calc. 2.247 eV	**5**.48b	[Iak00a, Zai01]
LE1808i	2.280	18.39	543.8	DAP11i	$^*(V_2Ni_1)N_x^{\circ}$	PLE of DAP11a-g; $s = 6$, calc. 2.289 eV	**5**.84b	[Iak00a, Zai01]
LE1808j	2.350	18.96	527.6	DAP11j	$^*(V_2Ni_1)N_x^{\circ}$	PLE of DAP11a-g; $s = 5$, calc. 2.340 eV	**5**.84b	[Iak00a, Zai01]
LE1785m	2.350	18.96	527.6	2Si-DAP53m	$^*(V_3Si_2)^+$	PLE of ZPL at 1.682 eV; $s = 4$, calc. 2.340 eV	**5**.48b	[Iak00a, Zai01]
LE1808k	2.400	19.36	516.6	DAP11k	$^*(V_2Ni_1)N_x^{\circ}$	PLE of DAP11a-g; $s = 4$, calc. 2.382 eV	**5**.84b	[Iak00a, Zai01]
LE1808-l	2.475	19.96	500.9	DAP11-l	$^*(V_2Ni_1)N_x^{\circ}$	PLE of DAP11a-g; $s = 3$, calc. 2.476 eV	**5**.84b	[Iak00a, Zai01]
LE1808m	2.520	20.33	492.0	DAP11m	$^*(V_2Ni_1)N_x^{\circ}$	PLE of DAP11a-g; $s = 2$, calc. 2.527 eV	**5**.84b	[Iak00a, Zai01]

Table 4.2.2 Low pressure synthetic diamond (CVD): excitation of photo luminescence lines (PLE), visible: ultramarine, violet (2.58–3.10 eV)

Line-label	Energy (eV)	Frequ. (10^3 cm^{-1})	Wavel. (nm)	Name	Impur./defect	Comment	Figs. **3–6.** [Zai01]; Fig. **7.** [Zai98]	References
LE1808n	2.620	21.13	473.2	DAP11n	$^*(V_2Ni_1)N_x^\circ$	PLE of DAP11a-g; $s = 1$, calc. 2.620 eV	**5.**84b	[Iak00a, Zai01]
LE2646 = LE2785z	2.646	21.34	468.5	DAP109z	$^*V_1Si_1^\circ$	var. 2.646–2.651 eV; PLE of DAP110 (see LL); ZPL of DAP109a-h (3.054 eV); QLVM at +74 meV (calc. 1Si = 74.5 meV)	**5.**128b	[Iak00a, Zai01]
LE2785a	2.785	22.46	445.2	DAP109a	$^*V_1Si_1^\circ$	PLE of DAP110; $s = 86$, calc. 2.785 eV	**5.**128b	[Iak00a, Zai01]
LE2785b	2.802	22.60	442.5	DAP109b	$^*V_1Si_1^\circ$	PLE of DAP110; $s = 72$, calc. 2.798 eV	**5.**128b	[Iak00a, Zai01]
LE2785c	2.828	22.81	438.4	DAP109c	$^*V_1Si_1^\circ$	PLE of DAP110; $s = 50$, calc. 2.828 eV	**5.**128b	[Iak00a, Zai01]
LE2785d	2.851	23.00	434.9	DAP109d	$^*V_1Si_1^\circ$	PLE of DAP110; $s = 42$, calc. 2.846 eV	**5.**128b	[Iak00a, Zai01]
LE2785e	2.884	23.26	429.9	DAP109e	$^*V_1Si_1^\circ$	PLE of DAP110; $s = 30$, calc. 2.881 eV	**5.**128b	[Iak00a, Zai01]
LE2785f	2.922	23.57	424.3	DAP109f	$^*V_1Si_1^\circ$	PLE of DAP110; $s = 22$, calc. 2.922 eV	**5.**128b	[Iak00a, Zai01]
LE2785g	2.982	24.05	415.8	DAP109g	$^*V_1Si_1^\circ$	PLE of DAP110; $s = 14$, calc. 2.992 eV	**5.**128b	[Iak00a, Zai01]
LE2785h	3.054	24.63	406.0	DAP109h	$^*V_1Si_1^\circ$	PLE of DAP110; $s = 10$, calc. 3.054 eV	**5.**128b	[Iak00a, Zai01]

4.3 Luminescence Lines (LL)

Table 4.3.1.1 Low pressure synthetic diamond (CVD): luminescence lines, near infrared, part 1 (1.24–1.69 eV)

Line-label	Energy (eV)	Frequ. (10^3 cm^{-1})	Wavel. (nm)	Name	Impur./ defect	Comment	Figs. **3–6.** [Zai01]; Fig. **7.** [Zai98]	References
LL1401a = DAP26z	1.401	11.30	884.9	1Ni-1.40a	$*V_1Ni_1^+$	2 lines a, b from 2.7 meV ground state splitting; named 1.4-eV cent. ZPL of 1Ni-DAP26*a*-k(1.000)	...	[Wol-10]
LL1401b = DAP26z	1.404	11.32	883.2	1Ni-1.40b	$*V_1Ni_1^+$	Main line.(intensities a:b = 4:6), named Ni-D_0, Ni_1 from isotopes (see Table 7.5)	...	[Wol-10]
LL1486a	1.486	11.99	834.3	DAP54a	$*(V_3Si_2)^+$	7 lines a–g (1.567 eV); $s = 50$, calc. 1.487 eV; ZPL at 1.682; 2Si from isotopes	**7.**69a	[Fen93, Zai01]
LL1486b	1.519	12.25	816.2	DAP54ba	$*(V_3Si_2)^+$	$s = 72$, calc. 1.519 eV	**5.**44	[Iak00a, Zai01]
LL1486c	1.524	12.29	813.5	DAP54c	$*(V_3Si_2)^+$	$s = 78$, calc. 1.525 eV	...	[Zai01]
LL1486d	1.527	12.32	811.9	DAP54d	$*(V_3Si_2)^+$	$s = 82$, calc. 1.529 eV	5.44	[Iak00a, Zai01]
LL1486e	1.534	12.37	808.2	DAP54e	$*(V_3Si_2)^+$	$s = 86$, calc. 1.533 eV	**5.**44	[Iak00a, Zai01]
LL1486f	1.556	12.55	796.8	DAP54f	$*(V_3Si_2)^+$	$s = 118$, calc. 1.554 eV	**5.**44, **7.**69	[Iak00a, Fen93]
LL1556a	1.556	12.55	796.8	DAP8a	$*(V_2Ni_1)N_x^-$	7 lines a–g (1.637 eV); $s = 198$, calc. 1.556 eV	**7.**67	[McC94a, Zai01]

(continued)

Table 4.3.1.1 (continued)

Line-label	Energy (eV)	Frequ. (10^3 cm^{-1})	Wavel. (nm)	Name	Impur./defect	Comment	Figs. **3–6.** [Zai01]; Fig. **7.** [Zai98]	References
LL1563	1.563	12.60	793.4	***S5**-DAP60a	$^*(V_2Ni_1)N_x^{\circ}$	11 lines a–k (see MA/ME/ML), $s = 162$, calc. 1.564 eV	...	[Wol-10]
LL1564a	1.564	12.62	792.7	DAP49a	$^*V_1N_1^-$	15 lines a–o (1.870 eV); $s = 8$, calc. 1.564 eV; lines (b, d, e, h, I, l, m. o) see ML	**5.**57	[Dol95, Zai01]
LL1486g	1.567	12.64	791.2	DAP54g	$^*(V_3Si_2)^+$	$s = 162$, calc. 1.573 eV	...	[Iak00a, Zai01]
LL1556b	1.575	12.70	787.2	DAP8b	$^*(V_2Ni_1)N_x^-$	$s = 130$, calc. 1.575 eV	**7.**67	[McC94a, Zai01]
LL1556c	1.586	12.79	781.7	DAP8c	$^*(V_2Ni_1)N_x^-$	$s = 106$, calc. 1.586 eV	**7.**67	[McC94a, Zai01]
LL1599	1.599	12.90	775.3	...	Ni?	...	**5.**21	[McC95, Zai01]
LL1556d	1.601	12.91	774.4	DAP8d	$^*(V_2Ni_1)N_x^-$	$s = 86$, calc. 1.601 eV	**7.**67	[McC94a, Zai01]
LL1556e	1.623	13.09	763.9	DAP8e	$^*(V_2Ni_1)N_x^-$	$s = 58$, calc. 1.625 eV	**7.**67	[McC94a, Zai01]
LL1556f	1.631	13.16	760.1	DAP8f	$^*(V_2Ni_1)N_x^-$	$s = 54$, calc. 1.631 eV	**7.**67	[McC94a, Zai01]
LL1564c	1.636	13.20	757.8	DAP49c	$^*V_1N_1^-$	$s = 12$, calc. 1.633 eV	**5.**57	[Dol95, Zai01]
LL1556g	1.637	13.20	757.3	DAP8g	$^*(V_2Ni_1)N_x^-$	$s = 50$, calc. 1.637 eV	**7.**67	[McC94a, Zai01]
LL1681	1.681	13.56	737.5	2Si-1.68 eV	$^*(V_3Si_2)^+$	ZPL of DAP54 a-g (1.486–1.567 eV); QLVM: −36, −45, LVM: −66, −83 meV	**5.**44, **5.**21	[Iak00a, McC95, Zai01]

Table 4.3.1.2 Low pressure synthetic diamond (CVD): luminescence lines, near infrared, part 2 (1.70–1.77 eV)

Line-label	Energy (eV)	Frequ. (10^3 cm^{-1})	Wavel. (nm)	Name	Impur./ defect	Comment	Figs. **3–6.** [Zai01]; Fig. **7.** [Zai98]	References
LL1564f	1.703	13.74	728.0	DAP49f	$*V_1N_1^-$	$s = 20$, calc. 1.702 eV	**5.**57	[Dol95, Zai01]
LL1723a	1.723	13.90	719.5	DAP6a	$*(V_2W_1)N_x^\circ$	6 lines a–f (1.759 eV); *shell* = 86; calc 1.723 eV	**5.**50	[Ste95, And97]
LL1564g	1.726	13.92	718.3	DAP49g	$*V_1N_1^-$	$s = 24$, calc. 1.724 eV	**5.**57	[Dol95, Zai01]
LL1723b	1.735	13.99	714.6	DAP6b	$*(V_2W_1)N_x^\circ$	$s = 72$; calc. 1.734 eV; named W5; dominant line; pseudo ZPL;3 SBs −24 meV; $L = 1.608$ eV	**5.**50	[Ste95, And97]
LL1723c	1.741	14.04	712.1	DAP6c	$*(V_2W_1)N_x^\circ$	$s = 64$; calc. 1.742 eV; named W4	**5.**50	[Ste95, And97]
LL1723d	1.750	14.12	708.4	DAP6d	$*(V_2W_1)N_x^\circ$	$s = 58$; calc. 1.749 eV; named W3	**5.**50	[Ste95, And97]
LL1723e	1.754	14.15	706.8	DAP6e	$*(V_2W_1)N_x^\circ$	$s = 54$; calc. 1.754 eV; named W2	**5.**50	[Ste95, And97]
LL1723f	1.759	14.19	704.8	DAP6f	$*(V_2W_1)N_x^\circ$	$s = 50$; calc. 1.759 eV; named W1	**5.**50	[Ste95, And97]
LL1748a	1.748	14.10	709.3	DAP7a	$*(V_2Ta_1)^\circ$	7 lines a–g (1.817 eV); $s = 162$, calc. 1.751 eV	**5.**51	[Har96, Zai01]

Table 4.3.2 Low pressure synthetic diamond (CVD): luminescence lines, visible: purple, red (1.77–2.03 eV)

Line-label	Energy (eV)	Frequ. (10^3 cm^{-1})	Wavel. (nm)	Name	Impur./ defect	Comment	Figs. **3–6.** [Zai01]; Fig. **7.** [Zai98]	References
LL1748b	1.771	14.28	700.0	DAP7b	$^*(V_2Ta_1)^{\circ}$	$s = 106$, calc. 1.770 eV	**5**.51	[Har96, Zai01]
LL1748c	1.774	14.31	698.9	DAP7c	$^*(V_2Ta_1)^{\circ}$	$s = 98$; calc. 1.774 eV; pseudo ZPL; 2 SBs −25 meV; $L =$ 1.667 eV	**5**.51	[Har96, Zai01]
LL1748d	1.783	14.38	695.3	DAP7d	$^*(V_2Ta_1)^{\circ}$	$s = 130$; calc. 1.785 eV	**5**.51	[Har96, Zai01]
LL1789	1.789	14.43	693.0	DAP65	$^*(V_3Si_2)^{\circ}$	$s = 22$, calc. 1.795; lines (a–e) see NL	. . .	[Zai01]
LL1564j	1.792	14.45	691.8	DAP49j	$^*V_1N_I^-$	$s = 50$, calc. 1.791 eV	**5**.57	[Dol95, Zai01]
LL1748e	1.794	14.47	691.1	DAP7e	$^*(V_2Ta_1)^{\circ}$	$s = 106$; calc. 1.795 eV	**5**.51	[Har96, Zai01]
LL1748f	1.807	14.58	686.1	DAP7f	$^*(V_2Ta_1)^{\circ}$	$s = 86$; calc. 1.807 eV	**5**.51	[Har96, Zai01]
Ll1808a	1.808	14.58	685.7	*S7-DAP11a	$^*(V_2Ni_1)N_x^{\circ}$	14 lines a–n (2.620 eV), lines (i–n) see LE sta-DAP; $s = 228$, calc. 1.815 eV; L at 1.724 eV	**5**.84a	[Iak00a, Zai01]
LL1564k	1.809	14.59	685.3	DAP49k	$^*V_1N_I^-$	$s = 64$, calc. 1.809 eV	**5**.57	[Dol95, Zai01]
LL1748g	1.817	14.66	682.3	DAP7g	$^*(V_2Ta_1)^{\circ}$	$s = 74$; calc. 1.817 eV	**5**.51	[Har96, Zai01]
LL1829	1.829	14.76	677.7		N	. . .	. . .	[Bac93, Zai01]
LL1808b	1.833	14.78	676.4	*S7-DAP11b	$^*(V_2Ni_1)N_x^{\circ}$	$s = 162$, calc. 1.833 eV	**5**.84a	[Iak00a, Zai01]
LL1564n	1.859	14.99	666.9	DAP49n	$^*V_1N_I^-$	$s = 162$, calc. 1.859 eV	**5**.57	[Dol95, Zai01]
LL1808c	1.901	15.33	652.2	*S7-DAP11c	$^*(V_2Ni_1)N_x^{\circ}$	$s = 64$, calc. 1.897 eV	**5**.84a	[Iak00a, Zai01]

LL1923	1.923	15.51	644.7	...	...	...	...	[Cla94, Zai01]
LL1943 = LL1564z	1.943	15.67	638.1	NV^-	$V_1N_1^-$	ZPL of DAP49a-o (1.567–1.870 eV); QLVM-61 meV (calc. 3C: 60.9); named **638 nm** center	5.57	[Dol95, Fre90, Zai01]
LL1808d	1.965	15.85	630.9	*S7-DAP11d	$^*(V_2Ni_1)N_x^\circ$	$s = 34$, calc. 1.961 eV	**5.**84a	[Iak00a, Zai01]
LL1968	1.968	15.87	630.0	...	...	Split into 3 lines (1.962, 1.968, 1.977 eV)	**5.**21, **7.**81	[McC95, Rua91b]
LL1910e	2.002	16.15	619.3	NV-DAP47e	$^*V_1N_1^\circ$	3 lines (e, f, h), lines a–d, g (1.910–2.059 eV) see ML, $s =$ 64, calc. 2.004 eV	**7.**95	[Jan91, Zai01]
LL1563h	2.005	16.17	618.3	*S5-DAP60h	$^*(V_2Ni_1)N_y^\circ$	3 lines (h–j), lines (a–g, k) see MA; $s = 6$, calc. 2.000 eV	...	[Iak00g]
LL2019	2.019	16.28	614.1	...	...	...	...	[Hei97, Zai01]
LL1808e	2.020	16.29	613.7	*S7-DAP11e	$^*(V_2Ni_1)N_x^\circ$	$s = 22$, calc. 2.019 eV	**5.**84a	[Iak00a, Zai01]
LL1910f	2.023	16.32	612.8	DAP47f	$^*V_1N_1^\circ$	$s = 86$, calc. 2.023 eV	**7.**95	[Jan91, Zai01]

Table 4.3.3 Low pressure synthetic diamond (CVD): luminescence lines, visible: orange, yellow (2.03–2.18 eV)

Line-label	Energy (eV)	Frequ. (cm^{-1})	Wavel. (nm)	Name	Impur./ defect	Comment	Figs. **3–6.** [Zai01]; Fig. **7.** [Zai98]	References
LL2055	2.055	16.58	603.2	...	...	...	...	[Hei97, Zai01]
LL1910h	2.059	16.61	602.1	DAP47h	$^*V_1N_1^\circ$	$s = 162$, calc. 2.059 eV	**7.**95	[Jan91, Zai01]
LL2070a	2.070	16.70	598.9	DAP1a	$^*N_1^\circ+B_1^\circ$	13 lines a–m (2.668), $s = 50$, calc. 2.067 eV	**7.**128	[Dis94a, Dis94b]
LL1808f	2.073	16.72	598.1	*S7-DAP11f	$^*(V_2Ni_1)N_x^\circ$	$s = 16$, calc. 2.070 eV	**5.**84a	[Iak00a, Zai01]
LL1808g	2.095	16.90	591.8	*S7-DAP11g	$^*(V_2Ni_1)N_x^\circ$	$s = 14$, calc. 2.095 eV	**5.**84a	[Iak00a, Zai01]
LL1563i	2.095	16.90	591.8	*S5-DAP60i	$^*(V_2Ni_1)N_y^\circ$	$s = 4$, calc. 2.089 eV	...	[Iak00g]
LL2070b	2.108	17.00	588.1	DAP1b	$^*N_1^\circ+B_1^\circ$	$s = 34$, calc. 2.105 eV	**7.**128	[Dis94a, Dis94b]
LL2154 = LL1910z	2.154	17.38	575.5	NV- DAP47z = **575 nm** center	$^*V_1N_1^\circ$	var. 2.153–2.156 eV; ZPL of DAP47a–h,, QLVM: −48 meV (calc. 1N + 3C:48.4)	**5.**57, **7.**95	[Dol95, Jan91, Zai01]

Table 4.3.4 Low pressure synthetic diamond (CVD): luminescence lines, visible: green (2.18–2.43 eV)

Line-label	Energy (eV)	Frequ. (cm^{-1})	Wavel. (nm)	Name	Impur./ defect	Comment	Figs. **3–6.** [Zai01]; Fig. **7.** [Zai98]	References
LL1808h	2.190	17.66	566.1	*S7-DAP11h	$^*(V_2Ni_1)N_x^{\circ}$	$s=9$, calc. 2.193 eV; lines (i–n) see LE	**5.**84a	[Iak00a, Zai01]
LL2070c	2.202	17.76	563.0	DAP1c	$^*N_1^{\circ}+B_1^{\circ}$	$s=16$; calc. 2.202 eV	**7.**128, **7.**129	[Dis94a, Dis94b, Col89d, Col90d]
LL1563j	2.220	17.91	558.5	*S5-DAP60j	$^*(V_2Ni_1)N_y^{\circ}$	$s=2$, calc. 2.228 eV	…	[Iak00g]
LL2070d	2.240	18.07	553.5	DAP1d	$^*N_1^{\circ}+B_1^{\circ}$	$s=13$; calc. 2.239 eV	**7.**128	[Dis94a, Dis94b, Kho93]
LL2241a	2.241	18.08	553.2	1Si-DAP110a	$^*V_1Si_1^{\circ}$	7 lines a–g (2.507 eV); $s=10$, calc. 2.238 eV	**5.**128a	[Iak00a, Zai01]
LL2246a	2.246	18.12	552.0	DAP113a	$^*V_1Si_1^{-}$	11 lines a–k (2.845 eV); $s=1$, calc. 2.246 eV	7.114	[Kho94, Zai98b]
LL2250	2.250	18.15	551.0	…	…	Main line of an unresolved doublet	**5.**21	[McC95, Zai01]
LL2241b	2.271	18.32	545.9	1Si-DAP110b	$^*V_1Si_1^{\circ}$	$s=12$, calc. 2.274 eV	**5.**128a	[Iak00a, Zai01]
LL2070e	2.284	18.42	542.8	DAP1e	$^*N_1^{\circ}+B_1^{\circ}$	$s=10$; calc. 2.285 eV	**7.**128	[Dis94a, Dis94b, Kho93]
LL2241c	2.291	17.75	563.3	1Si-DAP110c	$^*V_1Si_1^{\circ}$	$s=14$, calc. 2.300 eV	**5.**128a	[Iak00a, Zai01]
LL2329	2.329	18.79	532.3	…	N?	Single line in good quality, strain splitting in low quality (2.322 and 2.335 eV)	**7.**81, **7.**109, **7.**114, **7.**120, **7.**124, **7.**129	[Rua91b, Col92b, Kho94, Col89d, Zai01]

(continued)

Table 4.3.4 (continued)

Line-label	Energy (eV)	Frequ. (cm^{-1})	Wavel. (nm)	Name	Impur./defect	Comment	Figs. **3–6.** [Zai01]; Fig. **7.** [Zai98]	References
LL2330a	2.330	18.79	532.1	DAP101a	$^*(N_1C_1)_iB_1^-$	Lines a–d (2.687 eV); $s=1$, calc. 2.309 eV	...	[Ste96a]
LL2246b	2.330	18.79	532.1	DAP113b	$^*V_1Si_1^-$	$s=2$, calc. 2.323 eV; decay time 1.0 μs	**7.**114, **7.**120, **7.**135	[Kho94, Zai98b]
LL2070f	2.331	18.80	531.9	DAP1f	$^*N_1^\circ + B_1^\circ$	$s=8$; calc. 2.331 eV; also observed in boron doped CVD films [Hei97]	**7.**128, **7.**124, **7.**129	[Dis94a, Dis94b, Col89d, Col90d, Rua92b, Kho93]
LL2241d	2.362	19.05	524.9	1Si-DAP110d	$^*V_1Si_1^\circ$	$s=22$, calc. 2.360 eV	**5.**128a	[Iak00a, Zai01]
LL2070g	2.373	19.14	522.4	DAP1g	$^*N_1^\circ + B_1^\circ$	$s=7$; calc. 2.369 eV	**7.**128	[Dis94a, Dis94b, Kho93]
LL2070h	2.413	19.46	513.8	DAP1h	$^*N_1^\circ + B_1^\circ$	$s=6$; calc. 2.398 eV	**7.**128, **7.**124, **7.**129	[Dis94a, Dis94b, Col89d, Col90d, Rua92b, Kho93]
LL2417a	2.417	19.50	512.9	DAP99a	$^*(N_1C_1)_iB_1^\circ$	Lines a–h (2.696 eV); $s=8$, calc. 2.410 eV	...	[Ste96a]
LL2246c	2.430	19.60	510.2	DAP113c	$^*V_1Si_1^-$	$s=4$, calc. 2.444 eV	**7.**114, **7.**120, **7.**135	[Kho94, Zai98b]

Table 4.3.5 Low pressure synthetic diamond (CVD): luminescence lines, visible: blue (2.43–2.58 eV)

Line-label	Energy (eV)	Frequ. (cm^{-1})	Wavel. (nm)	Name	Impur./ defect	Comment	Figs. **3–6.** [Zai01]; Fig. **7.** [Zai98]	References
LL2241e	2.440	19.68	508.1	1Si-DAP110e	$^*V_1Si_1^\circ$	$s = 42$, calc. 2.446 eV; resonant 3 × QLVM?	**5.**128a	[Iak00a, Zai01]
LL2070-i	2.456	19.81	504.8	DAP1-i	$^*N_1^\circ + B_1^\circ$	$s = 5$; calc. 2.444 eV	**7.**128, **7.**124, **7.**129	[Dis94a, Dis94b, Col89d, Col90d, Kho93]
LL2464	2.464	19.88	503.2	**H3a**-DAP51	$V_1N_2^\circ$	var. 2.462–2.465 eV; ZPL of DAP51a-f (see ML)	**3.**109	[Zai01]
LL2417b	2.467	19.90	502.5	DAP99b	$^*(N_1C_1)_iB_1^\circ$	$s = 12$, calc. 2.480 eV	...	[Ste96a]
LL2473	2.473	19.95	501.3		B?	var. 2.466–2.476 eV	...	[Zai01]
LL2246d	2.480	20.00	499.9	DAP113d	$^*V_1Si_1^-$	$s = 5$, calc. 2.479 eV; decay time 1.0 μs	**7.**114, **7.**120, **7.**135	[Kho94, Zai98b]
LL2330b	2.487	20.06	498.5	DAP101b	$^*(N_1C_1)_iB_1^-$	$s = 4$, calc. 2.491 eV	...	[Ste96a]
LL2241f	2.490	20.08	497.9	1Si-DAP110f	$^*V_1Si_1^\circ$	$s = 72$, calc. 2.494 eV;	**5.**128a	[Iak00a, Zai01]
LL2070j	2.495	20.12	496.9	DAP1j	$^*N_1^\circ + B_1^\circ$	$s = 4$; calc. 2.481 eV	**7.**128, **7.**129	[Dis94a, Dis94b, Col89d, Col90d, Rua92b, Kho93]
LL2241g	2.507	20.22	494.5	1Si-DAP110g	$^*V_1Si_1^\circ$	$s = 86$, calc. 2.507 eV;	**5.**128a	[Iak00a, Zai01]

(continued)

Table 4.3.5 (continued)

Line-label	Energy (eV)	Frequ. (cm^{-1})	Wavel. (nm)	Name	Impur./ defect	Comment	Figs. **3–6.** [Zai01]; Fig. **7.** [Zai98]	References
LL2417c	2.512	20.26	493.5	DAP99c	$^*(N_1C_1)_i B_1^{\circ}$	$s = 16$, calc. 2.522 eV		[Ste96a]
LL2246e	2.537	20.46	488.7	DAP113e	$^*V_1Si_1^-$	$s = 6$, calc. 2.522 eV	**7.**120	[Kho94, Zai98b]
LL2246f	2.550	20.57	486.2	DAP113f	$^*V_1Si_1^-$	var. 1.547–1.557 eV; $s = 7$, calc. 2.548 eV; decay time 1.3 μs	**7.**114, **7.**120, **7.**135	[Kho94, Zai98b]
LL2567a	2.567	20.70	483.0	...	B?	2 lines a, b (2.572 eV); weak lines at 2.52–2.58 eV	**7.**109, **7.**131	[Col92b, Rua93a, Rua93b]
LL2070k	2.568	20.71	482.8	DAP1k	$^*N_1^{\circ} + B_1^{\circ}$	var. 2.561–2.575; $s = 3$; calc. 2.564 eV	**7.**128, **7.**124, **7.**129 **7.**131	[Dis94a, Dis94b, Col89d, Col90d, Kho93]
LL2567b	2.572	20.75	482.0	...	B?	Weak lines at 2.52–2.58 eV	**7.**109, **7.**131	[Col92b, Rua93a, Rua93b]
LL2246g	2.580	20.81	480.5	DAP113g	$^*V_1Si_1^-$	$s = 8$, calc. 2.584 eV; decay time 1.4 μs	**7.**114, **7.**120, **7.**135	[Kho94, Zai98b]

Table 4.3.6 Low pressure synthetic diamond (CVD): luminescence lines, visible: ultramarine (2.58–2.75 eV)

Line-label	Energy (eV)	Frequ. (10^3 cm^{-1})	Wavel. (nm)	Name	Impur./defect	Comment	Figs. **3–6.** [Zai01]; Fig. **7.** [Zai98]	References
LL2417d	2.582	20.83	480.2	DAP99d	$^*(N_1C_1)_iB_1^\circ$	$s = 26$, calc. 2.580 eV	...	[Ste96a]
LL2330c	2.597	20.95	477.4	DAP101c	$^*(N_1C_1)_iB_1^-$	$s = 7$, calc. 2.589 eV	...	[Ste96a]
LL2599	2.599	20.96	477.0	...	...	...	...	[Zai01]
LL2417e	2.612	21.07	474.6	DAP99e	$^*(N_1C_1)_iB_1^\circ$	$s = 34$, calc. 2.607 eV	...	[Ste96a]
LL2070-l	2.623	21.16	472.7	DAP1-l	$^*N_1^\circ + B_1^\circ$	$s = 2$; calc. 2.611 eV	**7.**128, **7.**129	[Dis94a, Dis94b, Col89d, Col90d]
LL2646 = LL2241z	2.646	21.34	468.5	DAP110z	$^*V_1Si_1^\circ$	var. 2.638–2.651 eV; ZPL of DAP110a-g; DAP109(LE); QLVM: 74 meV (calc. 1Si = 74.5)	**5.**128a	[Iak00a, Zai01, Rua91b]
LL2417f	2.657	21.43	466.6	DAP99f	$^*(N_1C_1)_iB_1^\circ$	$s = 72$, calc. 2.664 eV	...	[Ste96a]
LL2070m	2.668	21.52	464.7	DAP1m	$^*N_1^\circ + B_1^\circ$	$s = 1$; var. 2.60–2.68; calc. (incl. NSC) 2.694 eV	**7.**128, **7.**129	[Dis94a, Dis94b, Col89d, Col90d]
LL2676a	2.676	21.58	463.3	DAP17a	$^*O_1^\circ + B_1^\circ$	6 lines a–f (2.725 eV); $s = 62$, calc. 2.676 eV	**7.**131	[Rua93a, Zai01]
LL2676b	2.681	21.62	462.4	DAP17b	$^*O_1^\circ + B_1^\circ$	$s = 58$, calc. 2.676 eV	**7.**131	[Rua93a, Zai01]
LL2417g	2.687	21.67	461.4	DAP99g	$^*(N_1C_1)_iB_1^\circ$	$s = 106$, calc. 2.687 eV	...	[Ste96a]
LL2330d	2.687	21.67	461.4	DAP101d	$^*(N_1C_1)_iB_1^-$	$s = 8$, calc. 2.687 eV	...	[Ste96a]
LL2676c	2.692	21.71	460.5	DAP17c	$^*O_1^\circ + B_1^\circ$	$s = 50$, calc. 2.691 eV	**7.**131	[Rua93a, Zai01]
LL2608f	2.693	21.72	460.4	DAP98f	$^*(N_1C_1)_i^+$	Lines f–i (2.739 eV), a–e see ML; $s = 86$, calc. 2.691 eV	**7.**130b	[Dis94a]

(continued)

Table 4.3.6 (continued)

Line-label	Energy (eV)	Frequ. (10^3 cm^{-1})	Wavel. (nm)	Name	Impur./ defect	Comment	Figs. **3–6.** [Zai01]; Fig. **7.** [Zai98]	References
LL2676d	2.696	21.75	459.9	DAP17d	$^*O_1^\circ + B_1^\circ$	$s = 48$, calc. 2.695 eV	**7.**131	[Rua93a, Zai01]
LL2417h	2.696	21.75	459.9	DAP99h	$^*(N_1C_1)_i B_1^\circ$	$s = 130$, calc. 2.697 eV	...	[Ste96a]
LL2676e	2.702	21.79	458.8	DAP17e	$^*O_1^\circ + B_1^\circ$	$s = 44$, calc. 2.703 eV	**7.**131	[Rua93a, Zai01]
LL2246h	2.709	21.85	457.6	DAP113h	$^*V_1Si_1^-$	$s = 16$, calc. 2.704 eV	...	[Kho94, Zai98b]
LL2608g	2.713	21.88	457.0	DAP98g	$^*(N_1C_1)_i^+$	$s = 130$, calc. 2.712 eV	**7.**130b	[Dis94a, Zai01]
LL2715a	2.715	21.90	456.6	DAP100a	$^*(N_1C_1)_i B_1^+$	9 lines a–i (2.987 eV); $s = 8$, calc. 2.703 eV	**7.**129b	[Col89d]
LL2676f	2.725	21.98	455.0	DAP17f	$^*O_1^\circ + B_1^\circ$	$s = 34$, calc. 2.725 eV	**7.**131	[Rua93a, Zai01]
LL2608h	2.725	21.98	455.0	DAP98h	$^*(N_1C_1)_i^+$	$s = 162$, calc. 2.722 eV	**7.**130b	[Dis94a]
LL2608i	2.739	22.09	452.6	DAP98i	$^*(N_1C_1)_i^+$	$s = 228$, calc. 2.736 eV	**7.**130b	[Dis94a]
LL2440f	2.741	22.11	452.3	N3c-DAP24f	$V_1N_3^\circ$	Lines f, j, n, o; others see NL; $s = 22$, calc. 2.739 eV	**7.**109	[Col92b, Zai01]
LL2748	2.748	22.16	451.2	...	...	...	...	[Hei97]

Table 4.3.7.1 Low pressure synthetic diamond (CVD): luminescence lines, visible: violet, part 1 (2.75–2.97 eV)

Line-label	Energy (eV)	Frequ. ($10^3\,cm^{-1}$)	Wavel. (nm)	Name	Impur./ defect	Comment	Figs. **3–6.** [Zai01]; Fig. **7.** [Zai98]	References
LL2715b	2.756	22.23	449.8	DAP100b	$^*(N_1C_1)_iB_1^+$	$s=10$, calc. 2.744 eV	**7.**129b	[Col89d, Zai01]
LL2246i	2.760	22.26	449.2	DAP113i	$^*V_1Si_1^-$	$s=24$, calc. 2.756 eV	...	[Kho94, Zai98b]
LL2792 = LL2417z	2.792	22.51	444.0	DAP99z	$^*(N_1C_1)_iB_1^o$	ZPL of DAP99a-h(2.417–2.696 eV)	...	[Ste96a]
LL2807 = LL2608z	2.807	22.64	441.7	DAP98z	$^*(N_1C_1)_i^+$	var.2.786–2.807 eV; ZPL of DAP98a-i	**7.**129a, **7.**130b	[Col89d, Dis94a]
LL2715c	2.811	22.67	441.0	DAP100c	$^*(N_1C_1)_iB_1^+$	$s=16$, calc. 2.818 eV	**7.**130b	[Dis94a]
LL2246j	2.815	22.71	440.4	DAP113j	$^*V_1Si_1^-$	$s=42$, calc. 2.813 eV	...	[Kho94, Zai98b]
LL2715d	2.831	22.83	437.9	DAP100d	$^*(N_1C_1)_iB_1^+$	$s=18$, calc. 2.833 eV	**7.**129b, **7.**130b	[Col89d, Dis94a]
LL2440j	2.834	22.86	437.5	N3c-DAP24j	$V_1N_3^o$	$s=58$, calc. 2.833 eV	**7.**109, **7.**113	[Col92b, Col89e]
LL2246k	2.845	22.95	435.8	DAP113k	$^*V_1Si_1^-$	$s=64$, calc. 2.847 eV	...	[Kho94, Zai98b]
LL2440n	2.853	23.01	434.6	N3c-DAP24n	$V_1N_3^o$	$s=74$, calc. 2.851 eV	**7.**109	[Col92b, Zai01]
LL2246-l	2.855	23.03	434.2	DAP113-l	$^*V_1Si_1^-$	$s=72$, calc. 2.855 eV	**7.**135	[Kho94]
LL2440-o	2.872	23.17	431.7	N3c-DAP24-o	$V_1N_3^o$	$s=106$, calc. 2.873 eV	**7.**109	[Col92b, Zai01]

(continued)

Table 4.3.7.1 (continued)

Line-label	Energy (eV)	Frequ. (10^3 cm^{-1})	Wavel. (nm)	Name	Impur./defect	Comment	Figs. **3–6.** [Zai01]; Fig. **7.** [Zai98]	References
LL2246m	2.874	23.18	431.4	DAP113m	$^*V_1Si_1^-$	$s = 86$, calc. 2.867 eV	**7.**135	[Kho94, Zai01]
LL2715e	2.903	23.42	427.1	DAP100e	$^*(N_1C_1)_iB_1^+$	$s = 34$, calc. 2.903 eV	**7.**129b, **7.**130b	[Col89d, Dis94a]
LL2913	2.913	23.50	425.6	...	...	...	...	[Mel96, Zai01]
LL2920	2.920	23.55	424.6	...	B?	...	...	[Zai01]
LL2715f	2.927	23.61	423.6	DAP100f	$^*(N_1C_1)_iB_1^+$	$s = 44$, calc. 2.926 eV	**7.**130b	[Dis94a]
LL2715g	2.938	23.70	422.0	DAP100g	$^*(N_1C_1)_iB_1^+$	$s = 50$, calc. 2.937 eV	**7.**129b, **7.**130b	[Col89d, Dis94a]
LL2941	2.941	23.72	421.5	...	...	...	...	[Mel96, Zai01]
LL3188-s1	2.950	23.80	420.3	3.188-s1	$(^*(C_2)_iN_1^-$	LVM-SB(−238 meV), see Table 11.1	**7.**129a, **7.**130a	[Col89d, Dis94a]
LL2950	2.950	23.80	420.3	...	Si?	...	...	[Yac91, Zai01]
LL2715h	2.962	23.89	418.6	DAP100h	$^*(N_1C_1)_iB_1^+$	$s = 72$, calc. 2.962 eV	**7.**130b	[Dis94a]

Table 4.3.7.2 Low pressure synthetic diamond (CVD): luminescence lines, visible: violet, part 1 (2.98–3.10 eV)

Line-label	Energy (eV)	Frequ. (10^3 cm^{-1})	Wavel. (nm)	Name	Impur./defect	Comment	Figs. **3–6.** [Zai01]; Fig. **7.** [Zai98]	References
LL2985 = LL2440z	2.985	24.08	415.3	**N3c**-DAP24z	$V_1N_3^{\circ}$	ZPL of **N3c**-DAP24a-q(2.440–2897 eV); 0.59 meV excited state splitting	**7.**109, **7.**113	[Col92b, Col89e]
LL2715i	2.987	24.09	415.1	DAP100i	$^*(N_1C_1)_iB_1^+$	$s = 106$, calc. 2.985 eV	**7.**129b, **7.**130b	[Col89d, Dis94a]
LL2480 = LL2246z	2.991	24.13	414.5	DAP113z	$^*V_1Si_1^-$	ZPL of DAP113a-k(2.246–2.845 eV); in normal and time delayed spectra; QLVM: −74 meV (1Si)	**7.**114, **7.**135	[Kho94, Zai98b]
LL2992 = LL2330z	2.992	24.13	414.4	DAP101z	$^*(N_1C_1)_iB_1^-$	ZPL of DAP101a-d	...	[Ste96a]
LL3188-s2	2.997	24.17	413.7	3.188-s2	$^*(C_2)_iN_1^-$	LVM-SB(−190 meV), see Table 11.1	**7.**129a, **7.**130a	[Col89d, Dis94a]
LL3188-s3	3.009	24.27	412.0	3.188-s3	$^*(C_2)_iN_1^-$	LVM-SB(−179 meV), see Table 11.1	**7.**129a, **7.**130a	[Col89d, Dis94a]
LL3188-s4	3.026	24.41	409.7	3.188-s4	$^*(C_2)_iN_1^-$	LVM: −161 meV, C–C stretch (Table 11.1)	**7.**129a, **7.**130a	[Col89d, Dis94a]
LL3092 = LL2715z	3.092	24.94	401.0	DAP100z	$^*(N_1C_1)_iB_1^+$	var. 3.092–3.100 eV; ZPL of DAP100a-I; QLVM: −61 meV (calc. 3C = 60.9 meV)	**7.**129b, **7.**130b	[Col89d, Dis94a]

Table 4.3.8 Low pressure synthetic diamond (CVD): luminescence lines, near ultraviolet (3.10–3.94 eV)

Line-label	Energy (eV)	Frequ. ($10^3\,cm^{-1}$)	Wavel. (nm)	Name	Impur./ defect	Comment	Figs. **3–6.** [Zai01]; Fig. **7.** [Zai98]	References
LL3188-s5	3.111	25.09	398.5	3.188-s5	$^*(C_2)_iN_1^-$	QLVM: −77 meV, (calc. 1N + 1C = 79.7)	**7.**129a, **7.**130a	[Col89d, Dis94a]
LL3188 = LL2673z	3.188	25.71	388.9	**3.188 eV** = DAP72z	$^*(C_2)_iN_1^-$	var. 3.188–3.204 eV; ZPL of DAP72a-o(2.672–3.116 eV),see ML	**7.**129a, **7.**130a	[Col89d, Dis94a, Zai01]
LL3272	3.272	26.39	378.9	...	...	...	**5.**145	[Zai00a], [Zai01]
LL3487	3.487	28.13	355.5	...	...	...	...	[Mel96]
LL3853	3.853	31.08	321.8	...	...	...	...	[Fie92, Zai01]

Table 4.3.9 Low pressure synthetic diamond (CVD): luminescence lines, far ultraviolet (4.43–5.90 eV)

Line-label	Energy (eV)	Frequ. (10^3 cm^{-1})	Wavel. (nm)	Name	Impur./ defect	Comment	Figs. **3–6.** [Zai01]; Fig. **7.** [Zai98]	References
LL4582	4.582	36.96	270.6	**5RL**	$*(C_2)_i N_1^{\circ}$	ZPL of DAP80a-f(4.204–4.481), see ML	**7.**129a	[Col89d, Zai01]
LL5283-s1	4.617	37.24	268.5	*BE-Li-s1	Li	6 sidebands of BE-Li NPL(5.283 eV), observed in LL and ML, SB −666 meV (=2× TO(282) + TA(87) + LO(163)), see also Table A.1.3	**7.**158	[Rob93]
LL4711a	4.711	38.00	263.2	DAP2a	$P_1 + B_1$	23 LL lines a–w (5.215 eV), all lines are listed at HL; shell = 50; calc. 4.711 eV	...	[Ste99a, Ste99b]
LL5283-s2	4.757	38.37	260.6	*BE-Li-s2	Li	SB −526 meV (=2× TO(282) + TA(87) + LO(163))	**7.**158	[Rob93]
LL4760	4.760	38.39	260.5	...	...	...	...	[Rua91b, Zai01]
LL5283-s3	4.832	38.97	256.6	*BE-Li-s3	Li	SB −451 meV (=2× TO(282) + 2× TA(174))	**7.**158	[Rob93]
LL4862a	4.862	39.22	255.0	...	B?	2 lines a, b (4.999 eV)	**5.**153	[Yok99]
LL5283-s4	4.950	39.93	250.4	*BE-Li-s4	Li	SB −333 meV (=2×TA(174) + LO(163))	**7.**158	[Rob93]
LL4862b	4.999	40.32	248.0	...	B	...	**5.**153	[Yok99]
LL5316a-s1	5.020	40.49	247.0	BE-P(a)-s1	P	SB −296 meV (=2×148(TO)?)	**5.**155a	[Ste96b, Tan01]
LL5356a-s4	5.048	40.72	245.6	BE-B(a)-s4	B	SB −308 meV (=TO + LA/O?); named D_2	**5.**157	[Ste96b]

(continued)

Table 4.3.9 (continued)

Line-label	Energy (eV)	Frequ. (10^3 cm^{-1})	Wavel. (nm)	Name	Impur./ defect	Comment	Figs. **3–6.** [Zai01]; Fig. **7.** [Zai98]	References
LL5356b-s2	5.060	40.81	245.0	BE-B(b)-s2	B	SB −307 meV (= TO + LA/O?); named D_2′	**5.**157	[Ste96b]
LL5283-s5	5.120	41.30	242.1	*BE-Li-s5	Li	SB −163 meV(= LO(163))	**7.**158	[Rob93]
LL5316a-s2	5.170	41.70	239.8	BE-P(a)-s2	P	SB(−146 meV) = TO?	**5.**155a	[Ste96b, Tan01]
LL5316b-s1	5.190	41.86	238.9	BE-P(b)-s1	P	SB(−143 meV) = TO?	**5.**155a	[Ste96b, Tan01]
LL5356a-s5	5.193	41.89	238.7	BE-B(a)-s5	B	SB(−163 meV) = LA/O?; named D_1”	**5.**155a	[Ste96b, Tan01]
LL5356a-s6	5.215	42.06	237.7	BE-B(a)-s6	B	SB(−141 meV) = TO?; named D_1	**5.**155a	[Ste96b, Tan01]
LL5316a-s3	5.245	42.31	236.4	BE-P(a)-s3	P	SB(−71 meV)	**5.**155a	[Ste96b, Tan01]
LL5283	5.283	42.61	234.7	*BE-Li	Li	NPL of *BE-Li	**7.**158	[Rob93]
LL5316a	5.316	42.88	233.2	BE-P(a)	P	2 lines a, b (5.333 eV); NPL (upper VB)	…	[Tan01]
LL5316b	5.333	43.02	232.5	BE-P(b)	P	NPL (lower VB)	…	[Tan01]
LL5356a	5.356	43.20	231.5	BE-B(a)	B	NPL (upper VB); 2 lines a, b (5.367 eV)	**5.**159	[Ste97b]
LL5356b	5.367	43.29	231.0	BE-B(b)	B	NPL (lower VB)	**5.**159	[Ste97b]
LL5500	5.500	44.36	225.4	…	O?	…	…	[Hei97, Zai01]

4.4 Broad Bands (LB)

Table 4.4.1 Low pressure synthetic diamond (CVD): broad bands, mid infrared (0.18–1.24 eV)

Line-label	Energy (eV)	Frequ. ($10^3\,cm^{-1}$)	Wavel. (nm)	Name	Impur./defect	Comment	Figs. **3–6.** [Zai01]; Fig. **7.***x*[Zai98]	References
LB0391	0.3905	3.150	3175	...	B	**A**; *W* =45%, high B doping	**7.**49...	[Enc94]
LB0399a	0.3986	3.215	3110	3LP(a)	...	**A**; *W* =23%,3 lattice phonons, var. 0.396–0.399 eV; 2 bands a, b (0.4500); abs. coeff. =1.7–3.0 cm^{-1}	**3.**6...	[Fuc95a]
LB0399b	0.4500	3.630	2755	3LP(b)	...	**A**; *W* =8%, 3 lattice phonons; var. 0.444–0.450 eV; abs. coeff. =1.7–3.0 cm^{-1}	**3.**6...	[Fuc95a]

Observation: *A* absorption; *CL, EL, IL, PL, XL* luminescence; *PLE* photoluminescence excitation

Table 4.4.2 Low pressure synthetic diamond (CVD): broad bands, visible: purple to violet (1.77–3.10 eV)

Line-label	Energy (eV)	Frequ. ($10^3\,cm^{-1}$)	Wavel. (nm)	Name	Impur./ defect	Comment	Figs. **3–6.** [Zai01]; Fig. **7.**x[Zai98]	References
LB1800	1.800	14.52	688.8	B band	B?	**A, CL, PL**; W =20%; structure of 14 lines (1.742–2.138 eV, see DAP9=NL1743);	**7.**75	[Den92]
LB1880	1.880	15.16	659.5	...	Si?	**CL, PL**; W =21%	**7.**79	[Den93]
LB1900	1.900	15.33	652.4	...	P?	**CL**; W =10%	...	[Nij97]
LB1975	1.975	15.93	627.7	...	N, B?	**PL**; W =14%; var. 1.95–2.00 eV	...	[Ber93, Fre94a]
LB2020	2.020	16.29	613.7	...	W, Ta?	**PL**; W =30%	...	[Har96]
LB2100a	2.100	16.94	590.4	Green(a)	B?	**CL, PL, XL**; W =17–34%, 3 bands a–c (2.390 eV)	**5.**95	[Zai01]
LB2300	2.300	18.55	539.0	*C(α 1)	N_1°	**CL, PL**; W =8–16%, ZPL at 4.059 eV	**7.**79, **7.**124	[Den93, Col89d]
LB2100b	2.340	18.87	529.8	Green(b)	B?	**PL**; W =21%, named Green A, PLE at 2.41–2.95 eV; var. 2.15–2.34 eV	**5.**95, **5.**96, **7.**100, **7.**149	[Zai01, Iak00d, Law95]
LB2350	2.350	18.96	527.6	...	Si?	**CL**; W =36%	**7.**79	[Den93]
LB2360	2.360	19.04	525.3	...	Cr, Fe, Ni, Cu?	**Ion Beam Lum**	...	[Bet94]
LB2100c	2.390	19.28	518.7	Green(c)	B?	**PL**; W =21%, named Green B, PLE at >3.40 eV; var. 2.39–2.58 eV	**5.**96, **7.**100, **7.**149	[Iak00d, Law95]

LB2407	2.407	19.42	515.1	...	...	**PL**; intense excimer laser excitation	...	[Cre96]
LB2450	2.450	19.76	505.9	...	...	**PL**; W =13%; ZPL at 2.651 eV; PLE=LB2850	**5**.128, **7**.124	[Iak00a, Col89d]
LB2730a	2.730	22.02	454.1	Blue(a)	B?	**CL, PL, XL**; W =12%, 5 bands a–e (4.650 eV), var. 2.56–2.86 eV	**7**.109, **7**.123, **7**.124	[Col92b, Buh94, Col89d]
LB2740	2.740	22.10	452.5	...	...	**PL, XL**; W =18%; excited by LB3310	**5**.135, **7**.75	[Iak00a, Den92]
LB2850	2.850	22.99	435.0	...	...	**PLE** of LB2450; W =12%; ZPL at 2.651 eV	**5**.128	[Iak00a]
LB2880	2.880	23.23	430.5	*C(β 1)	N_1°	**CL, PL**; W =8–16%, var. 2.75–3.0 eV, ZPL at 4.059 eV	**7**.79, **7**.113, **7**.124	[Den93, Col89e, Col89d]
LB2950	2.950	23.79	420.3	...	B?	**PLE** of LB2100b=Green(b)	**5**.96	[Iak00d]
LB3100	3.100	25.01	399.9	"N"-band	N?	**PL, EL, IL**; W =13%	**7**.124	[Col89d]

Observation: *A* absorption; *CL, EL, IL, PL, XL* luminescence; *PLE* photoluminescence excitation

Table 4.4.3 Low pressure synthetic diamond (CVD): broad bands, ultraviolet (3.10–5.90 eV)

Line-label	Energy (eV)	Frequ. (10^3 cm^{-1})	Wavel. (nm)	Name	Impur./ defect	Comment	Figs. **3–6.** [Zai01]; Fig. **7.**x[Zai98]	References
LB3260	3.260	26.30	380.3	...	N?	**EL**; $W = 3\%$	...	[Zha96a]
LB3310	3.310	26.70	374.6	...	...	**PLE** of LB2740	**5.**135	[Iak00a]
LB3400	3.400	27.40	364.6	...	B?	**PLE** of LB2100c = Green(c)	**5.**96	[Iak00d]
LB3470	3.470	27.99	357.3	"Blue"(b)	B?	**CL**; $W = 10–17\%$; var. 3.35–3.50 eV	**5.**149, **7.**100, **7.**149	[Ste96a, Law95]
LB3850	3.850	31.05	322.0	...	...	**EL**; $W = 4\%$	...	[Man95]
LB4350a	4.350	35.09	285.0	"Blue"(c)	B?	**CL**; $W = 10\%$	**5.**149, **7.**149	[Ste96a, Law95]
LB4350b	4.500	36.30	275.5	"Blue"(d)	B?	**CL**; $W = 6\%$, dominant "Blue" band	**5.**149, **7.**149	[Ste96a, Law95]
LB4350c	4.650	37.51	266.6	"Blue"(e)	B?	**CL**; $W = 9\%$	**5.**149, **7.**149	[Ste96a, Law95]

Observation: *A* absorption; *CL, EL, IL, PL, XL* luminescence; *PLE* photoluminescence excitation

Chapter 5
Spectral Lines in Modified Diamond (Irradiation, Heat, etc.)

In this chapter ca. 1,000 lines (and bands), which are observed in modified diamond (see Sect. 1.2.4), are presented in 59 tables. The entries are assigned to 305 centers. The corresponding defect structure is established for 29 centers, is almost certain for 107 centers, and is incomplete or unknown for 169 centers.

Chapter 5 is subdivided into *absorption lines* (Sect. 5.1: 360 **MA** lines in 21 tables), into *photoluminescence excitation (PLE) lines* (Sect. 5.2: 96 **ME** lines in 7 tables), into *luminescence lines* (Sect. 5.3: 487 **ML** lines in 27 tables), and into *broad bands* (Sect. 5.4: 57 **MB** bands in 4 tables).

B. Dischler, *Handbook of Spectral Lines in Diamond*,
DOI 10.1007/978-3-642-22215-3_5, © Springer-Verlag Berlin Heidelberg 2012

5.1 Absorption Lines (MA)

Table 5.1.1.1 Modified diamond (irradiation, heat, etc.): absorption lines, far infrared, part 1 (0.04–0.14 eV)

Line-label	Energy (meV)	Frequ. (cm^{-1})	Wavel. (μm)	Name	Impur./defect	Treatment/comment	Figs. **3–6.** [Zai01]; Fig. **7.** [Zai98]	References
MA0060a′	60.00	484.0	20.66	**A**-cent.(a′)	N_2°	**HT**(1,700°C) or **IG+HT**(1,500°C); 5 IR lines a′–e′ (0.1589 eV), see Table 9.1.3	...	[Zai01]
MA0077	76.87	620.0	16.13	...	B, O?	**PO**(in KBr)	**3.**13	[Che94b]
MA0082	81.82	660.0	15.15	...	O?	**HT**(700°C, powder in O_2)	**3.**3b	[Zai01]
MA0097a′	93.50	754.2	13.26	**B**-cent.(a′)	$V_1N_4^\circ$	**HT**(2,500°C)or **IG+HT**(2,200°C); see Table 8.1.3.5; 7 IR lines a′–g′ (165.1 meV)	...	[Zai01]
MA0100a	99.80	805.0	12.42	...	B?	**II**(C+B)+**HT**(1,100°C); 4 lines a–d(173.6 meV)	...	[San89, Zai01]
MA0069c	102.8	829.0	12.06	1LP(c)		**II**(Xe, 5.6 GeV), **IE**(2 MeV); 1LP(a, b) see LA	**3.**3b, **3.**15	[Zai01, Col88c]
MA0100b	106.6	860.0	11.63	...	B?	**II**(C+B)+**HT**(1,100°C)	...	[San89, Zai01]
MA0117a′	117.8	950.0	10.52	**E**(a′) =**X**(a′)	N_1^+	**IE;** 7 lines a′–g′ (165.1 meV)	**3.**15	[Col88c]
MA0125	125.1	1,009	9.910	...		**IN**	...	...
MA0097b′	125.2	1,010	9.902	**B**-cent.(b′)	$V_1N_4^\circ$	**HT**(2,500°C)or **IG+HT**(2,200°C); see Table 8.1.3.5	...	...
MA0069g	126.3	1,019	9.814	1LP(g)		**II**(Xe, 5.6 GeV); **IN;** 1LP(a–q) see LA	**3.**3b, **7.**28	[Zai01, Bok86]

MA0126a	126.5	1,020	9.804	. . .	B?	**PO**(in KBr); 2 lines a, b (133.9 eV);	. . .	. . .
MA0117b′	129.7	1,046	9.559	**E**(b′)=**X**(b′)	N_1^+	**IE**(2 MeV)	**3**.15	[Col88c, Zai01]
MA0100c	131.4	1,060	9.434	. . .	B?	**II**(C+B)+**HT**(1,100°C)	. . ..	[San89, Zai01]
MA0132	131.9	1,064	9.399	. . .		**IN**(10^{18} cm^{-2})	. . ..	[Bok86, Zai01]
MA0069j	132.9	1,072	9.329	1LP(j)		**II**(Xe, 5.6 GeV)	**3**.3b	[Zai01]
MA0126b	133.9	1,080	9.259	. . .	B?	**PO**(in KBr)	**3**.13	[Che94b]
MA0060b′	135.1	1,090	9.177	**A**-cent.(b′)	N_2	**HT**(1,700°C) or **IG**+**HT**(1,500°C), see Table 9.1.3	. . .	[Zai01]
MA0105c′	136.0	1,097	9.116	**C**-cent.(c′)	$N_1{}^\circ$	**IE**(2 MeV);other c lines see NA; see Table 9.1.1	**3**.15	[Col88c]
MA0097c′	136.4	1,100	9.095	**B**-cent.(c′)	$V_1N_4^\circ$	**HT**(2,500°C)or **IG**+**HT**(2,200°C); see Table 8.1.3.5	. . .	[Zai01]
MA0069k	137.7	1,111	9.003	1LP(k)		**II**(Xe, 5.6 GeV), **IN**	**3**.3b, **7**.28	[Zai01, Bok86]
MA0117c′	138.2	1,115	8.971	**E**(c′)=**X**(c′)	N_1^+	**IE**(2 MeV); see Table 9.1.2	**3**.15	[Col88c]
MA0105d′	140.1	1,130	8.849	**C**-cent.(d′)	$N_1{}^\circ$	**IE**(2 MeV); **dominant** C-line; var.(139–141 meV)	**3**.15	[Col88c]

Treatment: *IE* irradiation, electrons; *IG* irradiation, general; *II* irradiation, ions; *IN* irradiation, neutrons; *IX* irradiation, X-ray; *HP* hydrogen plasma; *HT* heat treatment; *PD* plastic deformation; *PO* powder, *PO(KBr) compressed with KBr*

Table 5.1.1.2 Modified diamond (irradiation, heat etc.): absorption lines, far infrared, part 2 (0.14–0.18 eV)

Line-label	Energy (meV)	Frequ. (cm^{-1})	Wavel. (μm)	Name	Impur./defect	Treatment/comment	Figs. **3–6.** [Zai01]; Fig. **7.** [Zai98]	References
MA0097d/	145.7	1,175	8.509	**B**-cent.(d/)	$V_1N_4{}^{\circ}$	**HT**(2,500°C)or **IG+HT**(2,200°C); see Table 8.1.3.5	...	[Zai01]
MA0117d/	146.9	1185	8.440	**E**(d/) = **X**(d/)	N_1^+	**IE**(2 MeV); see Table 9.1.1.2	**3**.15	[Col88c]
MA0147	147.0	1186	8.434	**B**-cent.(irr.)	$^*V_1N_4V_1{}^{\circ}$	**IE+HT**(1,500°C); (indir.) **pseudo**-ZPL of DAP31	(see **5**.a, **5**.b)	[Col97, Kif99]
MA0069m	147.7	1191	8.394	1LP(m)		**II**(Xe, 5.6 GeV), **IN**	**3**.3b, **7**.28	[Zai01, Bok86]
MA0149	148.8	1200	8.333	...	B?	**II**(C+B)+**HT**(1,100°C)	...	[San89, Zai01]
MA0060d/	150.6	1215	8232	**A**-cent.(d/)	$N_2{}^{\circ}$	**HT**...	...	[Zai01]
MA0117e/	152.5	1230	8.130	**E**(e/)=**X**(e/)	N_1^+	**IE**(2 MeV); see Table 9.1.1.2	**3**.15	[Col88c]
MA0069-o	153.6	1239	8.071	1LP(o)		**II**(Xe, 5.6 GeV)	**3**.3b	[Zai01]
MA0154	154.0	1242	8.050	**A**-cent.(irr)	$^*V_1C_4N_2{}^{\circ}$	**IE**(1,000°C); (indirect) **pseudo**-ZPL of -DAP30	(see **5**.a, **5**.b)	[Col97, Kif99]
MA0156	156.2	1260	7.937	...	B?	**II**(C+B), **IN+HT**(>800°C)	...	[San89, Iak01]
MA0159	158.7	1280	7.813	...		**IN**(fast neutrons)	....	[Zai01]
MA0060d/	158.9	1282	7.802	**A**-cent.(d/)	N_2	**HT,** last A-center-line, **dominant**, FWHM= 4%	...	[Zai01]

MA0117f′	159.9	1290	7.753	**E**(f′)=**X**(f′)	N_1^+	**IE**(2 MeV); see Table 9.1.1.2	**3**.15	[Col88c]
MA0165	164.7	1328	7.527	...		**IN**+**HT**(800°C)	**3**.16	[Col88c]
MA0097e′	165.1	1332	7.509	**B**-cent.(e′)	V_1N_4	**HT**(2,500°C)or **IG**+**HT**(2,200°C); see Table 8.1.3.5	...	[Zai01]
MA0117g′	165.1	1332	7.509	**E**(g′)=**X**(g′)	N_1^+	**IE**(2 MeV); sharp, last E-line; see Table 9.1.1.2	**3**.15	[Col88c, Zai01]
MA0069q	165.2	1333	7.505	1LP(q)		**II**(Xe, 5.6 GeV), **IN**	**3**.3b, **7**.21a	[Zai01, Col88c]
MA0105g′	166.6	1344	7.442	C-cent.(g′)	$N_1{}^\circ$	**IE**(2 MeV), **IN** +HT(800°C);sharp(W= 0.2%) last C-center line; see Table 9.1.1.1	**3**.15, **3**.16 **7**.21a	[Col88c] [Col88c]
MA0041e′	169.8	1370	7.301	**D**(e′)=**platel.**	$^*(C_{2n})_i$	**HT**, var.0.158 eV (large) −0.173 eV (small platelets); lines (a′–f′) see NA and Table 8.2.2	...	[Zai01]
MA0174	173.6	1400	7.143	...	B?	**II**(C+B)+**HT**(1,100°C)	...	[San89, Zai01]
MA0176	176.0	1420	7.044	...	...	**IN**	...	[Zai01]
MA0180	179.8	1450	6.895	H1a1	$^*V_1C_4N_1^\circ$	**IG**+**HT**(750°C), LVM(N-C); i13, i15 isotope shifts (Tables 7.2, 7.3)	**3**.16, **7**.21c	[Col88c]

Treatment: IE irradiation, electrons; *IG* irradiation, general; *II* irradiation, ions; *IN* irradiation, neutrons; *IX* irradiation, X-ray; *HP* hydrogen plasma; *HT* heat treatment; *PD* plastic deformation; *PO* powder, *PO(KBr) compressed with KBr*

Table 5.1.2.1 Modified diamond (irradiation, heat etc.): absorption lines, mid infrared, part 1 (0.18–0.34 eV)

Line-label	Energy (meV)	Frequ. (cm^{-1})	Wavel. (μm)	Name	Impur./defect	Treatment/comment	Figs. **3–6.** [Zai01]; Fig. **7.** [Zai98]	References
MA0186	185.5	1496	6.684	...	$^*(C_2)_i^+$	**IE+HT**(450°C)	**3.**11	[Col88c]
MA0187	186.7	1506	6.640	H1a2	$^*(N_1C_1)_i^-$	**IG+HT**(750°C), var.186.2–187.2 meV, LVM(N-C); 188 meV in MA of 3.988 eV center (Table 8.2.4); i13, i15 shift, (Tables 7.2, 7.3)	**3.**16, **7.**13b	[Col88c, Bok86]
MA0190	190.0	1533	6.525	...	$^*(C_2)_i^-$	**IE**, var. 189–191 meV; 188 meV in MA of DAP35(ZPL at 2.638 eV); i13 (Table 7.2), no i15 shift	**3.**15	[Col88c]
MA0195	194.7	1570	6.368	...	$^*(C_2)_i^\circ$?	**IG (+HT),** i13 (Table 7.2), no i15 shift	**3.**11	[Col88c, Zai01]
MA0198	198.4	1600	6.250	N–H bend	H	**PO**(KBr), 2 lines a, b (390.5 meV)	**3.**13	[Che94b]
MA0205a	204.6	1650	6.061	...	H_2O?	**PO**(KBr), O–H bend vibration?	**3.**13	[Che94b, Zai01]
MA0212	211.5	1706	5.861	...	$^*(N_1C_1)_i^\circ$	**IE+HT**(450°C), LVM(N-C) 212 meV in MA of 2.897 eV center; i13 (Table 7.2); pseudo ZPL of DAP3	**3.**11, **7.**19b	[Col88c]
MA0230	230.1	1856	5.388	...	$^*(N_2)_i^\circ$	**IN+HT**(1,400°C), N_2 from isotope shift (Table 7.3)	**3.**12, **7.**20	2 × [Col87c]

MA0238	238.5	1924	5.198	C–C stretch	$^*(C_2)_iN_1^+$	**IE+HT**(450°C), LVM 240 meV in ML of 4.676 eV ce.	**3.**11	[Col88c]
MA0225c	244.0	1968	5.080	2LP(c)	lattice	**IN,** strong peak, var. 0.2440–0.2455; other 2LP lines see NA0225(a–q)	**7.**13	[Bok86]
MA0245	244.9	1975	5.063	**5RL**	$^*(C_2)_iN_1{}^+$	**IG+HT**(450°C), LVM(C–C) 245 meV in MA of DAP79 (ZPL at 4.582 eV); no i15 shift	**3.**11, **3.**12, **7.**20	[Col88c, Col87c] [Col87c]
MA0225e	251.1	2025	4.938	2LP(e)	lattice	**IN,** strong peak, var. 0.2511–0.2517 eV	**7.**13	[Bok86]
MA0260	259.7	2095	4.774	2LP(g)	lattice	**IN,** weak line	**7.**13	[Bok86]
MA0225i	267.0	2154	4.642	2LP(i)	lattice	**IN,** strong peak, var. 0.2670–0.2690 eV	**7.**13	[Bok86]
MA0225o	302.1	2.436	4.104	2LP(o)	lattice	**IN,** medium strong peak, var. 0.3013–0.3021 eV	**7.**13	[Bok86]
MA0225p	315.0	2.541	3.935	2LP(p)	lattice	**IN,** medium strong peak, var. 0.3150–0.3174 eV	**7.**13	[Bok86]
MA0331a	330.8	2668	3.748	DAP30a **H1d**	$^*V_1C_4N_2{}^\circ$	**IE+HT**(1,000°C)**,** 4 lines a–d(612.5 meV); $s = 50$; calc. 331 meV; precursor of H3=V_1N_2; SB (C–C bend): +64 meV; SB ^{13}C shift ca. −3–5% (calc. 3.9%)	**5.**1b1	[Kif99]

Treatment: IE irradiation, electrons; *IG* irradiation, general; *II* irradiation, ions; *IN* irradiation, neutrons; *IX* irradiation, X-ray; *HP* hydrogen plasma; *HT* heat treatment; *PD* plastic deformation; *PO* powder, *PO(KBr) compressed with KBr*

Table 5.1.2.2 Modified diamond (irradiation, heat etc.): absorption lines, mid infrared, part 2 (0.35–1.24 eV)

Line-label	Energy (meV)	Frequ. ($10^3 cm^{-1}$)	Wavel. (μ m)	Name	Impur. /defect	Comment	Figs. **3–6.** [Zai01]; Fig. **7.** [Zai98]	References
MA0155b	350.2	2.825	3.540	C–H stretch	H	**H covered** (111) surface, 1 × 1 reconstruction; lattice sp^3CH_1 stretch (Table 9.3.3); 349.9 meV in LA; MA observed by SFG (sum frequency generation)	**7.**15, **7.**17	[And94, Chi92]
MA0353a	353.4	2.851	3.507	DAP31a	$V_1N_4V_1{}^\circ$	**IE+HT**(1,500°C), 5 lines a–e(0.6408 eV); $s = 36$; calc. 356.0 meV; precursor of H4a, b = $N_3V_1N_1{}^\circ$; weak shoulder	**5.**1b2	[Kif99]
MA0085b	354.2	2.857	3.500	C–H stretch	H	**H covered** (111) surface, 2 × 1 reconstruction; sp^3CH_2 stretch (Table 9.3.3); 353.3 meV in RA	**7.**17	[Chi92]
MA0133b1	358.9	2.895	3.454	C–H stretch	H	**H covered** (100) surface, 2 × 1 reconstruction; sp^3CH_3 stretch (Table 9.3.3); 356.4 meV in RA	**7.**15	[And94]
MA0353b	361.8	2.918	3.427	DAP31b **H1e**	$V_1N_4V_1{}^\circ$	**IE+HT**(1,500°C), $s = 34$; calc. 362 meV; SB (C–C bend): +64 meV (see Table 11.1)	**5.**1b2	[Kif99]
MA0304p	362.7	2.926	3.418	**B-acc.**(p)	B_1	**II**(B), 1b → 8 (see Table 9.2.1)	...	[San89]
MA0133b2	366.4	2.955	3.384	C–H stretch	H	**H covered** (100) surface, 2 × 1 reconstruction; sp^3CH_3stretch (Table 9.3.3); 367.8 meV in RA	**7.**15	[And94]

MA0304r	372.0	3.001	3.333	**B-acc.**(r)	B_1	**II**(B), bound to free (see Table 9.2.1)	...	[San89]
MA0198b	390.5	3.150	3.175	N–H stretch	H	**PO**(KBr) or **NH_2terminated,** diffuse reflect. FTR	**3.**13	[Che94b, Mil95]
MA0353c	401.4	3.237	3.089	DAP31c	$V_1N_4V_1{}^{\circ}$	**IE+HT**(1,500°C), $s = 24$; calc. 404 meV	**5.**1b2	[Kif99]
MA0419	419.0	3.380	2.959	O–H stretch	H_2O?	**PO**(KBr), O–H stretch vibration?	**3.**13	[Che94b, Zai01]
MA0331b	487.6	3.933	2.543	DAP30b	$^{*}V_1C_4N_2{}^{\circ}$	**IE+HT**(1,000°C), $s = 18$; calc. 450 meV	...	
MA0353d	545.0	4.396	2.275	DAP31d **H1f**	$V_1N_4V_1{}^{\circ}$	**IE+HT**(1,500°C), $s = 10$; calc. 544 meV; SB (C–C bend): +64 meV (see Table 11.1)	**5.**1b2	[Kif99]
MA0331c	550.8	4.442	2.251	DAP30c	$^{*}V_1C_4N_2{}^{\circ}$	**IE+HT**(1,000°C), $s = 10$; calc. 551 meV; named **H1g**	**5.**1b1	[Kif99]
MA0331d	612.5	4.940	2.024	DAP30d	N_2?	**IE+HT**(1,000°C), $s = 8$;calc. 598 meV; named **H1b**	**5.**1b1	[Kif99]
MA0353e	640.8	5.169	1.935	DAP31e	$V_1N_4V_1^{\circ}$?	**IE+HT**(1,500°C), $s = 6$; calc. 659 meV; named **H1c**	**5.**1b2	[Kif99]

Treatment: IE irradiation, electrons; *IG* irradiation, general; *II* irradiation, ions; *IN* irradiation, neutrons; *IX* irradiation, X-ray; *HP* hydrogen plasma; *HT* heat treatment; *PD* plastic deformation; *PO* powder, *PO(KBr) compressed with KBr*

Table 5.1.3.1 Modified diamond (irradiation, heat etc.): absorption lines, near infrared, part 1 (1.24–1.67 eV)

Line-label	Energy (eV)	Frequ. ($10^3 cm^{-1}$)	Wavel. (nm)	Name	Impur. /defect	Treatment/comment	Figs. **3–6.** [Zai01]; Fig. **7.** [Zai98]	References
MA1245	1.245	10.04	995.8	…	Ni, N?	**HT**(>1,700°C)	…	[Yel95a]
MA1256z	1.256	10.13	987.1	H2 = DAP77z	$V_1N_2^-$	**HT**(1,700°C), ZPL of DAP77a–j; QLVM:+59, LVM:+77, +167 meV	**7.**78	[Col89c] [Law92b, Zai01]
MA1357	1.357	10.94	913.9	…	Ni, N?	**HT**(>1,700°C)	…	[Yel95a]
MA1374a	1.374	11.08	902.3	**H2**-DAP77a	$V_1N_2^-$	**HT**(1,700°C), $s = 106$, calc. 1.375 meV	…	[Law92b]
MA1374b	1.390	11.21	891.9	**H2**-DAP77b	$V_1N_2^-$	**HT**(1,700°C), $s = 86$, calc. 1.388 meV	**7.**78	[Col89c, Law92b]
MA1401a	1.401	11.30	884.9	**1Ni-1.4 eV**(a)	$^*V_1Ni_1^+$	**II**(Ni), 2 lines a, b (1.404 eV) from 2.7 meV ground state splitting; (see HA); ZPL of 1Ni-DAP25a–k	…	[Zai01]
MA1404	1.404	11.32	883.2	…	Ni, N?	**HT**(>1,700°C)	…	[Yel95a]
MA1374c	1.410	11.37	879.3	**H2**-DAP77c	$V_1N_2^-$	**HT**(1,700°C), $s = 64$, calc. 1.409 meV	…	[Law92b]
MA1441	1.441	11.62	860.2	…	Ni, N?	**HT**(>1,700°C)	…	[Yel95a]

MA1374d	1.480	11.94	837.7	**H2**-DAP77d	$V_1N_2^-$	**HT**(1,700°C), $s = 30$, calc. 1.480 meV	**7**.78	[Col89c, Law92b]
MA1521	1.521	12.27	815.1	...		**IG**	...	[Zai01]
MA1374e	1.530	12.34	810.3	**H2**-DAP77e	$V_1N_2^-$	**HT**(1,700°C), $s = 20$, calc. 1.531 meV	...	[Law92b]
MA1563a	1.563	12.61	793.2	***S5**-DAP60a	$*(V_2Ni_1)N_x^\circ$	**HT**(>1,600°C) or **IE+HT**(900°C), also HA, ME, 11 lines a–k(2.320 eV), L = 1.460 eV; $s = 162$, calc. 1.564 eV	...	[Kup99, Zai01]
MA1563b	1.600	12.91	774.9	***S5**-DAP60b	$*(V_2Ni_1)N_x^\circ$	**(see line a)**, $s = 86$, calc. 1.602 eV	...	[Kup99]
MA1563c	1.638	13.21	756.9	***S5**-DAP60c	$*(V_2Ni_1)N_x^\circ$	**(see line a)**, $s = 58$, calc. 1.634 eV	...	[Kup99]
MA1563d	1.648	13.29	752.3	***S5**-DAP60d	$*(V_2Ni_1)N_x^\circ$	**(see line a)**, $s = 50$, calc. 1.647 eV	...	[Kup99]
MA1374f	1.650	13.31	751.4	**H2**-DAP77f	$V_1N_2^-$	**HT**(1,700°C), $s = 10$, calc. 1.645 meV	...	[Law92b]
MA1665a	1.665	13.43	744.6	**GR1a**	V_1°	**IG,** 2 lines a, b (1.673 eV), see HL(a:b = 1:3)	...	[Zai01]
MA1665b = MA1771z	1.673	13.49	741.0	**GR1b** = DAP33z	V_1°	**IG,** ZPL of DAP33a–k(1.980 eV); 2 SBs at +41, +71 meV;	**5**.30a, **7**.82, **7**.68, **7**.145	[All98a, Col79a] [Dav77c]

Treatment: IE irradiation, electrons; *IG* irradiation, general; *II* irradiation, ions; *IN* irradiation, neutrons; *IX* irradiation, X-ray; *HP* hydrogen plasma; *HT* heat treatment; *PD* plastic deformation; *PO* powder, *PO(KBr) compressed with KBr*

Table 5.1.3.2 Modified diamond (irradiation, heat etc.): absorption lines, near infrared, part 2 (1.67–1.77 eV)

Line-label	Energy (eV)	Frequ. ($10^3 cm^{-1}$)	Wavel. (nm)	Name	Impur. /defect	Comment	Figs. **3–6.** [Zai01]; Fig. **7.** [Zai98]	References
MA1678a	1.678	13.53	738.8	***S9**-DAP4a	$*(V_2Ni_1)N_x^+$	**IN+HT**(950°C), 9 lines a–i(1.821 eV), $s = 72$, calc. 1.678 eV, lines(b–i) see ML [Sil95]	...	[Zai01]
MA1681	1.681	13.56	737.5	**2Si-1.68 eV** ce.= DAP53z	$*(V_3Si_2)^+$	**II**(Si) +HT or **HT**(1,900°C, Si doped); QLVM: 40 meV, calc. 2Si+1C:39.8; DAP53a-m, (LE)	...	[Zai01]
MA1685a= MA1825z	1.685	13.59	735.9	DAP71z	$(C_2)_i{}^\circ$	**IG,** 2 lines a, b (1.859 eV);(a) from upper ground state (+6.2 meV), ZPL of DAP71a-c	**5**.30a, **5**.30b	[All98a, Zai01] [Dav00]
MA1374g	1.690	13.63	733.6	**H2**-DAP77g	$V_1N_2^-$	**HT**(1,700°C), $s = 8$, calc. 1.691 meV	...	[Law92b]
MA1693a	1.693	13.66	732.3	**Ni-1.69 eV**-ce.	$*V_1Ni_2^+$	**HT**(1,600°C), 3 SBs:+51 meV (C–C bend, see Table 11.1); lines (b, c) at 1.940, 1.991 eV	**7**.70	[Col85b, Zai01]
MA1678d	1.697	13.69	730.7	***S9**-DAP4d	$*(V_2Ni_1)N_x^+$	**IN+HT**(950°C), 4 lines a, d, i, j(1.893 eV), other lines see ML1678; $s = 58$, calc. 1.698 eV	...	[Sil94a, Zai01]
MA1702= MA1834z	1.702	13.73	728.4	DAP45z	$*V_1C_4N_1^+$	**IG+HT**(800°C), ZPL of DAP45a-i (2.341 eV); precursor of V_1N_1	**5**.166	[Vin88b]
MA1563e	1.708	13.78	725.9	***S5**-DAP60e	$(V_2Ni_1)N_x{}^\circ$	**(see line a)**, $s = 30$, calc. 1.702 eV	...	[Kup99]
MA1374h	1.742	14.05	711.7	**H2**-DAP77h	$V_1N_2^-$	**HT**(1,700°C), $s = 6$, calc. 1.757 meV	...	[Law92b]

Treatment: IE irradiation, electrons; *IG* irradiation, general; *II* irradiation, ions; *IN* irradiation, neutrons; *IX* irradiation, X-ray; *HP* hydrogen plasma; *HT* heat treatment; *PD* plastic deformation; *PO* powder, *PO(KBr) compressed with KBr*

Table 5.1.4 Modified diamond (irradiation, heat etc.): absorption lines, visible: purple (1.77–1.88 eV)

Line-label	Energy (eV)	Frequ. ($10^3 cm^{-1}$)	Wavel. (nm)	Name	Impur. /defect	Treatment/comment	Figs. **3–6.** [Zai01]; Fig. **7.** [Zai98]	References
MA1771a	1.771	14.29	700.0	**GR1**-DAP33a	V_1^o	**IG,** 11 lines a–k, $s = 118$, calc. 1.769 eV	**5**.30	[All98a, Zai01]
MA1771b	1.793	14.46	691.5	**GR1**-DAP33b	V_1^o	**IG,** $s = 78$, calc. 1.791 eV	(Table 5.2)	[Zai01, p. 164]
MA1771c	1.805	14.56	686.9	**GR1**-DAP33c	V_1^o	**IG,** $s = 62$, calc. 1.805 eV	(Table 5.2)	[Zai01, p. 164]
MA1771d	1.809	14.59	685.3	**GR1**-DAP33d	V_1^o	**IG,** $s = 58$, calc. 1.810 eV	(Table 5.2)	[Zai01, p. 164]
MA1771e	1.814	14.63	683.4	**GR1**-DAP33e	V_1^o	**IG,** $s = 54$, calc. 1.815 eV	(Table 5.2)	[Zai01, p. 164]
MA1771f	1.819	14.67	681.6	**GR1**-DAP33f	V_1^o	**IG,** $s = 50$, calc. 1.819 eV	(Table 5.2)	[Zai01, p. 164]
MA1678i	1.821	14.69	680.8	***S9**-DAP4i	$*(V_2Ni_1)N_x^+$	**IN+HT**(950°C), $s = 22$, calc. 1.816 eV	...	[Sil94a, Zai01]
MA1771g	1.824	14.71	679.7	**GR1**-DAP33g	V_1^o	**IG,** $s = 48$, calc. 1.823 eV		[Dav00]

(continued)

Table 5.1.4 (continued)

Line-label	Energy (eV)	Frequ. ($10^3 cm^{-1}$)	Wavel. (nm)	Name	Impur. /defect	Treatment/comment	Figs. **3–6.** [Zai01]; Fig. **7.** [Zai98]	References
MA1825a	1.825	14.72	679.3	DAP71a	$(C_2)_i{}^\circ$	**IG,** 3 lines a-c (2.010 eV) $s = 64$, calc. 1.825 eV	**7.**121	[Cla56c]
MA1771h	1.828	14.74	678.2	**GR1**-DAP33h	V_1°	**IG,** $s = 44$, calc. 1.830 eV	(Table 5.2)	[Zai01, p. 164]
MA1834a	1.834	14.79	676.0	DAP45a	$V_1C_4N_1^\circ$	**IG+HT**(800°C), 9 lines a–i(2.331 eV); $s = 106$, calc. 1.831 eV	**5.**166	[Vin88b]
MA1771i	1.838	14.82	674.5	**GR1**-DAP33i	V_1°	**IG,** $s = 40$, calc. 1.837 eV	**5.**30	[All98a, Zai01]
MA1852 = MA1992z	1.852	14.94	669.3	**Co-1.85 eV =** DAP102z	$*V_1Co_2^+$	**HT**(1,500°C), Co melt, 3 SBs at +44 meV; ZPL of DAP102a-e(2.140 eV)	**5.**52	[Law96, Zai01]
MA1685b	1.859	15.00	666.8	pseudo ZPL	$(C_2)_i{}^\circ$	**IG,** 168 meV SB of 1.691 eV forbidden line	**5.**30a, **5.**30b	[All98a, Dav00]

Treatment: IE irradiation, electrons; *IG* irradiation, general; *II* irradiation, ions; *IN* irradiation, neutrons; *IX* irradiation, X-ray; *HP* hydrogen plasma; *HT* heat treatment; *PD* plastic deformation; *PO* powder, *PO(KBr) compressed with KBr*

Table 5.1.5 Modified diamond (irradiation, heat etc.): absorption lines, visible: red (1.88–2.03 eV)

Line-label	Energy (eV)	Frequ. (10^3cm^{-1})	Wavel. (nm)	Name	Impur. /defect	Treatment/comment	Figs. **3–6.** [Zai01]; Fig. **7.** [Zai98]	References
MA1374i	1.890	15.24	656.0	**H2-**DAP77i	$V_1N_2^-$	**HT**(1,700°C), s = 3, calc. 1.923 meV	...	[Law92b]
MA1678j	1.893	15.27	654.9	***S9-**DAP4j	$^*(V_2Ni_1)N_x^+$	**IN+HT**(950°C), s = 14, calc. 1.894 eV	...	[Sil94a, Zai01]
MA1834b	1.903	15.35	651.5	DAP45b	$^*V_1C_4N_1^\circ$	**IG+HT**(800°C), s = 42, calc. 1.907 eV	**5.**54	[Sil95]
MA1834c	1.909	15.40	649.4	DAP45c	$^*V_1C_4N_1^\circ$	**IG+HT**(800°C), s = 40, calc. 1.912 eV	**5.**54	[Sil95]
MA1834d	1.925	15.53	644.0	DAP45d	$^*N_1C_4V_1{}^\circ$	**IG+HT**(800°C), s = 34, calc. 1.928 eV	**5.**54, **5.**166	[Sil95, Vin88b]
MA1563f	1.930	15.57	642.4	***S5-**DAP60f	$^*(V_2Ni_1)N_x{}^\circ$	**(see line a)**, s = 8, calc. 1.928 eV	...	[Kup99]
MA1771j	1.939	15.64	639.4	**GR1**-DAP33j	V_1°	**IG,** s = 16, calc. 1.933 eV	**5.**30	[All98a], [Zai01]
MA1693b	1.940	15.65	639.1	**Ni-1.69 eV**-ce.	$^*V_1Ni_2^+$	**HT**(1,500°C), 7 SBs:+50 meV (C–C bend)	...	[Zai01]
MA1943= MA2019z	1.943	15.67	638.1	DAP48z=**NV**$^-$ = 638 **nm** center	$V_1N_1^-$(a)	**IG+HT**(>550°C), line (b) at 4.328 eV ZPL of DAP48a-j(2.315 eV) SB:+63 meV	**5.**59b, **7.**138	[Vin88b, Dav77c]
MA1825b	1.950	15.73	635.8	DAP71b	$(C_2)_i{}^\circ$	**IG,** s = 18, calc. 1.950 eV	**7.**121	[Cla56c]
MA1834e	1.962	15.83	631.9	DAP45e	$^*N_1C_4V_1{}^\circ$	**IG+HT**(800°C), s = 26, calc. 1.962 eV	**5.**166	[Vin88b, Zai01]
MA1563g	1.970	15.89	629.3	***S5-**DAP60g	$^*(V_2Ni_1)N_x{}^\circ$	**(see line a)**, s = 7, calc. 1.969 eV	...	[Kup99]

(continued)

Table 5.1.5 (continued)

Line-label	Energy (eV)	Frequ. ($10^3 cm^{-1}$)	Wavel. (nm)	Name	Impur. /defect	Treatment/comment	Figs. **3–6.** [Zai01]; Fig. **7.** [Zai98]	References
MA1979	1.979	15.96	626.5	...	$*(C_2)_iN_2^{\circ}$	**IE,** appears after bleaching of the 2.367 eV line	**7.**82	[Col79c, Zai01]
MA1771k	1.980	15.97	626.1	**GR1**-DAP33k	$*V_1^-$	**IG,** $s = 12$, calc. 1.973 eV	**5.**30	[All98], [Zai01]
MA1693c	1.991	16.06	622.7	**Ni-1.69 eV**-ce.	$*V_1Ni_2^+$	**HT**(1,500°C), 6 SBs:+50 meV (C–C bend)	**5.**56	[Kup99, Zai01]
MA1992a	1.992	16.07	622.4	DAP102a	$*V_1Co_2{}^{\circ}$	**HT**(1,600°C), pure Co melt; 5 lines a–e(2.140 eV), $s = 64$, calc. 1.997 eV	**5.**52	[Law96]
MA1563h	2.005	16.17	618.3	****S5**-DAP60h	$*(V_2Ni_1)N_x{}^{\circ}$	**(see line a)**, $s = 6$, calc. 2.000 eV	...	[Kup99]
MA1825c	2.010	16.21	616.8	DAP71c	$(C_2)_i{}^{\circ}$	**IG,** $s = 12$, calc. 2.010 eV	**7.**121	[Cla56c]
MA1834f	2.016	16.26	615.0	DAP45f	$*N_1C_4V_1{}^{\circ}$	**IG+HT**(800°C), $s = 18$, calc. 2.014 eV	**5.**166	[Vin88b, Zai01]
MA2019a	2.019	16.29	614.1	DAP48a	$V_1N_1^-$	**IG+HT**(>550°C), 10 lines a–j(2.315 eV), $s = 198$, calc. 2.019 eV	**5.**59b, **7.**138	[Vin88b, Dav77c]
MA1374j	2.020	16.29	613.7	**H2**-DAP77j	$V_1N_2^-$	**HT**(1,700°C), $s = 1$, calc. 2.051 meV	...	[Law92b]

Treatment: IE irradiation, electrons; *IG* irradiation, general; *II* irradiation, ions; *IN* irradiation, neutrons; *IX* irradiation, X-ray; *HP* hydrogen plasma; *HT* heat treatment; *PD* plastic deformation; *PO* powder, *PO(KBr) compressed with KBr*

Table 5.1.6 Modified diamond (irradiation, heat etc.): absorption lines, visible: orange, yellow (2.03–2.18 eV)

Line-label	Energy (eV)	Frequ. ($10^3 cm^{-1}$)	Wavel. (nm)	Name	Impur. /defect	Treatment/comment	Figs. **3–6.** [Zai01]; Fig. **7.** [Zai98]	References
MA2031	2.031	16.38	610.4	...	...	**IG+HT**(500°C)	...	[Zai01]
MA1992b	2.035	16.41	609.2	DAP102b	$^*V_1Co_2{}^\circ$	**HT**(1,600°C), $s = 42$, calc. 2.032 eV	**5.**52	[Law96]
MA1992c	2.060	16.62	601.8	DAP102c	$^*V_1Co_2{}^\circ$	**HT**(1,600°C), $s = 30$, calc. 2.062 eV	**5.**52	[Law96]
MA2019b	2.068	16.68	599.5	DAP48b	$^*V_1N_1^-$	**IG+HT**(>550°C), $s = 74$, calc. 2.068 eV	**7.**138	[Dav77c, Kif99]
MA2019c	2.073	16.72	598.1	DAP48c	$^*V_1N_1^-$	**IG+HT**(>550°C), $s = 64$, calc. 2.077 eV	**7.**138	[Dav77c]
MA1834g	2.086	16.83	594.3	DAP45g	$^*N_1C_4V_1{}^\circ$	**IG+HT**(800°C),$s = 12$, calc. 2.084 eV; LVM:165 meV	**5.**166	[Vin88b]
MA2086	2.086	16.83	594.3	**595 nm-** center	$^*V_2N_2^\circ$	**IG+HT**(700°C), QLVM: 75(2N), calc. 74.5 meV	**7.**88, **7.**137	[Dav77c, Dav80c]
MA2019d	2.092	16.87	592.6	DAP48d	$^*V_1N_1^-$	**IG+HT**(>550°C), $s = 50$, calc. 2.095 eV	**5.**59b, **7.**138	[Vin88b, Dav77c]
MA1563i	2.095	16.90	591.8	*__S5__-DAP60i	$^*(V_2Ni_1)N_x{}^\circ$	**(see line a)**, $s = 4$, calc. 2.089 eV	...	[Kup99]

(continued)

Table 5.1.6 (continued)

Line-label	Energy (eV)	Frequ. ($10^3 cm^{-1}$)	Wavel. (nm)	Name	Impur. /defect	Treatment/comment	Figs. **3–6.** [Zai01]; Fig. **7.** [Zai98]	References
MA2019e	2.110	17.02	587.6	DAP48e	$V_1N_1^-$	**IG**+**HT**(>550°C), $s = 42$, calc. 2.109 eV	**7.**138	[Dav77c]
MA1992d	2.117	17.08	585.6	DAP102d	*V_1Co_2°	**HT**(1,600°C), $s = 20$, calc. 2.112 eV	**5.**52	[Law96]
MA1992e	2.140	17.26	579.3	DAP102e	*V_1Co_2°	**HT**(1,600°C), $s = 16$, calc. 2.141 eV	**5.**52	[Law96]
MA2154= MA2244z	2.154	17.38	575.5	DAP46z= NV=**575 nm** center	*$V_1N_1^\circ$	**IG**+**HT**(>500°C), ZPL of DAP46a-e(2.399 eV) 7 SBs with combinations of 48 and 83 meV	**5.**59a, **7.**93b	[Vin88b,Dav79b] [Zai01]
MA2019f	2.165	17.46	572.6	DAP48f	*$V_1N_1^-$	**IG**+**HT**(>550°C), $s = 24$, calc. 2.162 eV	**7.**138	[Dav77c]
MA1834h	2.171	17.51	571.1	DAP45h	*$N_1C_4V_1$°	**IG**+**HT**(800°C), $s = 8$, calc. 2.170 eV	**5.**166	[Vin88b, Zai01]
MA2019g	2.171	17.51	571.1	DAP48g	*$V_1N_1^-$	**IG**+**HT**(>550°C), $s = 22$, calc. 2.172 eV	**7.**138	[Dav77c]

Treatment: IE irradiation, electrons; *IG* irradiation, general; *II* irradiation, ions; *IN* irradiation, neutrons; *IX* irradiation, X-ray; *HP* hydrogen plasma; *HT* heat treatment; *PD* plastic deformation; *PO* powder, *PO(KBr) compressed with KBr*

Table 5.1.7.1 Modified diamond (irradiation, heat etc.): absorption lines, visible: green, part 1 (2.18–2.37 eV)

Line-label	Energy (eV)	Frequ. ($10^3 cm^{-1}$)	Wavel. (nm)	Name	Impur. /defect	Treatment/comment	Figs. **3–6.** [Zai01]; Fig. **7.** [Zai98]	References
MA2201	2.201	17.75	563.3	...	Ni?	**HT**(1,700°C)...	...	[Zai01]
MA1563j	2.220	17.91	558.5	***S5**-DAP60j	$*(V_2Ni_1)N_x{}^{\circ}$	**(see line a)**, $s = 2$, calc. 2.228 eV	...	[Kup99]
MA2019h	2.230	17.99	556.0	DAP48h	$*V_1N_1^-$	**IG+HT**(>550°C), $s = 14$, calc. 2.231 eV	**7.**138	[Dav77c]
MA2070d	2.240	18.07	553.5	DAP1d	$N_1{}^{\circ}+B_1{}^{\circ}$	**HT**(>1,500°C),10 lines d-m(2.668), (a–c) see LL; MA-DAP1 lines ±20 meV; $s = 13$, calc.2.239 eV	**5.**87, **7.**97	[Law93a, Col85b] [Kup99]
MA2244a	2.244	18.10	552.5	DAP46a	$*V_1N_1^+$	**IG+HT**(>500°C), 5 lines a–e(2.399 eV) $s = 162$, calc. 2.249 eV	**5.**59a, 7.93b, **7.**95	[Vin88b,Dav79b] [Jan91]
MA2019i	2.255	18.19	549.8	DAP48i	$*V_1N_1^-$	**IG+HT**(>550°C), $s = 12$, calc. 2.253 eV	**7.**138	[Dav77c]
MA2244b	2.282	18.41	543.3	DAP46b	$*V_1N_1^+$	**IG+HT**(>500°C), $s = 86$, calc. 2.284 eV	**7.**95	[Jan91]
MA2070e	2.284	18.42	542.8	DAP1e	$N_1{}^{\circ}+B_1{}^{\circ}$	**HT**(>1,500°C), $s = 10$, calc. 2.285 eV	**5.**87, **7.**97	[Law93a, Col85b] [Kup99]
MA2313	2.313	18.66	536.0	...	...	**IG+HT**(1,300°C)...	...	[Zai01]

(continued)

Table 5.1.7.1 (continued)

Line-label	Energy (eV)	Frequ. ($10^3 cm^{-1}$)	Wavel. (nm)	Name	Impur. /defect	Treatment/comment	Figs. **3–6.** [Zai01]; Fig. **7.** [Zai98]	References
MA2019j	2.315	18.67	535.5	DAP48j	$^*V_1N_1^-$	**IG+HT**(>550°C), $s=8$, calc. 2.322 eV	**7.**138	[Dav77c]
MA1563k	2.320	18.71	534.4	***S5**-DAP60k	$(V_2Ni_1)N_x{}^\circ$	**(see line a)**, $s=1$, calc. 2.317 eV	...	[Kup99]
MA2244c	2.323	18.74	533.7	DAP46c	$^*V_1N_1^+$	**IG+HT**(>500°C), $s=50$, calc. 2.324 eV	**5.**59, 7.93b	[Vin88b,Dav79b]
MA2070f	2.331	18.80	531.9	DAP1f	$N_1{}^\circ+B_1{}^\circ$	**HT**(>1,500°C), var. 2.331–2.351 eV; $s=8$, calc. 2.331 eV	**5.**86, **7.**97	[Law93b, Col85b, Kup99]
MA1834i	2.331	18.80	531.9	DAP45i	$^*V_1C_4N_1^+$	**IG+HT**(800°C), var. 3.321–3.341 eV; $s=4$, calc. 2.331 eV	**5.**166	[Vin88b, Zai01]
MA2367 = MA2442z	2.367	19.09	523.9	DAP88z	$^*(C_2)_iN_2^-$	**IG,** dominant line of a triplet (2.378, 2.385 eV) from exc. state splitting; ZPL of DAP88a-f	**7.**82	[Col79c, Zai01]
MA2370	2.370	19.12	523.1	"**A**" line = DAP91z	$^*V_1Ni_3{}^\circ$	**HT**(1,900°C), ZPL of DAP91a–g(2.641 eV); corresponding lines observed in ML(DAP92)	...	[Nad93, Zai01]

Treatment: IE irradiation, electrons; *IG* irradiation, general; *II* irradiation, ions; *IN* irradiation, neutrons; *IX* irradiation, X-ray; *HP* hydrogen plasma; *HT* heat treatment; *PD* plastic deformation; *PO* powder, *PO(KBr) compressed with KBr*

Table 5.1.7.2 Modified diamond (irradiation, heat etc.): absorption lines, visible: green, part 2 (2.37–2.43 eV)

Line-label	Energy (eV)	Frequ. ($10^3 cm^{-1}$)	Wavel. (nm)	Name	Impur. /defect	Treatment/comment	Figs. **3–6.** [Zai01]; Fig. **7.** [Zai98]	References
MA2244d	2.374	19.15	522.2	DAP46d	$*V_1N_1^+$	**IG+HT**(>500°C), $s = 30$, calc. 2.374 eV	7.93b	[Dav79b]
MA2374	2.374	19.15	522.2	...	Ni?	**HT**(1,700°C)	...	[Yel95a, Osv97]
MA2383	2.383	19.22	520.3	...	Ni?	**HT**(1,600°C)	...	[Zai01]
MA2390	2.390	19.28	518.7	...	Ni?	**HT**(1,800°C)	...	[Law93b]
MA2394	2.394	19.31	518.0	...	Ni?	**HT**(1,700°C)	...	[Zai01]
MA2244e	2.399	19.35	516.8	DAP46e	$*V_1N_1^+$	**IG+HT**(>500°C), $s = 24$, calc. 2.400 eV	**5.**59a	[Vin88b]
MA2400	2.400	19.36	516.6	**M1**	...	**IG,** SB at +21 meV	**7.**107	[Dav77c]
MA2070h	2.413	19.46	513.8	DAP1h	$N_1{}^\circ + B_1{}^\circ$	**HT**(>1,500°C), $s = 6$, calc.2.398 eV	**5.**86, **7.**97	[Law93b, Col85b, Kup99]
MA2417a	2.417	19.50	512.9	**H4a**	$*N_3V_2N_1{}^\circ$	**IG+HT**(600°C) or **HT**(1,900°C), 2 lines a, b (2.498 eV); intens. H4a:b = 1:10	**7.**101, **7.**108 **7.**116a	[Col82b, Dav77c] [DeS77]
MA2419 = MA2543z	2.419	19.51	512.5	**TH5** = DAP15z	$*V_2{}^\circ$	**IG+HT**(600°C), weak line, ZPL of DAP15a-l(2.918 eV), dominant: a, i, j, k, l	**7.**121	[Cla56c, Zai01]
MA1660b	2.427	19.58	510.8	Ni 1.66 b	$*V_1Ni_2{}^\circ$	HT(>1600°C) QLVM: 25, LVM: 3 × 49 meV; PLE of 1.66 a	**5.**86, **7.**97	[Law93b, Col85b]

Treatment: IE irradiation, electrons; *IG* irradiation, general; *II* irradiation, ions; *IN* irradiation, neutrons; *IX* irradiation, X-ray; *HP* hydrogen plasma; *HT* heat treatment; *PD* plastic deformation; *PO* powder, *PO(KBr) compressed with KBr*

Table 5.1.8.1 Modified diamond (irradiation, heat etc.): absorption lines, visible: blue, part 1 (2.43–2.53 eV)

Line-label	Energy (eV)	Frequ. ($10^3 cm^{-1}$)	Wavel. (nm)	Name	Impur. /defect	Treatment/comment	Fig. **3–6.** [Zai01] Fig. **7.** [Zai98]	References
MA2442a	2.442	19.70	507.7	DAP88a	$^*(C_2)_i N_2^-$	**IG,** 6 lines a-f (2.600 eV) $s = 198$, calc. 2.444 eV	**7.**82	[Col79c, Zai01]
MA2445	2.445	19.72	507.1	**M2**	...	**IG**, SB at +21 meV	**7.**107	[Dav77c, Dav99a]
MA2454	2.458	19.83	504.4	...	Ni?	**HT**(1,700°C)	...	[Yel95a]
MA2070-i	2.456	198.1	504.8	DAP1-i	$N_1°+B_1°$	**HT**(>1,500°C), $s = 5$, calc.2.444 eV	**5.**86, **7.**97	[Law93b, Col85b, Kup99]
MA2460a	2.460	19.84	504.0	DAP91a	$^*V_1Ni_3°$	**HT**(1,900°C), $s = 162$, calc. 2.461 eV	...	[Nad93, Zai01]
MA2442b	2.462	19.86	503.6	DAP88b	$^*(C_2)_i N_2^-$	**IG,** $s = 130$, calc. 2.462 eV	**7.**82	[Col79c, Zai01]
MA2462	2.462	19.86	503.6	**3H**	$^*(C_2)_i^-$	**IG,** SB at +67 meV	**5.**30, **5.**34a **7.**107	[All98a, Zai01] [Dav77c]
MA2464	2.464	19.88	503.2	**H3a**	$^*V_1N_2°$	**IG+HT**(500–800°C); 2 lines a, b (3.361 eV), ZPL of DAP50a-e(2.981 eV), ZPL also in Figs. **7.**101, **7.**111, **7.**116a, **7.**137, **7.**138	**7.**108	[Dav77c, Zai01]
MA2460b	2.472	19.94	501.5	DAP91b	$^*V_1Ni_3°$	**HT**(2,200°C), $s = 130$, calc. 2.472 eV	...	[Nad93, Zai01]
MA2477a	2.477	19.98	500.5	**S3**-DAP58a	$(V_2Ni_1)N_3^+$	**HT**(1,700°C), lines b-j(2.890 eV) see ME	...	[Zai01]

MA2460c	2.479	20.00	500.1	DAP91c	*$V_1Ni_3{}^{\circ}$	**HT**(1,900°C), $s = 118$, calc. 2.477 eV	...	[Nad93, Zai01]
MA2442c	2.485	20.04	498.9	DAP88c	*$(C_2)_iN_2^-$	**IG,** $s = 86$, calc. 2.483 eV	**7.**82	[Col79c, Zai01]
MA2070j	2.495	20.12	496.9	DAP1j	$N_1{}^{\circ}+B_1{}^{\circ}$	**HT**(>1,500°C), $s = 4$, calc.2.481 eV	**5.**86, **7.**97	[Law93b, Col85b, Kup99]
MA2417b = MA2652z	2.498	20.15	496.3	**H4b** = DAP37z	*$N_3V_2N_1{}^{\circ}$	**IG+HT**(600°C)or **HT**(1,900°C), ZPL of DAP37a-m(3.342 eV), SBs at + 38, 154 meV...	**7.**101, **7.**108 **7.**116a	[Col82b, Dav77c] [DeS77]
MA2442d	2.500	20.71	495.8	DAP88d	*$(C_2)_iN_2^-$	**IG,** $s = 64$, calc. 2.502 eV	**7.**82	[Col79c, Zai01]
MA2500	2.500	20.17	495.8	...	...	**IG,** doublet	...	[Dav77c]
MA2515a	2.515	20.29	493.0	**S2**-DAP57a	$(V_2Ni_1)N_3^+$	**HT**(1,700°C), 5 lines a, b, e, f, h, other lines see ME	**7.**126	[Nad93, Zai01]
MA2442e	2.523	20.35	491.4	DAP88e	*$(C_2)_iN_2^-$	**IG,** $s = 50$, calc. 2.520 eV	**7.**82	[Col79c, Zai01]
MA2526	2.526	20.38	490.8	...	$V_1N_3^-$	**IG;** also in carbonado, ZPL of DAP76, see NL	**7.**111	[Dav77c]

Treatment: IE irradiation, electrons; *IG* irradiation, general; *II* irradiation, ions; *IN* irradiation, neutrons; *IX* irradiation, X-ray; *HP* hydrogen plasma; *HT* heat treatment; *PD* plastic deformation; *PO* powder, *PO(KBr) compressed with KBr*

Table 5.1.8.2 Modified diamond (irradiation, heat etc.): absorption lines, visible: blue, part 2 (2.53–2.58 eV)

Line-label	Energy (eV)	Frequ. ($10^3 cm^{-1}$)	Wavel. (nm)	Name	Impur. /defect	Treatment/comment	Figs. **3–6.** [Zai01]; Fig. **7.** [Zai98]	References
MA2533	2.533	20.43	489.4	...	...	**IG+HT**(600°C), SBs with 2 x 37 meV	**7.**111, **7.**138	[Dav77c]
MA2535 = MA2676z	2.535	20.45	489.1	DAP90z	$^*(C_2)_iN_2^+$	**IG,** ZPL of DAP90a-c(2.747 eV); QLVM: +73 meV, calc. (2N) 74.5 meV	**7.**82	[Col79c, Zai01]
MA2498-s3	2.536	20.46	488.9	H4b-s3	$^*N_3V_2N_1{}^\circ$	**IG+HT**(600°C), QLVM: +38 meV(calc. 1N+5C = 37.9 meV); SBs s1,s2 see ML2498	**7.**101, **7.**116a	[Col82b, DeS77]
MA2515b	2.537	20.46	488.7	**S2**-DAP57b	$(V_2Ni_1)N_3$	**HT**(1,700°C), named line **B**	**7.**126	[Nad93, Zai01]
MA2543a	2.543	20.51	487.5	**TH5**-DAP15a	$^*V_2{}^\circ$	**IG+HT**(600°C), 12 lines a–l(2.909 eV); $s = 130$, calc. 2.543 eV	**7.**121	[Cla56c, Zai01]

Treatment: IE irradiation, electrons; *IG* irradiation, general; *II* irradiation, ions; *IN* irradiation, neutrons; *IX* irradiation, X-ray; *HP* hydrogen plasma; *HT* heat treatment; *PD* plastic deformation; *PO* powder, *PO(KBr) compressed with KBr*

Table 5.1.9.1 Modified diamond (irradiation, heat etc.): absorption lines, visible: ultramarine, part 1 (2.58–2.66 eV)

Line-label	Energy (eV)	Frequ. ($10^3 cm^{-1}$)	Wavel. (nm)	Name	Impur. /defect	Treatment/comment	Figs. **3–6.** [Zai01]; Fig. **7.** [Zai98]	References
MA2534e	2.583	20.83	480.0	**TH5**-DAP15e	$*V_2°$	**IG**+**HT**(600°C), $s = 74$, calc. 2.583 eV	text p. 289	[Zai01]
MA2477c	2.583	20.83	480.0	**S3**-DAP58c	$(V_2Ni_1)N_3^+$	**HT**(1,700°C), $s = 50$, calc. 2.582 eV	**7.**126	[Nad93, Zai01]
MA2534f	2.590	20.89	478.7	**TH5**-DAP15f	$*V_2°$	**IG**+**HT**(600°C), $s = 68$, calc. 2.590 eV	text p. 289	[Zai01]
MA2460d	2.596	20.94	477.6	DAP91d	$*V_1Ni_3°$	**HT**(1,900°C), $s = 26$, calc. 2.598 eV	...	[Nad93, Zai01]
MA2515c	2.597	20.95	477.4	**S2**-DAP57c	$(V_2Ni_1)N_3$	**HT**(1,700°C), named line **C**	**7.**126	[Nad93, Zai01]
MA2477e	2.599	20.97	477.0	**S3**-DAP58e	$(V_2Ni_1)N_3^+$	**HT**(1,700°C), $s = 42$, calc. 2.601 eV	**7.**126	[Nad93, Zai01]
MA2534g	2.602	20.99	476.5	**TH5**-DAP15g	$*V_2°$	**IG**+**HT**(600°C), $s = 58$, calc. 2.604 eV	text p. 289	[Zai01]
MA2534h	2.610	21.05	475.0	**TH5**-DAP15h	$*V_2°$	**IG**+**HT**(600°C), $s = 54$, calc. 2.611 eV	text p. 289	[Zai01]
MA2460e	2.610	21.05	475.0	DAP91e	$*V_1Ni_3°$	**HT**(1,900°C), $s = 24$, calc. 2.607 eV	...	[Nad93, Zai01]
MA2617a	2.617	21.11	473.7	**H3a**-DAP50a	$*V_1N_2°$	**IG**+**HT**(500–800°C); $s = 58$, calc. 2.617 eV	**7.**108	[Dav77c, Zai01]
MA2534-i	2.618	21.12	473.6	**TH5**-DAP15-i	$*V_2°$	**IG**+**HT**(600°C), $s = 50$, calc. 2.619 eV	**7.**121	[Cla56c, Zai01]
MA2477f	2.622	21.15	472.8	**S3**-DAP58f	$(V_2Ni_1)N_3^+$	**HT**(1,700°C), $s = 34$, calc. 2.624 eV	**7.**126	[Nad93, Zai01]
MA2070-l	2.623	21.16	472.7	DAP1-l	$N_1°+B_1°$	**HT**(>1,500°C), $s = 2$, calc.2.565 eV	**7.**97	[Col85b, Law93b, Kup99]

(continued)

Table 5.1.9.1 (continued)

Line-label	Energy (eV)	Frequ. ($10^3 cm^{-1}$)	Wavel. (nm)	Name	Impur. /defect	Treatment/comment	Figs. **3–6.** [Zai01]; Fig. **7.** [Zai98]	References
MA2515f	2.623	21.16	472.7	**S2**-DAP57f	$(V_2Ni_1)N_3$	**HT**(1,700°C), named line **D**	**7.**126	[Nad93, Zai01]
MA2460f	2.632	21.23	471.0	DAP91f	$^*V_1Ni_3{}^{\circ}$	**HT**(1,900°C), $s = 20$, calc. 2.630 eV	. . .	[Nad93, Zai01]
MA2515h	2.634	21.25	470.7	**S2**-DAP57h	$(V_2Ni_1)N_3$	**HT**(1,700°C), named line **E**	**7.**126	[Nad93, Zai01]
MA2638	2.638	21.28	470.0	**TR12** = DAP35z	$^*(C_2)_i^+$(a)	**IG(+HT**300–600°C**)**, var. 2.629–2.638 eV, line(b) at 2.671 ZPL of DAP35a-n(2.965 eV), lines (k–n) see ME	**5.**30, **7.**68 **7.**107, **7.**127b	[All98a, Dav77c] [Dav77c, Wal79, Dav81b, Zai01]
MA2759z						SBs: 67, 77, 188(TR17), 222 meV		
MA2460g	2.641	21.30	469.4	DAP91g	$^*V_1Ni_3{}^{\circ}$	**HT**(1,900°C), $s = 18$, calc. 2.643 eV	. . .	[Nad93, Zai01]
MA2652a	2.652	21.39	467.5	**H4b**-DAP37a	$^*N_3V_2N_1{}^{\circ}$	**IG+HT**(600°C),13 lines a–m(3.343 eV), $s = 86$, calc. 2.647 eV	**7.**101, **7.**108 **7.**116a	[Col82b, Dav77c] [DeS77]
MA2534j	2.658	21.44	466.4	**TH5**-DAP15j	$^*V_2{}^{\circ}$	**IG+HT**(600°C), $s = 34$, calc. 2.660 eV	**7.**121	[Cla56c, Zai01]
MA2617b	2.659	21.45	466.3	**H3a**-DAP50b	$^*V_1N_2{}^{\circ}$	**IG+HT**(500–800°C); $s = 34$, calc. 2.662 eV	**7.**108	[Dav77c, Zai01]

Treatment: IE irradiation, electrons; *IG* irradiation, general; *II* irradiation, ions; *IN* irradiation, neutrons; *IX* irradiation, X-ray; *HP* hydrogen plasma; *HT* heat treatment; *PD* plastic deformation; *PO* powder, *PO(KBr) compressed with KBr*

Table 5.1.9.2 Modified diamond (irradiation, heat etc.): absorption lines, visible: ultramarine, part 2 (2.66–2.75 eV)

Line-label	Energy (eV)	Frequ. ($10^3 cm^{-1}$)	Wavel. (nm)	Name	Impur. /defect	Treatment/comment	Figs. **3–6.** [Zai01]; Fig. **7.** [Zai98]	References
MA2652a	2.652	21.39	467.5	**H4b**-DAP37a	$^*N_3V_2N_1{}^\circ$	**IG+HT**(600°C),13 lines a–m(3.343 eV), $s = 86$, calc. 2.647 eV	**7.**101, **7.**108 **7.**116a	[Col82b, Dav77c] [DeS77]
MA2534j	2.658	21.44	466.4	**TH5**-DAP15j	$^*V_2{}^\circ$	**IG+HT**(600°C), $s = 34$, calc. 2.660 eV	**7.**121	[Cla56c, Zai01]
MA2617b	2.659	21.45	466.3	**H3a**-DAP50b	$^*V_1N_2{}^\circ$	**IG+HT**(500–800°C); $s = 34$, calc. 2.662 eV	**7.**108	[Dav77c, Zai01]
MA2477g	2.660	21.46	466.1	**S3**-DAP58g	$(V_2Ni_1)N_3^+$	**HT**(1,700°C), $s = 26$, calc. 2.660 eV	**7.**126	[Nad93, Zai01]
MA2070m	2.668	21.52	464.7	DAP1m	$N_1{}^\circ+B_1{}^\circ$	**HT**(>1,500°C), $s = 1$, calc.2.644 eV	**7.**97	[Col85b, Law93b, Kup99]
MA2638b	2.671	21.54	464.2	**TR13**	$^*(C_2)_i^+$(b)	**IG**(+**HT**300–600°C)**, TR13**, from excited state splitting, no luminescence	**7.**107, **7.**127b	[Dav77c, Wal79, Zai01]
MA2676a	2.676	21.58	463.3	DAP90a	$^*(C_2)_iN_2^+$	**IG,** s = 58, calc. 2.677 eV	**7.**82	[Col79c, Zai01]
MA2652b	2.692	21.71	460.5	**H4b**-DAP37b	$^*N_3V_2N_1{}^\circ$	**IG+HT**(600°C), $s = 54$, calc. 2.687 eV	**7.**101, **7.**116a	[Col82b, DeS77]

(continued)

Table 5.1.9.2 (continued)

Line-label	Energy (eV)	Frequ. ($10^3 cm^{-1}$)	Wavel. (nm)	Name	Impur. /defect	Treatment/comment	Figs. **3–6.** [Zai01]; Fig. **7.** [Zai98]	References
MA2676b	2.702	21.79	458.8	DAP90b	$^*(C_2)_iN_2^+$	**IG,** $s = 42$, calc. 2.702 eV	**7.**82	[Col79c, Zai01]
MA2638-s6	2.705	21.82	458.3	**TR12**-s6	$^*(C_2)_i^+$	**IG(+HT**300–600°C**),** SB: +67 meV	**7.**127b	[Wal79, Zai01]
MA2617c	2.709	21.85	457.6	**H3a**-DAP50c	$^*V_1N_2^\circ$	**IG+HT**(500–800°C); $s = 22$, calc. 2.712 eV	**7.**108	[Dav77c, Zai01]
MA2712a	2.712	21.88	457.1	...	$^*V_1C_4N_1^+$	**IG+HT**(800°C), related to DAP45(1.702 eV)	...	[Zai01]
MA2638-s7	2.715	21.90	456.6	**TR12**-s7	$^*(C_2)_i^+$	**IG(+HT**300–600°C**),** SB: +77 meV	**7.**127b	[Wal79, Zai01]
MA2712b	2.718	21.92	456.1	...	$^*V_1C_4N_1^+$	**IG+HT**(800°C), related to DAP45(1.702 eV)	...	[Zai01]
MA2652c	2.741	22.11	452.3	**H4b**-DAP37c	$^*N_3V_2N_1^\circ$	**IG+HT**(600°C), $s = 34$, calc. 2.735 eV	**7.**101, **7.**116a	[Col82b, DeS77]
MA2676c	2.747	22.16	451.3	DAP90c	$^*(C_2)_iN_2^+$	**IG,** $s = 26$, calc. 2.747 eV	**7.**82	[Col79c, Zai01]

Treatment: IE irradiation, electrons; *IG* irradiation, general; *II* irradiation, ions; *IN* irradiation, neutrons; *IX* irradiation, X-ray; *HP* hydrogen plasma; *HT* heat treatment; *PD* plastic deformation; *PO* powder, *PO(KBr) compressed with KBr*

Table 5.1.10.1 Modified diamond (irradiation, heat etc.): absorption lines, visible: violet, part 1 (2.75–2.90 eV)

Line-label	Energy (eV)	Frequ. ($10^3 cm^{-1}$)	Wavel. (nm)	Name	Impur. /defect	Treatment/comment	Figs. **3–6.** [Zai01]; Fig. **7.** [Zai98]	References
MA2759a	2.759	22.25	449.4	**TR12**-DAP35a	$*(C_2)_i^+$	**IG(+HT**300–600°C**)**, 14 lines a–n (2.965 eV), lines k–n see ME, $s = 86$, calc. 2.759 eV	**7.**127b	[Wal79, Zai01]
MA2759b	2.769	22.34	447.7	**TR12**-DAP35b	$*(C_2)_i^+$	**IG(+HT**300–600°C**)**, $s = 72$, calc. 2.771 eV	**7.**127b	[Wal79, Zai01]
MA2534k	2.775	22.38	446.8	**TH5**-DAP15k	$*V_2°$	$s = 16$, calc. 2.771 eV	**7.**121	[Cla56c, Zai01]
MA2759c	2.777	22.40	446.4	**TR12**-DAP35c	$*(C_2)_i^+$	**IG(+HT**300–600°C**)**, **TR14**, $s = 72$, calc. 2.771 eV	**7.**127b	[Wal79, Zai01]
MA2617d	2.777	22.40	446.4	**H3a**-DAP50d	$*V_1N_2°$	**IG+HT**(500–800°C); $s = 14$, calc. 2.775 eV	**7.**108	[Dav77c, Zai01]
MA2781 = MA2881z	2.781	22.43	445.8	**GR2a**-DAP32z	$*V_1^+$	**IG,** ZPL of DAP32a-l (3.065 eV), **GR2–8** weak line	**5.**30	[All98a]
MA2759d	2.792	22.52	444.0	**TR12**-DAP35d	$*(C_2)_i^+$	**IG(+HT**300–600°C**)**, **TR15**, $s = 54$, calc. 2.791 eV	**7.**127b	[Wal79, Zai01]
MA2759e	2.799	22.58	442.9	**TR12**-DAP35e	$*(C_2)_i^+$	**IG(+HT**300–600°C**)**, $s = 50$, calc. 2.797 eV	**7.**127b	[Wal79, Zai01]
MA2759f	2.812	22.68	440.9	**TR12**-DAP35f	$*(C_2)_i^+$	**IG(+HT**300–600°C**)**, **TR16**, $s = 42$, calc. 2.812 eV	**7.**127b	[Wal79, Zai01]

(continued)

Table 5.1.10.1 (continued)

Line-label	Energy (eV)	Frequ. ($10^3 cm^{-1}$)	Wavel. (nm)	Name	Impur. /defect	Treatment/comment	Figs. **3–6.** [Zai01]; Fig. **7.** [Zai98]	References
MA2759g	2.818	22.73	439.9	**TR12**-DAP35g	$*(C_2)_i^+$	**IG(+HT300–600°C),** $s = 40$, calc. 2.816 eV	**7.**127b	[Wal79, Zai01]
MA2638-s8	2.826	22.80	438.7	**TR17** = TR12-s8	$*(C_2)_i^+$	**IG(+HT300–600°C), TR17,** SB(+188 meV)	**7.**127b	[Wal79, Zai01]
MA2759h	2.834	22.86	437.5	**TR12**-DAP35h	$*(C_2)_i^+$	**IG(+HT300–600°C),** $s = 32$, calc. 2.836 eV	**7.**127b	[Wal79, Zai01]
MA2652d	2.845	22.95	435.8	**H4b**-DAP37d	$*N_3V_2N_1{}^\circ$	**IG+HT(600°C),** $s = 16$, calc. 2.844 eV	**7.**101, **7.**116a	[Col82b, DeS77]
MA2759-i	2.848	22.97	435.3	**TR12**-DAP35-i	$*(C_2)_i^+$	**IG(+HT300–600°C),** $s = 30$, calc. 2.843 eV	**7.**127b	[Wal79, Zai01]
MA2759j	2.860	23.07	433.5	**TR12**-DAP35j	$*(C_2)_i^+$	**IG(+HT300–600°C), TR18,** $s = 26$, calc. 2.859 eV	**7.**127b	[Wal79, Zai01]
MA2881a	2.881	23.24	430.4	**GR2**-DAP32a	$*V_1^+$	**IG,** 12 lines a–l (3.065 eV); **GR2b,** $s = 130$, calc. 2.881 eV	**5.**30, **7.**68	[All98a, Dav77c, Zai01]
MA2881b	2.887	23.28	429.5	**GR2**-DAP32b	$*V_1^+$	**IG,** $s = 118$; calc. 2.886 eV, **GR3**	**7.**68	[Dav77c, Zai01]

Treatment: IE irradiation, electrons; *IG* irradiation, general; *II* irradiation, ions; *IN* irradiation, neutrons; *IX* irradiation, X-ray; *HP* hydrogen plasma; *HT* heat treatment; *PD* plastic deformation; *PO* powder, *PO(KBr) compressed with KBr*

Table 5.1.10.2 Modified diamond (irradiation, heat etc.): absorption lines, visible: violet, part 2 (2.90–3.10 eV)

Line-label	Energy (eV)	Frequ. ($10^3 cm^{-1}$)	Wavel. (nm)	Name	Impur. /defect	Treatment/comment	Figs. **3–6.** [Zai01]; Fig. **7.** [Zai98]	References
MA2881c	2.902	23.41	427.2	**GR2**-DAP32c	$^*V_1^+$	**IG,** $s = 86$; calc. 2.903 eV, **GR4**	**7.**68	[Dav77c, Zai01]
MA2534-1	2.909	23.46	426.2	**TH5**-DAP15-1	$^*V_2{}^\circ$	**IG+HT**(600°C), $s = 8$, calc. 2.918 eV	**7.**121	[Cla56c, Zai01]
MA2916	2.916	23.52	425.2	...	$^*V_2N_3{}^\circ$	**IE+HT**(700°C)	**7.**137, **7.**145	[Dav80c, Dav77c]
MA2918	2.918	23.54	424.9	...	...	**IG+HT**(600°C)	**7.**138, **7.**145	[Dav77c, Zai01]
MA2652e	2.920	23.55	424.6	**H4b**-DAP37e	$^*N_3V_2N_1{}^\circ$	**IG+HT**(600°C), **H6,**, variable 2.916–2.925 eV $s = 11$, calc. 2.920 eV	**7.**108, **7.**116a, **7.**137, **7.**139	[Dav77c, DeS77] [Dav80c, Dav77c]
MA2652f	2.936	23.68	422.3	**H4b**-DAP37f	$^*N_3V_2N_1{}^\circ$	**IG+HT**(600°C), $s = 10$, calc. 2.936 eV	**7.**101, **7.**116a	[Col82b, DeS77]
MA2881d	2.938	23.70	422.0	**GR2**-DAP32d	$^*V_1^+$	**IG,** $s = 50$; calc. 2.941 eV, **GR5**	**7.**68	[Dav77c, Zai01]
MA2881e	2.958	23.86	419.1	**GR2**-DAP32e	$^*V_1^+$	**IG,** $s = 42$; calc. 2.957 eV, **GR6a**	**5.**30, **7.**68	[All98a, Dav77c]
MA2881f	2.960	23.88	418.8	**GR2**-DAP32f	$^*V_1^+$	**IG,** $s = 40$; calc. 2.960 eV, **GR6b**	**5.**30, **7.**68	[All98a, Dav77c]
MA2652g	2.975	24.00	416.7	**H4b**-DAP37g	$^*N_3V_2N_1{}^\circ$	**IG+HT**(600°C), **H7,** variable 2.960–2.975 eV $s = 9$, calc. 2.967 eV	**7.**101, **7.**108, **7.**116a, **7.**137	[Col82b, Dav77c, DeS77, Dav80c]

(continued)

Table 5.1.10.2 (continued)

Line-label	Energy (eV)	Frequ. ($10^3 cm^{-1}$)	Wavel. (nm)	Name	Impur. /defect	Treatment/comment	Figs. **3–6.** [Zai01]; Fig. **7.** [Zai98]	References
MA2881g	2.976	24.00	416.6	**GR2**-DAP32g	$*V_1^+$	**IG,** $s = 34$; calc. 2.975 eV, **GR7a**	**7.**68	[Dav77c, Zai01]
MA2617e	2.981	24.04	415.9	**H3a**-DAP50e	$*V_1N_2°$	**IG+HT**(500–800°C); $s = 8$, calc. 2.981 eV	**7.**108	[Dav77c, Zai01]
MA2881h	2.982	24.05	415.8	**GR2**-DAP32h	$*V_1^+$	**IG,** $s = 32$; calc. 2.981 eV, **GR7b**	**7.**68	[Dav77c, Zai01]
MA2985	2.985	24.08	415.3	**N3c**-DAP23z	$V_1N_3°$	**IG**, ZPL of DAP23a-m(3.542 eV), lines (a–m), see NA	**6.**4a, **7.**139	[Wel94, Dav77c, Zai01]
MA2881-i	2.997	24.17	413.7	**GR2**-DAP32-i	$*V_1^+$	**IG,** $s = 27$a,b; calc. 3.000 eV, **GR8a,b**	**7.**68	[Dav77c, Zai01]
MA2881j	3.003	24.22	412.8	**GR2**-DAP32j	$*V_1^+$	**IG,** $s = 26$a,b; calc. 3.004 eV, **GR8c,d**	**5.**30, **7.**68	[All98a, Dav77c]
MA2652h	3.004	24.23	412.7	**H4b**-DAP37h	$*N_3V_2N_1°$	**IG+HT**(600°C), **H8,** $s = 8$, calc. 2.988 eV	**7.**116a	[DeS77]
MA2881k	3.006	24.25	412.4	**GR2**-DAP32k	$*V_1^+$	**IG,** $s = 25$; calc. 3.009 eV, **GR8e**	**7.**68	[Dav77c, Zai01]
MA3040	3.040	24.52	407.8	**R9**	...	**IG**...	...	[Zai01]
M°A2652i	3.062	24.70	404.9	**H4b**-DAP37i	$*N_3V_2N_1°$	**IG+HT**(600°C), **H9,** $s = 6$, calc. 3.063 eV	**7.**101, **7.**116a	[Col82b, DeS77]
MA2881l	3.065	24.72	404.5	**GR2**-DAP32-l	$*V_1^+$	**IG,** $s = 16$; calc. 3.064 eV	**5.**30	[All98a]

Treatment: IE irradiation, electrons; *IG* irradiation, general; *II* irradiation, ions; *IN* irradiation, neutrons; *IX* irradiation, X-ray; *HP* hydrogen plasma; *HT* heat treatment; *PD* plastic deformation; *PO* powder, *PO(KBr) compressed with KBr*

Table 5.1.11.1 Modified diamond (irradiation, heat etc.): absorption lines, near ultraviolet part 1 (3.10–3.53 eV)

Line-label	Energy (eV)	Frequ. ($10^3 cm^{-1}$)	Wavel. (nm)	Name	Impur. /defect	Comment	Figs. **3–6.** [Zai01]; Fig. **7.** [Zai98]	References
MA2652j	3.140	25.33	394.8	**H4b**-DAP37j	$^*N_3V_2N_1{}^\circ$	**IG+HT**(600°C), $s = 4$, calc. 3.156 eV	**7.**101	[Col82b]
MA3150	3.150	25.41	393.6	**ND1 = R10 =** DAP93z	$^*V_1^-$	**IG,** ZPL of DAP93a–f(3.650 eV), LVM: (4×) +80 meV; photochromic	**7.**145	[Dav77c, Wal79] [Zai01]
MA3188	3.188	25.71	388.9	**3.188 eV**-ce.-DAP77z	$^*(N_1C_1)_I{}^\circ$	var. 3.188–3.204; in ML ZPL of DAP77a–n	**7.**138	[Dav77c]
MA2652k	3.270	26.38	379.1	**H4b**-DAP37k	$^*N_3V_2N_1{}^\circ$	**IG+HT**(600°C), **H10,** $s = 3$, calc. 3.250 eV	**7.**147	[Col83a]
MA3290a	3.290	26.54	376.8	**ND1-**DAP93a	V_1^-	**IG,** 6 lines a–f(3.650 eV), $s = 78$, calc. 3.290 eV		[Wal79]
MA2652-l	3.311	26.71	374.4	**H4b**-DAP37-l	$^*N_3V_2N_1{}^\circ$	**IG+HT**(600°C), **H11,** $s = 2$, calc. 3.301 eV	**7.**147	[Col83a]
MA3333a	3.333	26.88	372.0	***S4-**DAP59a	$^*(V_2Ni_1)N_{1+2}$	**HT**(2,400°C), 12 lines a–l(3.820 eV), $s = 86$, calc. 3.333 eV	**7.**148	[Col85b]
MA2652m	3.343	26.96	370.9	**H4b**-DAP37m	$^*N_3V_2N_1{}^\circ$	**IG+HT**(600°C), **H12,** $s = 1$, calc. 3.394 eV	**7.**147	[Col83a]
MA3361 = MA3461z	3.361	27.11	368.9	**H3b**-DAP52z	$^*V_1N_2(b)\,^\circ$	**IG**+HT(500°C), **H13,** QLVM:+42 = H**14**, LVM: 112 meV = **H15b,** ZPL of DAP52a-i(4.042 eV)	**7.**147	[Col83a, Zai01]

(continued)

Table 5.1.11.1 (continued)

Line-label	Energy (eV)	Frequ. ($10^3 cm^{-1}$)	Wavel. (nm)	Name	Impur. /defect	Comment	Figs. **3–6.** [Zai01]; Fig. **7.** [Zai98]	References
MA3333b	3.380	27.26	366.8	***S4**-DAP59b	$^*(V_2Ni_1)N_{1+2}$	**HT**(2,400°C), $s=50$, calc. 3.378 eV	**7.**148	[Col85b]
MA3333c	3.412	27.52	363.4	***S4**-DAP59c	$^*(V_2Ni_1)N_{1+2}$	**HT**(2,400°C), $s=34$, calc. 3.417 eV	**7.**148	[Col85b]
MA3333d	3.443	27.77	360.1	***S4**-DAP59d	$^*(V_2Ni_1)N_{1+2}$	**HT**(2,400°C), $s=26$, calc. 3.451 eV	**7.**148	[Col85b]
MA3461a	3.461	27.92	358.2	**H3b**-DAP52a	$^*V_1N_2(b)\ ^\circ$	**IG**+HT(500°C), **H15a,** 9 lines a–i(4.042 eV), $s=136$, calc. 3.460 eV	**7.**147	[Col83a, Zai01]
MA3403	3.403	27.25	364.3	**H14**	$^*V_1N_2^\circ$	IG+HT(500°C), related to H3	**7.**147	[Col83a]
MA3333e	3.474	28.02	356.9	***S4**-DAP59e	$^*(V_2Ni_1)N_{1+2}$	**HT**(2,400°C), $s=22$, calc. 3.474 eV	**7.**148	[Col85b]
MA3461b	3.490	28.15	355.2	**H3b**-DAP52b	$^*V_1N_2(b)\ ^\circ$	**IG**+HT(500°C), **H16a,** $s=78$, calc. 3.497 eV	**7.**147	[Col83a]
MA3461c	3.504	28.26	353.8	**H3b**-DAP52c	$^*V_1N_2(b)\ ^\circ$	**IG**+HT(500°C), **H16b,** $s=64$, calc. 3.506 eV	**7.**147	[Col83a, Zai01]
MA3333f	3.509	28.30	353.3	***S4**-DAP59f	$^*(V_2Ni_1)N_{1+2}$	**HT**(2,400°C), $s=18$, calc. 3.503 eV	**7.**148	[Col85b]
MA3461d	3.523	28.42	351.9	**H3b**-DAP52d	$^*V_1N_2(b)\ ^\circ$	**IG**+HT(500°C), **H17a,** $s=50$, calc. 3.524 eV	**7.**147	[Col83a]

Treatment: IE irradiation, electrons; *IG* irradiation, general; *II* irradiation, ions; *IN* irradiation, neutrons; *IX* irradiation, X-ray; *HP* hydrogen plasma; *HT* heat treatment; *PD* plastic deformation; *PO* powder, *PO(KBr) compressed with KBr*

Table 5.1.11.2 Modified diamond (irradiation, heat etc.): absorption lines, near ultraviolet part 2 (3.54–3.94 eV)

Line-label	Energy (eV)	Frequ. (10^3cm^{-1})	Wavel. (nm)	Name	Impur. /defect	Comment	Figs. **3–6.** [Zai01]; Fig. **7.** [Zai98]	References
MA3290b	3.540	28.55	350.2	**ND1**-DAP93b	$^*V_1^-$	**IG,** $s = 10$, calc. 3.542 eV	. . .	[Wal79]
MA3333g	3.541	28.57	350.0	***S4**-DAP59g	$^*(V_2Ni_1)N_{1+2}$	**HT**(2,400°C), $s = 14$, calc. 3.546 eV	**7.**148	[Col85b]
MA3461e	3.543	28.58	349.9	**H3b**-DAP52e	$^*V_1N_2(b)$ °	**IG**+HT(500°C), **H17b,** $s = 42$, calc. 3.541 eV	**7.**147	[Col83a, Zai01]
MA3361f	3.560	28.72	34.83	**H3b**-DAP52f	$^*V_1N_2(b)$ °	**IG**+HT(500°C), **H18a,** $s = 34$, calc. 3.559 eV	**7.**147	[Col83a, Zai01]
MA3461g	3.570	28.80	347.3	**H3b**-DAP52g	$^*V_1N_2(b)$ °	**IG**+HT(500°C), **H18a,** $s = 30$, calc. 3.573 eV	**7.**147	[Col83a]
MA3333h	3.573	28.82	347.0	***S4**-DAP59h	$^*(V_2Ni_1)N_{1+2}$	**HT**(2,400°C), $s = 12$, calc. 3.573 eV	**7.**148	[Col85b]
MA3290c	3.590	28.96	345.3	**ND1**-DAP93c	$^*V_1^-$	**IG,** $s = 8$, calc. 3.588 eV	. . .	[Wal79]

(continued)

Table 5.1.11.2 (continued)

Line-label	Energy (eV)	Frequ. ($10^3 cm^{-1}$)	Wavel. (nm)	Name	Impur. /defect	Comment	Figs. **3–6.** [Zai01]; Fig. **7.** [Zai98]	References
MA3333i	3.612	29.13	343.2	***S4**-DAP59i	*$(V_2Ni_1)N_{1+2}$	**HT**(2,400°C), $s = 10$, calc. 3.610 eV	**7.**148	[Col85b]
MA3333j	3.650	29.44	339.7	***S4**-DAP59j	*$(V_2Ni_1)N_{1+2}$	**HT**(2,400°C), $s = 8$, calc. 3.659 eV	**7.**148	[Col85b]
MA3290d	3.650	29.44	339.7	**ND1**-DAP93d	*V_1^-	**IG,** $s = 10$, calc. 3.658 eV	...	[Wal79]
MA3290e	3.730	30.09	332.4	**ND1**-DAP93e	*V_1^-	**IG,** $s = 4$, calc. 3.738 eV	...	[Wal79]
MA3333k	3.743	30.19	331.2	***S4**-DAP59k	*$(V_2Ni_1)N_{1+2}$	**HT**(2,400°C), $s = 6$, calc. 3.731 eV	**7.**148	[Col85b]
MA3290f	3.650	29.44	339.7	**ND1**-DAP93f	*V_1^-	**IG,** $s = 2$, calc. 3.868 eV	...	[Wal79]
MA3333-l	3.820	30.81	324.5	***S4**-DAP59-l	*$(V_2Ni_1)N_{1+2}$	**HT**(2,400°C), $s = 4$, calc. 3.820 eV	**7.**148	[Col85b]
MA3461h	3.901	31.47	317.8	**H3b**-DAP52h	*V_1N_2(b) °	**IG**+HT(500°C), **N5b,** $s = 4$, calc. 3.911 eV	...	[Zai01]

Treatment: IE irradiation, electrons; *IG* irradiation, general; *II* irradiation, ions; *IN* irradiation, neutrons; *IX* irradiation, X-ray; *HP* hydrogen plasma; *HT* heat treatment; *PD* plastic deformation; *PO* powder, *PO(KBr) compressed with KBr*

Table 5.1.12 Modified diamond (irradiation, heat etc.): absorption lines, mid ultraviolet (3.94–4.43 eV)

Line-label	Energy (eV)	Frequ. (10^3cm^{-1})	Wavel. (nm)	Name	Impur. /defect	Comment	Fig. **3–6.** [Zai01]; Fig. **7.** [Zai98]	References
MA3988 = MA4060z	3.988	32.17	310.9	**R11**-DAP69z	$^*(N_1C_1)^-$	**IG,** anneals out at T > 400°C, ZPL of DAP69a-l, (4.488 eV), QLVM:+38, 79, LVM: 120 meV	**5.**148	[All98], [Zai01]
MA3461i	4.042	32.60	306.7	**H3b**-DAP52i	$^*V_1N_2(b)$ °	**IG**+HT(500°C), **N7,** $s = 2$, calc. 4.034 eV	...	[Zai01]
MA4060a	4.060	32.75	305.4	**R11**-DAP69a	$^*(N_1C_1)^-$	**IG,** $s = 228$, calc. 4.060 eV	**5.**148	[All98a, Zai01]
MA4060b	4.108	33.14	301.8	**R11**-DAP69b	$^*(N_1C_1)^-$	**IG,** $s = 86$, calc. 4.105 eV	**5.**148	[All98a, Zai01]
MA4060c	4.145	33.43	299.1	**R11**-DAP69c	$^*(N_1C_1)^-$	**IG,** $s = 50$, calc. 4.142 eV	**5.**148	[All98a, Zai01]
MA4060d	4.183	33.74	296.4	**R11**-DAP69d	$^*(N_1C_1)^-$	**IG,** $s = 32$, calc. 4.181 eV	**5.**148	[All98a, Zai01]
MA4060e	4.218	34.02	293.9	**R11**-DAP69e	$^*(N_1C_1)^-$	**IG,** $s = 22$, calc. 4.221 eV	**5.**148	[All98a, Zai01]
MA4060f	4.296	34.65	288.6	**R11**-DAP69f	$^*(N_1C_1)^-$	**IG,** $s = 12$, calc. 4.303 eV	**5.**148	[All98a, Zai01]
MA4060g	4.326	34.89	286.6	**R11**-DAP69g	$^*(N_1C_1)^-$	**IG,** $s = 10$, calc. 4.333 eV	**5.**148	[All98a, Zai01]
MA1942b	4.328	34.91	286.5	...	$^*V_1N_1^-(b)$	**IG+HT**(800°C)	...	[Zai01]
MA4060h	4.374	35.28	283.4	**R11**-DAP69h	$^*(N_1C_1)^-$	**IG,** $s = 8$, calc. 4.374 eV	**5.**148	[All98a, Zai01]
MA4380	4.380	35.33	283.1	...		**IN**	...	[Bie67, Zai01]
MA4060i	4.408	35.55	281.3	**R11**-DAP69i	$^*(N_1C_1)^-$	**IG,** $s = 7$, calc. 4.408 eV	**5.**148	[All98a, Zai01]

Treatment: IE irradiation, electrons; *IG* irradiation, general; *II* irradiation, ions; *IN* irradiation, neutrons; *IX* irradiation, X-ray; *HP* hydrogen plasma; *HT* heat treatment; *PD* plastic deformation; *PO* powder, *PO(KBr) compressed with KBr*

Table 5.1.13 Modified diamond (irradiation, heat etc.): absorption lines, far ultraviolet (4.43–5.90 eV)

Line-label	Energy (eV)	Frequ. ($10^3 cm^{-1}$)	Wavel. (nm)	Name	Impur. /defect	Comment	Figs. **3–6.** [Zai01]; Fig. **7.** [Zai98]	References
MA4060j	4.438	35.80	279.4	**R11**-DAP69j	$^*(N_1C_1)^-$	**IG**, $s = 6$, calc. 4.433 eV	**5.**148	[All98a, Zai01]
MA4060k	4.464	36.01	277.7	**R11**-DAP69k	$^*(N_1C_1)^-$	**IG**, $s = 5$, calc. 4.473 eV	**5.**148	[All98a, Zai01]
MA4060-l	4.488	36.20	276.2	**R11**-DAP69-l	$^*(N_1C_1)^-$	**IG**, $s = 4$, calc. 4.506 eV	**5.**148	[All98a, Zai01]
MA4582 = MA4653z	4.582	36.96	270.6	**5RL**-DAP79z	$^*(C_2)_iN_1{}^\circ$	**IG+HT**(500°C), ZPL of DAP79a-j(5.224 eV) QLVM:+42, LVM:+167, (2×)202, 246 meV	...	[Col86b, Zai01]
MA4653a	4.653	37.53	266.4	**5RL**-DAP79a	$^*(C_2)_iN_1{}^\circ$	**IG+HT**(500°C), $s = 228$, calc. 4.654 eV	...	[Col86b, Zai01]
MA4653b	4.666	37.64	265.7	**5RL**-DAP79b	$^*(C_2)_iN_1{}^\circ$	**IG+HT**(500°C), $s = 162$, calc. 4.668 eV	...	[Col86b, Zai01]
MA4698a	4.698	37.89	263.9	**2BD**(G)	$^*(C_2)_iB_2{}^\circ$	**IG** of type IIb, see ML	...	[Zai01]
MA4653c	4.702	37.93	363.7	**5RL**-DAP79c	$^*(C_2)_iN_1{}^\circ$	**IG+HT**(500°C), $s = 78$, calc. 4.705 eV	...	[Col86b, Zai01]
MA4653d	4.722	38.09	262.6	**5RL**-DAP79c	$^*(C_2)_iN_1{}^\circ$	**IG+HT**(500°C), $s = 58$, calc. 4.726 eV	...	[Col86b, Zai01]
MA4698b	4.777	38.53	259.5	**2BD**(F)	$^*(C_2)_iB_1v$	**IG** of type IIb, see ML(DAP97)	...	[Zai01]

MA4803	4.803	38.74	248.1	**2BD**(C)	*$(C_2)_iB_3$°	**IG** of type IIb, see ML	...	[Zai01]
MA4653e	4.872	39.30	254.5	**5RL**-DAP79e	*$(C_2)_iN_1$°	**IG**+**HT**(500°C), $s = 14$, calc. 4.874 eV	...	[Col86b, Zai01]
MA4653f	4.922	39.70	251.9	**5RL**-DAP79f	*$(C_2)_iN_1$°	**IG**+**HT**(500°C), $s = 10$, calc. 4.927 eV	...	[Col86b, Zai01]
MA4653g	5.022	40.51	246.9	**5RL**-DAP79g	*$(C_2)_iN_1$°	**IG**+**HT**(500°C), $s = 6$, calc. 5.027 eV	...	[Col86b, Zai01]
MA4653h	5.118	41.28	242.1	**5RL**-DAP79h	*$(C_2)_iN_1$°	**IG**+**HT**(500°C), $s = 4$, calc. 5.100 eV	...	[Col86b, Zai01]
MA4653i	5.180	41.78	239.3	**5RL**-DAP79i	*$(C_2)_iN_1$°	**IG**+**HT**(500°C), $s = 3$, calc. 5.174 eV	...	[Col86b, Zai01]
MA4653j	5.224	42.14	237.3	**5RL**-DAP79j	*$(C_2)_iN_1$°	**IG**+**HT**(500°C), $s = 2$, calc. 5.215 eV	...	[Col86b, Zai01]

Treatment: IE irradiation, electrons; *IG* irradiation, general; *II* irradiation, ions; *IN* irradiation, neutrons; *IX* irradiation, X-ray; *HP* hydrogen plasma; *HT* heat treatment; *PD* plastic deformation; *PO* powder, *PO(KBr) compressed with KBr*

5.2 Excitation of Photo Luminescence Lines (ME)

Table 5.2.1 Modified diamond (irradiation, heat etc.): excitation of photo luminescence lines (PLE), near infrared (1.24–1.77 eV)

Line-label	Energy (eV)	Frequ. ($10^3 cm^{-1}$)	Wavel. (nm)	Name	Impur. /defect	Comment	Figs. **3–6.** [Zai01]; Fig. **7.** [Zai98]	References
ME1563a	1.563	12.61	793.2	***S5**-DAP60a	*$(V_2Ni_1)N_x$	**HT**(>1,600°C) or **IE**+**HT**(900°C),lines a, d, f, h–k; other lines see MA, $s = 162$, calc. 1.564 eV; a PLE of S8-DAP16(1212–1489 eV)	**5.**19b	[Kup99, Zai01]
ME1563d	1.648	13.29	752.3	***S5**-DAP60d	*$(V_2Ni_1)N_x$	**HT**(>1,600°C) or **IE**+**HT**(900°C), $s = 50$, calc. 1.647 eV; **PLE** of *S5-DAP60a (1.563 eV)	**5.**19b	[Kup99]

Treatment: IE irradiation, electrons; *IG* irradiation, general; *II* irradiation, ions; *IN* irradiation, neutrons; *IX* irradiation, X-ray; *HP* hydrogen plasma; *HT* heat treatment; *PD* plastic deformation; *PO* powder, *PO(KBr) compressed with KBr*

Table 5.2.2 Modified diamond (irradiation, heat etc.): excitation of photo luminescence lines (PLE), visible: purple to green (1.77–2.43 eV)

Line-label	Energy (eV)	Frequ. ($10^3 cm^{-1}$)	Wavel. (nm)	Name	Impur. /defect	Comment	Figs. **3–6.** [Zai01]; Fig. **7.** [Zai98]	References
ME1563f	1.930	15.57	642.4	***S5**-DAP60f	$*(V_2Ni_1)N_x$	**HT**(>1,600°C) or **IE**+**HT**(900°C), $s=8$, calc. 1.928 eV; **PLE** of *S5-DAP60	**5.**19b	[Kup99]
ME1943	1.943	15.67	638.1	**NV**$^-$ or **638 nm** = DAP48z	$*V_1N_1^-$	**HT**(2,500°C) or **IG**+**HT**(>550°C), (DAP48a-j, see MA), 2 SBs +42 meV, PLE of N_1°	**5.**19b, **5.**59	[Kup99], [Vin88b] [Zai01]
ME1940b	1.991	16.06	622.7	***Ni1.69c**	$*V_1Ni_2^+$	**HT**(1,500°C), 6 SBs+50 meV; PLE of DAP111	**5.**56	[Kup99, Zai01]
ME1563h	2.005	16.17	618.3	***S5**-DAP60h	$*(V_2Ni_1)N_x$	**HT**(>1,600°C) or **IE**+**HT**(900°C), $s=6$, calc. 2.000 eV; **PLE** of *S5-DAP60	**5.**19b	[Kup99]
ME1563i	2.095	15.57	642.4	***S5**-DAP60f	$*(V_2Ni_1)N_x$	**HT**(>1,600°C) or **IE**+**HT**(900°C), $s=4$, calc. 2.089 eV; **PLE** of *S5-DAP60	**5.**19b	[Kup99]
ME2154	2.154	17.38	575.5	**NV**° or **575nm**	$*V_1N_1{}^\circ$	**IG**+**HT**(>550°C),-(DAP46a-f, see MA)	**5.**59	[Vin88], [Zai01]
ME1563j	2.220	17.91	558.5	***S5**-DAP60j	$*(V_2Ni_1)N_x$	**HT**(>1,600°C) or **IE**+**HT**(900°C), $s=2$, calc. 2.228 eV; PLE of *S5-DAP60a	**5.**19b	[Kup99]

(continued)

Table 5.2.2 (continued)

Line-label	Energy (eV)	Frequ. ($10^3 cm^{-1}$)	Wavel. (nm)	Name	Impur. /defect	Comment	Figs. **3–6.** [Zai01]; Fig. **7.** [Zai98]	References
ME2071b	2.298	18.54	539.5	*$\mathbf{Ni2.07b}$ = DAP27z	*$V_1Ni_2^-$ (b)	**HT**(1,500°C),PLE of Ni2.07a (DAP87, ML) and Ni2.07b(DAP28, ML); 3 SBs: QLVM:+30 (calc. 2 Ni: 29.0), LVM: +51, +80 meV	**5.**69b	[Kup99]
ME1563k	2.320	18.71	535.4	***S5**-DAP60k	*$(V_2Ni_1)N_x$	**HT**(>1,600°C) or **IE**+**HT**(900°C), $s = 1$, calc. 2.317 eV; PLE of *S5-DAP60a	**5.**19b	[Kup99]
ME2390a	2.390	19.28	518.7	DAP27a	*$V_1Ni_2^-$ (b)	**HT**(1,500°C), 10 lines a–j(2.555 eV), $s = 162$; calc. 2.389 eV, PLE see ME at 2.298 eV	**5.**19b	[Kup99]
ME1704b	2.401	19.37	516.4	***Ni1.70b** = DAP29z	*V_1Ni_1° (b)	**HT**(1,700°C), PLE of *Ni1.70a (DAP84), 2 SBs QLVM:+34(calc.Ni_1+C_3:32.8), LVM:+74 meV	**5.**49b	[Kup99]
ME2390b	2.410	19.44	514.4	DAP27b	*$V_1Ni_2^-$ (b)	**HT**(1,500°C), $s = 106$; calc. 2.411 eV, PLE see ME at 2.298 eV	**5.**19b	[Kup99]
ME2427	2.427	19.58	510.8	***Ni1.66b**	*V_1Ni_2° (b)	**HT**(1,500°C), 7 SBs: QLVM: +28 (calc. 2 Ni 29.0), LVM (6×) +49 meV, PLE of *Ni1.66a	**5.**27b	[Kup99, Zai01]

Treatment: IE irradiation, electrons; *IG* irradiation, general; *II* irradiation, ions; *IN* irradiation, neutrons; *IX* irradiation, X-ray; *HP* hydrogen plasma; *HT* heat treatment; *PD* plastic deformation; *PO* powder, *PO(KBr) compressed with KBr*

Table 5.2.3 Modified diamond (irradiation, heat etc.): excitation of photo luminescence lines (PLE), visible: blue (2.43–2.58 eV)

Line-label	Energy (eV)	Frequ. (10^3cm^{-1})	Wavel. (nm)	Name	Impur. /defect	Comment	Figs. **3–6.** [Zai01]; Fig. **7.** [Zai98]	References
ME2390c	2.453	19.79	505.4	DAP27c	$^*V_1Ni_2^-$(b)	**HT**(1,500°C), $s = 58$; calc. 2.451 eV	**5.**19b	[Kup99]
ME2462	2.462	19.86	503.6	**3H**	$^*(C_2)_i^+$	**IG**, PLE of 3H (DAP86, ML), photochromic	...	[Ste99a, Zai01]
ME2390d	2.471	19.93	501.7	DAP27d	$^*V_1Ni_2^-$(b)	**HT**(1,500°C), $s = 44$; calc. 2.474 eV	**5.**19b	[Kup99]
ME2496a	2.496	20.13	496.7	**S3**-DAP58a	$(V_2Ni_1)N_2^+$	**HT**(1,700°C), $s = 162$; calc. 2.494 eV; see ML	...	[Zai01]
ME2498	2.498	20.15	496.3	**H4b** = DAP37z	$^*N_3V_2N_1{}^\circ$	**HT**(>1,900°C) or **IG+HT**(>600°C), PLE of H4a (2.417 eV, ML), DAP37a-m, see MA	...	[Zai01]
ME2390e	2.500	20.71	495.8	DAP27e	$^*V_1Ni_2^-$(b)	**HT**(1,500°C), $s = 34$; calc. 2.496 eV	**5.**19b	[Kup99]
ME2390f	2.515	20.29	493.0	DAP27f	$^*V_1Ni_2^-$(b)	**HT**(1,500°C), $s = 30$; calc. 2.510 eV	**5.**19b	[Kup99]
ME2515a	2.515	20.29	493.0	**S2**-DAP57a	$(V_2Ni_1)N_3^+$	**HT**(1,700°C), $s = 228$; calc. 2.515 eV; 7 ME lines: DAP57e, g, h, j, m, n, o (3.341 eV)	(see HA,ML)	[Zai01]
ME2390g	2.530	20.41	490.0	DAP27g	$^*V_1Ni_2^-$(b)	**HT**(1,500°C), $s = 26$; calc. 2.526 eV	**5.**19b	[Kup99]
ME2390h	2.544	20.52	487.3	DAP27h	$^*V_1Ni_2^-$(b)	**HT**(1,500°C), $s = 22$; calc. 2.546 eV	**5.**19b	[Kup99]
ME2390-i	2.555	20.61	485.2	DAP27-i	$^*V_1Ni_2^-$(b)	**HT**(1,500°C), $s = 20$; calc. 2.558 eV	**5.**19b	[Kup99]
ME2390j	2.568	20.71	482.8	DAP27j	$^*V_1Ni_2^-$(b)	**HT**(1,500°C), $s = 18$, calc. 2.571 eV	**5.**19b	[Kup99]

Treatment: IE irradiation, electrons; *IG* irradiation, general; *II* irradiation, ions; *IN* irradiation, neutrons; *IX* irradiation, X-ray; *HP* hydrogen plasma; *HT* heat treatment; *PD* plastic deformation; *PO* powder, *PO(KBr) compressed with KBr*

Table 5.2.4 Modified diamond (irradiation, heat etc.): excitation of photo luminescence lines (PLE), visible: ultramarine (2.58–2.75 eV)

Line-label	Energy (eV)	Frequ. ($10^3 cm^{-1}$)	Wavel. (nm)	Name	Impur. /defect	Comment	Figs. **3–6.** [Zai01]; Fig. **7.** [Zai98]	References
ME2477c	2.583	20.83	480.0	**S3**-DAP58c	$(V_2Ni_1)N_2^+$	**HT**(1,700°C), 6 lines c, d, g–j; $s = 50$; calc. 2.582 eV;	**7.**104	[Yel92b, Zai01]
ME2477d	2.592	20.91	478.3	**S3**-DAP58d	$(V_2Ni_1)N_2^+$	**HT**(1,700°C), $s = 44$; calc. 2.597 eV;	**7.**104	[Yel92b, Zai01]
ME2515e	2.597	20.95	477.4	**S2**-DAP57e	$(V_2Ni_1)N_3$	**HT**(1,700°C), named **C;** $s = 64$, calc. 2.598 eV	7.126	[Nad93]
ME2515f	2.623	21.16	472.7	**S2**-DAP57f	$(V_2Ni_1)N_3$	**HT**(1,700°C), named **D;** $s = 50$, calc. 2.621 eV	...	[Zai01]
ME2515g	2.629	21.21	471.6	**S2**-DAP57g	$(V_2Ni_1)N_3$	**HT**(1,700°C); $s = 48$, calc. 2.626 eV	**7.**126	[Nad93]
ME2630a	2.630	21.21	471.4	***Ni1.70b** = DAP29a	$*V_1Ni_1°$ (b)	**HT**(1,700°C), 6 lines a–f(2.965 eV), $s = 26$, calc. 2.629 eV	**5.**49b	[Kup99]
ME2515h	2.634	21.25	470.7	**S2**-DAP57h	$(V_2Ni_1)N_3$	**HT**(1,700°C); named **E;** $s = 44$, calc. 2.636 eV	**7.**126	[Nad93, Zai01]
ME2638a	2.638	21.28	470.0	**TR12**-DAP35z	$*(C_2)_i^+$ (a)	**IG,** var. 2.629–2.638 eV; ZPL of DAP35c, e-j, l-m; (Fig. 1b in [Dav81b]), other lines see MA; PLE of DAP36a-n, SBs: +69, +76, +188 meV	(see comment)	[Dav81b, Zai01]

ME2515j	2.658	21.44	466.4	**S2**-DAP57j	$(V_2Ni_1)N_3$	**HT**(1,700°C)**;** $s = 34$, calc. 2.663 eV	**7**.104	[Yel92b]
ME2496f	2.660	21.46	466.1	**S3**-DAP58f	$(V_2Ni_1)N_2^+$	**HT**(1,700°C), $s = 26$; calc. 2.660 eV;	**7**.104	[Yel92b]
ME2638b	2.670	21.54	464.3	**TR13**	$^*(C_2)_i^+$ (b)	**IG,** "hot band" (absent in luminescence)	(see MA)	[Dav81b, Zai01]
ME2515-l	2.701	21.79	459.0	**S2**-DAP57-l	$(V_2Ni_1)N_3$	**HT**(1,700°C)**;** $s = 26$, calc. 2.699 eV	**7**.104	[Yel92b]
ME2630b	2.713	21.88	457.0	DAP29b	$^*V_1Ni_1{}^\circ$ (b)	**HT**(1,700°C), $s = 14$, calc. 2.711 eV	**5**.49b	[Kup99]
ME2515m	2.730	22.02	454.1	**S2**-DAP57m	$(V_2Ni_1)N_3$	**HT**(1,700°C)**;** $s = 22$, calc. 2.724 eV	**7**.104	[Yel92b]
ME2750a	2.750	22.18	450.8	***S6**-DAP61a	$^*(V_2Ni_1)N_x$	**HT**(2,200°C);PLE of *S5-DAP60; $s = 162$, calc. 2.748 eV; 3 SBs with 42 meV(LVM)	**5**.19b, **5**.26	[Kup99, Zai01]

Treatment: IE irradiation, electrons; *IG* irradiation, general; *II* irradiation, ions; *IN* irradiation, neutrons; *IX* irradiation, X-ray; *HP* hydrogen plasma; *HT* heat treatment; *PD* plastic deformation; *PO* powder, *PO(KBr) compressed with KBr*

Table 5.2.5.1 Modified diamond (irradiation, heat etc.): excitation of photo luminescence lines (PLE), visible: violet, part 1 (2.75–2.91 eV)

Line-label	Energy (eV)	Frequ. ($10^3 cm^{-1}$)	Wavel. (nm)	Name	Impur. /defect	Comment	Figs. **3–6.** [Zai01]; Fig. **7.** [Zai98]	References
ME2630c	2.765	22.30	448.4	DAP29c	$^*V_1Ni_1{}^\circ$ (b)	**HT**(1,700°C), $s = 10$, calc. 2.768 eV	**5.**49b	[Kup99]
ME2638c	2.777	22.40	446.4	DAP35c	$^*(C_2)_i^+$ (a)	**IG,** $s = 64$, calc. 2.778 eV; named **TR14**	(see ZPL)	[Dav81b, Zai01]
ME2496g	2.780	22.42	446.0	**S3**-DAP58g	$(V_2Ni_1)N_2^+$	**HT**(1,700°C), $s = 12$; calc. 2.791 eV	**7.**104	[Yel92b]
ME2781	2.781	22.43	445.8	DAP32z	$^*V_1^+$	**IG,** ZPL of **GR2**-DAP32a-l, PLE of GR1	**5.**30 (MA)	[All98b]
ME2630d	2.792	22.52	444.0	DAP29d	$^*V_1Ni_1{}^\circ$ (b)	**HT**(1,700°C), $s = 9$, calc. 2.794 eV	**5.**49b	[Kup99]
ME2638e	2.799	22.58	442.9	DAP35e	$^*(C_2)_i^+$ (a)	**IG,** $s = 50$, calc. 2.797 eV	(see ZPL)	[Dav81b, Zai01]
ME2638f	2.812	22.68	440.9	DAP35f	$^*(C_2)_i^+$ (a)	**IG,** $s = 42$, calc. 2.812 eV; named **TR16**	(see ZPL)	[Dav81b, Zai01]
ME2638g	2.818	22.73	439.9	DAP35g	$^*(C_2)_i^+$ (a)	**IG,** $s = 40$, calc. 2.816 eV;	(see ZPL)	[Dav81b, Zai01]
ME2496h	2.820	22.75	439.6	**S3**-DAP58h	$(V_2Ni_1)N_2^+$	**HT**(1,700°C), $s = 10$; calc. 2.830 eV;	**7.**104	[Yel92b]
ME2638h	2.834	22.86	437.5	DAP35h	$^*(C_2)_i^+$ (a)	**IG,** $s = 40$, calc. 2.816 eV;	(see ZPL)	[Dav81b, Zai01]

ME2630e	2.841	22.92	436.4	DAP29e	$*V_1Ni_1{}^\circ$ (b)	**HT**(1,700°C), $s = 7$, calc. 2.847 eV	**5.**49b	[Kup99]
ME2638-i	2.848	22.97	435.3	DAP35-i	$*(C_2)_i^+$ (a)	**IG,** $s = 30$, calc. 2.843 eV;	(see ZPL)	[Dav81b, Zai01]
ME2638j	2.860	23.07	433.5	DAP35j	$*(C_2)_i^+$ (a)	**IG,** $s = 26$, calc. 2.859 eV;	(see ZPL)	[Dav81b, Zai01]
ME2881a	2.881	23.24	430.4	**GR2**-DAP32a	$*V_1^+$	**IG,** 11 lines a–k (3.006 eV); $s = 130$; calc. 2.881 eV; PLE of GR1-DAP34; also in photocond. [Wal79]	**5.**40	[Col88a]
ME2881b	2.887	23.28	429.5	DAP32b	$*V_1^+$	**IG,** $s = 118$; calc. 2.886 eV, named **GR3**	**5.**40	[Col88a]
ME2496i	2.890	23.31	429.0	**S3**-DAP58i	$(V_2Ni_1)N_2^+$	**HT**(1,700°C), $s = 8$; calc. 2.882 eV	**7.**104	[Yel92b]
ME2881c	2.902	23.41	427.2	DAP32c	$*V_1^+$	**IG,** $s = 86$; calc. 2.903 eV, named **GR4**	**5.**40	[Col88a]
ME2638-l	2.910	23.47	426.0	DAP35-l	$*(C_2)_i^+$ (a)	**IG,** $s = 16$, calc. 2.919 eV;	(see ZPL)	[Dav81b, Zai01]

Treatment: IE irradiation, electrons; *IG* irradiation, general; *II* irradiation, ions; *IN* irradiation, neutrons; *IX* irradiation, X-ray; *HP* hydrogen plasma; *HT* heat treatment; *PD* plastic deformation; *PO* powder, *PO(KBr) compressed with KBr*

Table 5.2.5.2 Modified diamond (irradiation, heat etc.): excitation of photo luminescence lines (PLE), visible: violet, part 2 (2.93–3.10 eV)

Line-label	Energy (eV)	Frequ. ($10^3 cm^{-1}$)	Wavel. (nm)	Name	Impur. /defect	Comment	Figs. **3–6.** [Zai01]; Fig. **7.** [Zai98]	References
ME2881d	2.938	23.70	422.0	DAP32d	$^*V_1^+$	**IG,** $s = 50$; calc. 2.941 eV, named **GR5**	**5**.40	[Col88a]
ME2750b	2.942	23.73	421.4	***S6**-DAP61b	$^*(V_2Ni_1)N_x$	**HT**(2,200°C);PLE of *S5; $s = 22$, calc. 2.939 eV	**5**.26	[Kup99]
ME2638m	2.945	23.75	421.0	DAP35m	$^*(C_2)_i^+$ (a)	**IG,** $s = 14$, calc. 2.938 eV;	(see ZPL)	[Dav81b, Zai01]
ME2881e	2.958	23.86	419.1	DAP32e	$^*V_1^+$	**IG,** $s = 42$; calc. 2.957 eV, named **GR6a**	**5**.40	[Col88a]
ME2881f	2.960	23.88	418.8	DAP32f	$^*V_1^+$	**IG,** $s = 40$; calc. 2.960 eV, named **GR6b**	**5**.40	[Col88a]
ME2630f	2.965	23.92	418.1	DAP29f	$^*V_1Ni_1^\circ$ (b)	**HT**(1,700°C), $s = 4$, calc. 2.952 eV	**5**.49b	[Kup99]
ME2759n	2.965	23.92	418.1	DAP35n	$^*(C_2)_i^+$ (a)	**IG,** $s = 12$, calc. 2.963 eV;	(see ZPL)	[Dav81b, Zai01]
ME2881g	2.976	24.00	416.6	DAP32g	$^*V_1^+$	**IG,** $s = 34$; calc. 2.975 eV, named **GR7a**	**5**.40	[Col88a]
ME2881h	2.982	24.05	415.8	DAP32h	$^*V_1^+$	**IG,** $s = 32$; calc. 2.981 eV, named **GR7b**	**5**.40	[Col88a]
ME2750c	2.990	24.12	414.6	***S6**-DAP61c	$^*(V_2Ni_1)N_x$	**HT**(2,200°C);PLE of *S5; $s = 16$, calc. 2.989 eV	**5**.26	[Kup99]

ME2881-i	2.997	24.17	413.7	DAP32-i	$^*V_1^+$	**IG,** $s = 27$a,b; calc. 3.000 eV, named **GR8a,b**	**5.**40	[Col88a]
ME2881j	3.003	24.22	412.8	DAP32j	$^*V_1^+$	**IG,** $s = 26$a,b; calc. 3.004 eV, named **GR8c,d**	**5.**40	[Col88a]
ME2881k	3.006	24.25	412.4	DAP32k	$^*V_1^+$	**IG,** $s = 25$; calc. 3.009 eV, named **GR8e**	**5.**40	[Col88a]
ME2750d	3.015	24.32	411.2	***S6**-DAP61d	$^*(V_2Ni_1)N_x$	**HT**(2,200°C);PLE of *S5; $s = 14$, calc. 3.015 eV	**5.**26	[Kup99]
ME2750e	3.042	24.54	407.6	***S6**-DAP61e	$^*(V_2Ni_1)N_x$	**HT**(2,200°C);PLE of *S5; $s = 12$, calc. 3.045 eV	**5.**26	[Kup99]
ME3075a	3.075	24.80	403.2	**N3c**-DAP23a	$V_1N_3{}^\circ$ (c)	**IG**, ZPL at 2.985 eV; other lines see NA/NE; $s = 162$, calc. 3.075 eV, PLE of N3c-DAP24(a–q,z)	...	[Zai01]

Treatment: IE irradiation, electrons; *IG* irradiation, general; *II* irradiation, ions; *IN* irradiation, neutrons; *IX* irradiation, X-ray; *HP* hydrogen plasma; *HT* heat treatment; *PD* plastic deformation; *PO* powder, *PO(KBr) compressed with KBr*

Table 5.2.6 Modified diamond (irradiation, heat etc.): excitation of photo luminescence lines (PLE), near ultraviolet (3.10–3.94 eV)

Line-label	Energy (eV)	Frequ. ($10^3 cm^{-1}$)	Wavel. (nm)	Name	Impur. /defect	Comment	Figs. **3–6.** [Zai01]; Fig. **7.** [Zai98]	References
ME2750f	3.110	25.09	398.6	***S6**-DAP61f	$^*(V_2Ni_1)N_x$	**HT**(2,200°C);PLE of *S5; $s = 9$, calc. 3.115 eV	**5.**26	[Kup99]
ME2750g	3.140	25.33	394.8	***S6**-DAP61g	$^*(V_2Ni_1)N_x$	**HT**(2,200°C);PLE of *S5; $s = 8$, calc. 3.136 eV	**5.**26	[Kup99]
ME2750h	3.170	25.57	391.1	***S6**-DAP61h	$^*(V_2Ni_1)N_x$	**HT**(2,200°C);PLE of *S5; $s = 7$, calc. 3.180 eV	**5.**26	[Kup99]
ME2750i	3.210	25.89	386.2	***S6**-DAP61i	$^*(V_2Ni_1)N_x$	**HT**(2,200°C);PLE of *S5; $s = 6$, calc. 3.212 eV	**5.**26	[Kup99]
ME2515-o	3.234	26.09	383.4	**S2**-DAP57-o	$(V_2Ni_1)N_3$	**HT**(1,700°C)**;** $s = 2$, calc. 3.240 eV	**7.**104	[Yel92b]
ME2750j	3.260	26.30	380.3	***S6**-DAP61j	$^*(V_2Ni_1)N_x$	**HT**(2,200°C);PLE of *S5; $s = 5$, calc. 3.265 eV	**5.**26	[Kup99]
ME3270	3.270	26.38	379.1	**H4b**-DAP37-l named**H10**	$^*N_3V_2N_1{}^\circ$ (b)	**IE+HT**(500–1,500°C)**,** PLE of H3, H4a,b; other lines see MA; $s = 3$, calc. 3.250 eV	**7.**147	[Col83a, Zai01]
ME2750k	3.300	26.62	375.7	***S6**-DAP61k	$^*(V_2Ni_1)N_x^+$	**HT**(2,200°C);PLE of *S5; $s = 4$, calc. 3.307 eV	**5.**26	[Kup99]
ME3333a	3.333	26.88	372.0	***S4-**DAP59a	$^*(V_2Ni_1)N_3^\circ$	**HT**(2,500°C), 7 lines a–c, f, h, j, k; $s = 86$, calc. 3.333 eV, PLE of *Ni237-DAP92 and *S8-DAP16	**5.**19b	[Kup99]
ME2515-p	3.341	26.95	371.1	**S2**-DAP57-p	$(V_2Ni_1)N_3^+$	**HT**(1,700°C)**;** $s = 1$, calc. 3.335 eV;6SBs 34 meV	**7.**104	[Yel92b]

ME3361z	3.361	27.11	368.9	**H13**-DAP52z	$^*V_1N_2{}^\circ$ (b)	**IG+HT**(500–1,500°C), ZPL of H13, other lines see MA; SB +42 meV	**7.**147	[Col83a, Zai01]
ME3333b	3.380	27.26	366.8	*__S4__-DAP59b	$^*(V_2Ni_1)N_3^\circ$	**HT**(2,500°C), $s = 50$, calc. 3.378 eV	**5.**19b	[Kup99]
ME3333c	3.412	27.52	363.4	*__S4__-DAP59c	$^*(V_2Ni_1)N_3^\circ$	**HT**(2,500°C), $s = 34$, calc. 3.417 eV	**5.**19b	[Kup99]
ME2750-l	3.460	27.91	358.3	*__S6__-DAP61-l	$^*(V_2Ni_1)N_x^+$	**HT**(2,200°C);PLE of *S5; $s = 2$, calc. 3.455 eV	**5.**26	[Kup99]
ME3333f	3.509	28.30	353.3	*__S4__-DAP59f	$^*(V_2Ni_1)N_3^\circ$	**HT**(2,500°C), $s = 18$, calc. 3.503 eV	**5.**19b	[Kup99]
ME2750m	3.530	28.47	351.2	*__S6__-DAP61m	$^*(V_2Ni_1)N_x$	**HT**(2,200°C);PLE of *S5; $s = 1$, calc. 3.550 eV	**5.**26	[Kup99]
ME3333g	3.541	28.56	350.1	*__S4__-DAP59g	$^*(V_2Ni_1)N_3^\circ$	**HT**(1,700°C); $s = 14$, calc. 3.546 eV	**7.**104	[Yel92b]
ME3333h	3.573	28.82	347.0	*__S4__-DAP59h	$^*(V_2Ni_1)N_3^\circ$	**HT**(2,500°C), $s = 12$, calc. 3.573 eV	**5.**19b	[Kup99]
ME3333j	3.650	29.44	339.7	*__S4__-DAP59j	$^*(V_2Ni_1)N_3^\circ$	**HT**(2,500°C), $s = 8$, calc. 3.659 eV	**5.**19b	[Kup99]
ME3333k	3.743	30.19	331.2	*__S4__-DAP59k	$^*(V_2Ni_1)N_3^\circ$	**HT**(2,500°C), $s = 6$, calc. 3.731 eV	**5.**19b	[Kup99]

Treatment: IE irradiation, electrons; *IG* irradiation, general; *II* irradiation, ions; *IN* irradiation, neutrons; *IX* irradiation, X-ray; *HP* hydrogen plasma; *HT* heat treatment; *PD* plastic deformation; *PO* powder, *PO(KBr) compressed with KBr*

5.3 Luminescence Lines (ML)

Table 5.3.1 Modified diamond (irradiation, heat etc.): luminescence lines, mid infrared (0.18–1.24 eV)

Line-label	Energy (eV)	Frequ. ($10^3 cm^{-1}$)	Wavel. (nm)	Name	Impur. /defect	Treatment/comment	Figs. **3–6.** [Zai01]; Fig. **7.** [Zai98]	References
ML1020a	1.020	8.227	1215	**H2**-DAP78a	$^*V_1N_2^-$	**HT**(1,700°C); DAP78a-d; $s = 26$, calc. 1.015 eV	**5.**10, **7.**54	[Law92b]
ML1020b	1.078	8.695	1150	**H2**-DAP78b	$^*V_1N_2^-$	**HT**(1,700°C); $s = 48$, calc. 1.078 eV	**5.**10, **7.**54	[Law92b]
ML1020c	1.107	8.929	1120	**H2**-DAP78c	$^*V_1N_2^-$	**HT**(1,700°C); $s = 64$, calc. 1.103 eV	**5.**10, **7.**54	[Law92b]
ML1118	1.118	9.022	1108	...	...	**IG**, **IG**+**HT**(800°C)	...	[Gip95b, Zai01]
ML1020d	1.139	9.187	1088	**H2**-DAP78d	$^*V_1N_2^-$	**HT**(1,700°C); $s = 106$, calc. 1.137 eV	**5.**10, **7.**54	[Law92b]
ML1150	1.150	9.278	1078	...	...	**IG**, **IG**+**HT**(800°C)	...	[Gip95b, Zai01]
ML1177	1.177	9.494	1053	...	$^*V_1C_4N_1{}^\circ$	**IN**+**HT**(800°C); mirror of DAP45, $s = 6$	...	[Vin88], [Zai01]
ML1235a	1.235	9.962	1004	**Xe**-DAP42a	$^*(V_2Xe_1)^\circ$	**II**(Xe)+**HT**(1400C); $s = 16$, calc. 1.241 eV	**7.**62a	[Zai92a]

Treatment: IE irradiation, electrons; *IG* irradiation, general; *II* irradiation, ions; *IN* irradiation, neutrons; *IX* irradiation, X-ray; *HP* hydrogen plasma; *HT* heat treatment; *PD* plastic deformation; *PO* powder, *PO(KBr) compressed with KBr*

Table 5.3.2.1 Modified diamond (irradiation, heat etc.): luminescence lines, near infrared, part 1 (1.24–1.49 eV)

Line-label	Energy (eV)	Frequ. ($10^3 cm^{-1}$)	Wavel. (nm)	Name	Impur. /defect	Treatment/comment	Figs. **3–6.** [Zai01]; Fig. **7.** [Zai98]	References
ML1235b	1.246	10.05	995.0	**Xe**-DAP42b	$^*(V_2Xe_1)^\circ$	**II**(Xe)+**HT**(1400C); $s = 17$, calc. 1.246 eV	**7.**62a	[Zai92a]
ML1249	1.249	10.07	992.6	...	$^*(V_2Ti_1)^\circ$	**II**(Ti)+**HT**(1,400°C); 3SBs: −32, −89, −121 meV	**5.**8a = **7.**52	[Gip93b, Zai01]
ML1256 = ML1020z	1.256	10.13	987.1	**H2** = DAP78z	$^*V_1N_2^-$	**HT**(1,700°C); ZPL of H2-DAP78a-d, 4 SBs with −59 meV(QLVM: calc. 1N+2C:58.5).	**5.**10 = **7.**54	[Law92b, Zai01]
ML1277	1.277	10.30	970.8	...	Ti?	**II or HT**(1,500°C)	...	[Zai01]
ML1235c	1.281	10.33	967.8	**Xe**-DAP42c	$^*(V_2Xe_1)^\circ$	**II**(Xe)+**HT**(1400C); $s = 22$, calc. 1.282 eV	**7.**62a	[Zai92a]
ML1235d	1.286	10.37	964.1	**Xe**-DAP42d	$^*(V_2Xe_1)^\circ$	**II**(Xe)+**HT**(1400C); $s = 23$, calc. 1.287 eV	**7.**62a	[Zai92a]
ML1235e	1.360	10.97	911.6	**Xe**-DAP42e	$^*(V_2Xe_1)^\circ$	**II**(Xe)+**HT**(1400C); $s = 48$, calc. 1.362 eV	**7.**62a	[Zai92a]
ML1235f	1.367	11.03	906.9	**Xe**-DAP42f	$^*(V_2Xe_1)^\circ$	**II**(Xe)+**HT**(1400C); $s = 50$, calc. 1.366 eV	**7.**62a	[Zai92a]
ML1235g	1.385	11.17	895.1	**Xe**-DAP42g	$^*(V_2Xe_1)^\circ$	**II**(Xe)+**HT**(1400C); $s = 64$, calc. 1.384 eV	**7.**62a	[Zai92a]
ML1235h	1.398	11.28	886.8	**Xe**-DAP42g	$^*(V_2Xe_1)^\circ$	**II**(Xe)+**HT**(1400C); $s = 78$, calc. 1.398 eV	**7.**62a	[Zai92a]
ML1401a	1.403	11.32	883.7	1Ni-DAP26z	$V_1Ni_1^+$	**HT**(2,200°C); ZPL of 1Ni-DAP26a-k (see HL)	**5.**13aII	[Kup99, Zai01]
ML1407a	1.407	11.35	881.1	**Li**-DAP68a	$^*V_1Li_1^\circ$	**II**(Li)+**HT**(1,400°C); 4 lines a–d(1.710 eV), $s = 8$, calc. 1.398 eV	**5.**23	[Zai01]

(continued)

Table 5.3.2.1 (continued)

Line-label	Energy (eV)	Frequ. ($10^3 cm^{-1}$)	Wavel. (nm)	Name	Impur. /defect	Treatment/comment	Figs. **3–6.** [Zai01]; Fig. **7.** [Zai98]	References
ML1413	1.413	11.40	877.2	...	Ni+N?	**HT**(1,500°C)	**5.**13aI	[Kup99, Zai01]
ML1235i	1.422	11.47	871.9	**Xe**-DAP42i	*(V_2Xe_1) °	**II**(Xe)+**HT**(1,400°C); $s = 130$, calc. 1.427 eV	**7.**62a	[Zai92a]
ML1428	1.428	11.52	868.2	...	Ni?	**HT**(1,700°C)	...	[Col93d, Zai01]
ML1235j	1.443	11.64	859.2	**Xe**-DAP42j	*(V_2Xe_1)°	**II**(Xe)+**HT**(1400C); $s = 198$, calc. 1.446 eV	**7.**62a	[Zai92a]
ML1448	1.448	11.68	856.2	...	Ni?	**HT**(1,700°C)	...	[Col93d, Zai01]
ML1468	1.466	11.82	845.7	...	O?	**II**(O)+HT(1,650°C); low nitrogen diamond	**7.**59	[Gip92, Zai01]
ML1483	1.482	11.95	836.6	...	O?	**II**(O)+HT(1,650°C); low nitrogen diamond	**7.**59	[Gip92, Zai01]
ML1489	1.489	12.01	832.4	...	...	**HT**(1,500°C)	**5.**13aI	[Kup99, Zai01]
ML1508a	1.508	12.16	822.1	**GR1**-DAP34a	V_1°	**IG** or **IG**+HT(500°C); 7 lines a–g(1.582 eV), $s = 40$, calc. 1.509 eV	Table 5.2 in Zai01, p. 164	[Ned72, Zai01]
ML1508b	1.517	12.24	817.3	**GR1**-DAP34b	V_1°	**IG** or **IG**+HT(500°C)$s = 44$, calc. 1.516 eV	see line(a)	[Ned72, Zai01]

Treatment: IE irradiation, electrons; *IG* irradiation, general; *II* irradiation, ions; *IN* irradiation, neutrons; *IX* irradiation, X-ray; *HP* hydrogen plasma; *HT* heat treatment; *PD* plastic deformation; *PO* powder, *PO(KBr) compressed with KBr*

Table 5.3.2.2 Modified diamond (irradiation, heat etc.): luminescence lines, near infrared, part 2 (1.52–1.64 eV)

Line-label	Energy (eV)	Frequ. ($10^3 cm^{-1}$)	Wavel. (nm)	Name	Impur. /defect	Treatment/comment	Figs. **3–6.** [Zai01]; Fig. **7.** [Zai98]	References
ML1525	1.525	12.30	813.0	...	$*V_1C_4N_1{}^\circ$	**IN**+**HT**(>800°C); mirror of DAP45, $s = 58$, calc. 1.528 eV, spectral hole burning [Sil94a]	**5**.15, **7**.65	[Sil95, Bau93, Zai01]
ML1528 = ML1235z	1.528	12.32	811.4	**Xe**-DAP42z	$*(V_2Xe_1){}^\circ$	**II**(Xe)+**HT**(1400C); ZPL of Xe-DAP42a-j 2 SBs: −28 (QLVM, calc. 1Xe:27.3), −43 meV	**5**.16, **7**.62	[Gor98, Zai92a] [Zai01]
ML1508c	1.528	12.33	811.4	**GR1**-DAP34c	$V_1{}^\circ$	**IG** or **IG**+HT(500°C)$s = 50$, calc. 1.527 eV	see line(a)	[Ned72, Zai01]
ML1508d	1.542	12.44	804.0	**GR1**-DAP34d	$V_1{}^\circ$	**IG** or **IG**+HT(500°C)$s = 62$, calc. 1.541 eV	see line(a)	[Ned72, Zai01]
ML1508e	1.544	12.45	803.0	**GR1**-DAP34e	$V_1{}^\circ$	**IG** or **IG**+HT(500°C); $s = 66$, calc. 1.545 eV	see line(a)	[Ned72, Zai01]
ML1508f	1.554	12.54	797.8	**GR1**-DAP34f	$V_1{}^\circ$	**IG** or **IG**+HT(500°C); $s = 78$, calc. 1.555 eV	see line(a)	[Ned72, Zai01]
ML1563a	1.563	12.61	793.2	***S5**-DAP60a	$*(V_2Ni_1)N_x$	**HT**(>1,500°C); 4 lines a–d(1.648 eV), lines e–k see MA; $s = 162$, calc. 1.564 eV, SBs: −28, −59 meV	**5**.13a, **5**.19a **5**.16c	[Kup99] [Gor98, Zai01]
ML1564a	1.564	12.62	792.7	DAP49a	$*V_1N_1^-$	**IG**+**HT**(>550°C); 15 lines a–o (1.870 eV), lines c, f, j, n see LL; $s = 8$, calc. 1.564 eV	**7**.80	[Dav77c]
ML1407b	1.569	12.66	790.2	**Li**-DAP68b	$*V_1Li_1{}^\circ$	**II**(Li)+**HT**(1,400°C); $s = 20$, calc. 1.574 eV	**5**.23	[Zai01]
ML1569a	1.569	12.66	790.2	DAP84a	$*V_1Ni_1{}^\circ$	**HT**(>1,500°C); 3 lines a–c, $s = 72$, calc. 1.567 eV	**5**.49a	[Kup99, Zai01]

(continued)

Table 5.3.2.2 (continued)

Line-label	Energy (eV)	Frequ. (10^3cm^{-1})	Wavel. (nm)	Name	Impur. /defect	Treatment/comment	Figs. **3–6.** [Zai01]; Fig. **7.** [Zai98]	References
ML1508g	1.582	12.76	783.7	**GR1**-DAP34g	V_1°	**IG** or **IG**+HT(500°C)$s = 118$, calc. 1.587 eV	see line(a)	[Ned72, Zai01]
ML1594	1.594	12.86	777.8	...	B?	**II**(B)+**HT**(1,000°C)	**5**.20a, b	[Zai95, Zai96b]
ML1597	1.597	12.88	776.3	...	...	**IN**+**HT**(800°C)	...	[Vin88], [Zai01]
ML1599a	1.599	12.90	775.3	...	...	**HT**(>1,700°C); line b at 1.600 eV	**5**.21	[McC95, Yel99]
ML1564b	1.600	12.91	774.9	DAP49b	*$V_1N_1^-$	**IG**+**HT**(>550°C); $s = 10$, calc. 1.603 eV	**7**.80	[Dav77c]
ML1599b	1.600	12.91	774.9	...	...	**HT**(>1,700°C)	**5**.21	[McC95, Yel99]
ML1602	1.602	12.92	773.9	...	*$V_1C_4N_1$°	**IN**+**HT**(>800°C), mirror of DAP45, $s = 162$	**5**.15, **7**.65	[Sil95, Bau93]
ML1563b	1.602	12.92	773.9	***S5**-DAP60b	*$(V_2Ni_1)N_x$	**HT**(>1,500°C); $s = 86$, calc. 1.602 eV	**5**.13a	[Kup99, Zai01]
ML1569b	1.621	13.07	764.8	DAP84b	*V_1Ni_1°	**HT**(>1,500°C); $s = 198$, calc. 1.621 eV	**5**.49a	[Kup99, Zai01]
ML1569c	1.627	13.12	762.0	DAP84c	*V_1Ni_1°	**HT**(>1,500°C); $s = 228$, calc. 1.627 eV	**5**.49a	[Kup99, Zai01]
ML1407c	1.631	13.16	760.1	**Li**-DAP68b	*V_1Li_1°	**II**(Li)+**HT**(1,400°C); $s = 30$, calc. 1.631 eV	**5**.23	[Zai01]
ML1563c	1.638	13.21	756.9	***S5**-DAP60c	*$(V_2Ni_1)N_x$	**HT**(>1,500°C); $s = 58$, calc. 1.638 eV	**5**.13a	[Kup99, Zai01]

Treatment: IE irradiation, electrons; *IG* irradiation, general; *II* irradiation, ions; *IN* irradiation, neutrons; *IX* irradiation, X-ray; *HP* hydrogen plasma; *HT* heat treatment; *PD* plastic deformation; *PO* powder, *PO(KBr) compressed with KBr*

Table 5.3.2.3 Modified diamond (irradiation, heat etc.): luminescence lines, near infrared, part 3 (1.64–1.70 eV)

Line-label	Energy (eV)	Frequ. ($10^3 cm^{-1}$)	Wavel. (nm)	Name	Impur. /defect	Comment	Figs. **3–6.** [Zai01]; Fig. **7.** [Zai98]	References
ML1563d	1.648	13.29	752.3	*$\mathbf{S5}$-DAP60d	$*(V_2Ni_1)N_x$	**HT**(>1,500°C); $s = 50$, calc. 1.648 eV	**5.**13a	[Kup99, Zai01]
ML1564d	1.650	13.31	751.4	DAP49d	$*V_1N_1^-$	**IG**+**HT**(>550°C); $s = 14$, calc. 1.655 eV	**7.**80	[Dav77c]
ML1660a	1.660	13.39	746.9	. ***Ni** 1.66a	$*V_1Ni_2°$ (a)	**HT**(>1,600°C); line (b) at 2.427 eV (see ME); 2 SBs: −66 meV	**5.**27a	[Kup99, Zai00a]
ML1665a	1.665	13.43	744.6	**GR1**a	$V_1°$	**IG**, or **IG**+HT(500°C)2 lines a, b (1.673 eV)	**5.**39, **7.**96	[Zai01, Zai92a]
ML1666	1.666	13.44	744.2	...	...	**IN**+**HT**(950°C)	...	[Zai01]
ML1672	1.672	13.49	741.5	...	$*V_1Cr_1°$	**II**(Cr)+**HT**(>1,000°C), 2 SBs(QLVM) −34 meV	**5.**28, **7.**143b	[Zai92a, Zai01]
ML1665b = ML1508z	1.673	13.49	741.0	**GR1**b = DAP34z	$V_1°$	**IG**, or **IG**+HT(500°C), ZPL of DAP34a-g; main line of (a, b) is (b); SBs: −24, −36, −48 meV	**5.**34, **7.**96	[Zai01, Zai92a, Zai01]
ML1564e	1.673	13.49	741.0	DAP49e	$*V_1N_1^-$	**IG**+**HT**(>550°C); $s = 16$, calc. 1.675 eV	**7.**80	[Dav77c]
ML1678a	1.678	13.54	738.8	***S9**-DAP4a	$*(V_2Ni_1)N_x$	**IN**+**HT**(950°C); 9 lines a–i(1.821 eV), $s = 72$, calc. 1.678 eV	**5.**15 = **7.**61	[Sil95, Zai01]
ML1680a	1.680	13.55	738.0	**Tl**-DAP94a	$*V_1Tl_1°$	**II**(Tl)+**HT**(1,400°C); 8 lines a–h(1.881 eV); $s = 11$, calc. 1.678 eV	**5.**63	[Zai00a], [Zai01]

(continued)

Table 5.3.2.3 (continued)

Line-label	Energy (eV)	Frequ. ($10^3 cm^{-1}$)	Wavel. (nm)	Name	Impur. /defect	Comment	Figs. **3–6.** [Zai01]; Fig. **7.** [Zai98]	References
ML1682 = ML1487z	1.682	13.57	737.1	**2Si**−DAP54z = 1.681 eV center	$*(V_3Si_2)^+$	**II**(Si)+**HT**(1,400°C); ZPL of 2Si-DAP54a-g, lines b–l see LL, Si_2 isotopes see Fig. 5.47	**5.**20b, **5.**21 **5.**45, **7.**102	[Zai96b, McC95] [Zai01, Rua91b]
ML1678b	1.689	13.62	734.0	***S9**-DAP4b	$*(V_2Ni_1)N_x$	**IN**+**HT**(950°C); $s = 64$, calc. 1.689 eV	**5.**15 = **7.**61	[Sil95, Zai01]
ML1693a	1.693	13.66	732.3	...	$*V_1Ni_2^+$(a)	**HT**(>1,500°C), inverse lum., SB: 2 × +51 meV, 3 lines a–c(1.991 eV)	**5.**56a	[Kup99, Zai01]
ML1678c	1.694	13.66	731.9	***S9**-DAP4c	$*(V_2Ni_1)N_x$	**IN**+**HT**(950°C); $s = 62$, calc. 1.692 eV	**5.**15 = **7.**61	[Sil95, Zai01]
ML1696a	1.696	13.68	731.0	DAP70a	$*N_1^{2+} + B_1{}^\circ$	**IE**(300 keV, B doped CVD), 17 lines DAP70a-q (2.417), L = 1.567 eV, $s = 106$, calc. 1.695 eV	...	[Ste99d]
ML1678d	1.697	13.69	730.7	***S9**-DAP4d	$*(V_2Ni_1)N_x$	**IN**+**HT**(950°C); $s = 58$, calc. 1.698 eV;	**5.**15 = **7.**61	[Sil95, Zai01]
ML1680b	1.697	13.69	730.7	**Tl**-DAP94b	$*V_1Tl_1{}^\circ$	**II**(Tl)+**HT**(1,400°C); $s = 12$, calc. 1.697 eV	**5.**63	[Zai00a], [Zai01]

Treatment: IE irradiation, electrons; *IG* irradiation, general; *II* irradiation, ions; *IN* irradiation, neutrons; *IX* irradiation, X-ray; *HP* hydrogen plasma; *HT* heat treatment; *PD* plastic deformation; *PO* powder, *PO(KBr) compressed with KBr*

Table 5.3.2.4 Modified diamond (irradiation, heat etc.): luminescence lines, near infrared, part 4 (1.70–1.77 eV)

Line-label	Energy (eV)	Frequ. ($10^3 cm^{-1}$)	Wavel. (nm)	Name	Impur. /defect	Treatment/comment	Figs. **3–6.** [Zai01]; Fig. **7.** [Zai98]	References
ML1678e	1.703	13.74	728.0	***S9**-DAP4e	*$(V_2Ni_1)N_x$	**IN**+**HT**(950°C); $s = 54$, calc. 1.705 eV	**5.**15 = **7.**61	[Sil95, Zai01]
ML1704 = ML1569z	1.704	13.74	727.6	*Ni 1.70a = DAP84z	*$V_1Ni_1{}^{\circ}$	**HT**(>1,500°C); (QLVM): −44 meV (calc. Ni_1:43.6), Ni 1.70 b-DAP29 2.401 eV = PLE of (a)	**5.**27a, **5.**49a	2× [Kup99; Zai01]
ML1678f	1.708	13.78	725.9	***S9**-DAP4f	*$(V_2Ni_1)N_x$	**IN**+**HT**(950°C); $s = 52$, calc. 1.708 eV	**5.**15 = **7.**61	[Sil95, Zai01]
ML1708a	1.708	13.78	725.9	DAP111a	*$V_1Ni_2^+$(c)	**HT**(>1,500°C), 6 lines DAP111a-f (1.898 eV), $s = 16$, calc. 1.702 eV	**5.**56a	[Kup99]
ML1407d	1.710	13.79	725.0	**Li**-DAP68d	*$V_1Li_1{}^{\circ}$	**II**(Li)+**HT**(1,400°C); $s = 64$, calc. 1.710 eV	**5.**23	[Zai01]
ML1678g	1.712	13.81	724.2	***S9**-DAP4g	*$(V_2Ni_1)N_x$	**IN**+**HT**(950°C); $s = 50$, calc. 1.712 eV	**5.**15 = **7.**61	[Sil95, Zai01]
ML1716a	1.716	13.84	722.6	...	Ni?	**IE**+**HT**(>900°C); 2 lines a, b (1.720 eV)	...	[Zai01]
ML1708b	1.718	13.86	721.6	DAP111b	*$V_1Ni_2^+$(c)	**HT**(>1,500°C), $s = 18$, calc. 1.718 eV	**5.**56a	[Kup99]
ML1716b	1.720	13.88	720.6	...	Ni?	**IE**+**HT**(>900°C); SB: −36 meV	...	[Zai01]
ML1723a	1.723	13.90	719.5	NG(a)	Ne?	**II**(Ne)+**HT**(700°C); 2 lines a, b (1.731 eV)	**7.**72a	[Zai92a, Zai01]
ML1724	1.724	13.91	719.1	...	...	**II**+**HT**(>700°C)	**5.**45, **7.**71	[Zai01, Gip82a]
ML1564g	1.726	13.92	718.3	DAP49g	*$V_1N_1^-$	**IG**+**HT**(>550°C); $s = 24$, calc. 1.724 eV	**7.**80	[Dav77c]

(continued)

Table 5.3.2.4 (continued)

Line-label	Energy (eV)	Frequ. ($10^3 cm^{-1}$)	Wavel. (nm)	Name	Impur. /defect	Treatment/comment	Figs. **3–6.** [Zai01]; Fig. **7.** [Zai98]	References
ML1723b	1.731	13.96	716.2	NG(b)	Ne?	**II**(Ne)+**HT**(700°C); 10 SBs: −25 meV	**7.**72a	[Zai92a, Zai01]
ML1678h	1.732	13.97	715.8	***S9**-DAP4h	*$(V_2Ni_1)N_x$	**IN**+**HT**(950°C); $s = 50$, calc. 1.712 eV	**5.**15 = **7.**61	[Sil95, Zai01]
ML1738	1.738	14.02	713.4	...	Ni?	**HT**(>1,700°C)	...	[Yel99, Zai01]
ML1680c	1.740	14.03	712.5	**Tl**-DAP94c	*$V_1Tl_1°$	**II**(Tl)+**HT**(1,400°C); $s = 16$, calc. 1.740 eV	**5.**63	[Zai00a], [Zai01]
ML1742	1.742	14.05	711.7	DAP14z	*$(C_2)_i$ $+Ni_1°$	**IN**+**HT**(>800°C); ZPL of DAP14a-i (2.424 eV)	**7.**90	[Zai92a, Zai01]
ML1564h	1.745	14.08	710.5	DAP49h	*$V_1N_1^-$	**IG**+**HT**(>550°C); $s = 30$, calc. 1.747 eV	**7.**80	[Dav77c]
ML1696b	1.746	14.08	710.1	DAP70b	*$N_1^{2+} + B_1°$	**IE**(see line a), $s = 50$, calc. 1.752 eV	...	[Ste99d]
ML1750a	1.750	14.12	708.4	DAP73a	*$V_1B_1°$ (a)	**IX** of type IIb**;** 6 lines DAP73a f (2.240 eV), also HL; $s = 2$, calc. 1.742; line (b) at 3.120 eV	...	[Dea65a]
ML1753	1.753	14.14	707.2	...	Ni?	**HT** (>1,700°C)	...	[Yel99, Zai01]
ML1680d	1.755	14.16	706.4	**Tl**-DAP94d	*$V_1Tl_1°$	**II**(Tl)+**HT**(1,400°C); $s = 18$, calc. 1.756 eV	**5.**63	[Zai00a], [Zai01]
ML1564i	1.767	14.25	701.6	DAP49i	*$V_1N_1^-$	**IG**+**HT**(>550°C); $s = 42$, calc. 1.777 eV	**7.**80	[Dav77c]
ML1708c	1.768	14.26	701.2	DAP111c	*$V_1Ni_2^+$ (c)	**HT**(>1,500°C), $s = 26$, calc. 1.763 eV	**5.**56a	[Kup99]

Treatment: *IE* Irradiation electrons; *IG* irradiation general; *II* irradiation ions; *IN* irradiation neutrons; *IX* irradiation X-ray; *HT* heat treatment

Table 5.3.3 Modified diamond (irradiation, heat etc.): luminescence lines, visible: purple (1.77–1.88 eV)

Line-label	Energy (eV)	Frequ. ($10^3 cm^{-1}$)	Wavel. (nm)	Name	Impur. /defect	Treatment/comment	Figs. **3–6.** [Zai01]; Fig. **7.** [Zai98]	References
ML1771	1.771	14.28	700.1	...	Ni?	**HT** (>1,700°C)	...	[Yel99, Zai01]
ML1787	1.787	14.42	693.7	...	Ni?	**HT** (>1,700°C)	...	[Nad99], [Zai01]
ML1708d	1.789	14.43	693.0	DAP111d	$^*V_1Ni_2^+$(c)	**HT**(>1,500°C), $s = 34$, calc. 1.793 eV	**5.**56a	[Kup99]
ML1680e	1.792	14.45	691.8	**Tl**-DAP94e	$^*V_1Tl_1{}^\circ$	**II**(Tl)+**HT**(1,400°C); $s = 24$, calc. 1.791 eV	**5.**63	[Zai00a], [Zai01]
ML1696c	1.797	14.49	689.9	DAP70c	$^*N_1^{2+} + B_1{}^\circ$	**IE**(see line a), $s = 34$, calc. 1.793 eV	...	[Ste99d]
ML1696d	1.805	14.56	686.9	DAP70d	$^*N_1^{2+} + B_1{}^\circ$	**IE**(see line a), $s = 30$, calc. 1.807 eV	...	[Ste99d]
ML1564k	1.809	14.59	685.3	DAP49k	$^*V_1N_1^-$	**IG**+**HT**(>550°C); $s = 64$, calc. 1.809 eV	**7.**80	[Dav77c]
ML1678i	1.821	14.69	680.8	*__S9__-DAP4i	$^*(V_2Ni_1)N_x$	**IN**+**HT**(950°C); $s = 50$, calc. 1.712 eV	**5.**15 = **7.**61	[Sil95, Zai01]
ML1564-l	1.822	14.70	680.4	DAP49-l	$^*V_1N_1^-$	**IG**+**HT**(>550°C); $s = 74$, calc. 1.818 eV	**7.**80	[Dav77c]
ML1696e	1.823	14.70	680.0	DAP70e	$^*N_1^{2+} + B_1{}^\circ$	**IE**(see line a), $s = 26$, calc. 1.825 eV	...	[Ste99d]

(continued)

Table 5.3.3 (continued)

Line-label	Energy (eV)	Frequ. ($10^3 cm^{-1}$)	Wavel. (nm)	Name	Impur. /defect	Treatment/comment	Figs. **3–6.** [Zai01]; Fig. **7.** [Zai98]	References
ML1829	1.829	14.76	677.7	...	N?	**HT**(1,700°C)	...	[Zai01]
ML1708e	1.830	14.76	677.5	DAP111e	$^*V_1Ni_2^+$(c)	**HT**(>1,500°C), $s = 50$, calc. 1.828 eV	**5**.56a, **5**.69a	[Kup99]
ML1564m	1.832	14.78	676.7	DAP49m	$^*V_1N_1^-$	**IG**+**HT**(>550°C); $s = 98$, calc. 1.834 eV	**7**.80	[Dav77c]
ML1834	1.834	14.79	676.0	...	...	**IN**+**HT**(950°C)	...	[Sil95, Zai01]
ML1696f	1.839	14.83	674.0	DAP70f	$^*N_1^{2+}+B_1{}^\circ$	**IE**(see line a), $s = 24$, calc. 1.836 eV	...	[Ste99d]
ML1680f	1.856	14.97	668.0	**Tl**-DAP94f	$^*V_1Tl_1{}^\circ$	**II**(Tl)+**HT**(1,400°C); $s = 48$, calc. 1.858 eV	**5**.63	[Zai00], [Zai01]
ML1750b	1.860	15.00	666.5	DAP73b	$^*V_1B_1{}^\circ$ (a)	**IX** of type IIb; $s = 4$, calc. 1.861	...	[Dea65a]
ML1696g	1.862	15.02	666.0	DAP70g	$^*N_1^{2+}+B_1{}^\circ$	**IE**(see line a), $s = 20$, calc. 1.862 eV	...	[Ste99d]
ML1680g	1.867	15.06	664.0	**Tl**-DAP94g	$^*V_1Tl_1{}^\circ$	**II**(Tl)+**HT**(1,400°C); $s = 54$, calc. 1.867 eV	**5**.63	[Zai00], [Zai01]
ML1564-o	1.870	15.08	663.0	DAP49m	$^*V_1N_1^-$	**IG**+**HT**(>550°C); $s = 198$, calc. 1.867 eV	**5**.53a, **7**.80	[Zai89, Dav77c]
ML1696h	1.876	15.13	661.0	DAP70h	$^*N_1^{2+}+B_1{}^\circ$	**IE**(see line a), $s = 18$, calc. 1.877 eV	...	[Ste99d]

Treatment: IE irradiation, electrons; *IG* irradiation, general; *II* irradiation, ions; *IN* irradiation, neutrons; *IX* irradiation, X-ray; *HP* hydrogen plasma; *HT* heat treatment; *PD* plastic deformation; *PO* powder, *PO(KBr) compressed with KBr*

Table 5.3.4.1 Modified diamond (irradiation, heat etc.): luminescence lines, visible: red, part 1(1.88–1.98 eV)

Line-label	Energy (eV)	Frequ. ($10^3 cm^{-1}$)	Wavel. (nm)	Name	Impur. /defect	Treatment/comment	Figs. **3–6.** [Zai01]; Fig. **7.** [Zai98]	References
ML1881 = ML1407z	1.881	15.17	659.1	**Li**-DAP68z	$*V_1Li_1°$	**II**(Li)+**HT**(1,400°C); ZPL of **Li**-DAP68a d; SB: −36 meV (QLVM, calc. 1Li+6C: 36.4)	**5**.23	[Zai01]
ML1680h	1.881	15.17	659.1	**Tl**-DAP94h	$*V_1Tl_1°$	**II**(Tl)+**HT**(1,400°C); $s = 64$, calc. 1.880 eV	**5**.63	[Zai00], [Zai01]
ML1881a	1.881	15.17	659.0	...	$*(V_2Ne_1)$ °?	2 lines a, b (1.884 eV), (a) is dominant	**7**.72c	[Zai92a]
ML1881b	1.884	15.20	658.0	...	$*(V_2Ne_1)$ °?	SB: QLVM 2× −32 meV, calc.1Ne+6C:33.3	**7**.72c	[Zai92a]
ML1425e	1.890	15.24	656.0	DAP62e	$*V_1Ni_2N_2^{\circ}$	**IN**(high dose), $s = 34$, calc. 1.891 eV	**5**.53b	[Zai89]
ML1708f	1.898	15.31	653.2	DAP111f	$*V_1Ni_2^{+}$(c)	**HT**(>1,500°C), $s = 162$, calc. 1.900 eV	**5**.56a, **5**.69a	[Kup99]
ML1696i	1.901	15.33	652.0	DAP70i	$*N_1^{2+}+B_1°$	**IE**(see line a), $s = 16$, calc. 1.895 eV	...	[Ste99d]
ML1905a	1.905	15.37	650.8	DAP28a	$*V_1Ni_2^{-}$(b)	**HT** (>1,500°C) or **II**+**HT**(1,400°C), 13 lines a–m (2.210 eV), $s = 9$, calc. 1.905 eV	**5**.56a, **5**.69a	2× [Kup99]
ML1696j	1.907	15.38	650.0	DAP70j	$*N_1^{2+}+B_1°$	**IE**(see line a), $s = 15$, calc. 1.911 eV	...	[Ste99d]
ML1910a	1.910	15.41	649.1	DAP47a	$*V_1N_1°$	**IG**+**HT**(>500°C), 8 lines DAP47a-h(2.059 eV), $s = 24$, calc. 1.908 eV	**5**.73, **5**.74 **7**.93	[Zai96b, Zai01] [Dav79b]

(continued)

Table 5.3.4.1 (continued)

Line-label	Energy (eV)	Frequ. ($10^3 cm^{-1}$)	Wavel. (nm)	Name	Impur. /defect	Treatment/comment	Figs. **3–6.** [Zai01]; Fig. **7.** [Zai98]	References
ML1696k	1.913	15.43	648.1	DAP70k	$*N_1^{2+}+B_1{}^\circ$	**IE**(see line a), $s = 14$, calc. 1.918 eV	...	[Ste99d]
ML1696-l	1.931	15.58	642.0	DAP70-l	$*N_1^{2+}+B_1{}^\circ$	**IE**(see line a), $s = 13$, calc. 1.935 eV	...	[Ste99d]
ML1910b	1.934	15.60	641.0	DAP47b	$*V_1N_1{}^\circ$	**IG+HT**(>500°C), $s = 30$, calc. 1.934 eV	**5.**73, **7.**93	[Zai96b, Dav79b]
ML1937	1.937	15.62	640.0	...	...	II(C),IN	**5.**53a, **5.**55	[Zai89, Zai96b]
ML1693b	1.940	15.65	639.1	...	$*V_1Ni_2^+$(b)	**HT**(>1,500°C), SB: 2× −59 meV	**5.**56a, **5.**69a	[Kup99, Zai01]
ML1943= ML1564z	1.943	15.67	638.1	NV⁻ or 638nm, DAP49z	$*V_1N_1^-$	**IG+HT**(>550°C); ZPL of DAP49a-o; 2 SBs: −61 meV(QLVM, calc. 3C:60.9), −123 meV	**5.**16a, **5.**112, **7.**80	[Gor98, Zai01] [Dav77c]
ML1905b	1.945	15.69	637.4	DAP28b	$*V_1Ni_2^-$(b)	**HT** (>1,500°C, see line a), $s = 11$, calc. 1.944 eV	**5.**56a, **5.**69a	[Kup99]
ML1696m	1.949	15.72	636.0	DAP70m	$*N_1^{2+}+B_1{}^\circ$	**IE**(see line a), $s = 12$, calc. 1.947 eV	...	[Ste99d]
ML1905c	1.962	15.82	631.9	DAP28c	$*V_1Ni_2^-$(b)	**HT** (>1,500°C, see line a), $s = 12$, calc. 1.963 eV	**5.**56a, **5.**92a	[Kup99], [Zai00]
ML1910c	1.968	15.87	630.0	DAP47c	$*V_1N_1{}^\circ$	**IG+HT**(>500°C), $s = 42$, calc. 1.968 eV	**5.**73, **7.**93	[Zai96b, Dav79b]
ML1696n	1.980	15.97	626.0	DAP70n	$*N_1^{2+}+B_1{}^\circ$	**IE**(see line a), $s = 10$, calc. 1.983 eV	...	[Ste99d]

Treatment: IE irradiation, electrons; *IG* irradiation, general; *II* irradiation, ions; *IN* irradiation, neutrons; *IX* irradiation, X-ray; *HP* hydrogen plasma; *HT* heat treatment; *PD* plastic deformation; *PO* powder, *PO(KBr) compressed with KBr*

Table 5.3.4.2 Modified diamond (irradiation, heat etc.): luminescence lines, visible: red, part 2 (1.98–2.03 eV)

Line-label	Energy (eV)	Frequ. ($10^3 cm^{-1}$)	Wavel. (nm)	Name	Impur. /defect	Comment	Figs. **3–6.** [Zai01]; Fig. **7.** [Zai98]	References
ML1910d	1.987	16.03	623.9	DAP47d	$^*V_1N_1{}^\circ$	**IG+HT**(>500°C), $s = 50$, calc. 1.984 eV	**5.**73, **7.**93	[Zai96b, Dav79b]
ML1905d	1.991	16.06	622.7	DAP28d	$^*V_1Ni_2^-$(b)	**HT** (>1,500°C, see line a), $s = 12$, calc. 1.963 eV	**5.**56a, **5.**92a	[Kup99], [Zai00a]
ML1693c = ML1708z	1.991	16.06	622.7	DAP111z	$^*V_1Ni_2^+$(c)	**HT**(>1,500°C), ZPL of DAP111a-f(1.898 eV), SB: (2×) −59 meV	**5.**56a, **5.**69a	[Kup99] [Zai01]
ML1750c	2.000	16.13	619.9	DAP73c	$^*V_1B_1{}^\circ$ (a)	**IX** of type IIb**;** $s = 8$, calc. 1.997	. . .	[Dea65a]
ML1910e	2.002	16.15	619.3	DAP47e	$^*V_1N_1{}^\circ$	**IG+HT**(>500°C), $s = 64$, calc. 2.004 eV	**5.**73, **7.**93	[Zai96b, Dav79b]
ML1905e	2.009	16.20	617.1	DAP28e	$^*V_1Ni_2^-$(b)	**HT** (>1,500°C, see line a), $s = 16$, calc. 2.009 eV	**5.**56a, **5.**92a	[Zai00a]

(continued)

Table 5.3.4.2 (continued)

Line-label	Energy (eV)	Frequ. ($10^3 cm^{-1}$)	Wavel. (nm)	Name	Impur. /defect	Comment	Figs. **3–6.** [Zai01]; Fig. **7.** [Zai98]	References
ML2012a	2.012	16.23	616.2	DAP10a	...	**II**+**HT**(>1,300°C), 12 lines a–l(2.769 eV); $s = 86$, calc. 2.012 eV	**7.**84	[Zai01]
ML2012b	2.016	16.26	615.0	DAP10b	...	**II**+**HT**(>1,300°C), $s = 82$, calc. 2.015 eV	**7.**84	[Zai92a]
ML2019 = ML1680z	2.019	16.29	614.1	**Tl**-DAP94a	$*V_1Tl_1{}^{\circ}$	**II**(Tl)+**HT**(1,400°C); ZPL of **Tl**-DAP94a-h, SBs: −21 meV(QLVM, calc. 1Tl: 21.5), −45, −159, −319 meV (resonant with DAP lines)	**5.**63, **7.**62b	[Zai96b, Zai92a]
ML1910f	2.023	16.32	612.8	DAP47f	$*V_1N_1{}^{\circ}$	**IG**+**HT**(>500°C), $s = 86$, calc. 2.024 eV	**5.**73, **7.**93	[Naz92, Zai01]
ML1905f	2.025	16.33	612.2	DAP28f	$*V_1Ni_2^-$(b)	**HT** (>1,500°C, see line a), $s = 18$, calc. 2.025 eV	**5.**56a	[Zai92a, Zai01]

Treatment: IE irradiation, electrons; *IG* irradiation, general; *II* irradiation, ions; *IN* irradiation, neutrons; *IX* irradiation, X-ray; *HP* hydrogen plasma; *HT* heat treatment; *PD* plastic deformation; *PO* powder, *PO(KBr) compressed with KBr*

Table 5.3.5 Modified diamond (irradiation, heat etc.): luminescence lines, visible: orange (2.03–2.10 eV)

Line-label	Energy (eV)	Frequ. ($10^3 cm^{-1}$)	Wavel. (nm)	Name	Impur. /defect	Treatment/comment	Figs. **3–6.** [Zai01]; Fig. **7.** [Zai98]	References
ML1905g	2.035	16.41	609.2	DAP28g	$*V_1Ni_2^-$(b)	**HT** (>1,500°C, see line a), $s = 20$, calc. 2.038 eV	**5.**56a	[Naz92, Zai01]
ML1910g	2.046	16.50	605.9	DAP47g	$*V_1N_1°$	**IG**+**HT**(>500°C), $s = 130$, calc. 2.048 eV	**5.**73, **7.**93	[Zai00a]
ML2049a	2.049	16.53	605.1	DAP95a	$*V_1He_1°$ (a)	**II**(He)+**HT**(>900°C), 6 lines a–f(2.049 eV), $s = 72$, calc. 2.052 eV	**5.**98	[Tka85b]
ML2054	2.054	16.56	603.7	...	...	**II**+**HT**(1,600°C)	**5.**65, **5.**66	[Zai89, Zai96b, Ste99a, Zai92a, Col90c]
ML2055	2.055	16.58	603.3	...	...	**II**+**HT**(1,700°C)	**5.**55, **7.**86	[Zai01]
ML2012c	2.059	16.61	602.1	DAP10c	...	**II**+**HT**(>1,300°C), $s = 50$, calc. 2.059 eV	**7.**84	[Var88, Mel98, Col92b, Zai01]
ML1910h	2.059	16.61	602.1	DAP47h	$*V_1N_1°$	**IG**+**HT**(>500°C), $s = 162$, calc. 2.059 eV	**5.**73, **7.**93	[Zai96b, Dav79b]
ML1750d	2.062	16.63	601.2	DAP73d	$*V_1B_1°$ (a)	**IX** of type IIb**;** $s = 12$, calc. 2.070	...	[Gip93b, Zai01]
ML2066a	2.066	16.66	600.1	DAP14a	$*(C_2)_iNi_1°$	**II**+**HT**(>700°C), 7 lines DAP14a-g (2.271 eV), $s = 8$, calc. 2.063 eV	**7.**90	[Zai92a]

(continued)

Table 5.3.5 (continued)

Line-label	Energy (eV)	Frequ. (10^3cm^{-1})	Wavel. (nm)	Name	Impur. /defect	Treatment/comment	Figs. **3–6.** [Zai01]; Fig. **7.** [Zai98]	References
ML2049b	2.066	16.66	600.1	DAP95b	$^*V_1He_1{}^\circ$ (a)	**II**(He)+**HT**(>900°C), $s=86$, calc. 2.066 eV	**5.**98, **7.**105	2× [Zai96b]
ML2070a	2.070	16.70	598.9	DAP1a	$^*N_1{}^\circ + B_1{}^\circ$	**II**+**HT**(1,400°C), CL or **p-i-n** diode (electrolum.) 13 lines a–m(2.668 eV), $s=50$, calc. 2.067 eV	**5.**68a, **5.**92a	[Dea65a]
ML1905h	2.071	16.70	598.6	DAP28h	$^*V_1Ni_2^-$ (b)	**HT** (>1,500°C, see line a), $s=26$, calc. 2.070 eV	**5.**56a	[Yel99, Zai01]
ML2119-s1	2.072	16.71	598.3	...	O?	SB(−37 meV) = QLVM	**7.**59	[Gip92]
ML2075a	2.075	16.74	597.5	DAP89a	$^*(C_2)_iN_2^-$	**HT**(>1,600°C), 7 lines a–g(2.288 eV), $s=14$, calc. 2.077 eV	...	[Law96]
ML2049c	2.075	16.74	597.5	DAP95c	$^*V_1He_1{}^\circ$ (a)	**II**(He)+**HT**(>900°C), $s=98$, calc. 2.075 eV	**5.**98, **7.**105	[Tka85b, Zai92a]
ML1905i	2.089	16.85	593.5	DAP28i	$^*V_1Ni_2^-$ (b)	**HT** (>1,500°C, see line a), $s=30$, calc. 2.086 eV	**5.**56a	[Kup99]
ML1905j	2.097	16.91	591.2	DAP28j	$^*V_1Ni_2^-$ (b)	**HT** (>1,500°C, see line a), $s=34$, calc. 2.100 eV	**5.**92a	[Zai00a]

Treatment: IE irradiation, electrons; *IG* irradiation, general; *II* irradiation, ions; *IN* irradiation, neutrons; *IX* irradiation, X-ray; *HP* hydrogen plasma; *HT* heat treatment; *PD* plastic deformation; *PO* powder, *PO(KBr) compressed with KBr*

Table 5.3.6.1 Modified diamond (irradiation, heat etc.): luminescence lines, visible: yellow, part 1 (2.10–2.15 eV)

Line-label	Energy (eV)	Frequ. ($10^3 cm^{-1}$)	Wavel. (nm)	Name	Impur. /defect	Comment	Figs. **3–6.** [Zai01]; Fig. **7.** [Zai98]	References
ML2066b	2.101	16.95	590.1	DAP14b	$*(C_2)_iNi_1{}^\circ$	**II**+**HT**(>700° C), $s = 10$, calc. 2.100 eV	**7.**90	[Zai92a]
ML1425z	2.101	16.95	590.0	DAP62z	$*V_1Ni_2N_2{}^\circ$	**IN**(high dose), ZPL of DAP62a-f with a(1.425), b(1.596), c(1.660), d(1.744), e(1.890), f(2.013 eV)	**5.**53b	[Zai89, Zai01]
ML2104	2.104	16.97	589.2	...	...	**II**(N)+**HT**(1600)	...	[Var86a, Zai01]
ML2106	2.106	16.99	588.6	...	...	**IE**(high doses)	...	[Ste99a, Zai01]
ML2049d	2.107	17.00	588.4	DAP95d	$*V_1He_1{}^\circ$ (a)	**II**(He)+**HT**(>900° C), $s = 162$, calc. 2.105 eV	**5.**98, **7.**105	[Tka85b, Zai92a]
ML2070b	2.108	17.00	588.1	DAP1b	$*N_1{}^\circ+B_1{}^\circ$	**II**(see line a), $s = 34$, calc. 2.105 eV	**5.**68a, **5.**92a	[Zai01], [Zai00a]
ML2075b	2.113	17.04	586.7	DAP89b	$*(C_2)_iN_2^-$	**HT**(>1,600° C), $s = 18$, calc. 2.112 eV	...	[Law96]
ML2049e	2.117	17.56	569.5	DAP95e	$*V_1He_1{}^\circ$ (a)	**II**(He)+**HT**(>900° C), $s = 198$, calc. 2.115 eV	**5.**98, **7.**105	[Tka85b, Zai92a]
ML2117a	2.117	17.08	585.5	**Zn**-DAP114a	$*V_1Zn_1{}^\circ$	**II**(Zn)+**HT**(1,400° C), 6 lines a–f(2.287 eV), $s = 20$, calc. 2.119 eV	**5.**101	[Zai00a]

(continued)

Table 5.3.6.1 (continued)

Line-label	Energy (eV)	Frequ. ($10^3 cm^{-1}$)	Wavel. (nm)	Name	Impur. /defect	Comment	Figs. **3–6.** [Zai01]; Fig. **7.** [Zai98]	References
ML2120	2.120	17.10	584.8	...	O?	**II**(O)+**HT**(>1,500°C)	**7**.59	[Gip92, Zai01]
ML2066c	2.123	17.12	584.0	DAP14c	$^*(C_2)_iNi_1{}^\circ$	**II**+**HT**(>700°C), $s = 12$, 2.131 eV	**7**.90	[Zai92a]
ML2049f	2.128	17.16	582.6	DAP95f	$^*V_1He_1{}^\circ$ (a)	**II**(He)+**HT**(>900°C), $s = 228$, calc. 2.123 eV	**5**.98, **7**.105	[Tka85b, Zai92a]
ML2129a	2.129	17.17	582.3	**TR12**-DAP36a	$(C_2)_i^-$	**IG**(+HT >300°C), 14 lines a–n(2.519 eV), $s = 5$, calc. 2.137 eV	**5**.55	[Zai96b, Dav81b]
ML1905k	2.133	17.20	581.2	DAP28k	$^*V_1Ni_2^-$ (b)	**HT** (>1,500°C, see line a), $s = 50$, calc. 2.135 eV	**5**.92a	[Zai00a]
ML2135a	2.135	17.22	580.7	**Co-S5** DAP104a	$^*(V_2Co_1)N_x$	**HT**(1,800°C), 9 lines a–i(2.820 eV), $s = 50$, calc. 2.135 eV; split-off line at 2.040 eV	**5**.72	[Law96]
ML2138a	2.138	17.25	579.9	DAP86a	$^*(C_2)_i^+$	**IG,** 9 lines a–i(2.356 eV), $s = 12$, calc. 2.134 eV	...	[Ste99a]
ML2135b	2.140	17.26	579.3	...	$^*(V_2Co_1)N_x$	weaker split-off line of (a)	...	
ML2143	2.143	17.28	578.6	...	...	**II**+**HT**(1,400°C)	**5**.45	[Zai01]

Treatment: IE irradiation, electrons; *IG* irradiation, general; *II* irradiation, ions; *IN* irradiation, neutrons; *IX* irradiation, X-ray; *HP* hydrogen plasma; *HT* heat treatment; *PD* plastic deformation; *PO* powder, *PO(KBr) compressed with KBr*

Table 5.3.6.2 Modified diamond (irradiation, heat etc.): luminescence lines, visible: yellow, part 2 (2.15–2.18 eV)

Line-label	Energy (eV)	Frequ. (10^3cm^{-1})	Wavel. (nm)	Name	Impur. /defect	Treatment/comment	Figs. **3–6.** [Zai01]; Fig. **7.** [Zai98]	References
ML2154 = ML1910z	2.154	17.38	575.5	**NV°, 575 nm,** DAP47z	$V_1N_1{}^{\circ}$	**IG+HT**(>500°C), ZPL of DAP47a-h, SB: −48 meV = QLVM, calc. 1N+3C:48.4	**5.**73, **5.**115 **7.**93, **7.**90, **7.**96	2× [Zai96b] [Dav79b, Zai92a]
ML2156	2.156	17.39	574.9	...	...	**PD**(damaged < *100* > surface by identation)	...	[Zai01]
ML2157	2.157	17.40	574.8	...	Ni, N?	**HT**(1,700°C)	...	[Zai01]
ML2075c	2.160	17.42	574.0	DAP89c	$^*(C_2)_iN_2^-$	**HT**(>1,600°C), $s = 26$, calc. 2.155 eV	...	[Law96]
ML2012d	2.161	17.43	573.7	DAP10d	...	**II+HT**(>1,300°C), $s = 22$, calc. 2.162 eV	**7.**84	[Var89, Zai01]
ML2170a	2.170	17.50	571.3	DAP92a	$^*V_1Ni_3^+$	**HT**(2,200°C), 6 lines a–f(2.280 eV), $s = 34$, calc. 2.172 eV	**5.**97	[Kup99]
ML2173	2.173	17.52	570.6	...	...	**II(c)**	**5.**55, **5.**66	2×[Zai96b]
ML2012e	2.177	17.56	569.5	DAP10e	...	**II+HT**(>1,300°C), $s = 20$, calc. 2.177 eV	**7.**84	[Var89, Zai01]
ML1905-l	2.178	17.57	569.2	DAP28-l	$^*V_1Ni_2^-$(b)	**HT** (>1,500°C, see line a), $s = 98$, calc. 2.181 eV	**5.**92a	[Zai00a]
ML2179a	2.179	17.58	569.0	DAP13a	$^*V_1He_1{}^{\circ}$ (b)	**II**(He)+**HT** (550°C), 12 lines a–l(2.375 eV), $s = 23$, calc. 2.177 eV	**5.**99, **7.**105	2× [Zai92a, Zai01]
ML2075d	2.179	17.58	569.0	DAP89d	$^*(C_2)_iN_2^-$	**HT**(>1,600°C), $s = 32$, calc. 2.176 eV	...	[Law96]

Treatment: IE irradiation, electrons; *IG* irradiation, general; *II* irradiation, ions; *IN* irradiation, neutrons; *IX* irradiation, X-ray; *HP* hydrogen plasma; *HT* heat treatment; *PD* plastic deformation; *PO* powder, *PO(KBr) compressed with KBr*

Table 5.3.7.1 Modified diamond (irradiation, heat etc.): luminescence lines, visible:green, part 1 (2.18–2.23 eV)

Line-label	Energy (eV)	Frequ. ($10^3 cm^{-1}$)	Wavel. (nm)	Name	Impur. /defect	Comment	Figs. **3–6.** [Zai01]; Fig. **7.** [Zai98]	References
ML2066d	2.181	17.59	568.4	DAP14d	$^*(C_2)_iNi_1{}^\circ$	**II**+**HT**(>700°C), $s = 18$, 2.187 eV	**7.**90	[Zai92a]
ML2117b	2.187	17.64	566.8	**Zn**-DAP114b	$^*V_1Zn_1{}^\circ$	**II**(Zn)+**HT**(1,400°C), $s = 36$, calc. 2.190 eV	**5.**101	[Zai00a]
ML2138b	2.195	17.70	564.8	DAP86b	$^*(C_2)_i^+$	**IG,** $s = 18$, calc. 2.195 eV	...	[Ste99a]
ML2179b	2.198	17.73	564.0	DAP13b	$^*V_1He_1{}^\circ$ (b)	**II**(He)+**HT** (550°C), $s = 26$, calc. 2.196 eV	**5.**99, **7.**105	2× [Zai92a, Zai01]
ML1750e	2.200	17.75	563.5	DAP73e	$^*V_1B_1{}^\circ$ (a)	**IX** of type IIb; $s = 34$, calc. 2.203	...	[Dea65a]
ML2170b	2.200	17.75	563.5	DAP92b	$^*V_1Ni_3^+$	**HT**(2,200°C), $s = 50$, calc. 2.207 eV	**5.**97	[Kup99]
ML2066e	2.202	17.76	563.0	DAP14e	$^*(C_2)_iNi_1{}^\circ$	**II**+**HT**(>700°C), $s = 20$, 2.199 eV	**7.**90	[Zai92a]
ML2070c	2.202	17.76	563.0	DAP1c	$^*N_1{}^\circ + B_1{}^\circ$	**II**(see line a), $s = 16$; calc. 2.202 eV	**5.**68a, **5.**92a	[Zai98d], [Zai00]
ML2179c	2.206	17.79	562.0	DAP13c	$^*V_1He_1{}^\circ$ (b)	**II**(He)+**HT** (550°C), $s = 30$, calc. 2.211 eV	**5.**99, **7.**105	2× [Zai92a, Zai01]
ML2117c	2.206	17.79	562.0	**Zn**-DAP114c	$^*V_1Zn_1{}^\circ$	**II**(Zn)+**HT**(1,400°C), $s = 42$, calc. 2.204 eV	**5.**101	[Zai00a]
ML2135b	2.207	17.80	561.7	DAP104b	$^*(V_2Co_1)N_x$	**HT**(1,800°C), split-off line at 2.211 eV from excited state splitting, $s = 26$, calc. 2.208 eV	**5.**72	[Law96]
ML1905m	2.210	17.83	561.0	DAP28m	$^*V_1Ni_2^-$(b)	**HT** (>1,500°C, see line a), $s = 162$, calc. 2.207 eV	**5.**92a	[Zai00a]

ML2212a = ML2049z	2.212	17.84	560.5	**He** 2.212 eV = DAP95z	*V_1He_1° (a)	**II**(He)+**HT**(>600° C), ZPL of DAP95a-f(2.128 eV), line (b) at 2.316 eV	**5**.98, **5**.99 **5**.100, **7**.105	[Tka85b, Zai92a] [Zai98b, Zai92a]
ML2012f	2.215	17.87	559.7	DAP10f	...	**II**+**HT**(>1,300° C), $s = 16$, calc. 2.215 eV	**7**.84	[Var89, Zai01]
ML2075e	2.220	17.91	558.5	DAP89e	*$(C_2)_iN_2^-$	**HT**(>1,600° C), $s = 54$, calc. 2.220 eV	...	[Law96]
ML2170c	2.220	17.91	558.5	DAP92c	*$V_1Ni_3^+$	**HT**(2,200° C), $s = 58$, calc. 2.217 eV	**5**.97	[Kup99]
ML2179c	°2.224	17.94	557.5	DAP13c	He+?	$s = 86$, calc. 2.222 eV	**5**.99, **7**.105	[Zai92a, Zai01]
ML2226a	2.226	17.96	556.9	DAP12a	...	**II**(C) +**HT**(500° C), or **IE**(no HT), low N diam.; 6 lines a–f (2.440 eV), $s = 228$, calc. 2.229 eV; standard DAP with L = 2.140 eV	**7**.96	[Zai92a, Zai01]
ML2370a-s1	2.226	17.96	556.9	S4a-s1	$(V_2Ni_1)N_3$	SB(−144 meV) = 31(QLVM) + 118(LVM)	...	
ML2179d	2.230	17.99	556.0	DAP13d	*V_1He_1° (b)	**II**(He)+**HT** (550° C), $s = 34$, calc. 2.231 eV	**5**.99, **7**.105	2× [Zai92a]

Treatment: IE irradiation, electrons; *IG* irradiation, general; *II* irradiation, ions; *IN* irradiation, neutrons; *IX* irradiation, X-ray; *HP* hydrogen plasma; *HT* heat treatment; *PD* plastic deformation; *PO* powder, *PO(KBr) compressed with KBr*

Table 5.3.7.2 Modified diamond (irradiation, heat etc.): luminescence lines, visible: green, part 2 (2.23–2.27 eV)

Line-label	Energy (eV)	Frequ. ($10^3 cm^{-1}$)	Wavel. (nm)	Name	Impur. /defect	Comment	Figs. **3–6.** [Zai01]; Fig. **7.** [Zai98]	References
ML2170d	2.232	18.00	555.5	DAP92d	$^*V_1Ni_3^+$	**HT**(2,200°C), $s = 72$, calc. 2.233 eV	**5.**97	[Kup99]
ML2210b	2.234	18.02	555.0	...	...	**II+HT**(>400)	...	[Var86a, Zai01]
ML2066f	2.234	18.02	555.0	DAP14f	$^*(C_2)_iNi_1{}^\circ$	**II+HT**(>700°C), $s = 26$, 2.229 eV	**7.**90	[Zai92a]
ML2234	2.234	18.02	554.9	...	...	**II**(H), also in LL	...	[Zai01]
ML2234	2.234	18.02	554.9	...	...	**IG+HT**(>400°C)	...	[Var86a, Nis89]
ML2237a	2.237	18.04	554.2	**H3a** = DAP51a	$^*V_1N_2{}^\circ$ (a)	**IG+HT**(>500°C), 6 lines a–f(2.323 eV), $s = 42$, calc. 2.238 eV	Table 5.5	[Zai01], p.268
ML2117d	2.238	18.05	554.0	**Zn**-DAP114d	$^*V_1Zn_1{}^\circ$	**II**(Zn)+**HT**(1,400°C), $s = 58$, calc. 2.232 eV	**5.**101	[Zai00a]
ML2012g	2.239	18.06	553.7	DAP10g	...	**II+HT**(>1,300°C), $s = 14$, calc. 2.240 eV	**7.**84	[Var89, Zai01]
ML2138c	2.239	18.06	553.7	DAP86c	$^*(C_2)_i^+$	**IG**, $s = 26$, calc. 2.239 eV	**5.**105	[Ste99a]
ML2070d	2.240	18.07	553.5	DAP1d	$^*N_1{}^\circ + B_1{}^\circ$	**II**(see line a), $s = 13$; calc. 2.239 eV	**5.**92a	[Zai00a]
ML1750f	2.240	18.07	553.5	DAP73f	$^*V_1B_1{}^\circ$ (a)	**IX** of type IIb; $s = 50$, calc. 2.237	...	[Dea65a]
ML2135c	2.244	18.10	552.5	DAP104c	$^*(V_2Co_1)N_x$	**HT**(1,800°C), $s = 20$, calc. 2.245 eV	**5.**72	[Law96]

ML2179e	2.249	18.14	551.2	DAP13e	$^*V_1He_1{}^\circ$ (b)	**II**(He)+**HT** (550°C), $s = 42$, calc. 2.253 eV	**5**.99, **7**.105	2× [Zai92a]
ML2250	2.250	18.15	551.0	...	...	**HT** (>1,800°C) CVD,	**5**.21	[McC95, Zai01]
ML2138c	2.252	18.16	550.5	DAP86c	$^*(C_2)_i^+$	**IG,** $s = 30$, calc. 2.255 eV	**5**.105	[Ste99a]
ML2170e	2.256	18.20	549.5	DAP92e	$^*V_1Ni_3^+$	**HT**(2,200°C), $s = 106$, calc. 2.257 eV	**5**.97	[Kup99]
ML2117e	2.260	18.23	548.5	**Zn**-DAP114e	$^*V_1Zn_1{}^\circ$	**II**(Zn)+**HT**(1,400°C), $s = 86$, calc. 2.261 eV	**5**.101	[Zai00a]
ML2261	2.261	18.24	548.3	...	...	**HT**(2,000°C)+**IN**	...	[Osv97, Zai01]
ML2226b	2.262	18.24	548.1	DAP12b	...	**II** (see line a); $s = 118$, calc. 2.264 eV	**7**.96	[Zai92a]
ML2075f	2.265	18.27	547.4	DAP89f	$^*(C_2)_iN_2^-$	**HT**(>1,600°C), $s = 106$, calc. 2.262 eV	...	[Law96]
ML2266	2.266	18.28	547.1	...	...	**IG**	...	[Ste99a, Zai01]
ML2267	2.267	18.29	546.9	...	...	**HT**(>1,300°C) or **II**	**5**.88a	[Zai96b, Zai01]
ML2138d	2.268	18.29	546.6	DAP86d	$^*(C_2)_i^+$	**IG,** $s = 34$, calc. 2.268 eV	**5**.105	[Ste99a]

Treatment: IE irradiation, electrons; *IG* irradiation, general; *II* irradiation, ions; *IN* irradiation, neutrons; *IX* irradiation, X-ray; *HP* hydrogen plasma; *HT* heat treatment; *PD* plastic deformation; *PO* powder, *PO(KBr) compressed with KBr*

Table 5.3.7.3 Modified diamond (irradiation, heat etc.): luminescence lines, visible: green, part 3 (2.27–2.30 eV)

Line-label	Energy (eV)	Frequ. ($10^3 cm^{-1}$)	Wavel. (nm)	Name	Impur. /defect	Comment	Figs. **3–6.** [Zai01]; Fig. **7.** [Zai98]	References
ML2179f	2.271	18.31	546.0	DAP13f	$^*V_1He_1{}^\circ$ (b)	**II**(He)+**HT** (550°C), $s = 50$, calc. 2.272 eV	**5**.99, **7**.105	2× [Zai92a, Zai01]
ML2066g	2.271	18.31	546.0	DAP14g	$^*(C_2)_iNi_1{}^\circ$	**II**+**HT**(>700°C), $s = 42$, 2.273 eV	**7**.90	[Zai92a]
ML2271a	2.271	18.32	545.9	**H4b-** DAP38a	$^*N_3V_2N_1{}^\circ$	**IG**+**HT**(>600°C), 7 lines a–g(2.378 eV), $s = 34$, calc. 2.261 eV	Table 5.6	[Zai01, p. 281]
ML2272	2.272	18.33	545.7	...	...	**II** (C or H), var. 2.269–2.274 eV,	**5**.55, **5**.88b, c **7**.98	2× [Zai96b] [Gip83]
ML2179g	2.274	18.34	545.2	DAP13g	$^*V_1He_1{}^\circ$ (b)	**II**(He)+**HT** (550°C), $s = 52$, calc. 2.275 eV	**5**.99, **7**.105	2× [Zai92a, Zai01]
ML2237b	2.275	18.35	545.0	**H3a-** DAP51b	$^*V_1N_2{}^\circ$ (a)	**IG**+**HT**(>500°C), $s = 54$, calc. 2.276 eV	Table 5.5	[Zai01]
ML2277d	2.277	18.37	544.5	DAP104d	$^*(V_2Co_1)N_x$	**HT**(1,800°C), $s = 16$, calc. 2.278 eV; split-off lines at 2.283 and 2.291 eV from exc. st. spl. [Law96]	**5**.72	[Law96]
ML2041d	2.280	18.39	543.8	DAP75d	$^*(V_3Si_2)^-$	**II**(Si), lines a–c see NL; $s = 20$, calc. 2.280 eV	**5**.120a	[Naz92, Zai01]
ML2170f	2.280	18.39	543.8	DAP92f	$^*V_1Ni_3^+$	**HT**(2,200°C), $s = 162$, calc. 2.279 eV	**5**.97	[Kup99]

ML2070e	2.284	18.42	542.8	DAP1e	$^*N_1{}^\circ + B_1{}^\circ$	**II**(see line a), $s = 10$; calc. 2.285 eV	**5**.68a, **5**.92a	[Zai98d], [Zai00a]
ML2179h	2.286	18.44	542.3	DAP13h	$^*V_1He_1{}^\circ$ (b)	**II**(He)+**HT** (550°C), $s = 58$, calc. 2.285 eV	**5**.99, **7**.105	2× [Zai92a, Zai01]
ML2286	2.286	18.44	542.3	...	...	**IG** (+**HT**(1,950°C)), var. 2.286–2.291 eV	**5**.55	[Zai96b]
ML1696-o	2.287	18.45	542.1	DAP70-o	$^*N_1^{2+} + B_1{}^\circ$	**IE**(see line a), $s = 3$, calc. 2.281 eV	...	[Ste99c]
ML2117f	2.287	18.54	542.0	**Zn**-DAP114f	$^*V_1Zn_1{}^\circ$	**II**(Zn)+**HT**(1,400°C), $s = 130$, calc. 2.286 eV	**5**.101	[Zai00]
ML2075g	2.288	18.46	541.9	DAP89g	$^*(C_2)_iN_2^-$	**HT**(>1,600°C), $s = 198$, calc. 2.289 eV	...	[Law96]
ML2012h	2.293	18.50	540.7	DAP10h	...	**II**+**HT**(>1,300°C), $s = 11$, calc. 2.294 eV	**7**.84	[Var89, Zai01]
ML2138e	2.297	18.53	539.7	DAP86e	$^*(C_2)_i^+$	**IG,** $s = 48$, calc. 2.296 eV	**5**.105	[Ste99a]
ML2298 = ML1905z	2.298	18.54	539.5	DAP28z	$^*V_1Ni_2^-$(b)	**HT** (>1,500°C) or **II**+**HT**(>600°C), var. 2.292–2.298 eV, ZPL of DAP28a-m (2.210 eV); SBs: −29, −55 meV	**5**.92a, **7**.66	[Zai00a], [Fil92]
ML2237c	2.299	18.54	539.3	**H3a**–DAP51c	$^*V_1N_2{}^\circ$ (a)	**IG**+**HT**(>500°C), $s = 78$, calc. 2.298 eV	Tables 5.5 **7**.71	[Zai01] [Gip82a]

Treatment: IE irradiation, electrons; *IG* irradiation, general; *II* irradiation, ions; *IN* irradiation, neutrons; *IX* irradiation, X-ray; *HP* hydrogen plasma; *HT* heat treatment; *PD* plastic deformation; *PO* powder, *PO(KBr) compressed with KBr*

Table 5.3.7.4 Modified diamond (irradiation, heat etc.): luminescence lines, visible: green, part 4 (2.30–2.34 eV)

Line-label	Energy (eV)	Frequ. (10^3cm^{-1})	Wavel. (nm)	Name	Impur. /defect	Comment	Figs. **3–6.** [Zai01]; Fig. **7.** [Zai98]	References
ML2129b	2.301	18.56	538.8	...	...	**IG**(+HT >300°C), $s = 12$, calc. 2.313 eV	...	[Dav81b]
ML2302	2.302	18.57	538.6	...	...	**IE** of CVD diamond	...	[Fie92, Zai01]
ML2237dc	2.307	18.61	537.4	**H3a**-DAP51d	*V_1N_2° (a)	**IG**+**HT**(>500°C), $s = 86$, calc. 2.307 eV	Table 5.5	[Zai01, (p. 268)]
ML2307	2.307	18.61	537.4	...	N?	**HT**(1,700°C)	...	[Yel99, Zai01]
ML2271b	2.310	18.63	536.7	**H4b**-DAP38b	*$N_3V_2N_1$°	**IG**+**HT**(>600°C), $s = 54$, calc. 2.309 eV	Table 5.6	[Zai01, (p.281)]
ML2179-i	2.311	18.64	536.5	DAP13-i	*V_1He_1° (b)	**II**(He)+**HT** (550°C), $s = 78$, calc. 2.310 eV	**5**.99, **7**.105	2× [Zai92a, Zai01]
ML2313	2.313	18.66	536.0	...	...	**IG**+**HT**(1,300°C)	...	[Zai01]
ML2237e	2.314	18.66	535.8	**H3a**-DAP51e	*V_1N_2° (a)	**IG**+**HT**(>500°C), $s = 98$, calc. 2.317 eV	Table 5.5	[Zai01]
ML2138g	2.314	18.66	535.8	DAP86g	*$(C_2)_i^+$	**IG,** $s = 58$, calc. 2.313 eV		[Ste99a]
ML2316	2.316	18.68	535.3	...	Ni?	**HT**(1,700°C)	...	[Yel99, Zai01]
ML2212b	2.316°	18.68	535.2	***He** 2.32 eV	He_1N_1?	**II**(He)+**HT**(>600°C)	**5**.98, **5**.99 **5**.100, **7**.105	[Tka85b, Zai92a] [Zai88a, Zai92a]
ML2317	2.317	18.69	535.1	...	...	**IN**+**HT**(400°C)	...	[Nis89, Zai01]

ML2320	2.320	18.71	534.4	...	...	**IX**(γ -ray)	...	[Vac75c]
ML2237f	2.323	18.74	533.7	**H3a**-DAP51f	$*V_1N_2^{\circ}$ (a)	**IG+HT**(>500°C), $s = 106$, calc. 2.322 eV	Table 5.5	[Zai01]
ML2329	2.329	18.79	532.3	...	...	**II** or **HT**(>500°C), var. 2.322–2.335 eV, 2 lines in some spectra	**5**.88b, c, **5**.115 **7**.81, **7**.102	2× [Zai96b], 2× [Rua91b]
ML2135e	2.330	18.79	532.1	DAP104e	$*(V_2Co_1)N_x$	**HT**(1,800°C), $s = 12$, calc. 2.330 eV	**5**.72	[Law96]
ML2226c	2.330	18.79	532.1	DAP12c	...	**II** (see line a); $s = 50$, calc. 2.330 eV	**7**.96	[Zai92a, Zai01]
ML2070f	2.331	18.80	531.9	DAP1f	$*N_1^{\circ}+B_1^{\circ}$	**II**(see line a), $s = 8$; calc. 2.331 eV	**5**.68a, **5**.92a	[Zai98c], [Zai00a]
ML2129c	2.331	18.80	531.9	**TR12**-DAP36c	$(C_2)_i^-$	**IG**(+HT >300°C), $s = 14$, calc. 2.338 eV	**5**.55	[Zai96b, Dav81b]
ML2271c	2.335	18.83	531.0	**H4b**-DAP38c	$*N_3V_2N_1^{\circ}$	**IG+HT**(>600°C), $s = 72$, calc. 2.335 eV	Table 5.6	[Zai01]
ML2041e	2.335	18.83	531.0	DAP75e	$*(V_3Si_2)^-$	**II**(Si), $s = 34$, calc. 2.338 eV	**5**.120a	[Naz92, Zai01]
ML2138h	2.340	18.87	529.8	DAP86h	$*(C_2)_i^+$	**IG,** $s = 86$, calc. 2.340 eV	...	[Ste99a]

Treatment: IE irradiation, electrons; *IG* irradiation, general; *II* irradiation, ions; *IN* irradiation, neutrons; *IX* irradiation, X-ray; *HP* hydrogen plasma; *HT* heat treatment; *PD* plastic deformation; *PO* powder, *PO(KBr) compressed with KBr*

Table 5.3.7.5 Modified diamond (irradiation, heat etc.): luminescence lines, visible: green, part 5 (2.34–2.38 eV)

Line-label	Energy (eV)	Frequ. ($10^3 cm^{-1}$)	Wavel. (nm)	Name	Impur. /defect	Comment	Figs. **3–6.** [Zai01]; Fig. **7.** [Zai98]	References
ML2179j	2.347	18.93	528.2	DAP13j	$*V_1He_1°$ (b)	**II**(He)+**HT** (550°C), $s = 130$, calc. 2.345 eV	**5**.99, **7**.105	2× [Zai92a, Zai01]
ML2271d	2.347	18.93	528.2	**H4b**-DAP38d	$*N_3V_2N_1°$	**IG**+**HT**(>600°C), $s = 86$, calc. 2.349 eV	Table 5.6	[Zai01]
ML1696p	2.348	18.94	528.0	DAP70p	$*N_1^{2+} + B_1°$	**IE**(see line a), $s = 2$, calc. 2.330 eV	...	[Ste99d]
ML2355	2.355	19.00	526.4	...	...	**IG**	...	[Ste99a, Zai01]
ML2138i	2.356	19.00	526.2	DAP86i	$*(C_2)_i^+$	**IG**, $s = 118$, calc. 2.357 eV	...	[Ste99a]
ML2041f	2.357	19.01	526.0	DAP75f	$*(V_3Si_2)^-$	**II**(Si), $s = 42$, calc. 2.355 eV	**5**.120a	[Naz92, Zai01]
ML2179k	2.358	19.02	525.7	DAP13k	$*V_1He_1°$ (b)	**II**(He)+**HT** (550°C), $s = 162$, calc. 2.358 eV	**5**.99, **7**.105	2× [Zai92a, Zai01]
ML2129d	2.360	19.04	525.3	**TR12**-DAP36d	$(C_2)_i^-$	**IG**(+HT >300°C), $s = 16$, calc. 2.357 eV	...	[Dav81b]
ML2271e	2.361	19.04	525.1	**H4b**-DAP38e	$*N_3V_2N_1°$	**IG**+**HT**(>600°C), $s = 98$, calc. 2.359 eV	Table 5.6	[Zai01]
ML2365	2.365	19.08	524.2	...	...	**IN** of type Ib diamond + **HT**(550°C)	...	[Nis89, Zai01]
ML2135f	2.367	19.09	523.9	DAP104f	$*(V_2Co_1)N_x$	**HT**(1,800°C), $s = 10$, calc. 2.367 eV; split-off lines at 2.378 and 2.385 eV; coincidence	**5**.72	[Law96]

ML2367 = ML2075z	2.367	19.09	523.9	DAP89z	$^{*}(C_2)_iN_2^-$	**HT**(>1,600°C), ZPL of DAP89a-g (2.288 eV), in Co free sample [Col85b], coincidence	. . .	[Col85b, Law96]
ML2271f	2.367	19.09	523.9	**H4b**-DAP38f	$^{*}N_3V_2N_1{}^{\circ}$	**IG**+**HT**(>600°C), $s = 106$, calc. 2.364 eV	Table 5.6	[Zai01]
ML2041g	2.368	19.10	523.6	DAP75g	$^{*}(V_3Si_2)^-$	**II**(Si), $s = 50$, calc. 2.371 eV	**5.**120a	[Naz92, Zai01]
ML2370a	2.370	19.12	523.1	DAP92f	$^{*}V_1Ni_3^+$	**HT**(2,200°C), ZPL of DAP92a-f (2.280 eV), formerly "A" line of S2, NE3 in EPR	**5.**97	[Kup99]
ML2226d	2.371	19.12	522.9	DAP12d	. . .	**II** (see line a); $s = 34$, calc. 2.370 eV	**7.**96	[Zai92a]
ML2070g	2.373	19.14	522.4	DAP1g	$^{*}N_1{}^{\circ} + B_1{}^{\circ}$	**II**(see line a), $s = 7$; calc. 2.369 eV	**5.**68a, **5.**92a	[Zai98d], [Zai00a]
ML2374	2.374	19.15	522.2	. . .	. . .	**HT**(1,700°C)	. . .	[Zai01]
ML2179-1	2.375	19.16	522.0	DAP13-1	$^{*}V_1He_1{}^{\circ}$ (b)	**II**(He)+**HT** (550°C), $s = 228$, calc. 2.375 eV	**5.**99, **7.**105	2× [Zai92a, Zai01]
ML2129e	2.376	19.16	521.8	**TR12**-DAP36e	$(C_2)_i^-$	**IG**(+HT >300°C), $s = 18$, calc. 2.373 eV	**5.**55	[Zai96b, Dav81b]
ML2271g	2.378	19.18	521.3	**H4b**-DAP38g	$^{*}N_3V_2N_1{}^{\circ}$	**IG**+**HT**(>600°C), $s = 130$, calc. 2.376 eV	Table 5.6	[Zai01]

Treatment: IE irradiation, electrons; *IG* irradiation, general; *II* irradiation, ions; *IN* irradiation, neutrons; *IX* irradiation, X-ray; *HP* hydrogen plasma; *HT* heat treatment; *PD* plastic deformation; *PO* powder, *PO(KBr) compressed with KBr*

Table 5.3.7.6 Modified diamond (irradiation, heat etc.): luminescence lines, visible: green, part 6 (2.38–2.43 eV)

Line-label	Energy (eV)	Frequ. ($10^3 cm^{-1}$)	Wavel. (nm)	Name	Impur. /defect	Comment	Figs. **3–6.** [Zai01]; Fig. **7.** [Zai98]	References
ML2393	2.393	19.30	518.1	...	Ne?	**II**(Ne) of type IIa diamond	**7.**72b	[Zai92a]
ML2394 = ML2117z	2.394	19.31	518.0	**Zn**-DAP114z	$^*V_1Zn_1{}^\circ$	**II**(Zn)+**HT**(1,400° C), ZPL of DAP114a-f(2.287 eV), SBs: QLVM 2× 32 meV, calc. 1Zn+2C: 31.8 meV	**5.**101, **7.**63b	[Zai00a], [Zai92a]
ML1750z	2.395	19.32	517.7	DAP73z	$^*V_1B_1{}^\circ$ (a)	**IX,** ZPL of DAP73a-f(2.240 eV)	...	[Dea65]
ML2129f	2.404	19.39	515.7	**TR12**-DAP36f	$(C_2)_i^-$	**IG**(+HT >300°C), $s = 24$, calc. 2.408 eV	...	[Dav81b]
ML2041h	2.410	19.44	514.4	DAP75h	$^*(V_3Si_2)^-$	**II**(Si), $s = 98$, calc. 2.414 eV	**5.**120a	[Naz92, Zai01]
ML2070h	2.413	19.46	513.8	DAP1h	$^*N_1{}^\circ + B_1{}^\circ$	**II**(see line a), $s = 6$; calc. 2.398 eV	**5.**68a, **5.**92a	[Zai98d], [Zai00a]
ML2415	2.415	19.48	513.3	**He** 2.42 eV	He	**II**(He)+**HT** (>700°C)	**5.**99, **5.**100	[Zai92a], [Zai88a]
ML1696q	2.417	19.49	513.0	DAP70q	$^*N_1^{2+} + B_1{}^\circ$	**IE**(see line a), $s = 1$, calc. 2.418 eV	...	[Ste99c]

ML2417a	2.417	19.50	512.9	**H4a**	$^*N_3V_2N_1{}^\circ$	**IG+HT**(>600°C), 10× weaker than line (b) 2.498 eV; ZPL	**5**.45, **7**.116b	[Zai01, DeS77]
ML2129g	2.420	19.52	512.3	**TR12**-DAP36g	$(C_2)_i^-$	**IG**(+HT >300°C), $s = 26$, calc. 2.417 eV	**5**.127	[Zai96b, Dav81b]
ML2226e	2.427	19.58	510.8	DAP12e	...	**II** (see line a); $s = 22$, calc. 2.426 eV	**7**.96	[Zai92a, Zai01]
ML2129h	2.428	19.58	510.6	**TR12**-DAP36h	$(C_2)_i^-$	**IG**(+HT >300°C), $s = 30$, calc. 2.433 eV	**5**.55, **5**.127	[Zai96b, Dav81b]
ML2428	2.428	19.58	510.6	Ni1.66b	$^*V_1Ni_2{}^\circ$(b)	**IG or HT**(1,600°C), var. 2.421–2.429 eV	**5**.92a, **5**.127, **7**.90	[Zai00a], [Zai96b], [Zai92a]
ML2429a	2.429	19.59	510.4	**S1a**	$^*V_1Ni_2^+$(a)	**HT**(1,900°C), 2 lines a, b (2.463 eV); a-line 20× weaker than b; NL in Fig. **7**.110 [Sob69c]	...	[Zai01]

Treatment: IE irradiation, electrons; *IG* irradiation, general; *II* irradiation, ions; *IN* irradiation, neutrons; *IX* irradiation, X-ray; *HP* hydrogen plasma; *HT* heat treatment; *PD* plastic deformation; *PO* powder, *PO(KBr) compressed with KBr*

Table 5.3.8.1 Modified diamond (irradiation, heat etc.): luminescence lines, visible: blue, part 1 (2.43–2.49 eV)

Line-label	Energy (eV)	Frequ. ($10^3 cm^{-1}$)	Wavel. (nm)	Name	Impur. /defect	Treatment/comment	Figs. **3–6.** [Zai01]; Fig. **7.** [Zai98]	References
ML2066z	2.436	19.64	509.0	DAP14z	$*(C_2)Ni_1{}^\circ$	**II**+**HT**(>700°C), ZPL of DAP 14a–g(2.066–2.271 eV)	**7.**90	[Zai92a]
ML2436	2.436	19.64	509.0	...	...	**II**(**C**)	**5.**55	[Zai96b]
ML2041i	2.437	19.66	508.7	DAP75i	$*(V_3Si_2)^-$	**II**(Si), $s = 162$, calc. 2.438 eV	**5.**120a	[Naz92, Zai01]
ML2226f	2.440	19.68	508.1	DAP12f	...	**II** (see line a); $s = 20$, calc. 2.440 eV	**7.**96	[Zai92a, Zai01]
ML2041j	2.450	19.76	506.0	DAP75j	$*(V_3Si_2)^-$	**II**(Si), $s = 228$, calc. 2.452 eV	**5.**120a	[Naz92, Zai01]
ML2070-i	2.456	19.81	504.8	DAP1-i	$*N_1{}^\circ + B_1{}^\circ$	**II**(see line a), $s = 5$; calc. (incl. NSC) 2.444 eV	**5.**92°	[Zai00a]
ML2129-i	2.461	19.85	503.8	**TR12**-DAP36-i	$(C_2)_i^-$	**IG**(+HT >300°C), $s = 40$, calc. 2.460 eV	...	[Dav81b]
ML2462 = ML2138z	2.462	19.86	503.6	**3H** (name indicates closeness to H3)	$*(C_2)_i^+$	**IG** (also **IE** at 6K), typical radiation center, ZPL of DAP86a-i (2.356 eV), SBs: −74, −137, −169, −182, −218(−208, −213, −218;50%^{13}C) meV	**5.**53a, **5.**55 **5.**88a, **5.**105 **7.**96, **7.**133	[Zai89, Zai96b] [Zai96b, Ste99a] [Zai92a], [Col90c]
ML2429b	2.463	19.87	503.4	**S1b**	$*V_1N_2^+(b)$	**HT**(1,900°C), ZPL of DAP43a-g (see NL)	...	[Zai01]

ML2464a = ML2237z	2.464	19.88	503.2	**H3(a)** = DAP51z	$^{*}V_1N_2{}^{\circ}$	**IG**+**HT**(>500°C), line b(MA) 3.361 eV; ZPL of DAP51a-f(2.323 eV), SBs:3× −41, −152 meV	**5**.106, **5**.110b **7**.112	[Zai82, Mel98] [Col92b, Zai01]
ML2129j	2.464	19.88	503.2	**TR12**-DAP36j	$(C_2)_i^-$	**IG**(+HT >300°C), $s = 42$, calc. 2.464 eV	...	[Dav81b]
ML2466a	2.466	19.89	502.7	...	O?	**II**(O)+**HT** (>1,000°C), 2 lines a, b (2.488 eV)	...	[Gip93b, Zai01]
ML2212b = ML2179z	2.466	19.89	502.7	DAP13z	$^{*}V_1He_1{}^{\circ}$ (b)	**II**(He)+**HT** (550°C), weak ZPL of DAP13a-l	**7**.105	[Zai92a]
ML2473	2.473	19.95	501.3	...	B?	**IG** in CVD films var. 2.466–2.474 eV	**5**.88b, c, **5**.115	2× [Zai96b]
ML2476a	2.476	19.97	500.7	DAP74a	$^{*}V_1B_1{}^{\circ}$ (b)	**IX** of type IIb, 6 lines a–f(2.966 eV), $s = 2$, calc. 2.467 eV (structure on broad band)	...	[Dea65a]
ML2481	2.481	20.01	499.7	...	Ni?	**HT**(1,700°C)	...	[Yel99, Zai01]
ML2129k	2.485	20.04	498.9	**TR12**-DAP36k	$(C_2)_i^-$	**IG**(+HT >300°C), $s = 54$, calc. 2.485 eV	...	[Dav81b]
ML2465b	2.488	20.06	498.4	...	O?	**II**(O)+**HT** (>1,000°C)	...	[Gip93b, Zai01]

Treatment: IE irradiation, electrons; *IG* irradiation, general; *II* irradiation, ions; *IN* irradiation, neutrons; *IX* irradiation, X-ray; *HP* hydrogen plasma; *HT* heat treatment; *PD* plastic deformation; *PO* powder, *PO(KBr) compressed with KBr*

Table 5.3.8.2 Modified diamond (irradiation, heat etc.): luminescence lines, visible: blue, part 2 (2.49–2.53 eV)

Line-label	Energy (eV)	Frequ. ($10^3 cm^{-1}$)	Wavel. (nm)	Name	Impur. /defect	Comment	Figs. **3–6.** [Zai01]; Fig. **7.** [Zai98]	References
ML2070j	2.495	20.12	496.9	DAP1j	$^*N_1{}^\circ + B_1{}^\circ$	**II**(see line a), $s = 4$; calc. (incl. NSC) 2.481 eV	**5**.68a, **5**.92a	[Zai98d], [Zai00a]
ML2496a	2.496	20.13	496.7	**S3**-DAP58a	$(V_2Ni_1)N_2$	**II**(Ni)+**HT**(1,400° C) or **HT** (>1,800° C), 4 lines a, d, e, f, others see ME; $s = 162$, calc. 2.494 eV	**5**.92a, **5**.97 **5**.116b	[Zai00a], [Kup99] [Kan98]
ML2417b = ML2271z	2.498	20.15	496.3	**H4b**-DAP38z	$^*N_3V_2N_1{}^\circ$	**IG**+**HT**(>600° C), line (a) 2.417 eV; ZPL of DAP38a-g (2.378 eV), SBs: −37, −154 meV	**5**.45, **7**.116b	[Zai01, DeS77]
ML2129-l	2.498	20.15	496.3	**TR12**-DAP36-l	$(C_2)_i^-$	**IG**(+HT >300° C), $s = 64$, calc. 2.498 eV	**5**.55	[Zai96b, Dav81b]
ML2499	2.499	20.16	496.1	…	…	…	…	
ML2500	2.500	20.17	495.8	…	…	gamma-ray-luminescence	…	[Vac75c]
ML2129m	2.505	20.20	494.9	**TR12**-DAP36m	$(C_2)_i^-$	**IG**(+HT >300° C), $s = 72$, calc. 2.505 eV	…	[Dav81b]
ML2507	2.507	20.22	494.5	…	Ti	…	…	

ML2515a	2.515	20.29	493.0	**S2**-DAP57a	$(V_2Ni_1)N_3^+$	**HT**(>1,800°C) or **II**(Ni)+**HT**(1,400°C),11 lines DAP57a-l (2.681 eV), L = 2.422 eV; lines (m–o) see ME; $s = 228$, calc. 2.516 eV	**5**.92b	[Zai00a]
ML2129n	2.519	20.32	492.2	**TR12**-DAP36n	$(C_2)_i^-$	**IG**(+HT >300°C), $s = 86$, calc. 2.517 eV	**5**.55	[Zai96b, Dav81b]
ML2520	2.520	20.33	492.0	...	...	**II**(**C**) of Type IIa diamond	**5**.88a	[Zai96b, Zai01]
ML2523 = ML2041z	2.523	20.35	491.4	DAP75z	$^*(V_3Si_2)^-$	**II**(Si), ZPL of DAP75a-j (2.450 eV), lines a–c see NL; natural (brown) diamond	**5**.120a	[Naz92, Zai01]
ML2524	2.524	20.36	491.2	...	...	**IN**+**HT** (600°C)	...	[Zai01]
ML2526 = ML2164z	2.526	20.38	490.8	DAP76z	$^*V_1N_3^-$	**IG (PD)**, var. 2.523–2.529 eV; ZPL of DAP76a-m(2.457 eV), see NL2164a-m; no ML spectra	...	[Zai01]
ML2530	2.530	20.41	490.0	...	...	**IX**, gamma-ray-luminescence	...	[Vac75c]

Treatment: IE irradiation, electrons; *IG* irradiation, general; *II* irradiation, ions; *IN* irradiation, neutrons; *IX* irradiation, X-ray; *HP* hydrogen plasma; *HT* heat treatment; *PD* plastic deformation; *PO* powder, *PO(KBr) compressed with KBr*

Table 5.3.8.3 Modified diamond (irradiation, heat etc.): luminescence lines, visible: blue, part 3 (2.53–2.58 eV)

Line-label	Energy (eV)	Frequ. (10^3cm^{-1})	Wavel. (nm)	Name	Impur. /defect	Comment	Figs. **3–6.** [Zai01]; Fig. **7.** [Zai98]	References
ML2535	2.535	20.45	489.1	DAP90z	$^*(C_2)_iN_2^+$	**HT** (1,700°C), in absorption ZPL of DAP90a-c	...	[Zai01]
ML2535g	2.535	20.45	489.1	DAP104g	$^*(V_2Co_1)N_x$	**HT**(1,700°C), $s = 5$, calc. 2.537 eV	...	[Law96, (Fig. 6)]
ML2515b	2.537	20.46	488.7	**B = S2-** DAP57b	$(V_2Ni_1)N_3^+$	**HT, II**(see line a); var. 2.534–2.537 eV; $s = 162$, calc. 2.533 eV	**5.**92b, **5.**97 **5.**116, **7.**86	[Zai00a], [Kup99] [Kan98, Col85b]
ML2012-i	2.538	20.47	488.5	DAP10-i	...	**II+HT**(>1,300°C), $s = 4$, calc. 2.538 eV	**7.**84	[Var89, Zai01]
ML2553	2.553	20.59	485.6	...	...	**IG** (heavy irradiation)	**5.**53b, **5.**88a	[Zai89, Zai96b]
ML2135g	2.556	20.62	485.0	DAP104g	$^*(V_2Co_1)N_x$	**HT**(1,700°C), $s = 26$, calc. 2.208 eV	...	[Law96, (Fig. 6)]
ML2562	2.562	20.67	484.0	DAP106z	$^*(V_3Ni_2)^+$	**II**(Ni)+**HT**(1,400°C), ZPL of DAP106a-f (see HL), named 2.56 eV or 484 nm center	**5.**92a	[Zai00a]
ML2515c	2.564	20.68	483.5	**S2-**DAP57c	$(V_2Ni_1)N_3^+$	**HT, II**(see line a); $s = 98$, calc. 2.565 eV	**5.**92b, **5.**97 **5.**116	[Zai00a], [Kup99] [Kan98, Zai01]
ML2566	2.566	20.70	483.2	...	V?	**II** (C) of CVD diamond	**5.**88b, c	[Zai96b, Zai01]
ML2070k	2.568	20.71	482.8	DAP1k	$^*N_1{}^\circ + B_1{}^\circ$	**II**(see line a), $s = 3$; calc. 2.565 eV	**5.**92a	[Zai00a]
ML2576	2.576	20.78	481.2	...	...	**II** (C) of CVD diamond	**5.**115	[Zai96b, Zai01]

Treatment: IE irradiation, electrons; *IG* irradiation, general; *II* irradiation, ions; *IN* irradiation, neutrons; *IX* irradiation, X-ray; *HP* hydrogen plasma; *HT* heat treatment; *PD* plastic deformation; *PO* powder, *PO(KBr) compressed with KBr*

Table 5.3.9.1 Modified diamond (irradiation, heat etc.): luminescence lines, visible: ultramarine, part 1 (2.58–2.64 eV)

Line-label	Energy (eV)	Frequ. ($10^3 cm^{-1}$)	Wavel. (nm)	Name	Impur. /defect	Treatment/comment	Figs. **3–6.** [Zai01]; Fig. **7.** [Zai98]	References
ML2476b	2.590	20.89	478.7	DAP74b	$^*V_1B_1{}^\circ$ (b)	**IX** of type IIb, $s = 4$, calc. 2.586 eV	...	[Dea65a]
ML2590	2.590	20.89	478.7	...	$^*V_1Co_2^-$	**II**(Co)+**HT**(600° C) or **HT**(1,500° C)	...	[Zai01]
ML2515d	2.591	20.90	478.5	**S2**-DAP57d	$(V_2Ni_1)N_3^+$	**HT, II**(see line a); $s = 72$, calc. 2.588 eV	**5.**92b	[Zai00a]
ML2515e	2.597	20.95	477.4	**C = S2**-DAP57e	$(V_2Ni_1)N_3^+$	**HT, II**(see line a); $s = 64$, calc. 2.598 eV	**5.**92b, **5.**97 **5.**116	[Zai00a], [Kup99] [Kan98, Zai01]
ML2496d	2.599	20.96	477.0	**S3**-DAP58d	$(V_2Ni_1)N_2$	**II**(Ni)+**HT**(1,400° C), $s = 42$, calc. 2.601	**5.**92a	[Zai00a]
ML2599	2.599	20.96	477.0	...	...	**II**(c):**CL**; **IX**: (X ray luminescence)	5.88	[Sob79a], [Che82]
ML2608a	2.608	21.04	475.4	DAP98a	$^*(N_1C_1)_i^+$	**IG** (+**HT**(600° C)), 9 lines a–i(2.739 eV), s = 30, calc. 2.609 eV; lines (f–i) see LL	...	[Col89b, Col93c] [Zai01]
ML2608b	2.616	21.10	473.9	DAP98b	$^*(N_1C_1)_i^+$	**IG** (+**HT**(600° C)), $s = 32$, calc. 2.616 eV	...	[Col89b, Col93c]
ML2496e	2.622	21.15	472.8	**S3**-DAP58e	$(V_2Ni_1)N_2$	**II**(Ni)+**HT**(1,400° C), $s = 34$, calc. 2.624	**5.**92a	[Zai00a]
ML2515f	2.623	21.16	472.7	**D = S2**-DAP57b	$(V_2Ni_1)N_3^+$	**HT, II**(see line a); $s = 50$, calc. 2.621 eV	**5.**92b, **5.**116	[Zai00a], [Kan98]

(continued)

Table 5.3.9.1 (continued)

Line-label	Energy (eV)	Frequ. ($10^3 cm^{-1}$)	Wavel. (nm)	Name	Impur. /defect	Treatment/comment	Figs. **3–6.** [Zai01]; Fig. **7.** [Zai98]	References
ML2070-l	2.623	21.16	472.7	DAP1-l	$*N_1{}^\circ + B_1{}^\circ$	**II**(see line a), $s = 2$; calc. 2.611 eV	**5.**92b	[Zai00a]
ML2626	2.626	21.18	472.1	...	B?	**IN** of B doped CVD diamond	...	[Pop96, Zai01]
ML2515g	2.629	21.21	471.6	**S2**-DAP57g	$(V_2Ni_1)N_3^+$	**HT, II**(see line a); $s = 48$, calc. 2.626 eV	**5.**92b	[Zai00a]
ML2608c	2.630	21.21	471.4	DAP98c	$*(N_1C_1)_i^+$	**IG** (+**HT**(600°C)), $s = 38$, calc. 2.631 eV	...	[Col89b, Col93c]
ML2012j	2.632	21.23	471.0	DAP10j	...	**II**+**HT**(>1,300°C), $s = 3$, calc. 2.636 eV	**7.**84	[Var89, Zai01]
ML2515h	2.634	21.25	470.7	**E** = **S2**-DAP57h	$(V_2Ni_1)N_3^+$	**HT, II**(see line a); $s = 44$, calc. 2.636 eV	**5.**92b	[Zai00a], [Zai01]
ML2608d	2.635	21.25	470.5	DAP98d	$*(N_1C_1)_i^+$	**IG** (+**HT**(600°C)), $s = 40$, calc. 2.636 eV	...	[Col89b, Col93c]
ML2638 = ML2129z	2.638	21.28	470.0	**TR12**-DAP36z	$(C_2)_i^-$	**IG**(+**HT**(>300°C)), ZPL of DAP36a-n(2.519 eV), SBs: −65, −80, −199 meV	**5.**53a, **5.**55, **5.**127, **5.**145a, **7.**127	[Zai89, Zai96b] [Zai96b], [Zai00a] [Wal79, Zai01]
ML2608e	2.639	21.29	469.8	DAP98e	$*(N_1C_1)_i^+$	**IG** (+**HT**(600°C)), $s = 42$, calc. 2.640 eV	...	[Col89b, Col93c]

Treatment: IE irradiation, electrons; *IG* irradiation, general; *II* irradiation, ions; *IN* irradiation, neutrons; *IX* irradiation, X-ray; *HP* hydrogen plasma; *HT* heat treatment; *PD* plastic deformation; *PO* powder, *PO(KBr) compressed with KBr*

Table 5.3.9.2 Modified diamond (irradiation, heat etc.): luminescence lines, visible: ultramarine, part 2 (2.65–2.75 eV)

Line-label	Energy (eV)	Frequ. ($10^3 cm^{-1}$)	Wavel. (nm)	Name	Impur. /defect	Treatment/comment	Figs. **3–6.** [Zai01]; Fig. **7.** [Zai98]	References
ML2650	2.650	21.38	467.8	. . .	. . .	**II**(C) of CVD diamond	**5.**88b, c	[Zai96b]
ML2515i	2.652	21.39	467.5	**S2-**DAP57i	$(V_2Ni_1)N_3^+$	**HT, II**(see line a); $s = 38$, calc. 2.650 eV	**5.**92b	[Zai00a]
ML2515j	2.658	21.44	466.4	**S2-**DAP57j	$(V_2Ni_1)N_3^+$	**HT, II**(see line a); $s = 36$, calc. 2.657 eV	**5.**92b	[Zai00a]
ML2496f	2.663	21.48	465.6	**S3-**DAP58f	$(V_2Ni_1)N_2$	**II**(Ni)+**HT**(1,400°C), $s = 26$, calc. 2.660	**5.**92a	[Zai00a]
ML2070m	2.668	21.52	464.7	DAP1m	$^*N_1{}^\circ + B_1{}^\circ$	**II**(see line a), $s = 1$; var. 2.50–2.68; calc. 2.694 eV	**5.**92a	[Zai00a]
ML2673a	2.673	21.56	463.8	**3.2eV-** DAP72a	$^*(N_1C_1)_i{}^\circ$	**IG** (+**HT**(500°C)), $s = 4$, calc. 2.670 eV	**5.**145	[Zai00a]
ML2515k	2.675	21.58	463.5	**S2-**DAP57k	$(V_2Ni_1)N_3^+$	**HT, II**(see line a); $s = 32$, calc. 2.671 eV	**5.**92b	[Zai00a]
ML2515-l	2.681	21.62	462.4	**S2-**DAP57-l	$(V_2Ni_1)N_3^+$	**HT, II**(see line a); $s = 30$, calc. 2.680 eV	**5.**92b	[Zai00a]
ML2535h	2.682	21.63	462.3	DAP104h	$^*(V_2Co_1)N_x$	**HT**(1,700°C), $s = 3$, calc. 2.667 eV	. . .	[Law96, (Fig. 6)]
ML2012k	2.689	21.69	461.1	DAP10k	. . .	**II**+**HT**(>1,300°C), $s = 2$, calc. 2.689 eV	**7.**84	[Var89, Zai01]
ML2704	2.704	21.81	458.5	. . .	. . .	**IX,** x-ray luminescence, type IIb	. . .	[Bie67]
ML2705a	2.705	21.81	458.3	. . .	. . .	**II**+**HT**(1,400°C); 3 sharp lines a–d(2.930 eV)	**5.**131	[Var86a, Zai01]
ML2708	2.708	21.84	457.8	. . .	Si?	**II**+**HT**(>800°C)	**5.**45	[Zai01]
ML2476c	2.720	21.94	455.8	DAP74c	$^*V_1B_1{}^\circ$ (b)	**IX** of type IIb, $s = 8$, calc. 2.722 eV	. . .	[Dea65a]
ML2440f	2.741	22.11	452.3	**N3c-** DAP24f	V_1N_3	**II** or**HT**(see line z), $s = 22$, calc. 2.739 eV	**7.**86	[Col85b]

Treatment: IE irradiation, electrons; *IG* irradiation, general; *II* irradiation, ions; *IN* irradiation, neutrons; *IX* irradiation, X-ray; *HP* hydrogen plasma; *HT* heat treatment; *PD* plastic deformation; *PO* powder, *PO(KBr) compressed with KBr*

Table 5.3.10.1 Modified diamond (irradiation, heat etc.): luminescence lines, visible: violet, part 1 (2.75–2.97 eV)

Line-label	Energy (eV)	°Frequ. ($10^3 cm^{-1}$)	Wavel. (nm)	Name	Impur. /defect	Treatment/comment	Figs. **3–6.** [Zai01]; Fig. **7.** [Zai98]	References
ML2012-l	2.769	22.33	447.7	DAP10-l	...	**II**+**HT**(>1,300°C), $s = 1$, calc. 2.786 eV	**7.**84	[Var89, Zai01]
ML2807 = ML2807z	2.807	22.64	441.7	DAP98z	*$(N_1C_1)_i^+$	**IG** (+**HT**(600°C)), ZPL of DAP98a-i(2.739), lines (f–i) see LL	**7.**133	[Col90c, Col89b] [Col93c, Zai01]
ML2535i	2.820	22.75	439.6	DAP104i	*$(V_2Co_1)N_x$	**HT**(1,700°C), $s = 1$, calc. 2.805 eV	...	[Law96]
ML2705b	2.823	22.77	439.2	...	...	**II**+**HT**(1,400°C),SBs: −92, −158 meV	**5.**131, **7.**71	[Var86a, Gip82a]
ML2440h	2.823	22.77	439.2	**N3c**-DAP24h	V_1N_3°	**II** or**HT**(see line z), $s = 50$, calc. 2.823 eV	**7.**86	[Col85b]
ML2842a	2.842	22.92	436.2	DAP85a	*(V_2Ag_1) °	**II**(Ag)+**HT**(1,400°C), 6 lines a–f(3.027 eV), $s = 18$, calc. 2.844 eV	**5.**142	[Zai00a]
ML2845	2.845	22.95	435.7	...	...	**II**(C) of CVD diamond	**5.**134, **7.**133	[Zai96b]
ML2673b	2.847	22.96	435.5	**3.2eV**-DAP72b	*$(N_1C_1)_i$°	**IG** (+**HT**(500°C)), $s = 10$, calc. 2.843 eV	**5.**145	[Zai00a]
ML2867	2.867	23.12	432.4	...	...	**IX**, x-ray luminescence in nat. type IIb diam.	...	[Bie67]
ML2476d	2.876	23.30	431.1	DAP74d	*V_1B_1° (b)	**IX** of type IIb, $s = 22$, calc. 2.880 eV	...	[Dea65a]
ML2896	2.896	23.36	428.1	...	B?	**II** of B doped CVD diamond	...	[Pop96, Zai01]

ML2897	2.897	23.37	428.0	...	...	**II**+**HT**(1400)	**5**.45, **5**.131	[Zai01, Var86a]
ML2440q	2.897	23.37	428.0	**N3c**-DAP24q	V_1N_3°	**II** or**HT**(see line z), *s* = 162, calc. 2.895 eV	**7**.86	[Col85b]
ML2900	2.900	23.39	427.5	...	...	**IX,** x-ray luminescence in nat. type IIb diam.	...	[Bie67]
ML2705c	2.900	23.39	427.5	...	...	**II**+**HT**(1,400°C)	**5**.45, **5**.131	[Zai01, Var86a]
ML2842b	2.912	23.49	425.7	DAP85b	*(V_2Ag_1) °	**II**(Ag)+**HT**(1,400°C), *s* = 32, calc. 2.911 eV	**5**.142	[Zai00a]
ML2705d	2.930	23.63	423.1	...	...	**II**+**HT**(1,400°C)	**5**.45, **5**.131	[Zai01, Var86a]
ML2476e	2.930	23.63	423.1	DAP74e	*V_1B_1° (b)	**IX** of type IIb diamond, *s* = 34, calc. 2.928 eV	...	[Dea65a]
ML2842c	2.937	23.69	422.1	DAP85c	*(V_2Ag_1) °	**II**(Ag)+**HT**(1,400°C), *s* = 42, calc. 2.935 eV	**5**.142	[Zai00a]
ML2842d	2.957	23.85	419.3	DAP85d	*(V_2Ag_1) °	**II**(Ag)+**HT**(1,400°C), *s* = 54, calc. 2.956 eV	**5**.142	[Zai00a]
ML2476f	2.966	23.92	418.0	DAP74f	*V_1B_1° (b)	**IX** of type IIb diamond, *s* = 50, calc. 2.962 eV	...	[Dea65a]

Treatment: IE irradiation, electrons; *IG* irradiation, general; *II* irradiation, ions; *IN* irradiation, neutrons; *IX* irradiation, X-ray; *HP* hydrogen plasma; *HT* heat treatment; *PD* plastic deformation; *PO* powder, *PO(KBr) compressed with KBr*

Table 5.3.10.2 Modified diamond (irradiation, heat etc.): luminescence lines, visible: violet, part 2 (2.97–3.10 eV)

Line-label	Energy (eV)	Frequ. (10^3cm^{-1})	Wavel. (nm)	Name	Impur. /defect	Treatment/comment	Figs. **3–6.** [Zai01]; Fig. **7.** [Zai98]	References
ML2970	2.970	23.96	417.4	...	...	**IX,** X-ray lumin. in nat. type I and IIb diam.	...	[Zai01]
ML2972	2.972	23.97	417.2	...	Ni?	**HT**(1,400°C)	**5.**139	[Kan98, Zai01]
ML2673c	2.974	23.99	416.9	**3.2eV**-DAP72c	*$(N_1C_1)_i$°	**IG (+HT**(500°C)**),** $s = 26$, calc. 2.974 eV	**5.**145	[Zai00a]
ML2842e	2.977	24.01	416.5	DAP85e	*(V_2Ag_1) °	**II**(Ag)+**HT**(1,400°C), $s = 72$, calc. 2.977 eV	**5.**142	[Zai00a]
ML2985 = DAP2440z	2.985	24.08	415.3	**N3c**-DAP24z	V_1N_3°	**II+HT**(1,200°C) or **HT**(1,700°C), ZPL of DAP24a-q(2.897 eV), 3 DAP lines: f, h, q; other lines see NL	**5.**22a, **5.**45 **5.**131, **5.**134, **7.**86	2× [Zai01] [Var86a, Zai96b] [Col85b]
ML2673d	3.005	24.24	412.6	**3.2eV**-DAP72d	*$(N_1C_1)_i$°	**IG (+HT**(500°C)**),** $s = 36$, calc. 3.007 eV	**5.**145	[Zai00a]
ML2673e	3.011	24.29	411.7	**3.2eV**-DAP72e	*$(N_1C_1)_i$°	**IG (+HT**(500°C)**),** $s = 38$, calc. 3.012 eV	**5.**145	[Zai00a]
ML2673f	3.018	24.34	410.8	**3.2eV**-DAP72f	*$(N_1C_1)_i$°	**IG (+HT**(500°C)**),** $s = 42$, calc. 3.019 eV	**5.**145	[Zai00a]
ML2673g	3.023	24.38	410.1	**3.2eV**-DAP72g	*$(N_1C_1)_i$°	**IG (+HT**(500°C)**),** $s = 44$, calc. 3.023 eV	**5.**145	[Zai00a]

ML2842f	3.027	24.42	409.6	DAP85f	*$(V_2Ag_1)^\circ$	**II**(Ag)+**HT**(1,400° C), $s = 198$, calc. 3.030 eV	**5**.142	[Zai00a]
ML2673h	3.031	24.45	409.0	**3.2eV**-DAP72h	*$(N_1C_1)_i^\circ$	**IG** (+**HT**(500° C)), $s = 48$, calc. 3.030 eV	**5**.145	[Zai00a]
ML2673i	3.036	24.49	408.4	**3.2eV**-DAP72i	*$(N_1C_1)_i^\circ$	**IG** (+**HT**(500° C)), $s = 50$, calc. 3.034 eV	**5**.145	[Zai00a]
ML2673j	3.042	24.54	407.6	**3.2eV**-DAP72j	*$(N_1C_1)_i^\circ$	**IG** (+**HT**(500° C)), $s = 58$, calc. 3.044 eV	**5**.145	[Zai00a]
ML3053	3.053	24.62	406.1	...	...	**IX,** x-ray luminescence in nat. diamond	...	[Sob79a], [Zai01]
ML2673k	3.059	24.67	405.3	**3.2eV**-DAP72k	*$(N_1C_1)_i^\circ$	**IG** (+**HT**(500° C)), $s = 72$, calc. 3.059 eV	**5**.145	[Zai00a]
ML2673-1	3.064	24.71	404.6	**3.2eV**-DAP72-1	*$(N_1C_1)_i^\circ$	**IG** (+**HT**(500° C)), $s = 78$, calc. 3.065 eV	**5**.145	[Zai00a]
ML2673m	3.073	24.79	403.4	**3.2eV**-DAP72m	*$(N_1C_1)_i^\circ$	**IG** (+**HT**(500° C)), $s = 86$, calc. 3.071 eV	**5**.145	[Zai00a]
ML2673n	3.090	24.92	401.2	**3.2eV**-DAP72n	*$(N_1C_1)_i^\circ$	**IG** (+**HT**(500° C)), $s = 130$, calc. 3.092 eV	**5**.145	[Zai00a]

Treatment: IE irradiation, electrons; *IG* irradiation, general; *II* irradiation, ions; *IN* irradiation, neutrons; *IX* irradiation, X-ray; *HP* hydrogen plasma; *HT* heat treatment; *PD* plastic deformation; *PO* powder, *PO(KBr) compressed with KBr*

Table 5.3.11 Modified diamond (irradiation, heat etc.): luminescence lines, near and mid ultraviolet (3.10–4.43 eV)

Line-label	Energy (eV)	Frequ. ($10^3 cm^{-1}$)	Wavel. (nm)	Name	Impur. /defect	Comment	Figs. **3–6.** [Zai01]; Fig. **7.** [Zai98]	References
ML3111 = ML2842z	3.111	25.09	398.5	DAP85f	*$(V_2Ag_1)^\circ$	**II**(Ag)+**HT**(1,400°C), ZPL of DAP85a-f(3.027 eV), SBs(QLVM): −21,(LVM) −44, −67 meV	**5**.142 **7**.143a	[Zai00a] [Zai92a]
ML2673-o	3.116	25.11	398.3	**3.2eV**-DAP72-o	*$(N_1C_1)i^\circ$	**IG** (+**HT**(500°C)), $s = 228$, calc. 3.116 eV	**5**.145	[Zai00a]
ML3120 ML2476z	3.120	25.17	397.4	DAP74z	*$V/1B_1{}^\circ$ (b)	**IX** of type IIb, (weak) ZPL of DAP74a-f(2.966 eV)	…	[Dea65a]
ML3162a	3.162	25.50	392.1	…	B?	**IG** of type IIb diam. , 4 lines a–d(3.225 eV), positive temperature shift at 4K to 90K	…	[Maz80, Pop96]
ML3162b	3.184	25.68	389.4	…	B?	**IG,** temperature shift	…	[Maz80, Pop96]
ML3188 = ML2673z	3.188	25.71	388.9	3.2eV = 389 nm-DAP72z	*$(N_1C_1)_i{}^\circ$	**IG,** var. 3.188–3.204 eV, ZPL of DAP72a-o(3.116eV), 8 SBs: QLVM:-58, −77; LVM: −108, −138, −161, −179, −190, −236 meV	**5**.143, **5**.145 **7**.71, **7**.96 **7**.133, **7**.146	[Zai96b], [Zai00a] [Gip82a, Zai92a] [Col90c, Col92b]
ML3162c	3.203	25.84	387.1	…	B?	**IG,** temp. shift, 41 meV exc. state splitting…	…	[Maz80, Pop96]

ML3162d	3.225	26.01	384.4	...	B?	**IG,** temp. shift, 22 meV ground state split.	...	[Maz80, Pop96]
ML3450	3.450	27.83	359.4	...	B?	**IN** of B doped CVD diamond	...	[Pop96, Zai01]
ML3462	3.462	27.92	358.1	...	...	**HT**(1,200°C) of CVD diamond	...	[Rua91b, Zai01]
ML3570	3.570	28.80	347.3	...	...	**HT**(>1,200°C)	**7.**150	[Rua91b, Zai01]
ML4130	4.130	33.31	300.2	...	C_1^+ ions	**II**(C), ion beam induced luminescence	...	[Tag87, Zai01]
ML4204a	4.204	33.91	294.9	**5RL**-DAP80a	$*(C_2)_iN_1{}^\circ$	**IG(+HT**(500°C)**),** 6 lines a–f(4.481 eV), $s = 8$, calc. 4.196 eV	**5.**150	[Col92b]
ML4204b	4.242	34.22	292.3	**5RL**-DAP80b	$*(C_2)_iN_1{}^\circ$	**IG(+HT**(500°C)**),** $s = 10$, calc. 4.237 eV	**5.**150	[Col92b]
ML4204c	4.290	34.60	289.0	**5RL**-DAP80c	$*(C_2)_iN_1{}^\circ$	**IG(+HT**(500°C)**),** $s = 14$, calc. 4.290 eV	**5.**150	[Col92b]
ML4301	4.301	34.69	288.3	...	Si_1?	**II**(Si), ion beam induced luminescence	...	[Tag87, Zai01]
ML4417a	4.417	35.63	280.7	**2BD(F)**-DAP97a	$*(C_2)_iB_1{}^\circ$	**IG** of type IIb diamond, 4 lines a–d (4.667 eV); $s = 9$, calc. 4.411 eV	**7.**156	[Wal79]

Treatment: IE irradiation, electrons; *IG* irradiation, general; *II* irradiation, ions; *IN* irradiation, neutrons; *IX* irradiation, X-ray; *HP* hydrogen plasma; *HT* heat treatment; *PD* plastic deformation; *PO* powder, *PO(KBr) compressed with KBr*

Table 5.3.12.1 Modified diamond (irradiation, heat etc.): luminescence lines, far ultraviolet, part 1 (4.43–5.10 eV)

Line-label	Energy (eV)	Frequ. ($10^3 cm^{-1}$)	Wavel. (nm)	Name	Impur. /defect	Comment	Figs. **3–6.** [Zai01]; Fig. **7.** [Zai98]	References
ML4676-s5	4.436	35.78	279.5	...	$^*(C_2)_iN_1^+$	**IG,** SB: −240 meV	**7.**153	[Dav77c, Zai01]
ML4204d	4.438	36.80	279.4	**5RL**-DAP80d	$^*(C_2)_iN_1{}^\circ$	**IG(+HT**(500°C)), $s = 58$, calc. 4.438 eV	**5.**150	[Col92b]
ML4204e	4.458	35.96	278.1	**5RL**-DAP80e	$^*(C_2)_iN_1{}^\circ$	**IG(+HT**(500°C)), $s = 78$, calc. 4.459 eV	**5.**150	[Col92b]
ML4204f	4.481	36.14	276.7	**5RL**-DAP80f	$^*(C_2)_iN_1{}^\circ$	**IG(+HT**(500°C)), $s = 118$, calc. 4.481 eV	**5.**150	[Col92b]
ML4417b	4.507	36.35	275.1	**2BD-**DAP97b	$^*(C_2)_iB_1{}^\circ$	**IG** of type IIb diamond, $s = 16$, calc. 4.507 eV	**7.**156	[Wal79]
ML4582 = ML4204z	4.582	36.96	270.6	**5RL**-DAP80z	$^*(C_2)_iN_1{}^\circ$	**IG(+HT**(500°C)), ZPL of DAP80a-f(4.481 eV), SBs: (QLVM)-58(calc. 1N+2C:58.5) (LVM): −175,−193, (4×) −237 meV	**5.**150 **7.**153	[Col92b] [Dav77c, Zai01]
ML4417c	4.623	37.29	268.2	**2BD-**DAP97c	$^*(C_2)_iB_1{}^\circ$	**IG** of type IIb diamond, $s = 50$, calc. 4.625 eV	**7.**156	[Wal79]
ML4676-s6	4.639	37.42	267.2	...	$^*(C_2)_iN_1^+$	**IG,** SB:(QLVM): −37 meV, (calc. 1N+5C:37.9)	**7.**153	[Dav77c, Zai01]
ML4417d	4.667	37.64	265.6	**2BD-**DAP97d	$^*(C_2)_iB_1{}^\circ$	**IG** of type IIb diamond, $s = 98$, calc. 4.668 eV	**7.**156	[Wal79]

ML4676	4.676	37.72	265.1	. . .	$*(C_2)_iN_1^+$	**IG,** ZPL, SBs(QLVM): −37, (calc. 1N+5C:37.9), −195, 4× (−231–240) meV	**7.**153	[Dav77c, Zai01]
ML4698	4.698	37.89	263.9	**2BD(G)**	$*(C_2)_iB_2{}^\circ$	**IG** of type IIb diam., SBs: 3× −220 meV	**7.**156	[Wal79, Zai01]
ML5280-s1	4.757	38.37	260.6	BE-Li-s1	Li?	**HT** (see 5.280), SB: −523 meV = 2×TO+TA+LO?	**5.**151	[Zai97a, Zai01]
ML4777 = ML4417z	4.777	38.53	259.5	**2BD(F)** = DAP97z	$*(C_2)_iB_1{}^\circ$	**IG** of type IIb diam., ZPL of DAP97a-d(4.667 eV), splitting (4 meV) into 3 lines SBs: −26, 3× −210 meV	**7.**156	[Wal79, Zai01]
ML4803	4.803	37.89	263.9	**2BD(C)**	$*(C_2)_iB_3{}^\circ$	**IG** of type IIb diam., SBs: 3× −36 meV	**7.**156	[Wal79]
ML5280-s2	4.832	38.98	256.6	BE-Li-s2	Li?	**HT** (see 5.280), SB: −448 meV = 2× TO+TA?	**5.**151	[Zai97a, Zai01]
ML5280-s3	4.950	39.93	250.5	BE-Li-s3	Li?	**HT** (see 5.280), SB: −330 meV = 2× TO?	**5.**151	[Zai97a, Zai01]
ML5356a-s4	5.048	40.72	245.6	**D_2**-BE-B(a)-s4	B	**II** (B)+HT(600°C), SB 308 meV (= 141+163?)	. . .	[Zai01]

Treatment: IE irradiation, electrons; *IG* irradiation, general; *II* irradiation, ions; *IN* irradiation, neutrons; *IX* irradiation, X-ray; *HP* hydrogen plasma; *HT* heat treatment; *PD* plastic deformation; *PO* powder, *PO(KBr) compressed with KBr*

Table 5.3.12.2 Modified diamond (irradiation, heat etc.): luminescence lines, far ultraviolet, part 2 (5.10–5.90 eV)

Line-label	Energy (eV)	Frequ. ($10^3 cm^{-1}$)	Wavel. (nm)	Name	Impur. /defect	Treatment/comment	Figs. **3–6.** [Zai01]; Fig. **7.** [Zai98]	References
ML5280-s4	5.120	41.30	242.1	BE-Li-s4	Li?	**HT** (see 5.280), SB: −160 meV = LO?	**5.**151	[Zai97a]
ML5147a	5.147	41.52	240.9	X1...	...	**II,** 2 lines a, b (5.160 eV)	...	[Ste96a, Zai01]
ML5147b	5.160	41.62	240.3	X2...	...	**II**	...	[Ste96a, Zai01]
ML5316a-s2	5.170	41.70	239.8	BE-P(a)-s2	P	**II** (P), SB(−146 meV) = TO?	...	[Ste96b]
ML5316b-s1	5.190	41.86	238.9	BE-P(b)-s1	P	**II** (P), SB(−143 meV) = TO?	...	[Ste96b]
ML5356a-s5	5.193	41.89	238.7	**D_1?**BE-B(a)-s5	B	**II** (B)+HT(600°C), SB(−163 meV) = LA/O?	...	[Zai01]
ML5356a-s6	5.215	42.06	237.7	**D_1**-BE-B(a)-s6	B	**II** (B)+HT(600°C), SB(−141 meV) = TO?	...	[Zai01]
ML5252	5.252	42.36	236.1	**N9a**-DAP64z	*$V_1N_4^-$	**HT**(>1,000°C), ZPL of DAP64a-i, see NL	**7.**150	[Rua91b]
ML5280	5.280	42.59	234.8	BE-Li	Li?	**HT** (thermo-chemical polishing on hot iron), NPL with 4 SBs: −160, −330, −448, −543 meV	**5.**151	[Zai97a]
ML5316a	5.316	42.88	233.2	BE-P(a)	P	**II** (P), 2 lines a, b (5.333 eV); NPL (upper VB)	...	[Ste96b]
ML5316b	5.333	43.02	232.5	BE-P(b)	P	**II** (P), NPL (lower VB)	...	[Ste96b]
ML5356a	5.356	43.20	231.5	**D_0**-BE-B(a)	B	**II** (B)+HT(600°C), NPL (upper VB); 2 lines a, b	...	[Zai01]
ML5356b	5.367	43.29	231.0	**$D_0\prime$**-BE-B(b)	B	**II** (B)+HT(600°C), NPL (lower VB)	...	[Zai01]

Treatment: IE irradiation, electrons; *IG* irradiation, general; *II* irradiation, ions; *IN* irradiation, neutrons; *IX* irradiation, X-ray; *HP* hydrogen plasma; *HT* heat treatment; *PD* plastic deformation; *PO* powder, *PO(KBr) compressed with KBr*

5.4 Broad Bands (MB)

Table 5.4.1 Modified diamond (irradiation, heat etc.): broad bands, mid and near infrared (0.18–1.77 eV)

Line-label	Energy (meV)	Frequ. (cm^{-1})	Wavel. (μ m)	Name	Impur. /defect	Treatment/observation/comment	Figs. **3–6.** [Zai01]; Fig. **7.** [Zai98]	References
MB0384	385.0	3,105	3.220	...	...	**IE; *A***	...	[Pha65]
MB0399a	398.6	3,215	3.110	3LP(a)	3× lattice phonons	**IG; *A*;** var. 0.396–0.399 eV; 2 bands a, b (450.0); $W = 23\%$; abs. coeff. $= 1.7–3.0\,cm^{-1}$	**7.**13 (p. 25)	[Bok86] [Zai01]
MB0399b	450.0	3,630	2.755	3LP(b)	3× lattice phonons	**IG; *A*;** var. 0.444–0.450 eV; $W = 8\%$; abs. coeff. $= 1.7–3.0\,cm^{-1}$	**7.**13 (p. 25)	[Bok86] [Zai01]
	(eV)	($10^3 cm^{-1}$)	(nm)					
MB1076	1.076	8.68	1152	**H2**	$*V_1N_2^-$	**HT**(1,700°C); ***PL;*** $W = 23\%$, ZPL(DAP78) at 1.256 eV	**5.**10 = **7.**54	[Law92b]
MB1360	1.360	10.97	911.6	***S5**	$*(V_2Ni_1)N_x{}^\circ$	**HT**(2,500°C); ***PL***; unresolved SB, $W = 18\%$	**5.**19a	[Kup99]
MB1582	1.582	12.76	783.7	**GR1**-band	$V_1{}^\circ$	**II+HT**(500°C);***CL;*** unresolved DAP34, $W = 9\%$	7.96	[Zai92a]
MB1610	1.610	12.99	770.0	Radiation **"B"**-band	...	**II+HT**(>1,000°C) ***CL, PL*** or **LD;*EL***(Fig. **5.**68)**;** var. 1.6–1.7 eV; $W = 15–30\%$	**5.**68, **7.**66, **7.**72a, **7.**90	[Zai88c, Fil92] 2× [Zai92a]
MB1750	1.750	14.12	708.4	**NV**$^-$**-band**	$*V_1N_1^-$	**IG+HT**(>550°C); ***PL;*** unresolved DAP49, $W = 17\%$	**7.**80	[Dav77c]

Treatment: *IE* irradiation, electrons; *IG* irradiation, general; *II* irradiation, ions; *IN* irradiation, neutrons; *IX* irradiation, X-ray; *HP* hydrogen plasma; *HT* heat treatment; *PD* plastic deformation; *PO* powder, *PO(KBr) compressed with KBr*; *LD* LED
Observation: *A* absorption; *CL, EL, IL, PL, XL* luminescence; *PLE* photoluminescence excitation

Table 5.4.2 Modified diamond (irradiation, heat etc.): broad bands, visible: purple to green (1.77–2.43 eV)

Line-label	Energy (meV)	Frequ. (cm^{-1})	Wavel. (μ m)	Name	Impur. /defect	Treatment /observation/comment	Figs. **3–6.** [Zai01]; Fig. **7.** [Zai98]	References
MB1563α	1.810	14.60	685.0	*$\mathbf{S5}\alpha$	*$(V_2Ni_1)N_x{}^{\circ}$	**HT**(2,500°C); ***PLE*** of S5-DAP60; $W = 22\%$	**5.**19b	[Kup99]
MB1950	1.950	15.73	635.8	...	...	**HP;CL;**$W = 19\%$	**5.**94	[Hay97]
MB1980	1.980	15.97	626.1	**GR1**-band	$V_1{}^{\circ}$	**II, IE; *A;*** unresolved DAP33, ZPL at 1.673 eV $W = 25\%$	**5.**30, **6.**4, **7.**68, **7.**82	[All98a, Zai01] [Dav77c, Col79c]
MB2030	2.030	16.37	610.7	**NV°**-band	*$V_1N_1{}^{\circ}$	**IG+HT**(>500°C); ***CL;*** unresolved DAP47, $W = 12\%$	**5.**74, **5.**100 **7.**93a	2×[Zai01], [Dav79b]
MB2000	2.050	16.54	604.8	**AS**(a)	*$V_1B_1^-$(a)	**IX; *XL;*** $W = 24\%$	...	[Dea65a]
MB2066	2.066	16.66	600.1	"Red" band	P?	**LD; *EL;*** $W = 22\%$	5.68	[Zai88c]
MB2100	2.100	16.94	590.4	...	...	**II**(heavy ions like Zr); ***CL;*** $W = 14\%$	**5.**70	[Zai01]
MB2140	2.140	17.26	579.3	$\mathbf{B}_s$-band	...	**IE; CL;** $W = 10\%$	...	[Zai80b], [Gip81]
MB2200	2.200	17.74	563.5	amber(α).	...	**HT**(2,200°C); ***A*;** lines (a, b): 0.52, 0.77eV, $W = 23\%$,	...	[Eva79, Zai01]
MB2200	2.200	17.74	563.5	**AS**(b)	*$V_1B_1{}^{\circ}$ (a)	**IX; *XL;*** unresolved DAP73, $W = 30\%$	...	[Dea65a]

MB1563β	2.200	17.75	563.5	*S5β	$*(V_2Ni_1)N_x{}^\circ$	**HT**(2,500°C); ***PLE*** of S5-DAP60; $W = 18–30\%$	**5.**19b, **5.**26	2× [Kup99]
MB2210	2.210	17.83	561.0	**NV**$^-$-band	$*V_1N_1^-$	**IG+HT**(>550°C); ***A;*** unresolved DAP48, $W = 20\%$	**7.**138	[Dav77c]
MB2339	2.339	18.87	530.0	...	...	**HP;*CL*;**$W = 19\%$	**5.**94...	[Hay97]
MB2340a	2.340	18.87	529.8	**"Green"** band("A")	...	**II**(B) or**HP** of CVD**; *PL*;** var. 2.15–2.34 eV, 3 lines a–c see HB, LB; PLE at 2.41–2.95 eV; $W = 21\%$;	...	[Zai01]
MB2350	2.350	18.96	527.6	**H3a**–band	$*V_1N_2{}^\circ$	**II+HT**(>1,400°C); ***CL;*** or **LD;*EL*** (Fig. **5.**110), var. 2.32–2.38 eV, unresolved DAP51, $W = 6\%$	**5.**106, **5.**110b, **5.**111a	[Zai82, Mel98] [Zai98d]
MB2380	2.380	19.20	520.9	**NV**°-band	$*V_1N_1{}^\circ$	**IG+HT**(>500°C); ***A;*** unresolved DAP46, $W = 15\%$	**7.**93b	[Dav79b]
MB2400	2.400	19.36	516.6	...	...	**IG; *CL;*** $W = 40\%$	...	[Zai01]

Treatment: IE irradiation, electrons; *IG* irradiation, general; *II* irradiation, ions; *IN* irradiation, neutrons; *IX* irradiation, X-ray; *HP* hydrogen plasma; *HT* heat treatment; *PD* plastic deformation; *PO* powder, *PO(KBr) compressed with KBr*; *LD* LED
Observation: *A* absorption; *CL, EL, IL, PL, XL* luminescence; *PLE* photoluminescence excitation

Table 5.4.3 Modified diamond (irradiation, heat etc.): broad bands, visible: blue to violet (2.43–3.10 eV)

Line-label	Energy (eV)	Frequ. ($10^3 cm^{-1}$)	Wavel. (nm)	Name	Impur. /defect	Treatment/observation/comment	Figs. **3–6.** [Zai01]; Fig. **7.** [Zai98]	References
MB2430	2.430	19.60	510.2	...	...	**LD;** ***EL***; var 2.40–2.43 eV, $W = 8–12\%$	**5.**102	[Nai98, Zai01]
MB2480	2.480	20.00	500.0	...	...	**LD;** ***EL***; $W = 24\%$	**5.**123	[Zai01]
MB2480	2.480	20.00	499.9	A_s-band	...	**HP**(800°C) or **IE;** ***CL;*** $W = 26\%$	**7.**155	[Mor92a, Zai01]
MB2500	2.500	20.17	495.9	**AS**(c)	$^*V_1B_1^-$ (b)	**IX;** ***XL;*** $W = 14\%$	...	[Dea65a]
MB1563γ	2.600	20.97	476.8	***S5**γ*	$^*(V_2Ni_1)N_x{}^\circ$	**HT**(2,500°C); ***PLE*** of S5-DAP60	**5.**19b	[Kup99]
MB2645	2.645	21.33	468.7	...	$^*V_1Ni_1{}^\circ$ (b)	**HT**(1,700°C); ***PLE*** of $^*V_1Ni_1{}^\circ$ (a); unresolved DAP29; $W = 9\%$	**5.**49b	[Kup99]
MB2370α	2.720	21.94	455.8	**A**α	$^*(V_2Ni_1)N_3$	**HT**(>1,800°C); ***PLE*** (via S2) of line “A”(2.370 eV)	**7.**104	[Yel92b], [Nad93]
MB2750	2.750	2.18	450.8	**N3c**-band	$V_1N_3{}^\circ$ (c)	**II;*CL;***ZPL(DAP24) at 2.985 eV, $W = 12–21\%$	7.86	[Col85b]
MB2515α	2.790	22.50	444.4	**S2**α	$(V_2Ni_1)N_3$	**HT**(>1,800°C)or **II+HT**(1400), ***PLE*** of S2-DAP57	**7.**104	[Yel92b], [Nad93]
MB2800	2.800	22.58	442.8	**AS**(d)	$^*V_1B_1{}^\circ$ (b)	**IX;** ***XL;*** unresolved DAP74, $W = 13\%$	...	[Dea65a]
MB2496α	2.850	22.99	435.0	**S3**α	$(V_2Ni_1)N_2$	**HT**(>1,800°C); ***PLE*** of S3-DAP58	**7.**104	[Yel92b], [Nad93]
MB2880	2.880	23.23	430.5	Radiation **“A”** band	...	**II+HT**(>1,000°C) ***CL, PL*** or **LD;*EL;*** (Fig. **5.**68) var. 2.75–3.0 eV; $W = 8–16\%$	**5.**68a, **7.**66 **7.**90	[Zai88c, Fil92] [Zai92a]
MB1563δ	3.000	24.20	413.3	***S5**δ*	$^*(V_2Ni_1)N_x{}^\circ$	**HT**(2,500°C); ***PLE*** of S5	**5.**19b	[Kup99]

Treatment: IE irradiation, electrons; *IG* irradiation, general; *II* irradiation, ions; *IN* irradiation, neutrons; *IX* irradiation, X-ray; *HP* hydrogen plasma; *HT* heat treatment; *PD* plastic deformation; *PO* powder, *PO(KBr) compressed with KBr*, *LD* LED
Observation: *A* absorption; *CL, EL, IL, PL, XL* luminescence; *PLE* photoluminescence excitation

Table 5.4.4 Modified diamond (irradiation, heat etc.): broad bands, near, mid, and far ultraviolet (3.10–5.90 eV)

Line-label	Energy (eV)	Frequ. ($10^3 cm^{-1}$)	Wavel. (nm)	Name	Impur. /defect	Treatment/*observation*/comment	Figs. **3–6.** [Zai01]; Fig. **7.** [Zai98]	References
MB3100	3.100	25.01	399.9	"N"-band	N?	**HT**(600°C); ***PL, EL, IL;*** $W = 13\%$	...	[Zai01]
MB3263	3.263	26.32	380.0	**N3c**-band	$V_1N_3{}^\circ$ (c)	**II;*A;***ZPL(DAP23) at 2.985 eV, $W = 12–21\%$	6.4	[Zai01]
MB3300	3.300	26.62	375.7	Amber(β).	...	**HT**(2,200°C); ***A*;** lines (a, b): 0.52, 0.77eV	...	[Eva79, Zai01]
MB1563ε	3.300	26.62	375.7	***S5**ε*	$*(V_2Ni_1)N_x{}^\circ$	**HT**(2,500°C); ***PLE*** of *S5; $W = 9\%$	**5.**19b	[Kup99]
MB2370β	3.310	26.70	374.6	**A**β	$(V_2Ni_1)N_3$	**PLE**(via S2) of line "A"(2.370 eV) see DAP92f	**7.**104	[Yel92b], [Nad93]
MB3450	3.450	27.83	359.4	**ND1**-band	V_1^-	**IG; *A;*** unresolved DAP93, ZPL: 3.150 eV, $W = 13\%$	**7.**145	[Dav77c]
MB3470	3.470	27.99	357.3	...	B?	**II**(B)+**HT**1,200°C);***CL;*** var.3.35–3.50 eV; $W = 15\%$	...	[Zai01]
MB2515β	3.540	28.55	350.2	**S2**β	$(V_2Ni_1)N_3$	**HT**(>1,800°C) or **II**+**HT**(1400), ***PLE*** of S2-DAP57	**7.**104	[Yel92b], [Nad93]
MB2496β	3.600	29.04	344.4	**S3**β	$(V_2Ni_1)N_2$	**HT**(>1,800°C); ***PLE*** of S3-DAP58	**7.**104	[Yel92b], [Nad93]
MB3333α	3.650	29.44	339.7	***S4**α*	$(V_2Ni_1)N_3$	**HT**(>1,800°C); ***A, PLE*** of S4-DAP59	**7.**104, **7.**148	[Yel92b], [Col85b]
MB3750	3.750	30.25	330.6	...	O?	**HT** (oxygen, 200°C); ***CL;*** $W = 20\%$	**7.**155	[Mor92a, Zai01]

(continued)

Table 5.4.4 (continued)

Line-label	Energy (eV)	Frequ. (10^3cm^{-1})	Wavel. (nm)	Name	Impur. /defect	Treatment/*observation*/comment	Figs. **3–6.** [Zai01]; Fig. **7.** [Zai98]	References
MB2496γ	3.970	32.02	312.3	**S3γ**	$(V_2Ni_1)N_2$	**HT**(>1,800°C); ***PLE*** of S3-DAP58	**7.**104	[Yel92b], [Nad93]
MB4000	4.000	32.26	309.9	**"L"-band**	$^*(C_2)_iN_1{}^\circ$	**II** (+**HT**(1,200°C)) ; ***CL, IL.***, unresolved 5RL-DAP60, or **LD** (II(P)+HT1, 100°C)**;** ***EL***	**5.**150b(CL) **5.**102, **7.**152	[Col92b, Zai01, Nai98, Pri94]
MB4132	4.132	33.33	300.0	...	...	**II;** ***A;*** $W = 33\%$	...	[Zai01]
MB4450	4.450	35.89	278.6	...	P?	**LD** (II(P)+HT1, 300°C)**;** ***EL***	**5.**102	[Nai98, Zai01]
MB4350b	4.500	36.30	275.5	...	B?	**II** (B)+**HT** (1,200°C); ***CL;*** dominant of 3 bands; (4.35, 4.50, 4.65 eV), $W = 5\%$	**5.**149	[Ste96a, Zai01]
MB4560	4.560	36.78	271.9	...	...	**HT** (?);***A;***	...	[Zai01]
MB4640	4.640	37.43	267.2	...	O?	**HT** (oxygen, 200°C); ***CL;*** $W = 11\%$	**7.**155	[Mor92a, Zai01]
MB2515γ	4.680	37.75	264.9	**S2γ**	$(Ni_1V_2)N_3$	**HT**(>1,800°C) or **II**+**HT**(1400), ***PLE*** of S2-DAP57	**7.**104	[Yel92b], [Nad93]

Treatment: IE irradiation, electrons; *IG* irradiation, general; *II* irradiation, ions; *IN* irradiation, neutrons; *IX* irradiation, X-ray; *HP* hydrogen plasma; *HT* heat treatment; *PD* plastic deformation; *PO* powder, *PO(KBr) compressed with KBr*, *LD* LED *Observation*: *A* absorption; *CL, EL, IL, PL, XL* luminescence; *PLE* photoluminescence excitation

Chapter 6
Spectral Lines in Diamond-Related Materials: DLC, Lonsdaleite, etc.

In this chapter 84 lines (and bands), which are observed in diamond-related materials (see Sect. 1.2.5), are presented in 8 tables. The entries are assigned to 55 centers. The corresponding defect structure is established for seven centers, is almost certain for 12 centers, and is incomplete or unknown for 36 centers.

Chapter 6 is subdivided into *absorption lines* (Sect. 6.1: 64 **RA** lines in four tables), into *luminescence lines* (Sect. 6.2: 14 **RL** lines in two tables), and into *broad bands* (Sect. 6.3: six **RB** bands in two tables.

B. Dischler, *Handbook of Spectral Lines in Diamond*,
DOI 10.1007/978-3-642-22215-3_6, © Springer-Verlag Berlin Heidelberg 2012

6.1 Absorption Lines (RA)

Table 6.1.1.1 Diamond79.35-related materials: absorption lines, far infrared, part 1 (0.04–0.133 eV)

Line-label	Energy (meV)	Frequ. (cm^{-1})	Wavel. (μm)	Name	Impur./ defect	Material/ comment	Figs. **3–6.** [Zai01]; Fig. **7.** [Zai98]	References
RA0057	57.03	460.0	21.74	...	Si?	**SCI?**	**3.17, 7.30**	2 × [Ghe92b]
RA0079	79.35	640.0	156.2	...	Si, H?	**SCI?**	**3.17**	[Ghe92b]
RA0085a1	85.17	687.0	14.56	C–H_2 bend	H	**DGB, NGD;** 4 lines a1–a4 (178.5 meV); sp^3CH_2; C–H_2 bend, Table 9.3.2.1	**3.**8, **3.**17, **7.**173	[Rei98, Ghe92b],[Enc94, Dis87]
RA0087a	86.78	700.0	14.29	C–H_1 bend	H	**PAC;** sp_1CH_1; (only in PAC), Table 9.3.2.1, acetylenic C–H_1	...	[Dis87]
RA0088a1	88.02	710.0	14.09	C–H_2 bend	H	**DGB, NGD;** 4 lines a1–a4 (181.6 meV); olef. sp^2CH_2; C–H_2 bend, Table 9.3.2.1	**3.**17, **3.**26, **7.**5b	[Ghe92b, Fri91a], [Dis87]
RA0091a	90.75	732.0	13.66	...	N?	**NDC;** 2 lines a, b (102.9 meV); nitride inclusions	...	[Zai01]
RA0095a1	95.46	770.0	12.99	C–H_1 bend	H	**DGB, NGD;** 2 lines a1, 2 (177.3); arom. sp^2CH_1; Table 9.3.2.1	**3.**26, **7.**5b	[Fri91a, Dis87]
RA0099	99.18	800.0	12.50	...	Si?	**SCI;** Si–C interface, 10–100 nm thick, Si–C stretch?	**3.**17, **7.**30	2 × [Ghe92b]

RA0091b	102.9	830.0	12.05	...	N?	**NDC;** nitride inclusions	...	[Zai01]
RA0104	104.1	840.0	11.90	C–C bend	H	**NDC, DLC, PAC**; aromatic sp^2, Table 9.3.5	**3**.29	[Ang65, Dis87]
RA0108a1	107.9	870.0	11.48	$C–H_1$ bend	H	**DGB, NGD;** 2 lines a1, 2 (159.9); olefin.. sp^2CH_1; Table 9.3.2	**3**.26, **7**.5b **3**.29	[Fri91a, Ang65, Dis87]
RA0109a	109.1	880.0	11.40	...	...	**NDC;** 3 lines a–c (179.8 meV), carbonate inclusions?	**7**.9	[Bok86, Zai01]
RA0110	109.7	885.0	11.30	C–C bend	H	**DLC, PAC**; Table 9.3.5	...	[Dis87]
RA0088a2	114.1	920.0	10.87	$C–H_2$ bend	H	**DGB;** olef. sp^2CH_2 Table 9.3.2	**3**.17, **3**.18	[Ghe92b, Jan92]
RA0115a	114.8	926.0	10.80	...	...	**LON;** 5 lines a–e (165.1 meV); lonsdaleite inclusions in polycrystalline diamond	**7**.27	[Bok86, Zai01]
RA0120	120.3	970.0	10.31	C=C bend	H	**NDC, DLC, PAC**; olefinic sp^2, Table 9.3.5	**3**.29	[Ang65, Dis87]
RA0122	121.5	980.0	10.20	...	Si?	**SCI**	...	[Cel91, Zai01]

(continued)

Table 6.1.1.1 (continued)

Line-label	Energy (meV)	Frequ. (cm^{-1})	Wavel. (μm)	Name	Impur./ defect	Material/ comment	Figs. **3–6.** [Zai01]; Fig. **7.** [Zai98]	References
RA0085a2	125.2	1010	9.901	C–H_2 bend	H	**DGB, NGD;** aliphatic sp^3CH_2; Table 9.3.2	**3.**18, **3.**26 **7.**173	[Jan92, Fri91a] [Enc94, Dis87]
RA0115b	126.4	1020	9.808	...	...	**LON;** lonsdaleite inclusions in polycrystalline diamond	**7.**27	[Bok86, Zai01]
RA0133a1	133.3	1075	9.302	C–H_3 bend	H	**DGB, NGD;** 3 lines a1–a3 (185.3 meV); sp^3CH_3; Table 9.3.2	**3.**17, **3.**18 **3.**26, **7.**5b	[Ghe92b, Jan92] [Fri91a, Dis87]

BAL ballas; *CAR* carbonado; *DGB* diamond grain boundaries (in CVD diamond); *DLC* diamond-like carbon; *LON* lonsdaleite; *NBD* natural brown diamond; *NDC* natural diamond coat; *NGD* natural gray diamond; *PAC* polymer-like amorphous carbon; *SCI* Si–C interface; *PO* powder

Table 6.1.1.2 Diamond-related materials: absorption lines, far infrared, part 2 (0.135–0.18 eV)

Line-label	Energy (meV)	Frequ. (cm^{-1})	Wavel. (μm)	Name	Impur./ defect	Material/ comment	Figs. **3–6.** [Zai01]; Fig. **7.** [Zai98]	References
RA0109b	135.8	1095	9.130	...	...	**NDC;** carbonate inclusions?	**7.**9	[Bok86, Zai01]
RA0115c	136.2	1100	9.103	...	...	**LON;** lonsdaleite inclusions in polycrystalline diamond	**7.**27	[Bok86, Zai01]
RA0088a3	137.6	1110	9.009	C–H_2 bend	H	**DGB,** olef. sp^2CH_2 Table 9.3.2	**3.**17	[Ghe92b, Dis87]
RA0139a	138.5	1117	8.953	...	O?	**PO;** (fine grained) surface C=O?; 2 lines a, b (0.1569 eV)	...	[Nov68, Zai01]
RA0144	143.8	1160	8.621	C–C stretch	H	**DLC, PAC**; Table 9.3.5	...	[Dis87]
RA0085a3	145.1	1170	8.547	C–H_2 bend	H	**NBD, NGD, DGB;** sp^3CH_2; Table 9.3.2	**3.**8, **3.**14 **3.**18, **7.**173	[Rei98, Col82d] [Jan92, Enc94]
RA0115d	151.3	1220	8.194	...	...	**LON;** lonsdaleite inclusions in polycrystalline diamond	**7.**27	[Bok86, Zai01]
RA0154	154.4	1245	8.030	C–C stretch	H	**DLC, PAC**; Table 9.3.5	...	[Dis87]
RA0139b	156.9	1266	7.901	...	O?	**PO;** (fine grained) surface C=O vibration?	...	[Nov68, Zai01]
RA0108a2	159.9	1290	7.752	C–H_1 bend	H	**DGB;** olef. sp^2CH_1; Table 9.3.2; DLC: 159.9 meV	**3.**14	[Col82d]
RA0161	161.2	1300	7.692	...	O?	**PO;** surface diffuse reflection Fourier transform spectra	...	[Mil95, Zai01]

(continued)

Table 6.1.1.2 (continued)

Line-label	Energy (meV)	Frequ. (cm^{-1})	Wavel. (μm)	Name	Impur./defect	Material/comment	Figs. **3–6.** [Zai01]; Fig. **7.** [Zai98]	References
RA0161	161.2	1300	7.692	C–C stretch	H	**DLC, PAC**; Table 9.3.5	...	[Dis87]
RA0133a2	164.9	1330	7.518	C–H_3 bend	H	**DGB, NGD**; sp^3CH_3; Table 9.3.2	**3.**17, **3.**18 **3.**26, **7.**5b	[Ghe92b, Jan92] [Fri91a, Dis87]
RA0115e	165.1	1332	7.509	...	...	**LON;** lonsdaleite inclusions in polycrystalline diamond	**7.**27	[Bok86, Zai01]
RA0171a	171.1	1380	7.246	C–H_1 bend	H	**DGB, PO** [Che94b]; sp^3CH_1; Table 9.3.2	**3.**13, **3.**18	[Che94b, Jan92]
RA0175	175.3	1414	7..072	...	NH_2	**PO;** amin terminated, surface diffuse reflection FT spectra	...	[Mil95, Zai01]
RA0095a2	177.3	1430	6.993	C–H_1 bend	H	**DGB,** arom. sp^2CH_1; Table 9.3.2	...	[Dis87]
RA0178	178.5	1440	6.944	...	O?	**PO;** surface diffuse reflection Fourier transform spectra	...	[Mil95, Zai01]
RA0085a4	178.5	1440	6.944	C–H_2bend	H	**DGB, NGD**; sp^3CH_2; Table 9.3.2	**3.**14, **3.**18 **7.**173	[Col82d, Jan92] [Enc94, Dis87]
RA0109c	179.8	1450	6.900	...	...	**NDC;** carbonate inclusions?	**7.**9	[Bok86, Zai01]

BAL ballas; *CAR* carbonado; *DGB* diamond grain boundaries (in CVD diamond); *DLC* diamond-like carbon; *LON* lonsdaleite; *NBD* natural brown diamond; *NDC* natural diamond coat; *NGD* natural gray diamond; *PAC* polymer-like amorphous carbon; *SCI* Si–C interface; *PO* powder

Table 6.1.2 Diamond-related materials: absorption lines, mid-infrared (0.18–1.24 eV)

Line-label	Energy (meV)	Frequ. (cm^{-1})	Wavel. (μ m)	Name	Impur./ defect	Material/ comment	Figs. **3–6.** [Zai01]; Fig. **7.** [Zai98]	References
RA0088a4	181.6	1465	6.827	C–H_2 bend	H	**DGB**, **NBD**; olef. sp^2CH_2 Table 9.3.2	**3.**14	[Col82d, Dis87]
RA0133a3	185.3	1495	6.691	C–H_3 bend	H	**DGB**, **NGD**; sp^3CH_3; Table 9.3.2	**3.**17, **3.**26 **7.**5b	[Ghe92b, Fri91a] [Dis87]
RA0188	187.8	1515	6.602	C–C stretch	H	**DLC**, **PAC**; Table 9.3.5	...	[Dis87]
RA0191	191.5	1545	6.474	C–C stretch	H	**DLC**, **PAC**; Table 9.3.5	...	[Dis87]
RA0196	195.9	1580	6.329	C–C stretch	H	**DLC**, **PAC**; Table 9.3.5	...	[Dis87]
RA0200	199.6	1610	6.211	...	O?	**PO**(KBr)	**3.**13	[Che94b]
RA0201	200.8	1620	6.173	C–C stretch	H	**DLC**, **PAC**; Table 9.3.5	...	[Dis87]
RA0205a	204.6	1650	6.061	O–H bend?	H_2O?	**NDC**, **PO**(KBr); 2 lines a, b (419 meV)	**3.**13, **7.**9	[Che94b, Bok86]
RA0216	215.6	1739	5.750	...	O?	**DGB** (flame grown CVD) or **PO** (fine grained)	**3.**10	[Jan91, Zai01]
RA0270	270.3	2180	4.587	C–C stretch	H	**PAC**; acetylenic sp^1, Table 9.3.5	...	[Dis87]
RA0304	303.7	2450	4.082	...	CO_2?	**PO**(KBr); var. 285–304 meV	**3.**13	[Che94b, Zai01]
RA0085b1	353.3	2850	3.509	C–H_2 stretch	H	**DGB, NGD;** 2 lines b1, 2 (361.6 meV); sp^3CH_2; Table 9.3.3	**3.**10 = **7.**7 **7.**11, **7.**12 **7.**18, **7.**173	[Jan91] 2 × [Dis93] [Phi92, Enc94]

(continued)

Table 6.1.2 (continued)

Line-label	Energy (meV)	Frequ. (cm^{-1})	Wavel. (μm)	Name	Impur./ defect	Material/ comment	Figs. **3–6.** [Zai01]; Fig. **7.** [Zai98]	References
RA0133b1	356.4	2875	3.478	C–H_3 stretch	H	**DGB, PO;**2 lines b1, 2 (0.3678 eV); sp^3CH_3; Table 9.3.3	**3.**13, **7.**11	[Che94b, Dis93]
RA0171b	361.3	2914	3.432	C–H_1 stretch	H	**DGB;** sp^3CH_1; Table 9.3.3	**7.**11, **7.**12	2 × [Dis93]
RA0085b2	361.6	2917	3.429	C–H_2stretch	H	**DGB, NGD;** sp^3CH_2; Table 9.3.3	See line b1	
RA0133b2	367.8	2967	3.370	C–H_3 stretch	H	**DGB, PO;**sp^3CH_3; Table 9.3.3	**3.**13, **7.**11	[Che94b, Dis93]
RA0088b1	368.5	2972	3.365	C–H_2 stretch	H	**DGB;** 2 lines b1, 2 (375.0 meV); olef. sp^2CH_2; Table 9.3.3	**7.**11, **7.**12	2 × [Dis93, Zai01]
RA0108b	371.9	3000	3.333	C–H_1 stretch	H	**DGB;** olef. sp^2CH_1; Table 9.3.3	...	[Dis93, Zai01]
RA0088b2	375.0	3025	3.306	C–H_2 stretch	H	**DGB;** olef. sp^2CH_2; Table 9.3.3	**7.**11, **7.**12	2 × [Dis93, Zai01]
RA0095b	378.1	3050	3.279	C–H_1 stretch	H	**DGB;** arom. sp^2CH_1; Table 9.3.3	...	[Dis93, Zai01]
RA0087b	412.0	3323	3.009	C–H_1 stretch	H	**DGB;** sp^1CH_1; Table 9.3.3	...	[Dis93]
RA0205b	419.0	3380	2.958	O–H stretch?	H_2O?	NDC, PO(KBr)	**3.**13, **7.**9	[Che94b, Bok86]

BAL ballas; *CAR* carbonado; *DGB* diamond grain boundaries (in CVD diamond); *DLC* diamond-like carbon; *LON* lonsdaleite; *NBD* natural brown diamond; *NDC* natural diamond coat; *NGD* natural gray diamond; *PAC* polymer-like amorphous carbon; *SCI* Si–C interface; *PO* powder

Table 6.1.3 Diamond-related materials: absorption lines, visible: green and blue (2.18–2.58 eV)

Line-label	Energy (eV)	Frequ. (10^3**cm**$^{-1}$)	Wavel. (nm)	Name	Impur./ defect	Comment	Figs. **3–6.** [Zai01]; Fig. **7.** [Zai98]	References
RA2308	2.308	18.62	537.2	...	...	**NBD**	...	[Bie67, Zai01]
RA2526	2.526	20.38	490.8	...	...	**CAR;** coincides with ZPL of DAP76 ($V_1N_3^-$), see NL	...	[Zai01]

BAL ballas; *CAR* carbonado; *DGB* diamond grain boundaries (in CVD diamond); *DLC* diamond-like carbon; *LON* lonsdaleite; *NBD* natural brown diamond; *NDC* natural diamond coat; *NGD* natural gray diamond; *PAC* polymer-like amorphous carbon; *SCI* Si–C interface; *PO* powder

6.2 Luminescence Lines (RL)

Table 6.2.1 Diamond-related materials: luminescence lines, near infrared (1.24–1.77 eV)

Line-label	Energy (eV)	Frequ. ($10^3 cm^{-1}$)	Wavel. (nm)	Name	Impur./ defect	Material/ comment	Figs. **3–6.** [Zai01]; Fig. **7.** [Zai98]	References
RL1593	1.593	12.85	778.3	...	...	**LON;** lonsdaleite in natural polycryst. diam.	**7.**64	[Bok86, Zai01]
RL1635	1.635	13.19	758.3	...	...	**LON;** lonsdaleite in natural polycryst. diam.	**7.**64	[Bok86, Zai01]
RL1702	1.702	13.73	728.3	...	...	**DGB;** coating of laser assisted CVD diam.	...	[Bad97a, Zai01]
RL1736	1.736	14.01	714.0	...	...	**LON;** lonsdaleite in natural polycryst. diam.	**7.**64	[Bok86, Zai01]

BAL ballas; *CAR* carbonado; *DGB* diamond grain boundaries (in CVD diamond); *DLC* diamond-like carbon; *LON* lonsdaleite; *NBD* natural brown diamond; *NDC* natural diamond coat; *NGD* natural gray diamond; *PAC* polymer-like amorphous carbon; *SCI* Si–C interface; *PO* powder

Table 6.2.2 Diamond-related materials: luminescence lines, visible: purple to blue (1.77–2.58 eV)

Line-label	Energy (eV)	Frequ. ($10^3 cm^{-1}$)	Wavel. (nm)	Name	Impur./ defect	Material/comment	Figs. **3–6.** [Zai01]; Fig. **7.** [Zai98]	References
RL1773	1.773	14.30	699.3	...	...	**LON;** lonsdaleite in natural polycryst. diam.	**7.**64	[Bok86, Zai01]
RL1855	1.855	14.96	668.3	...	...	**LON;** lonsdaleite in natural polycryst. diam.	**7.**64	[Bok86, Zai01]
RL1884	1.884	15.20	658.1	...	...	**LON;** lonsdaleite in natural polycryst. diam.	**7.**64	[Bok86, Zai01]
RL1949	1.949	15.72	636.1	...	...	**BAL**	...	[Bok86, Zai01]
RL1980	1.980	15.97	626.1	...	...	**BAL**	...	[Bok86, Zai01]
RL2010	2.010	16.21	616.8	...	...	**BAL**	...	[Bok86, Zai01]
RL2501	2.501	20.17	495.7	...	...	**BAL**	...	[Bok86, Zai01]
RL2523	2.523	20.35	491.4	...	N?	**NBD;** coincides with ZPL of $*(V_3Si_2)^-$-DAP75	...	[Zai01]
RL2526	2.526	20.38	490.8	...	N?	**CAR**; coincides with ZPL of $*(V_1N_3)^-$-DAP76	...	[Zai01]
RL2537	2.537	20.46	488.7	...	...	**CAR**; main line of a group of narrow lines	...	[Bok86, Zai01]

BAL ballas; *CAR* carbonado; *DGB* diamond grain boundaries (in CVD diamond); *DLC* diamond-like carbon; *LON* lonsdaleite; *NBD* natural brown diamond; *NDC* natural diamond coat; *NGD* natural gray diamond; *PAC* polymer-like amorphous carbon; *SCI* Si–C interface; *PO* powder

6.3 Broad Bands (RB)

Table 6.3.1 Diamond-related materials: broad bands, near infrared (1.24–1.77 eV)

Line-label	Energy (eV)	Frequ. ($10^3 cm^{-1}$)	Wavel. (nm)	Name	Impur./ defect	Material/observation/ comment	Figs. **3–6.** [Zai01]; Fig. **7.** [Zai98]	References
RB1713	1.713	13.82	723.7	...	...	**DGB;***PL;* coating of laser assisted CVD diam.	...	[Bad97a, Zai01]
RB1734	1.734	13.99	715.0	...	...	**BAL, LON;** ***PL, XL***	...	[Bok86, Zai01]
RB1746	1.746	14.08	710.0	...	...	**LON;** ***PL;*** lonsdaleite in nat. polycryst. diam.	**7.**64	[Bok86, Zai01]

BAL ballas; *CAR* carbonado; *DGB* diamond grain boundaries (in CVD diamond); *DLC* diamond-like carbon; *LON* lonsdaleite; *NBD* natural brown diamond; *NDC* natural diamond coat; *NGD* natural gray diamond; *PAC* polymer-like amorphous carbon; *SCI* Si–C interface; *PO* powder
Observation: *A* absorption; *CL, EL, IL, PL, XL* luminescence; *PLE* photoluminescence excitation

Table 6.3.2 Diamond-related materials: broad bands, visible: purple to blue (1.77–2.58 eV)

Line-label	Energy (eV)	Frequ. ($10^3 cm^{-1}$)	Wavel. (nm)	Name	Impur./ defect	Comment	Figs. **3–6.** [Zai01]; Fig. **7.** [Zai98]	References
RB1771	1.771	14.29	700.0	...	...	**NBD;** ***A;*** "chameleon" diamond	...	[Fri91a, Zai01]
RB1975	1.975	15.93	627.7	...	N, B?	**BAL;*PL***	...	[Bok86, Zai01]
RB2460	2.460	19.84	504.0	...	...	**DLC;*PL;*** nondiamond phases, W=12%	...	[Ber93, Zai01]

BAL ballas; *CAR* carbonado; *DGB* diamond grain boundaries (in CVD diamond); *DLC* diamond-like carbon; *LON* lonsdaleite; *NBD* natural brown diamond; *NDC* natural diamond coat; *NGD* natural gray diamond; *PAC* polymer-like amorphous carbon; *SCI* Si–C interface; *PO* powder
Observation: *A* absorption; *CL, EL, IL, PL, XL* luminescence; *PLE* photoluminescence excitation

Chapter 7
Spectral Line Shifts from Substituted and Natural Isotopes

In this chapter, 235 lines for which isotope shifts are measured are presented in ten tables.

Chapter 7 is subdivided into $^{2}H_1,^{4}H_2,^{6}H_3,^{8}H_4$ shifts (Sect. 7.1: 74 lines (i2-, i4-, i6-, i8-) in four pages), into $^{13}C_1,^{26}C_2$ shifts (Sect. 7.2: 104 lines (i13, i26-) in 11 pages), into $^{15}N_1$ shifts (Sect. 7.3: 26 lines (i15) in three pages); into $^{57}Si_2,^{58}Si_2$ shifts (Sect. 7.4: eight lines (i57-, i58-) in one pages); into $^{60}Ni_1,^{61}Ni_1,^{62}Ni_1,^{64}Ni_1 or$ $^{118}Ni_2,^{120}Ni_2,^{122}Ni_2,^{124}Ni_2 shifts$ *(Sect. 7.5: 23 lines (i60-, i61-, i62-, i64- or i118-, i120, i122-, i124-)* in two pages).

For vibrational lines, the theoretical isotope shift can be calculated from the mass difference (see Sects. 13.2, 13.3, and Tables 7.1–7.3 captions), the isotopic line shifts at zero phonon lines are small (see Sect. 13.1). They can be negative or positive.

B. Dischler, *Handbook of Spectral Lines in Diamond*,
DOI 10.1007/978-3-642-22215-3_7,

7.1 Isotopic Shifts from ^{2}H (Deuterium) (i2, i4, i6, i8)

Table 7.1 Isotopic shifts (i2, i4, i6, i8) from one to four ^{2}H=D atoms (theoretical vibrational shift is−29.3%)

Line-label	Energy (meV)	Frequ. (cm^{-1})	Wavel. (μ m)	Name	Impur.	$^2H/^1H$ shift (%), comment	Figs. **3–6.** [Zai01]; Fig. **7.** [Zai98]	References
i2-RA0088a2	830.6	670.0	14.93	C–D_2 bend	D	−26.4%; 3 lines a2–a4 (0.1308 eV); olef. sp^2CD_2 Table 9.3.2	...	[Dis85]
i2-RA0085a2	892.6	720.0	13.89	C–D_2 bend	D	−30.1%; 3 lines a2–a4 (0.1302 eV); sp^3CD_2; Table 9.3.2	...	[Dis85]
i2-RA0133a1	929.8	750.0	13.33	C–D_3 bend	D	−30.2%; 3 lines a1–a3 (0.1333 eV); sp^3CD_3; Table 9.3.2	...	[Dis85]
i2-RA0088a3	979.4	790.0	12.66	C–D_2 bend	D	−28.8%; olef. sp^2CD_2; Table 9.3.2	...	[Dis85]
i2-RA0104	104.1	840.0	11.90	C–C stretch	D	*No shift*, DLC, PAC; Table 9.3.5	...	[Dis85]
i2-RA0110	109.7	885.0	11.30	C–C stretch	D	*No shift*, DLC, PAC; Table 9.3.5	...	[Dis85]
i2-RA0085a3	110.3	890.0	11.24	C–D_2 bend	D	−23.9%; sp^3CD_2; Table 9.3.2	...	[Dis85]
i2-RA0108a2	112.8	910.0	10.99	C–D_1 bend	D	−29.5%; olef. sp^2CD_1; Table 9.3.2	...	[Dis85]
i2-RA0133a2	117.8	950.0	10.53	C–D_3 bend	D	−28.3%; sp^3CD_3; Table 9.3.2	...	[Dis85]
i2-RA0171a	117.8	950.0	10.53	C–D_1 bend	D	−30.7%; sp^3CD_1; Table 9.3.2	...	[Dis85]
i2-RA0120	119.0	960.0	10.42	C–C stretch	D	*Small shift*: −1.0%; DLC, PAC; Table 9.3.5	...	[Dis85]
i2-RA0096a2	124.0	1,000	10.00	C–D_1 bend	D	−30.3%; arom. sp^2cd_1; Table 9.3.2	...	[Dis85]

i2-RA0085a4	130.2	1,050	9.524	C–D_2 bend	D	−27.1%; sp^{3}CD$_2$; Table 9.3.2	...	[Dis85]
i2-RA0088a4	130.8	1,055	9.479	C–D_2 bend	D	−27.2%; olef. sp^{2}CD$_2$; Table 9.3.2	...	[Dis85]
i2-RA0133a3	133.3	1,075	9.302	C–D_3 bend	D	−27.9%; sp^{3}CD$_3$; Table 9.3.2	...	[Dis85]
i2-RA0144	143.8	1,160	8.621	C–C stretch	D	*No shift*, DLC, PAC; Table 9.3.5	...	[Dis85]
i2-RA0154	154.4	1,245	8.030	C–C stretch	D	*No shift*, DLC, PAC; Table 9.3.5	...	[Dis85]
i2-RA0161	161.2	1,300	7.692	C–C stretch	D	*No shift*, DLC, PAC; Table 9.3.5	...	[Dis85]
i2-RA0188	187.8	1,515	6.602	C–C stretch	D	*No shift*, DLC, PAC; Table 9.3.5	...	[Dis85]
i2-RA0196	193.4	1,560	6.410	C–C stretch	D	*Small shift*: −1.3%; DLC, PAC; Table 9.3.5	...	[Dis85]
i2-RA0201	198.4	1,600	6.250	C–C stretch	D	*Small shift*: −1.2%; DLC, PAC; Table 9.3.5	...	[Dis85]
i2-LA0155b	244.9	1,975	5.063	C–D_1 stretch	D	−30.0%; **lattice** sp^{3}CD$_1$; Table 9.3.3	...	[Jan92]
i2-RA0085b1	254.8	2,055	4.866	C–D_2 stretch	D	−27.9%; sp^{3}CD$_2$; Table 9.3.3	...	[Dis85]
i2-RA0133b1	260.4	2,100	4.762	C–D_3 stretch	D	−27.0%; sp^{3}CD$_3$; Table 9.3.3	...	[Dis85]
i2-RA0085b2	266.6	2,150	4.651	C–D_2 stretch	D	−26.4%; sp^{3}CD$_2$; Table 9.3.3	...	[Dis85]
i2-RA0133b2	274.6	2,215	4.515	C–D_3 stretch	D	−25.4%; sp^{3}CD$_3$; Table 9.3.3	...	[Dis85]
i2-RA0171b	275.2	2,220	4.505	C–D_1 stretch	D	−24.0%; sp^{3}CD$_1$; Table 9.3.3	...	[Dis85]

(continued)

Table 7.1 (continued)

Line-label	Energy (meV)	Frequ. (cm^{-1})	Wavel. (μ m)	Name	Impur.	$^2H/^1H$ shift (%), comment	Figs. **3–6.** [Zai01]; Fig. **7.** [Zai98]	References
i2-RA0088b1	275.2	2,275	4.396	C–D_2 stretch	D	−22.8%; olef. sp^2CD_2; Table 9.3.3	...	[Dis85]
i2-RA0108b	287.6	2,320	4.310	C–D_1 stretch	D	−22.7%; olef. sp^2CD_1; Table 9.3.3	...	[Dis85]
i2-RA0088b2	293.2	2,365	4.228	C–D_2 stretch	D	−21.8%; olef. sp^2CD_2; Table 9.3.3	...	[Dis85]
i2-RA0095b	293.8	2,370	4.220	C–D_1 stretch	D	−22.2%; arom. sp^2CD_1; Table 9.3.3	...	[Dis85]
I2-RA0087b	320.5	2,585	3.868	C–D_1 stretch	D	−21.7%; sp^1CD_1; Table 9.3.3	...	[Dis85]
i8-LA0599z	(410.0)			(V_4D_4)-DAP40z	(V_4D_4)	*Small shift*, −0.5%; ZPL of i2-DAP40a-g indirect from lines (a–g), homo epitaxial films with 50% or 100% D	...	[Fuc95a], [Fuc95b]
i8-LA0599a	599.4	4,835	2.068	(V_4D_4)-DAP40a	(V_4D_4)	*No shift*; shell = 42; calc. 599 meV[a]	...	[Fuc95a], [Fuc95b]
i8-LA0599b	616.3	4,971	2.012	(V_4D_4)-DAP40b	(V_4D_4)	−0.02%; $s = 36$; calc. 613 meV[a]	...	[Fuc95a], [Fuc95b]
i8-LA0599c1	675.6	5,449	1.835	(V_4D_4)-DAP40c	(V_4D_4)	−2.2%; $s = 20$; calc. 684 meV[a]	...	[Fuc95a], [Fuc95b]
i6-LA0599c2	679.9	5,484	1.823	(V_4D_3)-DAP40c	($V_4D_3H_1$))	−1.6%; $s = 20$; calc. 684 meV[a], 50% D	**5.**6	[Fuc95a], [Fuc95b]

i4-LA0599c3	683.6	5,514	1.814	(V_4D_2)-DAP40c	$(V_4D_2H_2))$	−1.0%; $s = 20$; calc. 684 meV[a], 50% D	**5.6**	[Fuc95a, Fuc95b]
i2-LA0599c4	687.3	5,544	1.804	(V_4D_1)-DAP40c	$(V_4D_1H_3))$	−0.5%; $s = 20$; calc. 684 meV[a], 50% D	**5.6**	[Fuc95a, Fuc95b]
i8-LA0599d	845.9	6,823	1.466	(V_4D_4)-DAP40d	(V_4D_4)	−0.7%; $s = 8$; calc. 841 meV[a]	. . .	[Fuc95a, Fuc95b]
i4-LA0599d	847.1	6,833	1.464	(V_4D_2)-DAP40d	$(V_4D_2H_2)$	−0.5%; $s = 8$; calc. 841 meV[a], 50% D	**5.6**	[Fuc95a, Fuc95b]
i8-LA0599e	890.8	7,185	1.392	(V_4D_4)-DAP40e	(V_4D_4)	−0.7%; $s = 7$; calc. 880 meV[a]	. . .	[Fuc95a, Fuc95b]
i4-LA0599e	891.5	7,191	1.391	(V_4D_2)-DAP40e	$(V_4D_2H_2)$	−0.6%; $s = 7$; calc. 880 meV[a], 50% D	**5.6**	[Fuc95a, Fuc95b]
i8-LA0599f	908.4	7,327	1.365	(V_4D_4)-DAP40f	(V_4D_4)	−0.5%; $s = 6$; calc. 908 meV[a]	. . .	[Fuc95a, Fuc95b]
i4-LA0599f	910.7	7,346	1.361	(V_4D_2)-DAP40f	$(V_4D_2H_2)$	−0.3%; $s = 6$; calc. 908 meV[a], 50% D	**5.6**	[Fuc95a, Fuc95b]
i8-LA0599g	980.9	7,912	1.264	(V_4D_4)-DAP40g	(V_4D_4)	−0.5%; $s = 4$; calc. 990 meV[a]	. . .	[Fuc95a, Fuc95b]
i2-ML2272	2.274	18.34	545.2		D + ?	*Small shift*, +0.1%	7.98b	[Gip83, Dav94d]

[a]Calculated DAP transition energies with $L = 0.410$ eV and $D = 1.41$ eV

7.2 Isotopic Shifts from ^{13}C (i13, i26)

Table 7.2 Isotopic shifts (i13) from ^{13}C substitution (theoretical C–C vibrational shift is −3.9%)

Line-label	Energy (meV)	Frequ. (cm^{-1})	Wavel. (μ m)	Name[a]	Impur./defect	^{13}C/^{12}C shift (%)[b], comment	Figs. **3–6.** [Zai01]; Fig. **7.** [Zai98]	References
i13-HA0105b′	124.2	1,003	9.970	C(b′)	N_1°	−4.1%, Table 9.1.1.1	. . .	[Col88c]
i13-HA0105c′	130.2	1,050	9.524	C(c′)	N_1°	−4.3%, Table 9.1.1.1	. . .	[Col88c]
i13-HA0105d′	136.1	1,098	9.107	C(d′)	N_1°	−2.9%, Table 9.1.1.1	. . .	[Col88c]
i13-MA0147	(142.0)	1,145	8.734	H_1 b, d, g; type IaBirr.	$V_1N_4V_1$	−3.4%; (indirect) ZPL of DAP31	. . .	[Kif99]
i13-HA0105e′	143.2	1,155	8.658	C(e′)	N_1°	−4.4%, Table 9.1.1.1	. . .	[Col88c]
i13-MA0154	(152.0)	1,226	8.157	H_1 c,e,f; type IaAirr.	$N_1–N_1–V_1$	−1.3%; (indirect) ZPL of DAP30	. . .	[Kif99]
i13-HA0105f′	154.4	1,245	8.032	C(f′)	N_1°	−4.4%, Table 9.1.1.1	. . .	[Col88c]
i13-LA0157	156.8	1,265	7.905	C–H bend?	H?	No ^{12}C line; homoepitaxial film	**3**.19b	[Fuc95b]
i13-LA0165=i13-LA0240z	158.8	1,281	7.806	DAP39z	*$(V_4H_4)^\circ$	−3.8%; ZPL of DAP39a-j(438 meV); C–C stretch	**3**.6, **3**.19 b	[Fuc95a], [Fuc95b]
i13-MA0069q	158.9	1,282	7.800	1LP(q)	C	−3.9%, lattice Raman line	**4**.27	[Zai96a, Zai01]
i13-HA0105g′	160.2	1,292	7.740	C(g′)	N_1°	−3.9%; Table 9.1.1.1	. . .	[Col88c]
i13-LA0165	164.9	1,330	7.519	C–H bend?	H?	No ^{12}C line	**3**.19b	[Fuc95b]

i13-HA0041e/	165.6	1,336	7.485	D(e/) = platelets	$(C_{2n})_i$	−3.3 to −4.3%; Table 8.2.1	7.22c	[Sum88]
i13-LA0166	166.1	1,340	7.463	C–H bend?	H?	No ^{12}C line	3.19b	[Fuc95b]
i13-LA0168	168.1	1,355	7.375	C–H bend?	H?	No ^{12}C line	3.19b	[Fuc95b]
i13-LA0173	173.2	1,397	7.158	C–H bend?	H?	No ^{12}C line	3.19b	[Fuc95b]
i13-MA0180	176.5	1,423	7.025	C–N stretch	$*(C_2)_iN_1^-$	−2.2%; LVM of 3.188 eV center;176/180	. . .	[Zai01]
i13-MA0190	182.3	1,471	6.800	C–C stretch?	$*(C_2)_iN_1^-$	−3.9%; LVM of 3.188 eV center;182/190	7.19a	[Col88c]
i13-MA0187	182.6	1,473	6.789	C–N stretch?	N, C?	−1.9%	. . .	[Zai01]
i13-LA0185	185.3	1,495	6.689	C–H bend?	H?	No ^{12}C line	3.6, 7.6	[Fuc95a, Fuc95b]
i13-MA0195	187.1	1,509	6.628	C–C stretch?	C?	−3.9%	7.19a	[Col88c]
i13-MA0212	208.0	1,678	5.960	C–N stretch	N, C	−1.6%	7.19a	[Col88c]
i13-MA0238	229.2	1,849	5.408	C–C stretch	$*(C_2)_iN_1^-$	−3.9%; LVM of 3.188 eV center;229/238	7.19a	[Col88c]
i13-MA0245	235.1	1,896	5.274	C–C stretch?	C?	−4.0%	7.19	[Col88c]
i13-LA0225c	235.6	1,900	5.263	2LP(c)	C	−3.8%	3.4c	[Ant92]
i13-LA0240a	236.8	1,910	5.236	(V_4H_4)a-DAP39a	$*(V_4H_4)$	No ^{12}C line; s = 228, calc. 236.3 meV	3.6, 7.6	[Fuc95a, Fuc95b]

(continued)

Table 7.2 (continued)

Line-label	Energy (meV)	Frequ. (cm^{-1})	Wavel. (μ m)	Name[a]	Impur./defect	$^{13}C/^{12}C$ shift (%)[b], comment	Figs. **3–6.** [Zai01]; Fig. **7.** [Zai98]	References
i13-LA0225e	241.7	1,950	5.128	2LP(e)	C	−4.0%	**3**.4c	[Ant92]
i13-LA0237b	258.5	2,085	4.796	(V_4H_4)a-DAP39b	$^*(V_4H_4)^\circ$	No ^{12}C line; $s = 162$, calc. 251.0 meV	3.6, 7.6	[Fuc95a, Fuc95b]
i13-LA0225i	260.4	2,100	4.762	2LP(i)	C	−2.5%	3.4c	[Ant92]
i13-LA0237c	267.7	2,159	4.631	(V_4H_4)a-DAP39c	$^*(V_4H_4)^\circ$	No ^{12}C line; $s = 118$, calc. 268.0 meV	3.6, 7.6	[Fuc95a, Fuc95b]
i13-LA0225-o	292.6	2,360	4.237	2LP(o)	C	−3.1%	3.4c	[Ant92]
i13-LA0225p	306.2	2,470	4.049	2LP(p)	C	−2.8%	3.4c	[Ant92]
i13-MA0331a	329.7	2,659	3.761	H_1d=DAP30a	$^*V_1C_4N_2^\circ$	−0.3%; $s = 50$, calc. 330.0 meV; SB(62) −3.0%	...	[Kif99]
i13-LA0237d	334.1	2,695	3.711	(V_4H_4)a-DAP39d	$^*(V_4H_4)^\circ$	No ^{12}C line; $s = 46$, calc. 333.0 meV	3.6, 7.6	[Fuc95a, Fuc95b]
i13-LA0237e	343.2	2,768	3.613	(V_4H_4)a-DAP39e	$^*(V_4H_4)^\circ$	No ^{12}C line; $s = 42$, calc. 341.0 meV	3.6, 7.6	[Fuc95a, Fuc95b]
i13-LA0237f	346.4	2,794	3.579	(V_4H_4)a-DAP39f	$^*(V_4H_4)^\circ$	No ^{12}C line; $s = 40$, calc. 345.0 meV	3.6, 7.6	[Fuc95a, Fuc95b]
i13-LA0155 b	349.6	2,820	3.546	C–H_1 stretch	H	−0.4%; *lattice* sp^3CH_1; Table 9.3.3	...	[Dis93]
i13-LA0085 b1	351.4	2,834	3.529	C–H_2 stretch	H	−0.4%; sp^3CH_2; Table 9.3.3	...	[Dis85]
i13-LA0133 b1	355.4	2,867	3.488	C–H_3 stretch	H	−0.3%; sp^3CH_3; Table 9.3.3	...	[Dis85]

i13-MA0353 b	359.3	2, 898	3.451	H_1e=DAP31b	*$V_1N_4V_1$	−0.7%; $s = 34$, calc. 359.0 meV; B-center Irradiated	...	[Kif99]
i13-LA0085 b2	360.5	2, 908	3.439	C–H_2 stretch	H	−0.3%; sp^3CH_2; Table 9.3.3	...	[Dis85]
i13-LA0133 b2	366.4	2, 955	3.384	C–H_3 stretch	H	−0.4%; sp^3CH_3; Table 9.3.3	...	[Dis85]
i13-LA0237 g	374.9	3, 024	3.307	(V_4H_4)a-DAP39g	*$(V_4H_4)^\circ$	No ^{12}C line; $s = 30$, calc. 374.0 meV	3.6, 7.6	[Fuc95a, Fuc95b]
i13-LA0237 h	377.6	3, 046	3.283	(V_4H_4)a-DAP39h	*$(V_4H_4)^\circ$	No ^{12}C line; $s = 29$, calc. 379.0 meV	3.6, 7.6	[Fuc95a, Fuc95b]
i13-NA0385 b	384.1	3, 098	3.228	Nat. C–H stretch	*$(V_1H_1)^\circ$	−0.3%; sp^3CH_1; Table 9.3.1	...	[Woo83]
i13-HB0322 a	384.3	3, 100	3.226	3LP(a)	C	−3.6%, broad band	3.4 c	[Ant92]
i13-LA0237 i	386.1	3, 115	3.211	(V_4H_4)a-DAP39i	*$(V_4H_4)^\circ$	No ^{12}C line; $s = 26$, calc. 390.0 meV	3.6, 7.6	[Fuc95a, Fuc95b]
i13-LA0599z	411.2	3, 317	3.015	(V_4H_4)b-DAP40z	*$(V_4H_4)^\circ$	−0.2%; ZPL of DAP40a-g, homoepitaxial film	3.6, 7.6	[Fuc95a, Fuc95b]
i13-HB 0322 b	436.4	3, 520	2.841	3LP(b)	C	−3.0%, broad band	3.4c	[Ant92]
i13-LA0237 j	437.9	3, 532	2.831	(V_1H_4)a-DAP39j	*$(V_4H_4)^\circ$	No ^{12}C line; $s = 18$, calc. 436.0 meV	3.6, 7.6	[Fuc95a, Fuc95b]
i13-MA0353 d	544.1	4, 388	2.279	H_1f=DAP31d	*$V_1N_4V_1$	−0.2%; $s = 10$; calc. 544.0 meV, B-center irradiated	...	[Zai01]
i13-MA0331 c	552.0	4, 452	2.246	H_1g=DAP30c	*$V_1C_4N_2^\circ$	+0.2%; $s = 10$; calc. 552.0 meV; A-center irradiated	...	[Zai01]

(continued)

Table 7.2 (continued)

Line-label	Energy (meV)	Frequ. (cm^{-1})	Wavel. (μ m)	Name[a]	Impur./defect	$^{13}C/^{12}C$ shift (%)[b], comment	Figs. **3–6.** [Zai01]; Fig. **7.** [Zai98]	References
i13-LA599 a	598.8	4,830	2.070	(V_4H_4)b-DAP40a	$^*(V_4H_4)^{\circ}$	−0.1%; shell= 42; calc. 604.0 meV; homoepit.	3.6, 7.6	[Fuc95a, Fuc95b]
i13-MA0331 d	612.5	4,940	2.024	H_1b=DAP30d	$^*V_1C_4N_2^{\circ}$	*No shift*; $s = 8$; calc. 593.0 meV; A-center Irradiated	. . .	[Dav99b]
i13-LA0599 b	616.4	4.972	2.011	$^*(V_4H_4)$b-DAP40b	$^*(V_4H_4)$	*No shift*; $s = 36$; calc. 619.0 meV	3.6, 7.6	[Fuc95a, Fuc95b]
i13-LA0599 c	690.6	5.570	1.795	$^*(V_4H_4)$b-DAP40c	$^*(V_4H_4)$	−0.03%; $s = 20$; calc. 691.0 meV	3.6, 7.6	[Fuc95a, Fuc95b]
i13-LA0599 d	852.1	6.873	1.455	$^*(V_4H_4)$b-DAP40d	$^*(V_4H_4)$	+0.1%; $s = 8$; calc. 852.0 meV	3.6, 7.6	[Fuc95a, Fuc95b]
i13-LA0599 e	898.0	7.243	1.381	$^*(V_4H_4)$b-DAP40e	$^*(V_4H_4)$	+0.1%; $s = 7$; calc. 891.0 meV	3.6, 7.6	[Fuc95a, Fuc95b]
i13-LA0599 f	914.6	7.377	1.356	$^*(V_4H_4)$b-DAP40f	$^*(V_4H_4)$	+0.2%; $s = 6$; calc. 919.0 meV	3.6, 7.6	[Fuc95a, Fuc95b]
i13-LA0599 g	969.5	7.820	1.279	$^*(V_4H_4)$b-DAP40g	$^*(V_4H_4)$	−1.6%; $s = 4$; calc. 1,003 meV; broad line	3.6, 7.6	[Fuc95a, Fuc95b]

i13-ML1256	1.257	10,140	0.9863	**H2**	$*V_1N_2^+$	+0.7%	...	[Dav99a]
i13-HL1401 b	1.403	11,320	0.8835	*1Ni1.40b	$*V_1Ni_1^+$	−0.04%; ^{58}Ni, main line. (intensities a:b= 0.4 : 0.6)	...	[Col88c]
i13-MA1665 a	1.668	13,450	0.7435	**GR1a**	$*V_1^\circ$	+0.17%; intensities GR1a:GR1b= 1:6.3	5.39	[Col88c]
i13-MA1665 b	1.676	13,520	0.7396	**GR1b**	$*V_1^\circ$	+0.17%; ZPL of i13-GR1b-DAP33a-h	5.39	[Col88c]
i13-MA1685 a	1.687	13,610	0.7349	**R2**	$*(C_2)_i^\circ$	+0.1%; ZPL of DAP 71	...	[Dav00]
i13-MA1771 a	1.770	14,280	0.7003	GR1b-DAP33a	$*V_1^\circ$	−0.1%, s = 118; calc. 1.772 eV	...	[Dav01a]
i13-MA1771 b	1.796	14,490	0.6903	GR1b-DAP33b	$*V_1^\circ$	+0.2%; s = 78; calc. 1.794 eV	...	[Dav01a]
i13-MA1771 c	1.808	14,580	0.6857	GR1b-DAP33c	$*V_1^\circ$	−0.1%; s = 62; calc. 1.808 eV	...	[Dav01a]
i13-MA1771 d	1.812	14,620	0.6842	GR1b-DAP33d	$*V_1^\circ$	−0.1%; s = 58; calc. 1.813 eV	...	[Dav01a]
i13-MA1771 f	1.821	14,690	0.6808	GR1b-DAP33f	$*V_1^\circ$	−0.2%; s = 50; calc. 1.822 eV	...	[Dav01a]
i13-MA1771 g	1.825	14,720	0.6793	GR1b-DAP33g	$*V_1^\circ$	−0.2%; s = 48; calc. 1.826 eV	...	[Dav01a]

(continued)

Table 7.2 (continued)

Line-label	Energy (meV)	Frequ. (cm^{-1})	Wavel. (μ m)	Name[a]	Impur./defect	$^{13}C/^{12}C$ shift (%)[b], comment	Figs. **3–6.** [Zai01]; Fig. **7.** [Zai98]	References
i13-MA1771 h	1.835	14.80	675.6	GR1b-DAP33h	$^*V_1^{\circ}$	−0.2%; $s = 42$; calc. 1.837 eV	...	[Dav01a]
i13-MA1685 b	1.854	1,495	668.9	R2 sideband	$^*(C_2)_i^{\circ}$	−0.28 (−3.8% of 168-meV LVM); pseudo ZPL	...	[Dav99a]
i13-HA1883	1.887	15.22	657.1	*Ni 1.88	$^*(V_3Ni_2)^{\circ}$	+0.2%; ZPL of DAP107	...	[Dav94a]
i13-MA1943	1.945	15.69	637.3	**638 nm**	$^*V_1N_1^-$	+0.11%, ZPL of DAP48/49	...	[Dav99a]
i13-MA2086	2.087	16.83	594.2	**595 nm**	$^*V_2N_2$	+0.02%	...	[Dav99a]
i13-MA2154	2.157	17.40	574.7	**575 nm=NV**	$^*V_1N_1^{\circ}$	+0.14%, ZPL of DAP46/47	...	[Dav99a]
i13-MA2400	2.406	19.40	515.3	**M1**	?	+0.02%	...	[Dav99a]
i13-MA2445	2.449	19.76	506.2	**M2**	?	+0.2%	...	[Dav99a]
i13-ML2462	2.467	19.89	502.8	**3H**	$^*(C_2)_i^+$	+0.15%; ZPL of DAP50	5.105a	[Ste99a]
i13-HL2464 a	2.469	19.92	502.1	**H3a**	$^*V_1N_2^{\circ}$	+0.20%, Table 8.2.1	...	[Dav99a]
i13-MA2543a	2.550	20.56	486.3	**TH5**-DAP15a	$^*V_2^{\circ}$	+0.3%	...	[Dav01a]
i13-MA2638	2.642	21.31	469.3	**TR12**	$^*(C_2)_i^-$	+0.2%	...	[Dav99a]
i13-MA1665c	(2.788)	22.49	444.6	(GR2)	$^*V_1^+$	+0.3%; ZPL (indirect) of i13-GR2-DAP32a-m	...	[Dav01a]
i13-HL2807	2.810	22.67	441.1	2.81 eV	$^*(N_1C_1)_i^+$	+0.1%	...	[Col89a]

i13-ML3188-s6	2.883	23.25	430.0	**3.188 eV**-s6	$^*(C_2)_iN_1^-$	(−3.6%) at 308/313 meV LVM 2× (C–C)	...	[Col87b]
i13-ME2881a	2.888	23.29	429.4	**GR2**-DAP32a	$^*V_1^+$	+0.24% (+7 meV), $s = 130$; calc. 2.888 eV, lines c, d, (k–m) not observed	...	[Dav01a]
i13-ME2881b	2.894	23.34	428.4	**GR3**-DAP32b	$^*V_1^+$	+0.25% (+7.2 meV); $s = 118$; calc. 2.893 eV	...	[Dav01a]
i13-ML3188-s9	2.963	23.90	418.4	3.188 eV-s9	$^*(C_2)_iN_1^-$	(−3.5%) at 228/236 meV LVM	...	[Col87b]
i13-ME2881e	2.965	23.92	418.1	**GR6a**-DAP32e	$^*V_1^+$	+0.24% (+7.2 meV); $s = 42$; calc. 2.965 eV	...	[Dav01a]
i13-ME2881f	2.967	23.93	417.9	**GR6b**-DAP32f	$^*V_1^+$	+0.24% (+7.0 meV); $s = 40$; calc. 2.967 eV	...	[Dav01a]
i13-ME2881g	2.983	24.06	415.6	**GR7a**-DAP32g	$^*V_1^+$	+0.24% (+7.1 meV); $s = 34$; calc. 2.982 eV	...	[Dav01a]
i13-ME2881h	2.989	24.11	414.8	**GR7b**-DAP32h	$^*V_1^+$	+0.25% (+7.4 meV); $s = 32$; calc. 2.988 eV	...	[Dav01a]
i13-HA2985	2.990	24.11	414.7	**N3**	$V_1N_3^\circ$	+0.2%	...	[Dav99a]
i13-ME2881i	3.004	24.23	412.7	**GR8a,b**-DAP32i	$^*V_1^+$	+0.23% (+6.9 meV); $s = 27$; calc. 3.007 eV	...	[Dav01a]
i13-ML3188-s10	3.008	24.26	412.2	3.188 eV-s10	$^*(C_2)_iN_1^-$	(−3.8%) at 183/190 meV LVM (C–C)	...	[Col87b]
i13-MA2881j	3.010	24.28	411.9	**GR8c,d**-DAP32j	$^*V_1^+$	+0.23% (+6.9 meV); $s = 26$; calc. 3.011 eV	...	[Dav01a]
i13-ML3188-s11	3.016	24.33	411.0	3.188 eV-s11	$^*(C_2)_iN_1^-$	(−2.2%) at 175/179-meV LVM (N–C)	...	[Col87c]
i13-ML3188-s12	3.035	24.48	408.5	3.188 eV-s12	$^*(C_2)_iN_1^-$	(−3.7%) at 156/162-meV LVM (C–C)	...	[Col87c]

(continued)

Table 7.2 (continued)

Line-label	Energy (meV)	Frequ. (cm^{-1})	Wavel. (μ m)	Name[a]	Impur./defect	$^{13}C/^{12}C$ shift (%)[b], comment	Figs. **3–6.** [Zai01]; Fig. **7.** [Zai98]	References
i13-ML3188-s13	3.056	24.65	405.7	3.188 eV-s13	$^*(C_2)_iN_1^-$	(−2.3%) at 153/138-meV LVM (N–C)	...	[Col87b]
i13-ML3188-s14	3.085	24.88	401.9	3.188 eV-s14	$^*(C_2)_iN_1^-$	(−1.9%) at 106/108-meV LVM (N–C)	...	[Col87b]
i13-ML3188-s15	3.116	25.13	397.9	3.188 eV-s15	$^*(C_2)_iN_1^-$	(−3.2%) at 76/78-meV QLVM (1N+1C)	...	[Col87c]
i13-ML3188-s16	3.136	25.29	395.4	3.188 eV-s16	$^*(C_2)_iN_1^-$	(−3.4%) at 56/59-meV QLVM (1N+2C)	...	[Col87c]
i13-MA3150	3.154	25.44	393.1	**ND1**	$^*V_1^-$	+0.1%	...	[Dav99a]
i13-ML3188	3.191	25.74	388.5	**3.188 eV**	$^*(C_2)_iN_1^-$	+(0.1–0.3)%, ZPL of i13-DAP72a-p	...	[Col87c]
i26-ML4204a	4.226	34.09	293.4	5RL-DAP80a	$^*(C_2)_iN_1^\circ$	+0.52%, (+22 meV), $s = 8$, calc. 4.217 eV	...	[Col88c]
i26-ML4204b	4.265	34.40	290.7	5RL-DAP80b	$^*(C_2)_iN_1^\circ$	+0.54%, (+23 meV), $s = 10$, calc. 4.255 eV	...	[Col88c]
i26-ML4204c	4.307	34.74	287.9	5RL-DAP80c	$^*(C_2)_iN_1^\circ$	+0.40%, (+17 meV), $s = 14$, calc. 4.307 eV	...	[Col88c]
i13p-ML4582-s1	4.354	35.12	284.7	5RL-s1	$^*(C_2)_iN_1^\circ$	(−2.6%) at −231/-237 meV LVM (C–C); **40%** ^{13}**C**	...	[Col93e]
i26 ML4582-s1	4.363	35.19	284.2	5RL-s1	$^*(C_2)_iN_1^\circ$	(−4.2%) at −227/-237 meV LVM (C–C); 100% 13C	...	[Col88c]
i26 ML4582-s2	4.404	35.52	281.5	5RL-s2	$^*(C_2)_iN_1^\circ$	(−3.6%) at −186/-193-meV LVM (1N+2C)	...	[Col88c]

i26 ML4582-s3	4.421	35.66	280.4	5RL-s3	$^*(C_2)_iN_1^\circ$	(−3.4%) at −169/-175-meV LVM (1N+2C)	...	[Col88c]
i26 ML4204d	4.438	35.80	279.4	5RL-DAP80d	$^*(C_2)_iN_1^\circ$	+0.23%, (+10 meV), $s = 58$, calc. 4.451 eV	...	[Col88c]
i26-ML4204f	4.492	36.23	276.0	5RL-DAP80f	$^*(C_2)_iN_1^\circ$	+0.25%, (+11 meV), $s = 118$, calc. 4.492 eV	...	[Col88c]
i26 ML4582-s4	4.534	36.57	273.4	5RL-s4	$^*(C_2)_iN_1^\circ$	(−4.2%) at −56/-58-meV QLVM (1N+2C), calc. 56.3	...	[Col88c]
i13p ML4582	4.585	36.98	270.4	**5RL**-DAP80z	$^*(C_2)_iN_1^\circ$	+0.07%; (+3 meV), ZPL of i13-DAP80, **40%** ^{13}**C**	...	[Col93e]
i26-ML4582	4.590	37.02	270.1	**5RL**-DAP80z	$^*(C_2)_iN_1^\circ$	+0.17%; (+8 meV), ZPL of i26-DAP80a-f (line e not observed)	...	[Dav99a]
i13-HL5356a-s4	5.075	40.94	244.3	BE-B(a)-s4	B	(−4.2%) at −295/-308 meV; named D_2 (see Table 9.2.6)	...	[Ruf98]
i13-HL5356b-s2	5.085	41.02	243.8	BE-B(b)-s2	B	(−3.3%) at –297–307 meV; named D_2'	...	[Ruf98]
i13-HL5409a-s4	5.135	41.42	241.4	FE(a)-s4		(−2.0%) at −288/-294 meV; named B_2	...	[Ruf98]
i13-HL5356a-s5	5.215	42.06	237.7	BE-B(a)-s5	B	(−4.9%) at (LA/LO) −155/-163 meV; named D_1''	...	[Col90b]
i13-HL5356a-s6	5.235	42.22	236.8	BE-B(a)-s6	B	(−3.5%) at (TO) −136/-141 meV; named D_1	...	[Col90b]
i13-HL5356b-s3	5.246	42.31	236.3	BE-B(b)-s3	B	(−3.5%) at (TO) −136/-141 meV; named D_1'	...	[Col90b]

(continued)

Table 7.2 (continued)

Line-label	Energy (meV)	Frequ. (cm^{-1})	Wavel. (μ m)	Name[a]	Impur. /defect	$^{13}C/^{12}C$ shift (%)[b], comment	Figs. **3–6.** [Zai01]; Fig. **7.** [Zai98]	References
i13-HL5409a-s6	5.266	42.48	235.4	**FE(a)-s6**		(−3.7%) at (LA/LO) −157/-163 meV; $\mathbf{C}_1$, *very weak*	...	[Col90b]
i13-HL5409a-s7	5.288	42.65	234.5	**FE(a)-s7**		(−4.3%) at (TO) −135/-141 meV; $\mathbf{B}_1$, *dominant*	...	[Col90b]
i13-HL5409a-s8	5.341	43.08	232.1	**FE(a)-s8**		(−5.7%) at (TA) −82/-87 meV; named $\mathbf{A}_1$, *weak*	...	[Ruf98]
i13-HL5356a	5.370	43.31	230.9	BE-B(a)	B	+0.26%, (+14 meV); ZPL, named D_0	...	[Col90b]
i13-HL5356a	5.382	43.41	230.4	BE-B(b)	B	+0.26%, (+14 meV); ZPL, named D_0'	...	[Col90b]
i13-HL5409a	5.423	43.74	228.6	**FE(a)**		+0.26%, (+14 meV); ZPL **(indirect)**	...	[Col90b]
i13-HL5409b	5.430	43.80	228.3	**FE(b)**		+0.26%; (+14 meV); ZPL **(indirect)**	...	[Col90b]
i13 band gap	5.504	44.40	225.2	Band-to-band		+0.25% (+13.6 meV)	...	[Col90b]

[a] FE=free exciton, BE=bound exciton (boron); (a), (b)=difference from split-off valence band: 7 meV (FE) or 12 meV (BE)
[b] For the exciton sidebands the uncertainty of the shift is ±1 meV

7.3 Isotopic Shifts from ^{15}N (i15, i30)

Table 7.3 Isotopic shifts from ^{15}N (theoretical vibrational shift is −3.4% (N–N) or −1.9% (N–C)

Line-label	Energy (meV)	Frequ. (cm^{-1})	Wavel. (μ m)	Name	Impur. /defect	$^{14}N/^{15}N$ Shift, comment	Figs. **3–6.** [Zai01]; Fig. **7.** [Zai98]	References
i15-LA0105 b/	129.6	1,045	9.569	**C**-center(b/)	N_1	*No shift* (C–C)	7.23	[Col82f]
i15-LA0105 c/	136.0	1,097	9.116	**C**-center(c/)	N_1	*No shift* (C–C)	7.23	[Col82f]
i15-HA0105 d/	137.7	1,111	9.004	**C**-center(d/)	N_1	−1.3%	7.23	[Col82f, Dav94a]
i15-MA0060 e/	157.2	1,268	7.886	**A**-center(e/)	N_2°	−1.1%	. . .	[Dav94b]
i15-LA0105 f/	162.4	1,310	7.634	**C**-center(f/)	N_1	*No shift* (C–C)	7.23	[Col82f, Dav94a]
i15-LA0105 g/	166.6	1,344	7.442	**C**-center(g/)	N_1	*No shift* (C–C)	7.23	[Col82f, Col88c]
i15-MA0168	167.5	1,351	7.402	?	N, C	−0.8% (−1.4 meV)	. . .	[Col87a]
i15-HA0041 e/	170.4	1,374	7.278	**D**-center(e/)	$^*(C_{2n})_i^{\circ}$	*No shift*; platelets (C–C)	7.22b	[Sum88, Dav94b]
i15-MA0180	176.8	1,426	7.013	$\mathbf{H}_1$ **a,**3.188 eV	$^*(N_1C_1)_i^{\circ}$	−1.7%; LVM (177/179 meV) of 3.188 eV ce	. . .	[Col87a]
i15-MA0187	183.4	1,479	6.761	C–N str.	N, C	−1.8% LVM (183/186 meV)	. . .	[Col87a], [Col88c]

(continued)

Table 7.3 (continued)

Line-label	Energy (meV)	Frequ. (cm^{-1})	Wavel. (μ m)	Name	Impur. /defect	$^{14}N/^{15}N$ Shift (%), comment	Figs. **3–6.** [Zai01]; Fig. **7.** [Zai98]	References
i15-MA0190	190.4	1,535	6.515	**3.188 eV**	$^{*}(N_1C_1)_i^{\circ}$	*No shift*, LVM (190/190 meV) (C–C)	...	[Col87b]
i15-MA0195	194.7	1,570	6.368	?	C	*No shift...*	...	[Col88c]
i15-MA0212	208.0	1,678	5.960	?	N, C	−1.7% (−3.5 meV), LVM C–N str.?	...	[Col88c]
i30-MA0230	225.6	1,820	5.494	0.23 eV ce.	$^{*}(N_2)_i$	−2.0% (−4.5 meV), ^{15}N–^{15}N line (60% ^{15}N)	**3.**12, 7.20	[Col87b]
I15-MA0230	227.3	1,833	5.456	0.23 eV ce.	$^{*}(N_2)_i$	−1.2% (−2.8 meV), ^{14}N–^{15}N line (60% ^{15}N)	**3.**12, 7.20	[Col87b]
i15-MA0238	238.5	1,924	5.198	**3.188 eV**	$^{*}(N_1C_1)_i^{\circ}$	*No shift* at (238/238 meV), LVM (C–C)	...	[Col88c]
i15-MA0245	244.9	1,975	5.063	**5RL**	$^{*}(C_2)_iN_1^{\circ}$	*No shift* at (245/245 meV), LVM(C–C)	3.12	[Col87b]

i15-ML1910e	1.989	16,040	0.6233	DAP47e	$^{*}V_1N_1^{\circ}$	−3.0% (162/167 meV), $s = 50$, calc. 1.989 eV	...	[Col87b]
i15-ML1910f	2.008	16,200	0.6174	DAP47f	$^{*}V_1N_1^{\circ}$	−6.0% (143/152 meV), $s = 64$, calc. 2.007 eV	...	[Col87b]
i15-ML1910g	2.026	16,340	0.6119	DAP47g	$^{*}V_1N_1^{\circ}$	−3.0% (125/129 meV), $s = 86$, calc. 2.027 eV	...	[Col87b]
i15-ML1910h	2.050	16,540	0.6048	DAP47h	$^{*}V_1N_1^{\circ}$	−6.0% (101/108 meV), $s = 130$, calc. 2.050 eV	...	[Col87b]
i15-ML1910i	2.057	16,590	0.6027	DAP47i	$^{*}V_1N_1^{\circ}$	−1.0% (94/95 meV), $s = 162$, calc. 2.061 eV	...	[Col87b]
i15-ML2154-s2	2.107	17,000	0.5884	NV°-s2	$^{*}V_1N_1^{\circ}$	−8.3% (44/48 meV), QLVM: 1N+3C	...	[Col87b]
i15-ML2154	2.151	17,350	0.5764	**NV°**,DAP47z	$^{*}V_1N_1^{\circ}$	−0.1% (−3.0 meV)	...	[Col87b]
i15-ML3188-s1	2.998	24,180	0.4136	3.188 eV-s1	$^{*}(C_2)_iN_1^{-}$	*No shift* at −190/-190 meV, LVM(C–C)	...	[Col87b]
i15-ML3188-s2	3.011	24,290	0.4117	3.188 eV-s2	$^{*}(C_2)_iN_1^{-}$	(−1.7%) at −176/-179 meV, LVM(N–C)	...	[Col87b]
i15-ML3188	3.187	25,710	0.3890	3.188 eV	$^{*}(C_2)_iN_1^{-1}$	−0.01%(−0.4 meV), ZPL of DAP72; 50% 15N	...	[Col87b]

7.4 Isotopic Shifts from Si_2 (^{28} Si+^{29}Si or ^{30}Si) (i57, i58)

Table 7.4 Isotopic shifts from Si_2 (i57, i58) natural silicon isotopes (92% ^{28}Si, 5% ^{29}Si, 3% ^{30}Si)

Line-label[a,b]	Energy (meV)	Frequ. (cm^{-1})	Wavel. (μ m)	Name[c]	Impur. /defect	$^{56}Si_2/^{57}Si_2$, $^{58}Si_2$isotope shift, comment	Figs. **3–6.** [Zai01]; Fig. **7.** [Zai98]	References
i58-HL1682 a	1.6814	13.562	737.4	*2Si 1.68 a	$*(V_3^{58}Si_2)^+$	−0.04%; (−0.7 meV), ZPL DAP54	5.47	[Cla95, Ste95]
i58-HL1682 b	1.6816	13.564	737.2	*2Si 1.68 b	$*(V_3^{58}Si_2)^+$	−0.04%, (−0.7 meV)	5.47	[Cla95, Ste95]
i57-HL1682 a	1.6817	13.565	737.2	*2Si 1.68 a	$*(V_3^{57}Si_2)^+$	−0.02%, (−0.4 meV)	5.47	[Cla95, Ste95]
i57-HL1682 b	1.6819	13.566	737.1	*2Si 1.68 b	$*(V_3^{57}Si_2)^+$	−0.02%, (−0.4 meV)	5.47	[Cla95, Ste95]
i58-HL1682 c	1.6825	13.571	736.9	*2Si 1.68 c	$*(V_3^{58}Si_2)^+$	−0.04%, (−0.7 meV)	5.47	[Cla95, Ste95]
i58-HL1682 d	1.6827	13.573	736.7	*2Si 1.68 d	$*(V_3^{58}Si_2)^+$	−0.04%, (−0.7 meV)	5.47	[Cla95, Ste95]
i57-HL1682 c	1.6828	13.574	736.7	*2Si 1.68 c	$*(V_3^{57}Si_2)^+$	−0.02%, (−0.4 meV)	5.47	[Cla95, Ste95]
i57-HL1682 d	1.6831	13.576	736.6	*2Si 1.68 d	$*(V_3^{57}Si_2)^+$	−0.02%, (−0.3 meV)	5.47	[Cla95, Ste95]

[a] Also observed in absorption, see Table 9.4.2, where calculated energies (20 lines) and calculated and observed intensities are given

[b] Lines a, b, c, d arise from the splittings: 0.2 meV in the ground state and 1.1 meV in the excited state [c] *2Si 1.68 = $*(V_3Si_2)^+$ (see Table 8.3c)

7.5 Isotopic Shifts from Ni_1(i60, i61, i62, i64), or Ni_2(i118, i120, i122)

Table 7.5 Isotopic shifts from natural nickel Ni_1 (i60, i61, i62, i64) or Ni_2 (i118, i120, i122, i124) isotopes, natural abundances are (^{58}Ni (68%), ^{60}Ni (26%), ^{61}Ni *(1%)*, ^{62}Ni (4%), and ^{64}Ni (1%))

Line-label	Energy (eV)	Frequ. (10^3 cm^{-1})	Wavel. (nm)	Name[a]	Impur. /defect	Isotope shift (%)/comment	Figs. **3–6.** [Zai01]; Fig. **7.** [Zai98]	References
i62-HA1401 a	1.40063	11.297	885.2	*1Ni 1.40a	$^{62}Ni_1$	−0.024%; (−0.33 meV); line b at 1.40343 eV	...	[Dav89, Naz91b]
i60-HA1401 a	1.40096	11.299	885.1	*1Ni 1.40a	$^{60}Ni_1$	−0.012%, (−0.17 meV)	...	[Dav89, Naz91b]
i64-HA1401 b	1.40327	11.319	883.5	*1Ni 1.40b	$^{64}Ni_1$	−0.035%, (−0.49 meV)	**5.**12c	[Dav89, Naz91b]
i62-HA1401 b	1.40343	11.320	883.3	*1Ni 1.40b	$^{62}Ni_1$	−0.024%, (−0.33 meV)	**5.**12c	[Dav89, Naz91b]
i61-HA1401 b	1.40354	11.321	883.3	*1Ni 1.40b	$^{61}Ni_1$	−0.016%, (−0.22 meV)	**5.**12c	[Dav89, Naz91b]
i60-HA1401 b	1.40359	11.321	883.3	*1Ni 1.40b	$^{60}Ni_1$	−0.012%, (−0.17 meV)	**5.**12c	[Dav89, Naz91b]
i60-LL1802 b	1.842	14.858	673.0	*S7-DAP62b	$^{60}Ni_1$	+0.5%, (+9 meV); shell 162	...	[Iak00g]
i62-LL1802 b	1.851	14.930	669.8	*S7-DAP62b	$^{62}Ni_1$	+1.0%, (+18 meV), shell 162	...	[Iak00g]
i60 LL1802 c	1.913	15.430	648.1	*S7-DAP62c	$^{60}Ni_1$	+0.6%, (+12 meV); shell 64	...	[Iak00g]
i62 LL1802 c	1.925	15.527	644.0	*S7-DAP62c	$^{62}Ni_1$	+1.3%, (+24 meV); shell 64	...	[Iak00g]

(continued)

Table 7.5 (continued)

Line-label	Energy (eV)	Frequ. (10^3 cm^{-1})	Wavel. (nm)	Name[a]	Impur. /defect	Isotope shift (%)/comment	Figs. **3–6.** [Zai01]; Fig. **7.** [Zai98]	References
i60 LL1802 e	2.033	16.398	609.8	*S7-DAP62e	$^{60}Ni_1$	+0.6%, (+13 meV), shell 22	...	[Iak00g]
i62 LL1802 e	2.046	16.503	605.0	*S7-DAP62e	$^{62}Ni_1$	+1.3%, (+26 meV), shell 22	...	[Iak00g]
i60 LL1802 g	2.105	16.979	589.0	*S7-DAP62g	$^{60}Ni_1$	+0.5%, (+10 meV), shell 14	...	[Iak00g]
i62 LL1802 g	2.115	17.060	586.2	*S7-DAP62g	$^{62}Ni_1$	+1.0%, (+20 meV), shell 14	...	[Iak00g]
i60 LL1802 h	2.200	17.745	563.5	*S7-DAP62h	$^{60}Ni_1$	+0.6%, (+13 meV), shell 9	...	[Iak00g]
i62 LL1802 h	2.216	17.874	559.5	*S7-DAP62h	$^{62}Ni_1$	+1.2%, (+26 meV), shell 9	...	[Iak00g]
i118-HL2558a1	2.55925	20.643	484.4	*2Ni 2.56a	$^{118}Ni_2$	+0.06%, (+1.6 meV); (^{58}Ni+^{60}Ni)	**5.**124	[Col97]
i118-HL2558a2	2.55962	20.646	484.4	*2Ni 2.56a	$^{118}Ni_2$	+0.06%, (+1.6 meV), (^{58}Ni+^{60}Ni)	**5.**124	[Col97]
i120-HL2558a1	2.56085	20.656	484.1	*2Ni 2.56a	$^{120}Ni_2$	+0.12%, (+3.2 meV) (^{58}Ni+^{62}Ni, ^{60}Ni+^{60}Ni)	**5.**124	[Col97]
i120-HL2558a2	2.56120	20.659	484.1	*2Ni 2.56a	$^{120}Ni_2$	+0.12%, (+3.2 meV) (^{58}Ni+^{62}Ni, ^{60}Ni+^{60}Ni)	**5.**124	[Col97]
i122-HL2558a2	2.56274	20.671	483.8	*2Ni 2.56a	$^{122}Ni_2$	+0.18%, (+4.7 meV) (^{58}Ni+^{64}Ni, ^{60}Ni+^{62}Ni)	**5.**124	[Col97]
i122-HL2558a3	2.56304	20.673	483.7	*2Ni 2.56a	$^{122}Ni_2$	+0.18%, (+4.7 meV) (^{58}Ni+^{64}Ni, ^{60}Ni+^{62}Ni)	**5.**124	[Col97]
i124-HL2558a3	2.56449	20.685	483.4	*2Ni 2.56a	$^{124}Ni_2$	+0.26%, (+6.7 meV) (^{60}Ni+^{64}Ni, ^{62}Ni+^{62}Ni)	**5.**124	[Col97]

[a] *1Ni 1.40 =*$V_1Ni_1^+$-DAP25; *S7-DAP62= $(V_2Ni_1)N^{\circ}_{2+x+y}$; *2Ni 2.56 =*$(V_3Ni_2)^+$-DAP106; the i116 line is split by 0.35 meV into a1=2.55765, a2=2.55800, a3 (indirect)=2.55830 eV

Chapter 8
Intrinsic Defects and Their Associates in Diamond

The intrinsic defects (lattice vacancies, self-interstitials, and their associates with impurities) give rise to over 60 defects in diamond (see Tables 8.1.1–8.1.7.2 and 8.2.1–8.2.4). Many of the configurations in these tables are proposals which need confirmation (marked by a preceding *).

From the isolated lattice vacancy and divacancy, four optical centers are observed (see Table 8.1.1).

Associates of the single or double lattice vacancy with one to four nitrogen atoms are observed in 21 centers (see Table 8.1.3.1–8.1.3.6).

One to three nickel atoms form 16 associates with single, double, or triple lattice vacancies (see Tables 8.1.4–8.1.5, 9.5.1.1–9.5.1.2).

A total of 22 centers from other impurities with one or multiple lattice vacancies are listed in Table 8.1.7.1–8.1.7.2 (see also Sects. 9.1–9.7).

The isolated lattice vacancy and self-interstitial are produced by irradiation. These defects have limited thermal stability (see Sect. 8.3.1). Therefore, they are absent in natural diamond. However, associates of the vacancy with nitrogen, nickel, silicon, or hydrogen atoms are stable up to ca. $2,000\,^{\circ}$C, and are well-known defects in natural diamond, especially those with historical names: NV center for $V_1N_1^{\circ}$, N3 center for $V_1N_3^{\circ}$, and B center for $V_1N_4^{\circ}$.

8.1 Lattice Vacancies and Associates

8.1.1 The Isolated Single Vacancy

The isolated single vacancy is a typical radiation center and can be created by high energy irradiation (e.g., electrons, ions, neutrons, and γ-rays). The properties of this important defect have been investigated by theory, electron paramagnetic resonance (EPR), and optically.

B. Dischler, *Handbook of Spectral Lines in Diamond*,
DOI 10.1007/978-3-642-22215-3_8, © Springer-Verlag Berlin Heidelberg 2012

Theory. The results of theoretical calculations [Lan68, Bre95, Jou96, Mai97] are summarized in recent reviews [Dav01a, New01c]. The vacancy can exist in three charge states: V_1^-, V_1°, and V_1^+. There is a special situation, where the relative stability of the different charge states depends on the position of the Fermi level E_F. The most stable charge state are V_1^- for $E_F > (E_c - 2.2\,eV)$ (typical for type Ib), V_1° for $(E_v + 1.0\,eV) < E_F < (E_c - 2.2\,eV)$ (typical for types Ia and IIa), and V_1^+ for $E_F < (E_v + 1.0\,eV)$ (typical for type IIb).

The $\mathbf{V_1}^-$ center has stable T_d symmetry with a $^4A_2$ ground state, and allowed transitions to a calculated $^4T_1$ state at +3.3 eV.

The $\mathbf{V_1}^\circ$ center distorts from T_d to D_{2d} symmetry with a gain in Jahn–Teller energy of 0.36 eV. However, the Jahn–Teller effect is dynamic in producing effective T_d symmetry, and a splitting in the ^1E ground state of a few meV (observed is 8 meV).

The $\mathbf{V_1}^+$ center is stable in the Jahn–Teller distorted D_{2d} symmetry with a $^2T_2$ ground state. The excited states are "well above "the ground state.

The carbon atoms of the first shell relax radially outward by ca. 13% of the initial bond length (independent of the charge state).

EPR measurements: The EPR and ENDOR results confirm the theoretical description [New01c]. Fortunately, also the $\mathbf{V_1}^\circ$ center with its diamagnetic ^1E ground state has a paramagnetic $^5A_2$ excited state, which can be populated via ND1 absorption (see Table 8.1.1).

Optical measurements: The close correlation of EPR intensity and optical absorbance allows to assign the ND1 center to V_1^- and the GR1 center to V_1°, even with calibration factors [Twi99]. The GR2–GR8 lines, which are now interpreted as donor–acceptor pair transitions of V_1^+, were previously assigned to ZPL transitions of V_1°.

All three charge states of the vacancy have sideband systems from donor–acceptor pair transitions (V_1^-: DAP93, V_1°: DAP33/34, and V_1^+: DAP32, see Table 8.1.1). The V_1^+-DAP32 sidebands are very efficient in the excitation of the V_1°(a, b) luminescence, and have been named GR2 to GR8. The broad V_1°(b)-DAP33 absorption band, centered at 1.98 eV (630 nm), is (together with a blue-violet absorption band) responsible for the green color of some irradiated diamonds [All98a]. The green color of rare natural diamonds (e.g., the 41 karat volume colored green diamond gemstone of Dresden) is ascribed to irradiation from natural radioactivity [Bos89, Col97]. In a different case, a natural green skin (about 0.02-mm thick) has been reported [Wel94].

Table 8.1.1 The single and double lattice vacancy in diamond

Defect	Observed	ZPL (eV)	SB-DAP	D (eV)	Name/comment
V_1^-	MA, EPR(S1, S2) $S = 3/2$	3.150	93a–d	+1.10	**R10** = **ND1**/LVM: (4×) +80 meV; peak of broad band at 3.45 eV; *no luminescence (e^- emission in exc. st.)*; photochromic with GR1 at $1.1 < E < 2.8$ eV; stable T_d symmetry with $^4A_2$ gr. st.
V_1°	MA, EPR, $S = 4$, (only in $^5A_2$ exc. st.; obs. via ND1 absorption)	1.673	33a–k	+1.20	**GR1**b; line(a) at 1.665 eV/QLVM: +41, LVM: +68, +93 meV; gr. st. spl. = 8 meV from dynamic JT; effective T_d symmetry with ^1E ground state; peak of broad band at 1.95 eV; photochromic with ND1 at $E > 3.1$ eV
V_1°	ML	1.673	34a–g	−1.20	**GR1**b; line(a) at 1.665 eV/QLVM: −37, −91; LVM: −70 meV; gr. st. *spl.* = 8 meV from dynamic JT
$^*V_1^+$	MA, ME; EPR (NIRIM3) $S = 1/2$	2.781	32a–m	+1.31	**GR2b-GR8e** (at 2.881–3.060 eV), very weak ZPL (GR2a), PLE of GR1, observed only in irradiated type IIb or boron doped CVD diamonds; *no luminescence (e^-capture in exc. st.)*; resolved spectrum also in photoconductivity [Ver74]; D_{2d} symmetry from JT distortion, $^2T_2$ ground state ($S = 1/2$)
(V_2^-)	EPR (W29)	–	–	–	**W29**(EPR), $S = 3/2$; *no optical lines*; C_{2h} symmetry from JT distortion
V_2°	MA, EPR(R4 = W6) $S = 1$	2.419	15a–l	+1.63	**TH5**/weak ZPL; DAP lines at 2.543 (former ZPL) to 2.909 eV/no QLVM or LVM SBs, C_{2h} symmetry from JT distortion

EPR see [New01c]

SB-DAP = sideband DAP with ZPL = L = limiting energy, and D = dielectric factor; gr. st. spl. = ground state splitting, exc. st. = excited state; JT = Jahn–Teller effect

8.1.2 The Isolated Divacancy

The isolated divacancy occurs in the neutral state ($V_2{}^{\circ}$ = optical TH5 center). The negative charge state ($V_2{}^{-}$) is observed only in EPR (W29) [New01b]. Divacancies are formed by aggregation of two single vacancies at anneal temperatures $T >$ 600 °C. The concentration of $V_2{}^{\circ}$ in irradiated and annealed diamond is always low because the migrating single vacancies are preferentially trapped by nitrogen centers, and seldom by other single vacancies. A group of weak absorption lines is observed in the 2.4–2.9-eV range and has been named TH5 [Cla56c]. The group of 13 lines arises from the sideband $V_2{}^{\circ}$-DAP15 (see Table 8.1.1) with a very weak ZPL at 2.419 eV (previously the first sideband at 2.543 eV was erroneously assigned to the ZPL).

8.1.3 Associates of Single or Double Vacancies with Nitrogen

When the vacancies begin to migrate (at a typical anneal temperature of 800 °C, E_A = 2.4 eV), most of them are trapped by impurities before reaching the surface. Some very stable (beyond T = 1,600–2,300 °C) associates with nitrogen are observed in natural, and also in as-grown synthetic HPHT or in as-grown synthetic CVD diamond. They are named: The NV center (= $V_1N_1{}^{\circ}$), the 638 nm center (= $V_1N_1{}^{-}$), the H3 center (= $V_1N_2{}^{\circ}$), the S1 center (= $V_1N_2{}^{-}$), the N3 center (= $V_1N_3{}^{\circ}$), the 2,526-eV center (= $V_1N_3{}^{-}$), the B-center (= $V_1N_4{}^{\circ}$), and the N9 center (= $V_1N_4{}^{-}$).

Metastable centers are H1(a) = $V_1C_4N_1{}^{\circ}$ (precursor of NV = $V_1N_1{}^{\circ}$, Table 8.1.3.1), H1(b, d, g) = $V_1C_4N_2{}^{\circ}$ (precursor of H3 = $V_1N_2{}^{\circ}$, Table 8.1.3.1), and H1(c, e, f) = $V_1N_4V_1{}^{\circ}$ (precursor of H4 = $N_3V_2N_1{}^{\circ}$, Table 8.1.3.4). Similar to the SB-DAPs with hydrogen centers (see Sect. 9.3), the precursor DAPs 30 and 31 are sidebands of a dominant infrared vibration. The DAP analysis yields L = 154 meV (from the A-center = N_2, see Table 9.1.1.3) for H1(b,d,g)-DAP30, and L = 147 meV (from the B-center = V_1N_4, see Table 8.1.3.5) for H1(c,e,f)-DAP31. The sharpness of the precursor DAP sidebands indicates a similar sharpness for the involved infrared vibrations (pseudo ZPL), which in direct observation are evidently broadened by resonance with lattice vibrations.

The trapping of a vacancy by nitrogen centers has been theoretically analyzed [Mai94a], see Sect. 9.1.6.2.

Table 8.1.3.1 Associates of the single vacancy with one or two nitrogens

Defect	Observed	ZPL (eV)	SB-DAP	*D* (eV)	Name/comment
$^{*}V_1N_1^-$(a)	NA, HA, LA, MA	1.943	48a–j	+1.24	**638 nm** or NV^- center/QLVM: +63; LVM: +75, +143; (combinations +213, +417, +520, +587) meV
$^{*}V_1N_1^-$(a)	NL, HL, LL, ML	1.943	49a–o	−1.24	**638 nm** or NV^- center/QLVM: −63; LVM: −123 meV
$^{*}V_1N_1^-$(b)	MA	4.328	...	...	NV^-(b) center; trigonal symmetry
$V_1N_1^{\circ}$	NA, HA, LA, MA	2.154	46a–f	+1.39	**NV = 575-nm** center/QLVM: +90; LVM: +164 meV
$V_1N_1^{\circ}$	NL, HL, LL, ML	2.154	47a–i	−1.39	**NV = 575-nm** center/QLVM: −48(1N+3C); LVM: −155 meV
$^{*}V_1C_4N_1^{\circ}$	MA	0.180	...	...	**H1(a)**/in irradiated and annealed diamond/precursor of $V_1N_1^{\circ}$ = NV
$^{*}V_1C_4N_1^{\circ}$	MA	1.702	45a–i	+1.53	In irradiated and annealed diamond/weak ZPL, precursor of $V_1N_1^{\circ}$ = NV; three luminescence lines at mirror positions; DAP lines at **1.909**, named **1.908 eV center** (hole burning [Sil95]); **2.087**(named **595 nm center**), QLVM: +70 (61 + 80?), LVM: +165 meV
$^{*}V_1N_2^-$	MA	1.256	77a–k	+1.35	**H2**/LVM: +65, +167 (sharp) meV
$^{*}V_1N_2^-$	ML	1.256	78a–g	−1.35	**H2**/LVM −62 meV
$^{*}V_1N_2^{\circ}$(a)	NA, HA, LA, MA	2.464	50a–h	+1.34	**H3(a)**/DAP lines **H6–H12** at 2.975–3.343 eV (*s* = 8 − −1)/QLVM: +46 meV
$^{*}V_1N_2^{\circ}$(a)	NL, HL, LL, ML	2.464	51a–f	−1.34	**H3(a)**/QLVM: −41; −81; LVM: −157 meV

(continued)

Table 8.1.3.1 (continued)

Defect	Observed	ZPL (eV)	SB-DAP	D (eV)	Name/comment
$^{*}V_1N_2{}^{\circ}$(b)	NA, HA, LA, MA	3.361	52°-g	+1.34	**H3(b)** = **H13**/DAP lines **H15-H18** at 3.457–3.560 eV (s = 228–34)
$^{*}V_1N_2{}^{+}$	NL, HL, LL, ML	2.429	…..	…..	**S1(a)**; line (a) 2.429 eV (20× weaker than (b))/34 meV exc. st. splitting
$^{*}V_1N_2{}^{+}$	NL, HL, LL, ML	2.463	43a–g	−1.33	**S1(b)**; line (a) 2.429 eV (20×weaker)/QLVM: −48; LVM: −67 meV; energy transfer between S1(b) and H3(a) centers
$^{*}V_1C_4N_2^{\circ}$	MA	0.154	30a–d	+1.45	In irradiated and annealed type IaA/ZPL indirect, precursor of H3(a, b) = $^{*}V_1N_2$; DAP lines **H1(d)** at 0.331 eV (s = 50), **H1(g)**, at 0.551 eV (s = 10), **H1(b)**, at 0.612 eV (s = 8)/QLVM: +64; LVM: +154 meV/ZPL resonant with A(e′), see Table 9.1.1.3

SB-DAP = *sideband* DAP with ZPL = L = limiting energy, D = dielectric factor, and s = shell number

Table 8.1.3.2 Associates of the single vacancy with three nitrogens

Defect	Observed	ZPL (eV)	SB-DAP	D (eV)	Name/comment
$^{*}V_1N_3{}^{-}$	NL, ML	2.526	76a–n	−1.19	QLVM: −48; LVM: −144 meV/dyn. JT
$V_1N_3{}^{\circ}$(a)	NL, HL, ML,	2.297	55a–e	−1.33	**N3(a)** (formerly **N3b**)/delayed luminescence; observed only in samples with *large* platelets (with *small* platelets see the 2.202 eV line below)/QLVM: −67 meV/dyn. JT

(continued)

Table 8.1.3.2 (continued)

Defect	Observed	ZPL (eV)	SB-DAP	*D* (eV)	Name/comment
$V_1N_3{}^{\circ}$(b)	NL, HL, ML	2.680	56a–d	−1.33	**N3(b)** (formerly **N3a**)/delayed luminescence/QLVM: −59 meV/dyn. JT
$V_1N_3{}^{\circ}$(c)	NA, HA, LA, MA, NE, HE, LE, ME	2.985	23a–o	+1.33	**N3(c)** (formerly **N3**); ZPL split by 0.59 meV; DAP lines N4 at 3.610 (s = 3) and N5 at 3.730 eV (s = 1)/PLE of N3(a, b, c)/QLVM: +36, +64 meV; IR: +177 meV/dyn. JT
$V_1N_3{}^{\circ}$(c)	NL, HL, LL, ML	2.985	24a–q	−1.33	**N3(c)** (formerly **N3**)/QLVM: −36, −69; LVM: −160 meV/dyn. JT
$^*V_1N_3{}^{+}$	NL	3.414	22a–l	−1.36	QLVM:-36, −74 meV
$^*(V_1H_1)N_3{}^{\circ}$	NA	0.178	–	–	nat. N–H bend
	NA	0.390	–	–	nat. N–H stretch
$^*V_1N_3(C_2)_i{}^{\circ}$	NA	2.202	–	–	(Named **563-nm** center) in natural brown diamonds with high hydrogen content/QLVM: +32, +72 meV
$^*V_1N_3(C_2)_i{}^{\circ}$	NL, HL, ML	2.202	–	–	Delayed luminescence/similarity with N3(a); observed only in samples with *small* platelets

SB-DAP = sideband DAP with ZPL = *L* = limiting energy, *D* = dielectric factor, and s = shell number; dyn. JT = dynamic Jahn–Teller effect

Table 8.1.3.3 Associates of the single vacancy with four nitrogens

Defect	Observed	ZPL (eV)	SB-DAP	*D* (eV)	Name/comment
$^*V_1N_4^-$	NA, NE	5.252	63a–g	+1.43	**N9**(a); lines (b) 5.262 eV, (c) 5.277 eV/QLVM:+75, LVM: +120, +140 meV
$^*V_1N_4^-$	NL	5.252	64a–i	−1.43	**N9**(a); lines (b) 5.262 eV, (c) 5.277 eV/LVM: −65, −76 meV

(continued)

Table 8.1.3.3 (continued)

Defect	Observed	ZPL (eV)	SB-DAP	*D* (eV)	Name/comment
$V_1N_4^{\circ}$	NA, MA	4.191	82a–g	+1.64	**B**(b); lines (a) 4.184, (c) 4.197 eV/QLVM: +36, +72 meV; IR: +94,+125, +134, **+146**, +165 meV (see Table 9.1.3.5)/dyn. JT
$^*V_1N_4^{+}$	HL	2.510	–	–	Four sidebands at +40 meV
$^*(V_1H_1)N_4^{\circ}$	MA	0.192	–	–	NH bend
$^*(V_1H_1)N_4^{\circ}$	MA	(0.424)	–	–	Expected NH stretch
$^*V_1N_4(C_2)_i^{\circ}$(a)	NA, HA	2.145	–	–	***F(a)**/QLVM: +23 meV/IR lines F(a′)–F(i′), see Table 8.1.7
$^*V_1N_4(C_2)_i^{\circ}$(a)	NL, HL	2.145	–	–	***F(a)**/ZPL of F(α1) = "Red band" with 1.80-eV peak, 14 SBs 31 meV
$^*V_1N_4(C_2)_i^{\circ}$(b)	NE, HE	2.721	–	–	***F(b)**/PLE of F(β1) = "Yellow band" with 2.25-eV peak
$^*V_1N_4(C_2)_i^{\circ}$(b)	NL, HL	2.721	–	–	***F(b)**/ZPL of F(β1) = "Yellow band" with 2.25-eV peak, seven SBs 34 meV
$^*V_1N_4(C_2)_i^{\circ}$(c)	NA, HA	4.567	81a–f	+1.67	***F(c)**/QLVM: +55; LVM: +136, +154 meV

SB-DAP = sideband DAP with *L* = ZPL = limiting energy, *D* = dielectric factor, and *s* = shell number; dyn. JT = dynamic Jahn–Teller effect

Table 8.1.3.4 Associates of the double vacancy with one to four nitrogens

Defect	Observed	ZPL (eV)	SB-DAP	*D* (eV)	Name/comment
$^*V_2N_1{}^{\circ}$	LL, ML	2.329	–	–	Very strain sensitive, variable 2.322–2.335 eV
$^*V_2N_2$	MA	2.086	–	–	**595 nm** center; in irradiated and annealed (275–800° C) type Ib diamond; resonant with DAP45, s = 12 line of $^*V_1C_4N_1^{\circ}$ at 2.086 eV; QLVM: +75 (2N), LVM: +165 meV
$^*V_2N_3{}^{\circ}$	MA	2.916	–	–	Very weak line

(continued)

Table 8.1.3.4 (continued)

Defect	Observed	ZPL (eV)	SB-DAP	*D* (eV)	Name/comment
$^*V_1N_4V_1{}^\circ$	MA	0.147	31a–e	+1.45	In irradiated and annealed type IaB/ZPL indirect, precursor of H4(a,b) = $^*N_3V_2N_1$; DAP lines **H1(e)** at 0.362 eV (s = 50), **H1(f)**, at 0.545 (s = 10), **H1(c)**, at 0.641 eV (s = 8)/QLVM: +64; LVM: +154 meV/ZPL resonant with B(d′), see Table 8.1.3.5
$^*N_3V_2N_1{}^\circ$	MA	2.498	37a–h	1.60	**H4**(b); line (a) at 2.417 eV (ten times weaker than H4(b)/QLVM: +38 meV
$^*N_3V_2N_1{}^\circ$	ML	2.498	38a–g	−1.60	**H4**(b);/QLVM: −42, −81 meV

SB-DAP = sideband DAP with L = ZPL = limiting energy, D = dielectric factor, and s = shell number

Table 8.1.3.5 IR absorption bands and vibronic UV absorption lines and bands from four nitrogens surrounding a vacancy: ($V_1N_4{}^\circ$ = **B center**)

Line/band	*E* (meV) *(cm⁻¹)*	Width (meV) *(%)*	Rel. Intens (%)	Comment
B(a′)	93.5 *754*	14 *15*	6	Variable 93.5–96.7 meV
B(b′)	125.2 *1,010*	8 *6*	7	–
B(c′)	134.5 *1,085*	3 *2*	2	–
B(d′)	145.7 *1,175*	10 *7*	53	Close to the ZPL of $^*V_1N_4V_1$-SB-DAP31 (see Table 8.1.5)
B(e′)	150.6 *1,215*	6 *4*	6	Overlapping band
B(f′)	158.9 *1,282*	14 *9*	14	Overlapping band
B(g′)	165.1 *1,332*	1 *0.6*	9	–
Line/band	*E* (eV)	Width (eV)		Comment
*B(al)	4.184	0.003	–	ZPL(al) of B(α2)
*B(a2)	4.191	0.003	–	ZPL(a2) of B(α2)
*B(a3)	4.197	0.003	–	ZPL(a3) of B(α2)

(continued)

Table 8.1.3.5 (continued)

Line/band	*E* (meV) *(cm⁻¹)*	Width (meV) *(%)*	Rel. Intens (%)	Comment
B(α2)	4.70	0.56	–	Absorption band, same band in photoconductivity [Den67]
DAP82a-d	4.307–4.431	–	–	SB-DAP; L = B(a1–3) ZPL (4.191 eV), *D* = 1.64

No luminescence from UV center
DAP: L = limiting energy, D = dielectric factor of donor–acceptor pair

Table 8.1.3.6 Absorption lines and bands from four nitrogens surrounding a vacancy associated with a self-interstitial: $V_1N_4(C_2)_i{}^{\circ}$ = **F center**

Line/band	*E* (meV) *(cm^{-1})*	Width (meV) *(%)*	Rel. Intens (%)	Comment
F(a′)	121.0 *976*	6 *x5*	4	–
F(b′)	126.7 *1,024*	6 *5*	2	Also observed as +127 meV LVM sideband at *F(c) (4.567 eV)
F(c′)	144.1 *1,162*	8 *6*	32	Close to B(d′) center line at 145.7 meV (see Table 8.2e)
F(d′)	154.0 *1,242*	7 *5*	23	Also observed as +154 meV LVM sideband at *F(c) (4.567 eV)
F(e′)	157.5 *1,270*	7 *5*	21	–
F(f′)	161.0 *1,299*	6 *4*	13	–
F(g′)	164.0 *1,323*	1.5 *1*	2	–
F(h′)	191.4 *1,544*	3 *2*	1	Typical for $(C_2)_i$, see Table 8.2.1
F(i′)	195.9 *1,580*	5 *3*	2	Typical for $(C_2)_i$, see Table 8.2.1

(continued)

Table 8.1.3.6 (continued)

Line/band	E (eV)	Width (eV)	Rel. Intens.(%)	Comment
*F(α1)	1.80	0.30	–	Lumin. "Red band"; 14 SBs 31 meV [Col82b]
*F(a)	2.145	–	–	ZPL(absorption and luminescence) of F(α1, α2); very weak [Col82b], SBs: +23, −31 meV
*F(α2)	2.55	0.27	–	Absorption band; SBs 23 meV; correlates with F center IR absorption [Col82b]
*F(β1)	2.25	0.55	–	lumin. "Yellow band"; 7 SBs 34 meV [Col82b]
*F(b)	2.721	0.001	–	ZPL (absorption and PLE) of F(β1, β2) [Col82b], SBs: −34, +38 meV
*F(β2)	3.05	0.45	–	Absorption band; nine SBs 38 meV [Col82b]
(*F(γ1))	(4.30)	–	–	(Expected luminescence band)
*F(c)	4.567	0.01	–	ZPL (absorption) of F(γ2) and of $*V_1N_4(C_2)_i{}^{\circ}$(c) DAP81a-g (D = 1.67, see Table 8.2.3),/LVM: +55 (harmonic 106), 126, 154 meV [Naz87]
*F(γ2)	5.00	0.60	–	Absorption band [Naz87]

DAP: ZPL = L = limiting energy, D = dielectric factor

8.1.4 Associates of the Single Vacancy with Ni_1, Ni_2, *and* Ni_3

Nickel is a frequent impurity in natural and HPHT diamonds (in the latter case from the melt). Of the six nickel centers in Table 8.1.4, two centers occur in natural diamond, especially the Ni 1.4 eV (=* $V_1Ni_1^+$) center. The number of associated Ni atoms is obtained from the Ni isotope shifts (Ni_1) or from the QLVM frequency (Ni_2 and Ni_3).

Table 8.1.4 Associates of the single vacancy with Ni_1, Ni_2, and Ni_3 (see also Tables 9.5.1.1 and 9.5.6)

Defect	Observed	ZPL (eV)	SB-DAP	*D* (eV)	Name/comment
*$V_1Ni_1^\circ$(a)	HA, MA	1.704	83a–e	+1.34	***Ni 1.70(a),** line (b) at 2.401eV/LVM: +61, +80 meV
*$V_1Ni_1^\circ$(a)	HL, ML	1.704	84a–c	−1.34	***Ni 1.70(a),** QLVM: −44 meV/Ni_1 from QLVM (calc. 1Ni: 44.0 meV)
*$V_1Ni_1^\circ$(b)	HA,MA, ME	2.401	29a–f	+1.34	***Ni 1.70(b),** QLVM: +34, LVM: +74 meV
*$V_1Ni_1^+$	NA, HA, MA/EPR: NIRIM2 Ni_1^+; $S = \frac{1}{2}$	1.404	25a–k	+1.33	***1Ni 1.40**; weaker line at 1.401 eV from 2.8 meV gr. st. spl.; trigonal center, ⟨111⟩ polarized line, photochromic/Ni_1 from isotope shifts/QLVM: +45; LVM: +80 meV
*$V_1Ni_1^+$	NL, HL, ML	1.404	26a–i	−1.33	***1Ni 1.40**/QLVM: −45, LVM: −80 meV
*$V_1Ni_2^-$ (a)	ML	2.071	87a–k	−1.34	***Ni 2.07(a),** line (b) at 2.298 eV/LVM: (3×) −55 meV
*$V_1Ni_2^-$ (b)	ME	2.298	27a–j	+1.34	***Ni 2.07(b),** QLVM: +30, LVM:+51 meV/Ni_2 from QLVM/PLE of *Ni 2.07(a)
*$V_1Ni_2^-$ (b)	ML	2.298	28a–h	−1.34	***Ni 2.07(b),** QLVM: −32; LVM: −55 meV
*$V_1Ni_2^\circ$ (a)	HL, ML	1.660	–	–	***Ni 1.66(a)**/LVM: (2×) −66 meV/slow luminescence (1.1 ms)

(continued)

Table 8.1.4 (continued)

Defect	Observed	ZPL (eV)	SB-DAP	*D* (eV)	Name/comment
$^*V_1Ni_2{}^{\circ}$ (b)	HA, MA, HE, ME	2.427	–	–	***Ni 1.66(b)**/QLVM: +28, LVM: (6×) +49 meV/Ni_2 from QLVM
$^*V_1Ni_2{}^{\circ}$ (b)	HL, ML	2.427	–	–	***Ni 1.66(b)**/LVM: (2×) −61 meV
$^*V_1Ni_2{}^{+}$ (a)	HA, MA	1.693	–	–	***Ni 1.69(a),** lines(b,c) at 1.940, 1.991 eV/LVM: (3×) +51 meV/C_{2v} symmetry
$^*V_1Ni_2{}^{+}$ (a)	HL, ML	1.693	–	–	***Ni 1.69(a)**
$^*V_1Ni_2{}^{+}$ (b)	HA, MA	1.940	–	–	***Ni 1.69(b)**/LVM: (7×) +50 meV
$^*V_1Ni_2{}^{+}$ (b)	HL, ML	1.940	–	–	***Ni 1.69(b)**/LVM: −59, −64 meV
$^*V_1Ni_2{}^{+}$ (c)	HA, MA/E	1.991	–	–	***Ni 1.69(c)**/LVM: (6×) +50 meV
$^*V_1Ni_2{}^{+}$ (c)	HL, ML	1.991	111a–f	−1.34	***Ni 1.69(c)**/LVM: (3×) −57 meV
$^*V_1Ni_2N_2^{\circ}$	NL, ML	2.101	62a-f	−1.42	***Ni 2.10;** intense DAP lines at 1.660 and 1.890 eV (see also Table 9.5.1.2)
$^*V_1Ni_3{}^{+}$	NA, MA	2.370	91a–g	+1.34	***Ni 2.37,** named**A line;** QLVM: +37, +49 meV
$^*V_1Ni_3{}^{+}$	NL, ML	2.370	92a–f	−1.34	***Ni 2.37,** named**A line;** QLVM: −23, −38; LVM: −70 meV/Ni_3 from QLVM (calc. 3Ni: 23.3 meV)

SB-DAP = sideband DAP with ZPL = *L* = limiting energy, and *D* = dielectric factor, gr. st. spl. = ground state splitting

8.1.5 *Centers of* Ni_1 *in the Double Vacancy and Association with Nitrogen Ligands (S2-*S9)*

Eight centers (S2–*S9) are known with a single nickel atom at the center of a divacancy, see Table 8.1.5. The DAP transitions of these centers coincide in absorption and luminescence, i.e., they are of the standard type (see Sect. 10.1). Additional lines are observed for S2 and *S4, which indicate the existence of a second electronic transition for these centers.

Table 8.1.5 Centers of interstitial Ni_1 in the double vacancy and association with nitrogen ligands (S2, S3, *S4–*S9), see also Sect. 9.5.12

Defect[a]	Observed	Lines (eV)	Stand.-DAP[b]	L (eV)	D (eV)	Name/comment
$(V_2Ni_1)N_{2+0+1}{}^{+}$	NA, HA, MA, ME, NL, HL, ML EPR(**NE2**)	2.515–3.341	57a–m	2.422	+1.63	**S2**/DAP lines: ***B*** = 2.537, ***C*** = 2.597, ***D*** = 2.623, ***E*** = 2.634 eV/QLVM: +29, LVM: +34, −39, +41, −46; (3×) +63, (2×)+71, −90, ±102 meV/inactive in CL
$(V_2Ni_1)N_{2+0+0}{}^{+}$	NA, HA, MA, ME, NL, HL, ML EPR(**NE1**)	2.496– 2.800	58a–h	2.383	+1.63	**S3**/QLVM: +30, LVM: −41; −79, ±117 meV/inactive in CL
$^*(V_2Ni_1)N_{1+2+0}{}^{\circ}$	NA, NE, HA, HE, MA,ME EPR(**NE3**)	3.333– 3.820	59a–i	3.191	+1.53	***S4**/PLE of *Ni 2.37(V_1Ni^{3+}, ZPL = line A, see Table 8.3a)/QLVM: −31, LVM: −38; −97, ±117, −144 meV
$^*(V_2Ni_1)N_{x+y+z}{}^{\circ}$	NA, HA, MA, ME, NL, HL, LL, ML	1.563–2.320	60a–g	1.460	+1.53	***S5**/QLVM: −33, LVM: +43, −44, −61; −70, −113, −135, −153 meV
$^*(V_2Ni_1)N_{x+y+z}{}^{+}$	LE, ME, EPR(NE8)?	2.750–3.530	61a–m	2.637	+1.63	***S6**/QLVM: +42; LVM: +99 meV/PLE of *S5
$^*(V_2Ni_1)N_{x+y+z}{}^{+}$	LE, LL, EPR(S = 1 center.)	1.802–2.620	11a–i	1.724	+1.60	***S7**/PLE of *S5/Ni_1 from Ni isotopes (9-meV splitting); PLE in own spectrum
$^*(V_2Ni_1)^{-}$	HAEPR(NE4)?	1.212–1.383	16a–l	1.036	+1.45	***S8**/broad band at 1.40 eV (named Ni 1.4 eV band)/in some samples resonance DAP 16 with 12 lines, peaking at 1.22 eV; photochromic/LVM: +64, +113, +152 meV
$^*(V_2Ni_1)N_{x+y+z}{}^{+}$	NL, MA, ML	1.678–1.812	4a–i	1.509	+1.66	***S9**/spectral hole burning in absorption [Sil94a]

Standard DAP transitions (abs. = lum., *no* ZPL), for further DAP information see Table 9.5.10

[a] N_{a+b+c} indicates the number of nitrogen atoms in the first (a), second (b), or third shell (c), (as obtained from hyperfine interaction [Nad93]; see Sect. 9.5)

[b] Donor–acceptor pairs: (see Chap. 10) standard DAP57-61, 11, 16, 4 with L = limiting energy and D = dielectric factor

Four centers (S2–*S5) occur in *natural* diamond. The structure of S2–*S4 has been determined by analysis of the hyperfine interaction (HFI) between the nickel nucleus and the unpaired electron spin, which is partially located on the nitrogen ligands. The correspondence of EPR and optical centers is NE1 = S3 = $(V_2Ni_1)N_{2+0+0}{}^+$, NE2 = S2 = $(V_2Ni_1)N_{2+0+1}{}^+$, and NE3 = *S4 = $(V_2Ni_1)N_{1+2+0}{}^+$ [Nad93]. In the notation $(V_2Ni_1)N_{x+y+z}{}^+$, the nitrogen subscripts N_{x+y+z} mean: N_x in the first shell (average HFI 15.5 G), N_y in the second shell (average HFI 5.6 G), and N_z in the third shell (average HFI 2.6 G). These HFI values follow closely a $1/r^3$ rule ($x : y : z = 1.0 : 0.34 : 0.22$), where r is the distance in the unperturbed lattice (see also Sect. 9.5.10).

The nucleus for nitrogen trapping is the nitrogen-free *S8 center $(V_2Ni_1)^-$. The charge state of the center influences the dielectric factor D (see Table 8.1.5; positive: neutral: negative): $N_2 : N_1 : N_0 = 1.63 : 1.53 : 1.45$ eV (see also Sect. 8.1.5).

8.1.6 Centers of Si_2, Ni_2, or Co_2 in the Triple Vacancy

$(V_3Si_2)^\circ$: For the 1.681-eV silicon center, the line shifts and intensities for the natural isotopes clearly reveal two equivalent Si atoms [Ste95]. From theory and detailed experimental analysis (observed $\langle 110 \rangle$ oriented dipole), it is known that each of the two equivalent Si atoms is situated midway between two vacancies of a triple vacancy, forming a $\langle 110 \rangle$ oriented defect [Bro95]. The 1.681-eV silicon center is a prominent defect in HTHP and CVP synthetic diamonds, and is stable above 2,200 °C. The other two charge states (*$(V_3Si_2)^-$ at 2.523 eV and *$(V_3Si_2)^\circ$ at 2.052 eV) are also thermally stable, as indicated by their presence in natural diamond (see Table 8.1.6). It appears that this unusual but very stable structure is the preferred incorporation of Si in natural diamond, see Sect. 9.4.

$(V_3Ni_2)^\circ$: This center is observed in as-grown HTHP synthetic diamonds, grown from a nickel melt, and with a nitrogen content > 50 ppm. The only LVM frequency (+61 meV) is unchanged in ^{13}C diamond, indicating a Ni–Ni vibration (see Table 8.1.6).

$(V_3Co_2)^\circ$: This center is observed in HPHT synthetic diamond, grown from a cobalt melt, see Tables 8.1.6 and 9.6.

Table 8.1.6 Associates of the triple vacancy with two equivalent interstitial silicon, cobalt, or nickel atoms

Defect	Observed	ZPL (eV)	SB-DAP	*D* (eV)	Name/comment
$^*(V_3Si_2)^-$	NE, NL, Si impl. ME, ML	2.523	75a–j	−1.25	*2Si 2.52/QLVM: −31(w = 25), 44 (w = 40) meV, Si_2 from QLVM (calc. 2Si = 44.9 meV)/PLE of *2Si 1.68/decay time 1.0–1.8 μs
$^*(V_3Si_2)^\circ$	NL	2.052	65a–e	−1.39	*2Si 2.05/QLVM: −45(w = 23); LVM: −70 meV, Si_2 from QLVM (calc. 2Si = 44.9 meV)/PLE of *2Si 1.68
$^*(V_3Si_2)^+$	HA, LA Si impl. MA	1.682	53a–m	1.60	*2Si 1.68/QLVM: +43(w = 30); LVM: +67 meV, Si_2 from QLVM (calc. 2Si = 44.9 meV)/ photochromic
$^*(V_3Si_2)^+$	HL, LL Si impl. ML	1.682	54a–g	−1.60	*2Si 1.68/QLVM: −42(w = 25); LVM: −65 meV/gr. st. spl. = 0.2 meV; exc. st. spl. = 1.1 meV/Si_2 from shifts of two equivalent Si isotopes (see Table 9.4.2)/⟨110⟩ oriented dipole, decay time from 2.4 ns to 105 μs, photochromic
$^*(V_3Co_2)^+$	HL	1.989	112a–f	−1.37	*2Co 1.99/QLVM: −25, LVM: −56 meV
$^*(V_3Ni_2)^\circ$(a)	HA, LA	1.883	107a–g	+1.36	*2Ni 1.88a/LVM: (2×)+61 meV/lines (b, c) from exc. st. spl., see below
$^*(V_3Ni_2)^\circ$(a)	HL	1.883	108a,b	−1.36	*2Ni 1.88a/QLVM: −26, LVM: −65 meV
$^*(V_3Ni_2)^\circ$(b)	HA,HL	1.905	–	–	*2Ni 1.88b/LVM: (2×)+61 meV
$^*(V_3Ni_2)^\circ$(c)	HA,HL	1.913	–	–	*2Ni 1.88c/LVM: (2×) +61 meV
$^*(V_3Ni_2)^+$(a)	NA,HA MA	2.562	105a–c	+1.55	*2Ni 2.56a/weak ZPL/line (b) at 2.588 eV from 26 meV exc. st. spl.

(continued)

Table 8.1.6 (continued)

Defect	Observed	ZPL (eV)	SB-DAP	D (eV)	Name/comment
$^*(V_3Ni_2)^+$(a)	NL,HLML	2.562	106a–f	−1.55	*2Ni 2.56a/line (b) at 2.588 eV from 26 meV exc. st. spl./QLVM: −25, −36 meV/Ni_2 from shifts of two equiv. Ni isotopes (Table 7.5)/gr. st. spl. 0.36 and 0.66 meV/slow luminescence (140 μs at 2.3 K)
$^*(V_3Ni_2)^+$(c)	NA,HA HE,MA	3.064	–	–	*2Ni 2.56c/line (d) at 3.076 eV from 12 meV exc. st. spl./QLVM: +26 meV/PLE of line (a)

SB-DAP = sideband DAP with ZPL = L = limiting energy and D = dielectric factor; gr. st. spl. = ground state splitting; exc. st. spl. = excited state splitting

8.1.7 Associates of the Vacancies with H, (H + N), He, Li, B, N, Al, Si, Ti, Cr, Co, Ni, Zn, As, Zr, Ag, Xe, Ta, W, and Tl

Only the associates of the vacancy with hydrogen, boron, aluminum, and silicon occur in natural diamond, the others being produced by implantation or doping (see Tables 8.1.7.1 and 8.1.7.2).

The single lattice vacancy can be filled with one hydrogen atom (with concentrations up to 1 at. %). Optical absorption is observed from two fundamental vibrations: C–H bend at 174 meV and C–H stretch at 385 meV, see Table 8.1.7.1. There are six harmonic and combination sidebands [Dav84b], see Table 9.3.1. A five-line sideband system from DAP transitions is observed for the C–H stretch line at 0.385 eV.

In homoepitaxial CVD diamond, a center with T_d symmetry (as evidenced by isotopic deuterium substitution [Fuc95a, b]) is observed. In a tentative assignment, a fourfold vacancy is filled by four hydrogen atoms. There are two ZPL lines at 165 meV (C–C stretch?) and 412 meV (C–H stretch?). For both lines, a sideband DAP system (DAP39, 40) is observed, see Table 8.1.7.2 and Sect. 9.3.

The $^*V_1B_1$(a, b) centers occur only in boron-rich type IIb diamonds, while the $^*V_1Al_1{}^\circ$ is found in HPHT synthetic diamonds (with Al in the melt) and in rare natural diamonds.

The other centers in Tables 8.1.7.1 and 8.1.7.2 are observed after implantation of the respective ions. The QLVM frequencies have the expected values (see Tables 11.2 and A.3). For these centers, the dielectric factor (D = 1.25–1.60 eV) is similar to those for $V_1{}^\circ$ and $V_2{}^\circ$ (1.24–1.63 eV) or V_1N_n (1.33–1.69 eV).

Table 8.1.7.1 Associates of the single vacancy with H_1, H_1+N_1, H_1+N_3, H_1 + N_4, He, Li, B, Al, Cr, Zn, and As

Defect	Atomic Radius (nm) [1)]	Observed	ZPL (eV)	SB DAP	*D* (eV)	Name/comment
*$(V_1H_1)^\circ$	(140)	NA	0.174	–	–	nat. CH bend (see Table 9.3.1)
*$(V_1H_1)^\circ$	(140)	NA	0.385	41a–e	1.54	nat. CH stretch (see Table 9.3.1)
*$(V_1H_1)N_1^\circ$	(140)	NA	0.182	–	–	nat. NH bend, harmonic at 0.362 eV, (see Table 9.3.1)
*$(V_1H_1)N_1^\circ$	(140)	NA	0.401	–	–	nat. NH stretch, combination line at 0.583 eV (=0.401 + 0.182 eV)
*$(V_1H_1)N_3^\circ$	(140)	MA	0.198	–	–	NH bend (see Table 9.3.1)
*$(V_1H_1)N_3^\circ$	(140)	MA	0.390	–	–	NH stretch (see Table 9.3.1)
*$(V_1H_1)N_4^\circ$	(140)	MA	0.192	–	–	NH bend (see Table 9.3.1)
*$(V_1H_1)N_4^\circ$	(140)	MA	(0.424)	–	–	Expected NH stretch (see Table 9.3.1)
*$V_1He_1(a)^\circ$	(90)	ML (He impl.)	2.212	95a–f	−1.57	Line (b) 2.316 eV/QLVM: −38; LVM: −91 meV
*$V_1Li_1^\circ$	77	ML (Li impl.)	1.881	68a–g	−1.58	QLVM: −36 meV
*$V_1B_1(a)^\circ$	(86)	NL	2.395	73a–f	−1.30	Also broad band at 2.20 eV
*$V_1B_1(b)^\circ$	(86)	NL	3.120	74°-f	−1.30	Also broad band at 2.85 eV
V_1N_n	***70***	–	–	–	–	***See Tables 8.1.3.1–8.1.3.6***
*$V_1Al_1^\circ$	141	NL,HL (Al dop.)	2.974	66a–e	−1.52	QLVM: −42 meV

***V_mSi_n**	***119***	–	–	–	–	***See Tables 8.1.6 and Sect. 9.4***
***V_mNi_n**	***127***	–	–	–	–	***See Tables 8.1.4–8.1.6 and Sect. 9.5***
*V_1Cr_1°	126	ML (Cr impl.)	1.673	–	–	QLVM: −31 meV
***V_mCo_n**	***128***	–	–	–	–	***See Tables 8.1.6 and Sect. 9.6***
*V_1Zn_1°	134	ML (Zn impl.)	2.394	114a–f	−1.42	QLVM: (2×) −30 meV
*V_1As_1°	120	MA (As doped)	0.113	115a–f	+1.32	QLVM: +25 meV

[a][Fin56], Fig. 1, with (interpolated) values, SB-DAP = sideband DAP with ZPL = L = limiting energy and D = dielectric factor; QLVM see Table 11.2

Table 8.1.7.2 Associates of the multiple vacancy with H_4, N, Si, Ni, Ti, Co, Zr, Ag, Xe, Ta, W, and Tl

Defect	Atomic radius (nm) [a]	Observed	ZPL (eV)	SB/*sta.* DAP	D (eV)	Name/comment
$*(V_4H_4)^-$	(140)	LA	0.165	39a–j	+1.31	homo-epi-CVD: *C–C stretch
$*(V_4H_4)^\circ$	(140)	LA	0.412	40a–g	+1.43	homo-epi-CVD: *C–H stretch
$*\boldsymbol{V_mN_n}$	***70***	–	–	–	–	***See Tables 8.1.3.4 and Sect. 9.1***
$*\boldsymbol{V_mSi_n}$	***119***	–	–	–	–	***See Tables 8.1.6 and Sect. 9.4***
$*(V_2Ti_1)^\circ$	150	ML (Ti impl.)	1.249	–	–	QLVM: −28, 85; LVM: 32, 119 meV
$*\boldsymbol{V_mCo_n}$	***128***	–	–	–	–	***See Tables 8.1.6 and Sect. 9.6***
$*\boldsymbol{V_mNi_n}$	***127***	–	–	–	–	***See Tables 8.1.5, 8.1.6 and Sect. 9.5***
$*(V_2Zr_1)^\circ$	163	HL (Zr melt)	$L = 1.364$	*stand., 5a–j*	+1.39	Zr–Ni–Si melt
$*(V_2Ag_1)(a)^\circ$	144	ML (Ag impl.)	3.111	85a–f	−1.31	Line (b) 3.117 eV/QLVM: (2×) −21(1Ag+8C); LVM: −67 meV/gr. st. spl. = 6 meV
$*(V_2Xe_1)$, distorted°	233	ML (Xe impl.)	1.528	42a–j	−1.33	QLVM: −29 meV(1 Xe)
$*(V_2Ta_1)$	141	LL (hot filament)	$L = 1.667$	*stand, 7a–g*	+1.23	DAP7b with LVM: −48 meV
$*(V_2W_1)$	138	LL (hot filament)	$L = 1.608$	*stand, 6a–f*	+1.24	DAP6b with LVM: (5×) −24 meV(1 W)
$*(V_2Tl_1)^\circ$, distorted	199	ML (Tl impl.)	2.019	94a–f	−1.57	QLVM: −21(1 Tl); LVM: −45 meV

[a][Fin56], Fig. 1, with (interpolated) values, SB-DAP = sideband DAP/*sta. DAP* = *standard* DAP with $ZPL = L =$ *limiting energy*; D = dielectric factor; QLVM see Table 11.2

8.2 The ⟨100⟩ Split Self-Interstitial and Associates

In a fundamental theoretical analysis, it has been shown that the only stable structure for the self-interstitial in diamond (in the 1−, neutral, 1+, and 2+ charge state) is the ⟨100⟩ split interstitial [Bre95]. Earlier theoretical work [Wei73, Mai78] had already favored the ⟨100⟩ split self-interstitial configuration. The ⟨100⟩ split self-interstitial configuration has also been confirmed experimentally by detailed analysis of the ^{13}C isotope shifts [Ste99a, Dav00, Twi01] (see Sect. 8.3.6). A Jahn–Teller distortion from D_{2d} to D_2 is predicted for the 1−, neutral, and 1+ charge state [Bre95]. The calculated activation energy for migration of the neutral self-interstitial is 1.7 eV [Bre95]. The notation used here is $(C_2)_i$.

The isolated self-interstitial is optically observed in the negative charge state (3H center at 2.462 eV), in the neutral charge state (1.685-eV center = R2 in EPR), and in the positive charge state (TR12 center at 2.638 eV); see Sect. 8.2.1 and Table 8.2.1. For the neutral charge state (1.685-eV center), an energy level scheme was calculated and yielded good agreement with the experimental data, especially with the 6-meV splitting of the ground state [Dav01b, Gos01a, Mai01].

By association with one nitrogen ($(^*C_2)_iN_1{}^\circ = $ 5RL center at 4.582 eV), the ⟨100⟩ split self-interstitial can be "stabilized" (up to $T = 1{,}000\,^\circ$C and thus observable in as-grown CVD diamond).

Association with two nitrogens is observed after heavy electron irradiation of nitrogen-rich diamond [Col79c]. The ZPLs of the three charge states are at 2.367 eV ($(^*C_2)_iN_2{}^-$), at 1.979 eV ($(^*C_2)_iN_2{}^\circ$), and at 2.535 eV ($(^*C_2)_iN_2{}^+$).

A very stable center (observed in as-grown HPHT and CVD diamond) consists of the split interstitial with one interstitial carbon replaced by nitrogen, here denoted by $(C_1N_1)_i$. This center is optically observed in the negative charge state (3.988-eV center = R11), in the neutral charge state (0.212-eV center), and in the positive charge state (*2.807-eV center); see Sect. 8.2.4 and Table 8.2.4.

8.2.1 The Isolated ⟨100⟩ Split Self-Interstitial

The isolated ⟨100⟩ split self-interstitial is observed in three charge states.

$\mathbf{(C_2)_i{}^+}$ = **TR12**: In irradiated diamond, optical spectra of the TR12 center with two ZPLs at 2.638 eV (TR12) and 2.670 eV (TR13, in absorption only) are observed. There are numerous sidebands (TR14-TR18 in absorption). With the exception of the TR17 LVM line at +188 (absorption) and −199 meV (luminescence), these absorption sidebands belong to $(C_2)_i{}^+$-DAP35 transitions (absorption) and $(C_2)_i{}^+$-DAP36 transitions (luminescence). The TR12 center anneals out above 700 °C.

Table 8.2.1 The isolated and multiple ⟨100⟩ split C_2 self-interstitial in diamond

Defect	Observed	ZPL (eV)	SB-DAP	*D* (eV)	Name/comment
$^*(C_2)_i^-$	ML	2.462	86a–i	−1.31	**3H**/LVM: −139, −165, −169, −182, −187, −218(C) meV/dyn. JT
$^*(C_2)_i^-$	MA	2.462	–	–	**3H**/LVM: +67 meV (see Table 11.1)
$^*(C_2)_i^+$(a)	MA, ME	2.638	35a–n	+1.30	**TR12**/LVM: +67, +77, +139 = **TR14**, +188 = **TR17**, +225 meV/IR: +190 meV/dyn. JT
$^*(C_2)_i^+$(a)	ML	2.638	36a–n	−1.30	**TR12**/LVM: −65, −80, −143, −199, −209 meV/dyn. JT
$^*(C_2)_i^+$(b)	MA, ME	2.670	–	–	**TR13**/from 32 meV excited state splitting/LVM: +188 meV/no luminescence
$^*(C_2)_i^\circ$	MA, EPR (R2)	1.685	71a–c	+1.30	**R2**/QLVM: +40, +80; LVM: +169 meV/IR: +167 meV/gr. st. spl. = 6 meV; ZPL starting from upper gr. st./dyn. JT
$^*(C_2)_i^\circ$	MA	((1.859)	–	–	**pseudo** ZPL, LVM sideband (169 meV) starting from lower gr. st. of R2 center (1.685 eV)
$(C_4)_i^\circ$	EPR (R1)	–	–	–	**R1**(EPR) = **I_2** [Twi01]; no optical lines
$(C_6)_i^\circ$	EPR (O3)	–	–	–	**O3**(EPR) = **I_3** [Twi01]; no optical lines
$^*(C_{2n})_i^\circ$	NL, ML	1.528	96a–g	−1.30	**D(a) = platelets**/weak ZPL; broad band at 1.25 eV/IR bands D(a′)–D(f′) see Table 8.2.2

SB-DAP = sideband DAP with L = ZPL = limiting energy and D = dielectric factor; IR: = observed IR bands (LVM absorption); From ^{13}C and ^{15}N isotope shifts [Col88c]; (C) = established as C–C vibration by full ^{13}C (and no ^{15}N) isotope shift; (N) = established as N–C vibration by ^{15}N (and reduced ^{13}C) isotope shift; dyn. JT = dynamic Jahn–Teller effect; gr. st. spl. = ground state splitting; exc. st. spl. = excited state splitting

$(C_2)_i{}^{\circ}$: The neutral charge state gives rise to a ZPL at 1.685 eV and an intense sideband line at 1.859 eV. The ground state is split by 6.2 meV and, due to selection rules, the ZPL starts at the upper ground state, while the sideband line starts at the lower ground state. The 6.2-meV ground state splitting is interpreted as a "tunneling splitting" arising from the movement between two equivalent equilibrium sites [Dav01b]. Only the excited state couples to a phonon with an energy of 169 meV. The 1.685-eV ZPL couples to the sideband $(C_2)_i{}^{\circ}$-DAP71 transitions with three lines. Thermal annealing of the 1.685 eV center = $(C_2)_i{}^{\circ}$ occurs at relatively low temperatures (typically 420–540 °C). In the first stage, the $(C_2)_i{}^{\circ}$ is quantitatively converted into $(C_2)_i{}^{+}$ = TR12 [All98a]. In a second stage ($T > 700$ °C), all isolated self-interstitials are annihilated by migration to vacancy centers (GR1 etc).

$(C_2)_i{}^{-}$ = 3H: From the singly negative charge state, a luminescence spectrum of the 3H center with the ZPL at 2.462 eV and several sidebands is observed. The LVM sidebands have energies of −169, −182, −187, and −218 meV. At least nine sidebands can be assigned to $(C_2)_i{}^{-}$-DAP86 transitions. The 3H center is not observed in p-type diamond. Strong optical (reversible) bleaching of the 3H center occurs for 488 nm or UV illuminations [Ste99a].

8.2.2 The Multiple ⟨100⟩ Split Self-Interstitial

Aggregates of self-interstitials are observed in EPR [Twi01], but not optically. These are (2 × $(C_2)_i{}^{\circ}$ = EPR-R1 named I_2) and (3 × $(C_2)_i{}^{\circ}$ = EPR-O3 named I_3).

$(C_{2n})_i$ = platelets = D center: Much larger self-interstitial aggregates (with dimension up to 100 μm) with orientation in (100) planes are observed in natural or

Table 8.2.2 Lines and bands from the D center (platelets, B′): *$(C_{2n})_i$ = *D(a′–f′)* and D(α), D(a)

Line/band[a]	*E* (meV)*(cm^{-1})*	Width (meV)*(%)*	Rel. Intens (%)	Comment/for vibrations see also Table 11.1
D(a′)	40.9 *331*	–	–	Out of plane bend sp^2–sp^2; [Zai01]
D(b′)	150.0 *1,210*	6 *4*	34	Mixed sp^2–sp^{3}stretch at the *surface* of the platelet; Figs. 3.20 [Woo83], 3.25 [Cla84b]
D(c′)	156.1 *1,258*	7 *4*	32	Mixed sp^2–sp^3 stretch at the *surface* of the platelet; Fig. 3.25 [Cla84b]
D(d′)	167.6 *1,352*	2 *1*	31	Mixed sp^2–sp^3 stretch at the *surface* of the platelet, very asymmetric line shape; Figs. 3.20 [Woo83], 3.25 [Cla84b]

(continued)

Table 8.2.2 (continued)

Line/band[a]	E (meV)(*cm*$^{-1}$)	Width (meV)(*%*)	Rel. Intens (*%*)	Comment/for vibrations see also Table 11.1
D(e′1)	168.0 *1,355*	0.7 *0.4*	1	**Large** platelets; mixed sp^2–sp^3 stretch at the *rim* of the platelet; Figs. 3.7 [Fer96], 3.20 [Woo83]
D(e′2)	172.0 *1,387*	3 *2*	2	**Small** platelets; mixed sp^2–sp^3 stretch at the *rim* of the platelet; Figs. 3.20 [Woo83], 3.26 [Fri91a]
D(f′)	177.3 *1,430*	–	–	Very weak; symmetric sp^2–sp^2 stretch; [Zai01]
Line/band	E (eV)	Width (eV)(*%*)	Rel. Intens (%)	Comment/for vibrations see also Table 11.1
D(α)	1.250	0.300 *24*		Strongly polarized E ∥ (100) platelet plane; Fig. 5.9 [Ruo91a]
D(a)	1.526	0.038 *2*		ZPL of D(α), Fig: 5.9 [Ruo91a], 5.15 [Sil95], 7.80 [Dav77c]; LVM: (2×) −45 meV (corresponding to IR line D(a′)), (2×) −163 meV (corresponding to IR lines D(b′), D(c′), D(d′)); ZPL of DAP97a-f

[a]D(a′–f′) observed in NA, HA, MA; D(α), D(a) observed in NL, HL, ML

annealed diamond (see Tables 8.2.1 and 8.2.2), and also in heavily neutron irradiated diamond. These aggregates have been named “platelets”, and can be observed as X-ray or electron microscopy spikes. Optically, there is characteristic IR absorption (with the historical name “D center”) and a ZPL at 1.525 with a broad sideband at 1.25 eV (see Table 8.2.1 and 8.2.2). In absorption, there are three infrared lines at 41, 158–173, and 177 meV. The position of the intense sharp line depends on the size of the platelets and varies from 173 meV (for small) to 158 meV (for large platelets). In luminescence, a broad band at 1.25 eV with a weak ZPL at 1.526 and seven DAP96 lines is observed (see Table 8.2.1 and 8.2.2).

8.2.3 *Associates of the* $(C_2)_i$ *Split Interstitial with* N_1, N_2, *with* B_1, B_2, B_3, *or with Ni*

Ten associates of $(C_2)_i$ with nitrogen, boron, or nickel are listed in Table 8.2.3.

Table 8.2.3 Associates of the $(C_2)_i$ ⟨100⟩ split interstitial with N, B, or Ni

Defect	Observed	ZPL (eV)	DAP[a]	*D* (eV)	Name/comment[b–d]
$^*(C_2)_iN_1{}^-$	LA, MA	3.188	–	–	**3.188 eV** center (weak absorption)/IR: +180 (51% C, 49% N), +190 (98% C), +232 (98% C) meV
$^*(C_2)_iN_1{}^-$	LL, ML	3.188	72 a-n	−1.26	**3.188 eV** ce./QLVM: −58,−80; LVM: −77 ? −108 (46% C), −138 (54% C, 46% N), −161 (78% C), −179 (54% C), −190 (95% C), −238 (100% C), (combinations: −290, −300, −322, −393, −413, −468, −491) meV
$^*(C_2)_iN_1{}^\circ$	LA, MA	4.582	79a–j	+1.26	**5RL**/QLVM: +42; LVM: +167, 195, 202 meV/IR: +167 (100% C), +186 (56% C, 44% N), +201 (61% C, 39% N), +245 (100% C) meV
$^*(C_2)_iN_1{}^\circ$	LL, ML	4.582	80a–e	−1.26	**5RL**/QLVM: −58; LVM: −146 (68% C, 32%N), −175 (88% C, 12% N), −193 (93% C, 7% N), (4×) −237 (100% C) meV
$^*(C_2)_iN_1{}^+$	ML	4.676	–	–	QLVM: −58; LVM: −195, (4×) −240 (C)/IR: +238 meV
$^*(C_2)_iN_2{}^-$	MA	2.367	88a–f	+1.25	More lines at 2.378, 2.385, 2.678 eV/QLVM: +38 meV; LVM combination: +314 meV/gr. st. spl. = 3.2 meV/N_2C_4 from QLVM (calc 37.3 meV)/ photochromic with $^*(C_2)_iN_2{}^\circ$ at 1.979 eV
$^*(C_2)_iN_2{}^-$	ML	2.367	89a–j	−1.25	QLVM: −38; LVM: −90, −160, −232 meV/N_2C_4 from QLVM (calc. 37.3 meV)
$^*(C_2)_iN_2{}^\circ$	MA	1.979	–	–	QLVM: +37, +67 meV/gr. st. spl. = 4.1 meV
$^*(C_2)_iN_2{}^+$	MA	2.535	90a–c	+1.25	QLVM: +70 meV (calc. 74.5 meV for N_2)
$^*(C_2)_iB_1{}^\circ$	ML (type IIb, irrad.)	4.777	97a–d	−1.25	**2BD**(F); split off lines at 4.773, 4.781 eV/LVM: (3×) −210 meV/dyn. JT?
$^*(C_2)_iB_2{}^\circ$	ML (type IIb, irrad.)	4.698	–	–	**2BD**(G)/LVM: (3×) −220 meV

(continued)

Table 8.2.3 (continued)

Defect	Observed	ZPL (eV)	DAP[a]	D (eV)	Name/comment[b–d]
*$(C_2)_iB_3{}^{\circ}$	ML (type IIb, irrad.)	4.803	–	–	**2BD(C)**/LVM: (3×) −236 meV
*$(C_2)_iNi_1{}^{\circ}$	ML	2.436	14a–f	−1.22	Resonance/SB DAP14(2.066–2.271 eV), resonance lines at 2.156 ($V_1N_1{}^{\circ}$) and 2.427 eV ($V_1Ni_2{}^{+}$(b)); QLVM: −75 meV

[a]All DAPs are of the sideband type with L = ZPL = limiting energy and D = dielectric factor
[b]***5RL**-IR: LVM absorption including ^{13}C and ^{15}N isotope line shifts [Col88c]; from precise data, the percentage of C or N vibrational involvement could be calculated for this table
[c]dyn. JT = dynamic Jahn–Teller effect
[d]gr. st. spl. = ground state splitting

***$(C_2)_iN_1{}^{-}$**: The 3.188 eV center is observed in luminescence with the ZPL at 3.188 eV, and with a rich sideband structure. This center occurs in as-grown HPHT and CVD diamond and in irradiated nitrogen containing diamond. Valuable information for the assignment of the LVM sidebands is obtained from the isotope shifts (see Sect. 8.3.6) The 238-meV LVM (central N–C vibration) is hidden under an intense combination sideband (see below). There are nine LVM combination bands at 238, 290, 300, 322, 393, 404, 413, 468, and 491 meV. It has been shown [Dis94a], that a complete assignment is possible, and this in turn confirms the frequencies of the one phonon LVM lines. If the latter are denoted by $a = 108$, $b = 138$, $c = 161$, $d = 179$, $e = 190$, and $f = 238$ meV, than the combinations are $a + b = 238$, $a + d = 290$, $a + e = b + c = 300$, $2c = 322$, $2a + d = 393$, $2a + e = c + f = 404$, $d + f = 413$, $2d = 468$, and $b + 2e = 491$ meV. Thanks to the high luminescence intensity of the 3.188-eV center, 14 weak lines (in addition to the phonon sidebands) have been identified [Zai01], which can be assigned to **$(C_2)_iN_1{}^{-}$**-DAP72a-n transitions.

***$(C_2)_iN_1{}^{\circ}$**: From this 5RL center, a luminescence spectrum is observed, both in irradiated diamond and in as-grown CVD diamond. The 5RL spectrum is also observed in absorption. The ZPL is at 4.582 eV. The luminescence sideband structure is very unusual and has been analyzed by isotopic substitution [Col93b]. The 237-meV phonon (typical for the vibration of the central C–C bond, and dipole enhanced by the associated nitrogen) arises from two almost equivalent carbon atoms. This vibration produces not only an intense first-order sideband (which is more intense than the ZPL), but also a second, third, and fourth harmonic sideband. Combination lines from the harmonics and the other phonons are also observed. Five weak lines can be assigned to $(C_2)_iN_1$-DAP79 transitions. The 5RL center anneals out at temperatures from 800 to 1,200 °C [Zai01].

***$(C_2)_iN_1{}^{+}$**: The QLVM frequency (−58 meV in luminescence) is the same as for the neutral charge state, and both are in excellent agreement with the theoretical value of 58.5 meV, calculated for N_1C_2.

$^*(C_2)_iN_2{}^-$: This 2.367-eV center is observed in absorption (DAP88, QLVM at +38 meV from N_2C_4) and in luminescence (DAP89, QLVM at −38 meV from N_2C_4, LVM at −232 meV from $(C_2)_i$). Cathodoluminescence from the 2.367-eV line is observed in cobalt grown synthetic HPHT diamond after heating at 1,500 °C [Col85b].

$^*(C_2)_iN_2{}^\circ$: The 1.979-eV center is created after optical bleaching of the 2.367 eV center ($^*(C_2)_iN_2{}^-$) [Col79c]. The reverse process occurs by thermal recovery with an activation energy of 0.7 eV [Col79c].

$^*(C_2)_iN_2{}^+$: The 2.535-eV center is relatively weak in samples showing also the 1.979 and 2.367-eV centers. Only three DAP90 sidebands and one QLVM at 73 meV (from N_2) are observed.

From thermal data, a ground state splitting of 3.25 meV for the 2.367 eV center (=* $(C_2)_iN_2{}^-$), and 6.7 meV for the 1.979 eV center (=* $(C_2)_iN_2{}^\circ$) have been determined [Col79c]. The above three centers disappear at $T > 350$ °C from the 1.6 to 2.9 eV spectral range [Col79c], possibly by transformation into $^*(C_2)_iN_2{}^{2+}$ with an expected ZPL in the 4.0–4.5 eV range.

$^*(C_2)_iB_1{}^\circ$, $^*(C_2)_iB_2{}^\circ$, $^*(C_2)_iB_3{}^\circ$: In irradiated boron-rich type IIb diamonds, three 2BD centers are observed (2BD stands for IIb damaged). These centers are also observed in as-grown boron doped CVD diamond [Yok92]. Similar to the 5RL = $(C_2)_i$ $N_1{}^\circ$ center, very intense high frequency sidebands (210–237 meV from $(C_2)_i$) are observed for the 2BD centers. In an attempt, to interpret the three similar spectra, the following assignment can be proposed: The 4.698 eV (2BD(G)) center has one boron ligand, the 4.777 eV (2BD(F)) center has two boron ligands, and the 4.803 eV (2BD(C)) center has three boron ligands (see Table 8.2.3). The ZPL of the 2BD(F) center is a triplet with 10-meV spitting (lines at 4.767, 4.777, and 4.787 eV at 100 K). At 50 K, the 4.767-eV component disappears. For the 2BD(F) center, weak sidebands from DAP97a-d can be identified.

$^*(C_2)_iNi_1{}^\circ$: In low-nitrogen natural or HPHT synthetic diamond after ion (e.g., Ni or C^+) implantation and > 700 °C anneal. The seven DAP lines at 2.066, 2.101, 2.123, 2.181, 2.202 2.234, and 2.271 (ZPL at 2.436 eV) are resonant with $V_1N_1{}^\circ$ at 2.154 eV and with $^*V_1Ni_2{}^\circ$(b) at 2.427 eV.

8.2.4 The ⟨100⟩ Split Nitrogen–Carbon Interstitial $(N_1C_1)_i$ *and Associates with* $B_1{}^-$, $B_1{}^\circ$, $B_1{}^+$

Substitution of nitrogen for carbon in the ⟨100⟩ self-interstitial leads to $(N_1C_1)_i$ (three charge states listed), and associates with boron (three charge states), in Table 8.2.4. The $(N_2)_i$ center (IR line at 0.230 eV = LVM) is not a truly "intrinsic" defect).

Table 8.2.4 The $(N_1C_1)_i$ split ⟨100⟩ interstitial and associates with boron, and the ⟨100⟩ split $(N_2)_i$ interstitial in diamond

Defect	Observed	ZPL (eV)	SB-DAP[a]	D (eV)	Name/comment[b]
*$(N_1C_1)_i^-$	MA	3.988	69a–l	+1.26	**R11** ce./QLVM:+38, +79; LVM: +120, +195?, +188, +218, +230?, (combinations: +308, +338) meV/IR: +186 (100% C), +195 (50% N)
*$(N_1C_1)_i^\circ$	NA	0.212	3a–i	+1.24	Pseudo ZPL from N–C stretch (51% C, 49% N) = first DAP-SB ($s = 38$, at 0.386 eV) in resonance with nat. C–H stretch (0.385 eV)
*$(N_1C_1)_i^+$	LA, MA	2.807	–	–	ZPL is weak in absorption/IR: +190 (100% C), +212 meV (51% C, 49% N)
*$(N_1C_1)_i^+$	LL, ML	2.807	98a–i	−1.25	QLVM: −58; LVM: −171 (98% C, 0% N), −175 (87% C, 13% N); N_1 from (50%) ^{15}N isotope doublet [Col89c]
*$(N_1C_1)_iB_1^-$	LL (B doped)	2.992	101a–h	−1.20	Hot filament CVD, [Ste96a]
*$(N_1C_1)_iB_1^\circ$	LL (B doped)	2.792	99a–h	−1.25	QLVM: −50 meV [Ste96a], [Dis94a, b]
*$(N_1C_1)_iB_1^+$	LL, ML (B doped)	3.092	100a–g	−1.27	QLVM: −58; LVM: −105, −165, −205, −216 meV [Col89d, Dis94a]
*$(N_2)_i^\circ$	MA	0.230	–	–	IR LVM: +230 meV/N_2 from isotopes (55%) ^{15}N: +225, +227, +230 meV [Col87c]/*not truly intrinsic!* see also Table 9.1.4

[a]SB-DAP = sideband DAP with L = ZPL = limiting energy and D = dielectric factor
[b]IR: LVM absorption including ^{13}C and ^{15}N isotope shifts [Col88c, Col89b]; from the precise data, the percentage of C or N vibrational involvement could be calculated for this table

$^*(N_1C_1)_i{}^-$: In the negative charge state (R11 center), an absorption spectrum with the ZPL at 3.988 eV and a rich sideband structure is observed. There are two fundamental differences in comparison with the $(N_1C_1)_i{}^+$ and $(N_1C_1)_i{}^\circ$ centers, which occur in as-grown samples. (1) The $(N_1C_1)_i{}^-$ center is observed only in irradiated diamond, and it starts to anneal out above 400 °C, while the $(N_1C_1)_i{}^+$ and $(N_1C_1)_i{}^\circ$ centers occur in as-grown samples; (2) the $(N_1C_1)_i{}^-$ center is observed only in absorption, while the other charge states are intense in luminescence and very weak in absorption. Unfortunately, no results from isotopic substitution are reported for the $(N_1C_1)_i{}^-$ center. It is therefore difficult to separate vibrational sidebands from the DAP69 transitions. Assuming that the slightly sharper sidebands at 120, 308, and 338 meV are LVM lines (the latter two as LVM combinations), LVM frequencies of 120, 188, and 218 meV can be deduced, which coincide with two weak lines at 188 and 218 meV in the spectrum. Twelve sidebands can be assigned to $(N_1C_1)_i{}^-$-DAP69 transitions.

$^*(N_1C_1)_i{}^+$: The 2.807 center is observed in irradiated diamond [Zai01] and in as-grown CVD diamond [Dis94a]. The 2.807-eV center has sharp sidebands from at least seven LVM frequencies. In a detailed ^{13}C and ^{15}N isotope shift analysis [Col89a], it has been found that five vibrations (168, 171, 172, 177, and 199 meV) are localized at carbon atoms, one vibration (191 meV) at a nitrogen atom, and one vibration (175 meV) mainly at a carbon and partly at the nitrogen atom. Remarkable are the DAP3 donor–acceptor transitions of the pseudo ZPL at 0.212 eV (N–C stretch vibration). This DAP is a mixture of sideband DAP and resonance DAP (observed are weak lines (0.386–0.421 eV) from nine shells ($s = 46$–32), and a sharp cut-off at the energy of the coexisting natural C–H stretch vibration at 0.385 eV). Effective in the creation of this center is the implantation of nitrogen or nickel, but normally the 2.807-eV center is 30 times weaker than, e.g., the $(N_1C_1)_i{}^\circ = 3.188$-eV center [Col89a]. The center anneals out at 1,200–1,400 °C.

$^*(N_1C_1)_iB_1{}^-$: This center with four DAP101 lines (ZPL at 2.992 eV) is observed in boron doped CVD diamond. The spectra of the boron associates are very similar to those of the respective isolated centers (see above and Table 8.2.4); however, the DAP sidebands are considerably enhanced by the presence of the boron.

$^*(N_1C_1)_iB_1{}^\circ$: In boron contaminated CVD diamond, this center with eight DAP99 lines (ZPL at 2.792 eV) is observed, together with $V_1N_1{}^\circ$ (2.156 eV), $^*(N_1C_1)_i{}^\circ$ (3.188 eV), and $^*(C_2)iN_1{}^\circ$ (4.582 eV).

$^*(N_1C_1)_iB_1{}^+$: This center with a ZPL at 3.092 eV is occasionally observed in the luminescence spectrum of as-grown CVD diamond [Col89c, Dis94a]. The low intensity is not sufficient for a detailed sideband analysis.

$^*(N_2)_i{}^\circ$: This center is not truly intrinsic, but closely related to the $^*(N_1C_1)_i$ centers. So far, only the infrared LVM line at 0.230 eV is observed (C–C stretch). The structure is revealed by the three-line splitting for 55% ^{15}N substitution [Col87c].

8.3 Discussion of Spectral Data from Intrinsic Defects

8.3.1 *Thermal Stability of the Intrinsic Defects*

The existing data on the thermal stability of intrinsic defects can be summarized as follows:

1. The isolated ⟨100⟩ split self-interstitial anneals out by migration at temperatures of 420–700 °C ($E_a = 1.7$ eV for $(C_2)i^{\circ}$).
2. The single vacancy anneals out at temperatures of 500–1,000 °C. The theoretical activation energy for V_1° migration is 2.4 eV [Mai94a]. A first reduction in V1° concentration occurs, when the self-interstitials (with the lower activation energy of 1.7 eV) start to recombine with the vacancies. The divacancy is created from two isolated vacancies by 1,000 °C anneal. While the isolated V1° is absent inside natural or as-grown HPHT synthetic diamonds, a significant concentration of V1° can be found near the growth surface of CVD diamond [All98b].
3. Increased thermal stability (evidently by reduced mobility) is observed for associates of the vacancies with impurities like nitrogen (see Tables 8.1.3.1–8.1.3.6), nickel (see Tables 8.1.4–8.1.6), hydrogen or boron (see Tables 8.1.7.1 and 8.1.7.2), and silicon (see Table 8.1.6). All these associates are found in natural diamond, which demonstrates their thermal stability.
4. For the ⟨100⟩ split self-interstitial, the gain in thermal stability by association (see Table 8.2.3) or substitution with one nitrogen ($(N_1C_1)_i$, see Table 8.2.4) is not sufficient to observe these centers in natural diamond. However, the thermal stability of the 3.188-eV center =* $(C_2)N_1{}^-$ (up to 1,500 °C) is sufficient for observation in as-grown HPHT and CVD diamond. In as-grown CVD diamond the centers 5RL =*$(C_2)_iN_1{}^{\circ}$ and 2.807-eV =*$(N_1C_1)_i{}^+$ are observed.
5. Confirmation of the stability data on the isolated defects $(C_2)_i{}^{\circ}$ and $V_1{}^{\circ}$ (taking into account the increased mobility in the damaged lattice), was obtained by a study, where the temperature dependence of the production rate of these defects during 2-MeV electron irradiation was measured. The results are: Zero production of $(C_2)_i{}^{\circ}$ at $T > 350$ °C and zero production of $V_1{}^{\circ}$ at $T > 800$ °C [Dav01b].

8.3.2 *Donor–Acceptor Pair Transitions at Intrinsic Defects*

For the majority of the intrinsic defect centers the ZPL couples to donor–acceptor pair transitions of the *sideband* type (Tables 8.1.1—8.2.4). Exceptions are the centers with a (V_2Ni_1) nucleus, which have DAP transitions of the *standard* type (S2–*S9, Table 8.1.5).

It is worth noting, that all 60 DAP centers of the *sideband* type in Table 10.1 can be ascribed to intrinsic defect centers. It is likely, that, that all 75 intrinsic defects in Tables 8.1.1–8.2.4 are accompanied by DAP transitions. However, the DAP transitions are often difficult to identify, and for 15 intrinsic centers in Tables 8.1.1—8.2.4 they are not yet found.

This special property (SB-DAPs) of intrinsic defects is probably due to the occurrence of *dangling bonds* (there are four dangling bonds at the lattice vacancy and two at the self-interstitial). In agreement with this interpretation would be the fact, that for the centers with a (V_2Ni_1) nucleus (S2–*S9) the DAP transitions are of the *standard* type, probably because the nickel "d" orbitals can saturate the carbon dangling bonds.

The DAPs of the sideband type have only one free parameter, namely the dielectric factor "D" for the Coulomb energy. The other parameter "L" for the limiting energy coincides with the ZPL.

In Tables 8.1.1–8.2.4, the dielectric factor D of the DAP analysis is included. For the centers with *vacancies*, the D values are in the range 1.10–1.69 eV (see Table 8.1.1–8.1.7.2). For centers with *self-interstitials*, the variation of the D values is very small (1.22–1.31 eV, see Tables 8.2.1–8.2.4 and Sect. 10.4).

If a defect exists in several charge states, there is a tendency for increasing D values when going from the negative to the neutral and to the positive charge states (see Sect. 10.5).

Characteristic for *sideband* DAPs is the "mirror image" behavior of the luminescence (low energy side of the ZPL) and absorption (high energy side of the ZPL). Consequently, the dielectric factor D is *negative* for luminescence, and *positive* for absorption. The N3(c) center at 2.985 eV is a well-known example for the mirror image (DAP23 and DAP24). Note that two different DAP numbers are required for absorption and luminescence of the same center.

Characteristic for DAPs of the *standard* type is the coincidence of absorption and luminescence lines, as observed for the S2–*S9 centers in Table 8.1.9. Also characteristic for standard DAPs is the photoluminescence excitation in the own spectrum (see *$(V_2Ni_1)N_{x+y+z}$-DAP11 in Table 8.1.5). Another characteristic occurs with the LVM frequencies of those phonons, which couple to the standard DAP transitions. As can be seen in Tables 8.1.4–8.1.5, these frequencies are identical in absorption and luminescence.

In the past, it was tempting to interpret standard DAP lines (which coincide in absorption and luminescence) as zero phonon lines, (e.g., the B, C, D, and E line of the S2 center). The present book reveals a non-ZPL character for these and many other lines.

8.3.3 *Quasilocal Vibrational Mode Sidebands at Intrinsic Defects*

In a pioneering study [Zai00a] it was shown, that the quasilocal vibrational mode (QLVM) frequency is directly determined by the vibrating mass (see Sect. 11.2).

Therefore, the observed frequencies are very helpful for the determination of the atomic composition of intrinsic defects. Examples are given in Tables 11.2 and A.3. The maximum deviations between observed and calculated frequencies are only a few percent, which is of the order of experimental uncertainty or variation.

A problem exists for centers when a *dynamic Jahn–Teller effect* occurs. Numerous QLVM frequencies are observed for the isolated vacancy and vacancy-nitrogen associate centers. However, the values in the range 36–46 and 55–72 meV cannot be assigned to the expected calculated frequencies. Exceptions are the associates of the vacancy with atoms, which have a mass sufficiently different from carbon (see Table 8.1.7.2 and 11.2).

For associates of the vacancy with one impurity atom, the QLVM cluster can include the three impurity ligands (C_3), plus the three vacancy ligands (total = C_6). With Jahn–Teller distortion, three impurity ligands (C_3) and one vacancy ligand (C_1) (total = C_4) may be involved:

8.3.4 Local Vibrational Mode Sidebands at Intrinsic Defects

Vibronic sidebands from local vibrational modes (LVM) are sharper than the QLVM sidebands. Six types of C–C vibrations can be distinguished (see Table 11.1). Very similar frequencies are expected for N–C vibrations, because there is little difference with respect to atomic mass and bonding force constant.

The observed LVM frequencies are in the expected range. Remarkable is the considerable reduction (ca. 25%) of the LVM frequencies in platelets = (C2n)i compared to the isolated self-interstitial = $(C_2)_i$.

Theoretical calculations for $(C_2)_i$ [Bre95] yield 198 meV for the *ligand-central* C–C bond vibration (observed 120–199 meV) and 251 meV for the *central* C–C bond vibration (observed 202–245 meV).

The results from Table 11.1 can be extrapolated for the LVM frequencies of the heavier metal atoms. Assuming comparable force constants, and replacing C–C by M–C in the vacancy associates, the predicted bending frequencies are for ^{48}Ti in the range 28–57 (observed 32) meV, for ^{59}Co 26–52 (observed 44) meV, for ^{108}Ag 20–40 (observed 21) meV, and for ^{204}Tl 15–30 (observed 21) meV. The predicted stretching frequencies are for ^{59}Ni in the range 70–97 (observed 70–80) meV, for ^{108}Ag 54–75 (observed 67) meV, and for ^{204}Tl 40–56 (observed 45) meV.

8.3.5 Infrared Vibrational Absorption Lines

Characteristic infrared vibrational absorption lines are observed from the B-center ($V_1N_4°$ see Tables 8.1.3.3 and 8.1.3.5) and from the platelets (see Tables 8.2.1 and 8.2.2). The vibrational lines are broadened to bands by resonance with the lattice phonons in the range 69–165 meV. At higher frequencies, the vibrational lines are sharp and allow to observe ^{13}C and ^{15}N isotope shifts [Col88c]. In Tables 8.2.1 and 8.2.3, these lines are marked by (IR:).

8.3.6 *Vibrational Isotope Shifts from ^{13}C and ^{15}N Substitution*

The observed isotope shifts after partial or full ^{13}C or ^{15}N substitution provide very important information on the structure of the intrinsic defects.

$^{+}(\mathbf{C_2})_i{}^{-}$ at 2.462 (see Table 8.2.1): The isotope shifts in a sample with 50% ^{13}C reveals localization of the vibrations on different sites [Ste99a]. The 218-meV sideband splits into a perfect triplet with 1:2:1 intensities, i.e., two equivalent interstitial carbon atoms vibrate with an energy, which is typical for $C = C$ double bonds (see Table 11.1). This observation strongly supports the theoretical split interstitial model. The 182- and 187-meV sidebands split into quartets with 1:3:3:1 intensities, indicating a vibration of *one* central carbon with *two* carbon ligands. The 169-meV sideband shifts to a bell-shaped band with its maximum at 50% isotope shift. This vibration is clearly located at the four carbon ligands and their 12 carbon lattice neighbors, i.e., 16 carbon atoms are involved.

$^{*}(\mathbf{C_2})_i\mathbf{N_1}{}^{-}$ at 3.188 eV (see Table 8.2.3): A detailed analysis of the LVM lines at 179 and 190 meV in luminescence spectra after partial substitution with 50% ^{15}N [Col87b] or with 40% ^{13}C [Col93b] enables the present author to determine the distribution of the vibrational energy on one central atom C(1) with ligands N and C(3), and the other central atom C(2) with the ligands C(4, 5). For the *178.5 meV ("N")* vibration, the result is: N: 47%, C(1): 34%, C(2): 13%, and C(3): 6%. For the *190.5 ("C")* vibration the result is: C(2): 50%, C(1): 32%, C(4): 8%, C(5): 7%, N: 3%. The LVMs at 168, 170, and 232 meV are pure carbon vibrations, because they are not influenced by ^{15}N substitution [Col87b].

8.3.7 *Isotope Shifts of the Zero Phonon Lines by ^{13}C and ^{15}N Substitution*

Frequency shifts of the zero phonon lines (see Tables 7.2–7.3) are reported for ten vacancy centers (Tables 8.1.1–8.1.7.2) in the range +0.5 (for $^{*}V_2N_2{}^{\circ}$ at 2.086 eV) to +6.6 meV (for $V_2{}^{\circ}$ at 2.543 eV), and for five interstitial centers (Tables 8.2.1–8.2.4) in the range +1.2 for ($(C_2)_i{}^{\circ}$ at 1.685 eV) to +8.0 meV (for $^{*}(C_2)_iN_1{}^{\circ}$ at 4.582 eV) [Dav99a].

In a theoretical treatment, *four* contributions to the ZPL isotope shifts have been discussed [Dav99a]. Typical values for these contributions are:

1. +0.9 meV from lattice compression (in case of 13C)
2. +1.5 meV for lower vibrational frequencies in the excited state
3. +0.4 meV for anharmonicities (if applicable)

4. −0.8 meV for a transition between a dynamic Jahn–Teller active ground state and a non-Jahn–Teller active excited state

Unexpected isotope shifts clearly reveal non ZPL transitions. Examples are the −5.2-meV shift for the luminescence pseudo ZPL line at 1.859 eV of $(C_2)_i{}^\circ$, and the +6.7 to +7.2 meV shifts for the GR2–GR8 PLE and absorption lines at 2.880–3.005 eV ($V_1{}^+$ SB-DAP32).

Chapter 9
Impurity Defects in Diamond

In the following seven sections, the spectral lines (and bands) from impurity defects are described.

Six important impurities are nitrogen (Sect. 9.1), boron (Sect. 9.2), hydrogen (Sect. 9.3), silicon (Sect. 9.4), nickel (Sect. 9.5), and cobalt (Sect. 9.6).

In Sect. 9.7, lines from 18 other impurities are listed, and a general discussion of impurity centers is given.

9.1 Nitrogen in Diamond

Nitrogen is the most common and also a very important impurity in natural and synthetic diamond (see the fundamental defects in Tables 1.1.1–1.1.2). Five of the six well-known infrared centers contain nitrogen (A–C, E, and F). The historical classification scheme for natural diamonds is based on the nitrogen content, i.e., type I for high and type II for low (<40 ppm) nitrogen concentration, as monitored by the visible light absorption in the 2.5–5.5 eV range (see Sect. 1.2.1 [Rob34]).

There are five types (1–5) of the 42 nitrogen centers in diamond. The majority (30 centers, 2–4) are associates with intrinsic defects (see Chap. 8). The properties of the associates are dominantly determined by the intrinsic defects.

1. Nitrogen only (4): isolated substitutional nitrogen

   $N_1{}^{\circ}$ = **C** center (see Table 9.1.1.1)
   $N_1{}^{+}$ = **E** or **X** center (see Table 9.1.1.2)
   N_2 = **A** center (see Table 9.1.1.3)
   $N_2{}^{+}$ (see Table 9.1.1.3)

2. Associates (18) with lattice vacancies (see Tables 8.1.3.1–8.1.5):

   (NV = $V_1N_1{}^{\circ}$, **H2** = $V_1N_2{}^{-}$, **H3** = $V_1N_2{}^{\circ}$, **N3** = $V_1N_3{}^{\circ}$, **N9** = $V_1N_4{}^{-}$, **B** = $V_1N_4{}^{\circ}$ (see Table 8.1.3.5), **F** = $V_1N_4(C_2)_i{}^{\circ}$ (see Table 8.1.3.6), **H4** = $V_2N_4{}^{\circ}$, etc.

DOI 10.1007/978-3-642-22215-3_9, © Springer-Verlag Berlin Heidelberg 2012

3. Associates (6) of N_1 or N_2 with the ⟨1 0 0⟩ split carbon self-interstitial: ($5RL = (C_2)_iN_1$, etc.) (see Table 8.2.3).
4. The ⟨1 0 0⟩ split $(N_1C_1)_i$ *interstitial* (4) in three charge states: (**R11**, etc.), and related centers, like $(N_2)_i$ (see Table 8.2.4).
5. Donor–acceptor pairs (DAP) (9) of nitrogen with boron, and of nitrogen with an unknown acceptor (see Table 9.1.5).

9.1.1 Isolated Nitrogen Centers

$N_1{}^\circ = $C center: The absorption lines and bands of the C center are listed in Table 9.1.1.1. In EPR, the paramagnetic $N_1{}^\circ$ center is known as the P1 center [New01a].

The properties of the single substitutional nitrogen (C center) have been the subject of numerous experimental and theoretical investigations. The C center is the dominant nitrogen center in type Ib diamonds. Type Ib is the common type in synthetic HPHT diamonds, but is seldom found among natural diamonds.

The original T_d symmetry is lowered to C_{3v} with a large gain in Jahn–Teller energy, and a considerable lowering of the donor level to a calculated position at 2.20 eV above the valence band, i.e., 3.30 eV below the conduction band [Mes73]. One of the four N–C bonds is elongated by ca. 26%, where the two atoms move apart by roughly the same amount [Mes73, Mai94a].

With partial ^{13}C substitution (40%, 50%, 70%, 95% ^{13}C), a splitting into four to five lines of the 167-meV line (100%C, 0%N) is observed [Col93e],[San93]. In the analysis by the present author, the vibration is located with 24% at C′ (distorted direct neighbor of N), $1\times20\%+2\times12\% = 44\%$ at 3 C″ (three ligands of C′, further distorted), and $9\times3.5\% = 32\%$ at C‴ (Nine ligands of 3 C″). With these parameters (and a slight adjustment of the partial ^{13}C content, i.e., 44%, 56%, 75%, and 89%), the line shifts can be reproduced within 1% of the total shift (−52 meV). This fit is much better than that obtained previously with the force constant model [San93].

For the C center, three broad vibronic absorption bands are observed: C(α2) at 3.30 eV, C(β2) at 3.90 eV, and C(γ2) at 4.55 eV. However, only for C(γ2) the corresponding ZPL is observed (C(c) at 4.059 eV), with quasi-local vibrational mode (QLVM) sidebands at 4.120 and 4.179 eV. The two other ZPLs are expected at the onset of the absorption and luminescence bands: *C(a) at 2.80 and *C(b) at 3.40 eV (see Table 9.1.1.1).

The DAP1 analysis (see Table 10.3) yields for the $N_1{}^\circ$ donor ground level a position at ***3.23 eV*** below the conduction band (close to the theoretical value of 3.30 eV [Mes73], see below). Therefore, the excited state of *C(a) would be in the energy gap, while the excited states of *C(b) and C(c) would be inside the conduction band. This interpretation is in agreement with the (very weak) photocurrent (PC) at 3.30–4.05 eV, with the PC peak = C(β2) at 3.90 eV, and the intense PC starting at 4.05 eV with the (not resolved) PC peak C(γ2) at 4.55 eV [Den67]. Structure on this band with onset at 4.070, 4.150, and 4.240 eV

Table 9.1.1.1 Transitions from the neutral isolated substitutional nitrogen atom in diamond ($N_1{}^{\circ}$ = C center; in EPR: P1 center)

Line/ band[a]	E (meV) (*cm*$^{-1}$)	Width (meV) (%)	Relat. int. (%)	i13 shift[b] $^{13}E/^{12}E$(%)	i15 shift[c] $^{15}E\perp^{14}E$(%)	Comment, percentage of C or N vibrational involvement from isotope shifts
C(a′)	105.4	25	36	–	–	N–C or C–C bend vibration
	850	*24*				
C(b′)	129.6	7	7	−4.1	–	100%C
	1,045	*5*				
C(c′)	136.0	3	1	−4.3	–	100%C
	1,097	*2*				
C(d′)	**140.1**	**8**	**28**	−2.9	−1.3	70%C – 30%N vibration
	1,130	***6***				var. 138.7–140.7 meV
C(e′)	149.5	14	23	−3.9	–	100%C
	1,206	*9*				
C(f′)	161.2	5	4	−4.4	±0	100%C
	1,300	*3*				
C(g′)	166.6	0.6	1	−3,9	±0	100%C (see also text Sect. 9.1.1)
	1,344	0.4				
	E (eV)	Width (eV)				
*C(α1)	2.30	0.40	–	–	–	Luminescence band
(*C(a))	(2.80)	–	–	–	–	Expected ZPL (onset of C(α1) and C(α2))
C(α2)	3.30	0.60	–	–	–	Polarized absorption band, E//trigonal axis
*C(β1)	2.88	0.35	–	–	–	Luminescence band
(*C(b))	(3.40)	–	–	–	–	Expected ZPL (onset of C(β1) and C(β2))
*C(β2)	3.90	0.80	–	–	–	Polarized absorption band, E/trigonal axis

(continued)

Table 9.1.1.1 (continued)

Line/ band[a]	E (meV) (cm^{-1})	Width (meV) (%)	Relat. int. (%)	i13 shift[b] $^{13}E/^{12}E(\%)$	i15 shift[c] $^{15}E/^{14}E(\%)$	Comment, percentage of C or N vibrational involvement from isotope shifts
C(c)	4.059	0.01	–	–	–	ZPL of C(γ); QLVM: +61 meV [Naz87]
*C(γ2)	4.550	0.6	–	–	–	Polarized absorption band, E//trigonal axis

[a]The C(a′–g′) IR and C(c) UV absorption lines are observed in NA, HA, and LA. The C(α1)–C(γ2) bands are observed in NB, HB, and LB. The C(c) line and the C(γ2) band are also observed in photoconductivity (PC)
[b]From [Col88c], Table 1 and Fig. 1;
[c]From [Col82], [Dav94a]

corresponds to the ZPL = C(c) at 4.059, with two QLVM sidebands at 4.120 and 4.179 eV (see Table 9.1.1.1).

The depth of the isolated nitrogen donor has been the subject of several theoretical and experimental investigations (summarized in [Naz94a]).

In a pioneering theoretical treatment, a dramatic energy gain by a ⟨1 1 1⟩ Jahn–Teller distortion (26% lengthening of one N–C bond) was calculated [Mes73]. While the effective mass treatment predicts a depth of 0.4 eV for T_d symmetry, the depth for the C_{3v} distorted nitrogen in the cluster calculation is 3.3 eV (2.2 eV above the valence band).

From PC measurements, different depth values of ***1.7 eV*** (natural type Ib, [Far69], [Ver75]), ***2.0 eV*** (HPHT synthetic type Ib [Far74]), or ***4.05 eV*** (natural type Ia, Ib, or natural intermediate type [Den67, Dea65a]) have been derived.

However, a great difficulty in PC data interpretation arises from the fact that all transitions from standard DAP are expected to be PC active. Two clear examples of PC spectra with well-resolved features are (a) DAP32 = GR2–8 [Ver75b] and (b) DAP63/64 = N9 [Den67].

With the present knowledge, it can be concluded that none of the above-mentioned depth values (***1.7, 2.0***, and ***4.05 eV***) represents the true ionization energy of the $N_I{}^{\circ}$ = C center (which is ***3.23 eV*** from DAP1). Three possible explanations can be offered: (a) the onset at ***1.7 eV*** arises from DAP9 (E or X center + boron, 1.74–2.18 eV); (b) the onset at ***2.0 eV*** arises from DAP1 (C center + boron, 2.02–2.67 eV); and (c) the onset at ***4.05 eV*** arises from transitions from the ground state in the gap to excited states inside the conduction band (C center: 4.059 eV; B center: 4.184, 4.191, and 4.197 eV; and A center: 4.470 eV).

An isolated PC peak from a type Ia sample occurs at 4.7 eV [Den67] and can be assigned to the B(α2) absorption peak at 4.7 eV. An unresolved PC peak at

Table 9.1.1.2 The positively charged isolated substitutional nitrogen atom in diamond: $N_1^+ = E$ (or X) center, IR absorption E(a′–g′), and PC bands *E(α2 − γ2)

Line/ band $N_1^+E=X$	E (meV) *(cm^{-1})*	Width (meV)*(%)*	Relat. Int. (%)	i13 shift	i15 shift	Comment
E(a′)	**117.8**	**12**	**26**	–	–	HA [Law93b]; Fig. 3.21
	950	***24***				MA [Col88c]; Fig. 3.15
E(b′)	129.7	7	19	–	–	NA[Woo83]; Fig. 3.20,
	1,046	*5*				HA[Law93b]; Fig. 3.21 MA[Col88c]; Fig 3.15
E(c′)	138.2	8	19	–	–	NA[Woo83]; Fig. 3.20,
	1,115	*2*				HA[Law93b]; Fig. 3.21, MA[Col88c]; Fig. 3.15
*E(d′)	146.9	7	9	–	–	MA[Col88c]; Fig. 3.15
	1,185	*6*				Overlapping band
*E(e′)	152.5	8	14	–	–	MA[Col88c]; Fig. 3.15
	1,230	*9*				Overlapping band
*E(f′)	159.9	7	8	–	–	MA[Col88c]; Fig. 3.15
	1,290	*3*				Overlapping band
E(g′)	165.1	0.7	5	–	–	HA[Law93b]; Fig. 3.21,
	1,332	*0.4*				MA[Col88c]; Fig. 3.15
	E (peak) (eV)	Width (eV)*(%)*				
*E(α2)?	2.50	0.30 *12*	–	–	–	From photoconductivity [Far74]
*E(β2)?	3.50	0.45 *13*	–	–	–	From photoconductivity [Far74]
*E(γ2)?	4.20	0.50 *12*	–	–	–	From photoconductivity [Far74]

Table 9.1.1.3 The neutral N_2° = A center (IR absorption and vibronic lines and bands), and the positively charged $^*N_2^+$ center (vibronic lines)

Line/band	E (meV) *(cm⁻¹)*	Width (meV) *(%)*	Relat. int. (%)	i15 shift $^{15}E/^{14}E$(%)	Comment
N_2°					
A(a′)	60.0	6	11	–	N–C bond
	484	*10*			
A(b′)	135.1	4	2	–	–
	1,090	*3*			
A(c′)	146.9	7	21	–	–
	1,185	*5*			
A(d′)	150.6	6	8	–	–
	1,215	*4*			
A(e′)	**158.9**	**7**	**58**	**−1.1** [Dav94b]	N(29%)–C(71%) stretch; close to the ZPL of SB-DAP30 from the $N_2 + V_1$ center: (L = 0.154, D = 1.45 eV) with lines H1 b, d, g
	1,282	***5***			
	E (eV)	Width (eV)			
*A(a)	3.928	0.015	–	–	Named **N6**, proposed LVM sideband at 4.088 eV (+160 meV = A(e′))
*A(α2)	> 3.9	–	–	–	Absorption continuum with onset at 3.9 eV; photoconductivity threshold at 4.0 eV
*A(b)	4.470	0.010	–	–	ZPL; SB (QLVM) at f = 58, w = 38 meV [Naz87]; (calc. f = 58.5, w = 36.3 meV for N_1C_2; see Table 11.2)
*A(β2)	5.00	0.60	–	–	Absorption (see Table 2.4.3) and PC [Den67] band
$^*N_2^+$(a)?	1.602	0.012	–	–	Observed in PL; sidebands: QLVM at f = –74 meV, w = 57 meV (calc. f = 74.5, w = 59 meV for N_2; see Table 11.2); LVM (overlap) at –77 meV
$^*N_2^+$(b)?	2.910	0.012	–	–	Observed in absorption and CL; LVM at –77 meV

5.00 eV from an intermediate type sample [Den67] corresponds probably to the $A(\beta 2)$ absorption peak at 5.00 eV.

$N_1^+ = E$ center or X lines: The assignment of the $(N_1^+ = E = X)$ center was accomplished by a detailed study of the interconversion of N_1° and N_1^+ [Law98]. First, a synthetic HPHT diamond was irradiated with $2 \times 10^{18}\,cm^{-2}$ 1.9-MeV electrons, which resulted in a ratio of 40%N_1° to 60%N_1^+. Second, photochromic studies showed that a quantitative conversion from N_1^+ to N_1° occurs with UV illumination of $E > 3.1$ eV and the reverse process with illumination of $2.1 < E < 3.1$ eV [Law98].

The absorption lines and bands of the E = X center are listed in Table 9.1.1.2. The IR spectrum of N_1^+ is similar to that of N_1°. Especially, the small shift in the sharp line from 166.6 (N_1°) to 165.1 meV (N_1^+) indicates similar distortions for both centers. The expected depth of N_1^+ is $E_D = 3.48$ eV (see DAP9, Table 9.1.5).

Also, the broad bands from N_1^+ transitions to the excited states (inside or near the conduction band) are similar to those of N_1° : $E(\alpha 2)/E(\beta 2)/E(\gamma 2) = 2.50/3.50/4.20$ eV, compared to that of $C(\alpha 2)/C(\beta 2)/C(\gamma 2) = 3.30/3.90/4.55$ eV.

$N_2^\circ = A$ center: The absorption lines and bands of the A center are listed in Table 9.1.3. The A center is paramagnetic in the ionized state.

From theory, it is known that the length of the N–N bond is increased by 30%, and the length of the N–C bond is decreased by 4% (compared to the lattice C–C bond) [Mai94a].

The DAP20 analysis (see Table 9.1.5) yields for the N_2° donor ground level a position at 3.71 eV below the conduction band. This is in good agreement with the observed onset of the absorption continuum at 3.9 eV. In the ultraviolet range, an absorption and PC band $A(\beta 2)$ at 5.00 eV with ZPL at 4.470 eV is observed (see Table 9.1.1.3).

$^*N_2^+$ center: This center with a depth of 1.44 eV (see Table 9.1.5) is indirectly observed in the DAP67 $(^*N_2^+ + B_1^\circ)$ transitions (3.86–4.28 eV). A luminescence line at 1.602 eV in high nitrogen diamonds can possibly be assigned to $^*N_2$(a). In addition, a luminescence (2.897 eV) and absorption (2.910–2.916 eV) line can be tentatively assigned to $^*N_2$(b). This line appears in spectra of DAP18, e.g., in luminescence after N^+ implantation (and 1,400 °C anneal) of type IIa diamonds [Zai01]. Both ZPLs have a local vibrational mode (LVM) sideband at −77 meV (see Table 9.1.1.3).

9.1.2 Associates of Nitrogen with Single or Double Vacancies

$V_1N_1^-$; $V_1N_1^\circ$(= NV center) and $V_1N_2^+$(= S1 center): Details of these centers are given in Table 8.1.3.1. All three centers are observed in natural, as grown HPHT, or CVD synthetic diamonds, and also in irradiated (and annealed) diamond.

$^*(V_1H_1)^\circ N_1$: This configuration is tentatively assigned to weak infrared lines from natural diamond (see Table 8.4a).The NH bend vibration is not directly observed, but can be indirectly determined from the stretch + bend combination line.

($V_1N_2^-$ = H2 center) and ($V_1N_2^\circ$ = H3 center): Details of these centers are given in Table 8.1.3.1. The **H2** center (1.256 eV, DAP77/78) appears after heating diamonds to 1,700 °C, or after irradiation and annealing at 500–1,500 °C. The **H3** center (2.464 eV, DAP51/52) is observed in natural, in HPHT or CVD synthetic, and in irradiated (and annealed) diamond. Photochromic interconversion of H2 and H3 is achieved by illumination below or above 600 nm (= 2.07 eV) [Mit90]. Uniaxial stress experiments on H2 [Law92b] and H3 [Dav76a,Dav76b] have shown that these centers have C_{2v} symmetry with the C_2 axis along ⟨0 0 1⟩ and two (1 1 0) reflection planes, i.e., the structure is N–V–N.

***$V_1N_3^-$($V_1N_3^\circ$ = N3 center) and *$V_1N_3^\pm$**: Details of these centers are given in Table 8.1.3.2. The well-known N3 center has three ZPL transitions: $V_1N_3^\circ$(a,b,c), at 2.297, 2.680, and 2.985 eV. There are luminescence sidebands (with structure) for $V_1N_3^\circ$(a,b,c), but there is an absorption sideband (with structure) only for $V_1N_3^\circ$(c). The transitions of $V_1N_3^\circ$(a) and $V_1N_3^\circ$(b) are observed only in delayed photoluminescence [Sob76].

***$V_1N_3^\circ(C_2)_i$**: When the vacancy of the $V_1N_3^\circ$ center is created, a $(C_2)_i$ self-interstitial is expelled, which is later gathered in the platelets (see the discussion in Sect. 9.1.6.1). In an intermediate state (only with *small* platelets, 5 nm), the self-interstitial is connected to V_1N_3, forming the *$V_1N_3^\circ(C_2)_i$ center (see Table 8.1.3.2). This center shows great similarity to $V_1N_3^\circ$(a) (both centers observed only in delayed photoluminescence). The ZPL of $V_1N_3^\circ(C_2)_i$ is shifted by −95 meV with respect to the ZPL of the isolated $V_1N_3^\circ$(a) (only with *large* platelets, 100 μm) [Sob76].

($V_1N_4^-$ = N9 center) and ($V_1N_4^\circ$ = B center): For details of these centers, see Table 8.1.3.2. The ZPL absorption lines of the B center (in the ultraviolet) are very weak [Naz87]. The ZPL splitting of 6–7 meV is typical for a dynamic Jahn–Teller effect. Four lines of the sideband DAP82 are superimposed on the B(α) absorption band. The B center is well known for its vibrational infrared absorption lines (see Table 8.1.3.5).

The ZPL of the N9 center is split into three lines: N9(a) at 5.252, N9(b) at 5.262, and N9(c) at 5.277 eV. This 10–15 meV splitting is similar to that of the B center, indicating again a dynamic Jahn–Teller effect. Well-resolved lines from the sideband DAP63 (absorption) and the sideband DAP64 (luminescence) are observed.

***$V_1N_4(C_2)_i^\circ$ = F center**: As described in Sect. 9.1.6.1, the self-interstitial is connected to ($V_1N_4^\circ$ = B center), forming the (*$V_1N_4(C_2)_i^\circ$ = F center). The ZPL of $V_1N_4(C_2)_i^\circ$ is shifted by +376 meV with respect to the ZPL of the isolated $V_1N_4^\circ$(b). Similar to the B center, the F center has well-known vibrational infrared absorption lines, which correlate with the ZPL F(a) (see Table 8.1.3.3). However, compared to the vibrations of the B center (highest frequency 165 meV), the F center vibrations extend to 191 and 196 meV, which are typical frequencies of $(C_2)_i$ vibrations. There are three ZPL transitions at F(a) = 2.145, F(b) = 2.721, and F(c) = 4.567 eV, each with broad sidebands. The structure on the 5.0-eV sideband of F(c) reveals the involvement of donor–acceptor transitions (DAP81a–f).

$V_2N_1^\circ$, $V_2N_3^\circ$, and ($N_3V_2N_1^\circ$ = H4 center): It has been shown by theory that the H4 center is formed when the (V_1N_4 = B center) captures a vacancy (see Eqs. (9.8a,b)). The infrared center with the lines ($H1(c, e, f) = V_1N_4{}^\circ + V_1{}^\circ$) is an intermediate state (see the discussion below). A similar process can be assumed for the formation of V_2N_1 and V_2N_3. Details of these centers are given in Table 8.1.3.4.

9.1.3 *Associates of the Split Self-interstitial $(C_2)_i$ with One or Two Nitrogens*

$^*(C_2)_iN_1{}^\circ$ = 5RL: The 5RL center (4.582 eV, DAP79/80) is prominent in as-irradiated diamonds, is enhanced by 500 °C anneal, and anneals out at 800–1,200 °C. The luminescence spectrum is unusual, showing four intense sidebands (partly exceeding the ZPL intensity) of the 237-meV C–C vibration. This gave rise to the name 5RL = "5 radiation lines." The proposed structure is in good agreement with the observed vibrational frequencies and their isotope shifts [Col88c], and also with the uniaxial stress results [Col86b]. Details of the 5RL center are given in Table 8.2.3.

$^*(C_2)_iN_1{}^+$: The positive charge state of the 5RL has properties similar to that of the neutral charge state (see Table 8.2.3).

$^*(C_2)_iN_2{}^-$, $^*(C_2)_iN_2{}^\circ$, $^*(C_2)_iN_2{}^+$: The three charge states of the $^*(C_2)_iN_2$ center have very similar properties (see Table 8.2.3).

9.1.4 *The ⟨1 0 0⟩ Split Nitrogen–Carbon Interstitial $(N_1C_1)_i$ in Three Charge States, Associates with Boron, and the $(N_2)_i{}^\circ$ center*

These seven centers are described in Sect. 8.2.4 and in Table 8.2.4.

9.1.5 *Donor–Acceptor Pairs of Nitrogen (N_1 and N_2) with a Single Boron and of Nitrogen (N_2) with an Unknown Acceptor*

$^*N_1{}^\circ + B_1{}^\circ$ = DAP1: For diamond, this is the first donor–acceptor pair system described in the literature [Dis94a,b]. The transitions of this standard DAP occur in the range 2.070–2.668 eV (shells $s = 50$ to $s = 1$). The DAP line of the nearest neighbor N–B pair ($s = 1$) is very intense, but no electronic transition from this center is observed. The notation is $^*N_1{}^\circ + B_1$, which emphasizes the DAP character, and the * indicates that the atomic composition is not yet fully confirmed.

Table 9.1.5 Donor–acceptor pairs (standard DAP) of a nitrogen donor (N_1 or N_2) with the boron acceptor

Defect	Observed	L (eV)	Lines (range) (eV)	Standard-DAP	D(eV)	Name/comment
$^*N_1^{\circ}+B_1^-$	NL	1.871	1.559–1.791	106 a–k	−1.43	shells: 16, 20, 22, 24, 26, 32, 34, 54, 58, 198, 228/$E_D + E_A = 3.619$; $E_A(B_1^-) = 0.392$ from $E_D = 3.227$ eV
$^*N_1^{\circ}+B_1^{\circ}$	HA, MA, NL, LL, ML	1.893	2.067–2.668	1a–m	1.43	Shells: 1–8, 10, 13, 16, 34, 50/$E_D + E_A = 3.597$; $E_D(N^{\circ}) = 3.227$ from $E_A = 0.370$ eV
$^*N_1^{+}+B_1^{\circ}$	NA, NL	1.641	1.743–2.180	9a–v	1.53	Shells: 6–10, 12–14, 16, 18, 20, 22, 24, 26, 34, 36, 50, 58, 74, 98, 106, 162/$E_D + E_A = 3.849$, $E_D(N^+) = 3.479$ from $E_A = 0.370$ eV
$^*N_1^{2+}+B_1^{\circ}$	ML	1.567	1.697–2.417	70a–q	1.52	Shells: 1, 2, 3, 10, 12, 13, 14, 15, 16, 18, 20, 24, 26, 30, 34, 50, 106/$E_D + E_A = 3.923$, $E_D(N^{2+}) = 3.553$ from $E_A = 0.370$ eV
$^*N_2^{\circ}+B_1^-$	NL	1.890	2.095–2.622	19a–j	1.30	Lines **A** ($s = 4$), **B** ($s = 3$) in yellow luminescence/shells: 1, 2, 3, 4, 5, 10, 12, 16, 20, 32/$E_D + E_A = 3.600$
$^*N_2^{\circ}+B_1^{\circ}$	NL	1.407	1.798–2.105	20a–g	1.38	Shells: 1, 3, 4, 5–8/$E_D + E_A = 4.070$; $E_D(N_2^{\circ}) = 3.700$ from $E_A = 0.370$ eV
$^*N_2^{+}+B_1^{\circ}$	NA, MA	3.684	3.865–4.280	67a–i	1.43	Shells: 4, 6, 10, 12, 14, 16, 26, 34, 50/lines of $s = 1$–3 are outside spectrum/$E_D + E_A = 1.806$; $E_D(N_2^+) = 1.436$ from $E_A = 0.370$ eV
$^*N_2^{2+}+B_1^{\circ}$	NL	2.495	2.649–3.204	18a–g	1.54	Lines C ($s = 50$), D ($s = 32$), E ($s = 20$), F ($s = 2$), H ($s = 1$) in yellow luminescence/shells: 1, 2, 4, 20, 26, 32, 50/$E_D + E_A = 3.156$; $E_D = 2.786$ from $E_A = 0.370$ eV

DAP: L = limiting energy (different from ZPL), D = dielectric factor

$^*N_1{}^+ + B_1{}^\circ$ = DAP9: A set of weak lines is superimposed on the luminescence "B" band of type IIb diamonds. A possible correlation with the $N_1{}^+$ = E = X center has not been investigated.

$^*N_2{}^\circ + B_1{}^\circ$ = DAP20: Some 2–5% of all gem-quality natural diamonds are brown in color. Some of these exhibit bright yellow photoluminescence when excited at 3.40 eV [Col82b]. The infrared absorption spectrum of these diamonds is dominated by the F center lines ($V_1N_4\ (C_2)_i{}^\circ$, see Table 8.1.3.6). In addition, lines from A centers ($N_2{}^\circ$), B centers (V_1N_4), and also from sp^2CH_2 (at 182 meV) and sp^3CH_2 (at 178 meV) are observed. These latter lines indicate the presence of hydrogen in grain boundaries (see Sect. 9.3). It can be speculated that the grain boundaries stabilize the F centers by blocking the reactions of (9.3b) and (9.4). Superimposed on the broad luminescence bands from the F center (but not related to those) are sets of sharp lines, which belong to DAP18, DAP19, and DAP20. Some lines of DAP18 and DAP19 have been studied under uniaxial stress, and the symmetry of the center (from the dipole orientation) has been determined [Moh82b]. No explanation for the different symmetries was given in the original publication, but a fascinating agreement (and confirmation of the DAP analysis) is obtained when the lattice vectors of the respective DAP are considered (see $^*N_2{}^+ + B_1$ = DAP18, below).

$^*N_2{}^+ + B_1{}^\circ$ = DAP67: For a general description, see DAP20 above. Two lines are given names [Moh82b]: A ($s = 4$) and B ($s = 3$). The symmetry of B is triclinic in agreement with the lattice vector $\langle 3\ 1\ 1\rangle$.

$^*N_2{}^+ + B_1{}^\circ$ = DAP18: Five lines are given names [Moh82b]: C ($s = 50$), D ($s = 32$), E ($s = 20$), F ($s = 2$), H ($s = 1$), and J (perturbed $s = 2$). The symmetry (obtained from uniaxial stress) for the lines C, D, E, and J is C_{2v} with the corresponding lattice vectors C$\langle 10,\ 10,\ 0\rangle$, D $\langle 8,\ 8,\ 0\rangle$, E $\langle 8,\ 4,\ 0\rangle$, and F, J $\langle 2,\ 2,\ 0\rangle$. The symmetry of line H is trigonal, as expected for the lattice vector $\langle 1,\ 1,\ 1\rangle$. Six lines from DAP18 are also observed in absorption (at the same energies as in luminescence, indicating a standard DAP).

$^*N_2{}^+$ = DAP4: When a natural type Ia diamond with 8×10^{19} cm^{-3} $N_2{}^\circ$ (A center) is irradiated with 1×10^{19} cm^{-3} neutrons, the DAP4 transitions together with the electronic transition at 1.602 eV of $^*N_2{}^+$ are observed. Vibrational sidebands of DAP4 are at 56 meV (typical for N_2, see Table 9.1.5). The acceptor is unknown.

$^*N_2{}^{2+}$ = DAP45: This center is observed in HTHP synthetic diamond with high nitrogen concentration (1.3×10^{19} cm^{-3}) after neutron irradiation and 12 h annealing at 800 °C [Vin88b]. The acceptor is unknown; however, the type of spectrum and the hole burning [Sil94a, Sil95] confirm the DAP interpretation.

9.1.6 Discussion

9.1.6.1 Thermally Induced Nitrogen Aggregation in Natural and As-Grown HPHT Synthetic Diamonds

In "newly grown" natural and as-grown HPHT synthetic diamonds, nitrogen is dominantly present in the single substitutional form ($N_1{}^\circ$ = C center, type Ib).

Natural type Ib diamonds are rare, because aggregation occurs under the conditions in the earth's crust.

The sequence of aggregation steps and the structure of the aggregates have been established by annealing experiments [Chr77, Woo86, Kif00] and by theoretical calculations [Mai94]. Triple and quadruple nitrogen centers are stable in natural diamonds only in association with a lattice vacancy [Mai94a].

Step 1: At $T > 1,500\,°C$, the single nitrogen starts to migrate, and nitrogen pairs ($N_2{}^\circ$ = A center) are formed with an energy gain of 0.5 eV.

$$N_1 + N_1 \rightarrow (N_2 = \text{A center}) \quad (9.1)$$

Step 2a: At $T > 1,700\,°C$, some of the nitrogen pairs dissociate, while other pairs capture one nitrogen. The triple nitrogen expels a lattice carbon, i.e., a vacancy/interstitial pair is formed.

$$N_2 + N_1 \rightarrow ({*}V_1N_3(C_2)_i = 563\,\text{nm center?}). \quad (9.2a)$$

Unfortunately, the $^*V_1N_3(C_2)_i$ center is not well documented in the literature. The 563-nm center with absorption and luminescence at 2.202 eV is a possible candidate, due to its great similarity to the N3(a) center at 2.297 eV (see Table 8.2.2).

Step 2b: At the same temperature ($T > 1,700\,°C$), the interstitial can leave the $^*V_1N_3(C_2)_i$ center, and the N3 center is formed.

$$^*V_1N_3(C_2)_i \rightarrow (V_1N_3 = \text{N3 center}) + (\text{distant})(C_2)_i. \quad (9.2b)$$

Step 3a: At $T > 1,900\,°C$, a further nitrogen is captured and the $^*V_1N_4(C_2)_i$ = F center is formed with a theoretical energy gain of ca. 3 eV [Mai94].

$$^*V_1N_3(C_2)_i + N_1 \rightarrow ({*}V_1N_4(C_2)_i = \text{F center}). \quad (9.3a)$$

The F center is named after its intense IR absorption bands (see Table 8.1.3.6). Correlated with these IR bands are three ZPL transitions at 2.145, 2.721, and 4.567 eV. The latter is similar to the 4.184–4.197 transitions of the B center (see Sect. 9.1.2 and Tables 8.2.3 and 9.1.4).

Step 3b: At the same temperature ($T > 1,900\,°C$), the interstitial can leave the $^*V_1N_4(C_2)_i$ center, and a (V_1N_4 = **B** center) is formed.

$$* \, V_1N_4(C_2)_i \rightarrow (V_1N_4 = \text{B center}) + (\text{distant})(C_2)_i. \quad (9.3b)$$

Step 3c: At the same temperature ($T > 1,900\,°C$), the V_1N_3 = N3 center can capture a single nitrogen, and a V_1N_4 = **B** center is formed with a theoretical energy gain of 3 eV [Mai94].

$$V_1N_3 + N_1 \rightarrow (V_1N_4 = \text{B center}) \quad (9.3c)$$

Evidently, there are two possible paths leading to the **B** center formation, depending on whether the $(C_2)_i$ has been released from the $^*V_1N_3(C_2)_i$ center ((9.2b), followed by (9.3c)) or not released ((9.3a), followed by (9.3b)). The centers with three nitrogens (**N3** and $^*V_1N_3(C_2)_i$) are metastable with respect to the centers with four nitrogens. For this reason, the concentration of the triple nitrogen centers is always very low compared to the centers with one, two, or four nitrogens.
Step 4: The migrating self-interstitials $(C_2)_i$ form planar aggregates called platelets.

$$n(C_2)_i \rightarrow (C_{2n})_i = \text{platelets} \tag{9.4}$$

The calibrated absorption of the platelets has been compared with the absorption from B centers. In "regular" type IaB diamonds, the absorptions are proportional, while in "irregular" type IaB diamonds, the platelet absorption is very low or absent. A possible explanation can be a partial decomposition (see (9.5a)). An alternate explanation would be that the platelets have never or incompletely been created [Col97].
Step 5a: At still higher temperatures ($T > 2{,}700\,°C$), the platelets decompose, and dislocation loops are formed [Kif00].

$$(C_{2n})_i = \text{platelets} \rightarrow \text{dislocation loops} \tag{9.5a}$$

Step 5b: At the same temperature ($T > 2{,}700\,°C$), the B centers decompose, and voidites are formed. These voidites are octahedral defects consisting of molecular nitrogen in a crystalline form, with dimensions between 2 and 50 nm [Kif00].

$$(nV_1N_4 = n \text{ B centers}) \rightarrow \text{voidites} = N_{4n}(\text{crystalline}) \tag{9.5b}$$

In natural diamonds, all intermediate states of thermal nitrogen aggregation are observed: With single nitrogen (type Ib), with nitrogen pairs (type IaA), with partial transformation from A to B (type IaAB), or with full transformation into the very stable B centers (type IaB). Voidites are very rare in natural diamonds.

9.1.6.2 Thermally Induced Nitrogen Aggregation in the Presence of Isolated Vacancies in Irradiated Diamonds

In irradiated diamond, the concentration of isolated *vacancies* can be comparable to the concentration of *nitrogen*. At $T > 600\,°C$, the vacancies start to migrate, and can be captured by nitrogen centers with a considerable gain in energy (3.0–6.0 eV) [Mai94a]. The respective reactions are denoted in (9.6a–9.8a). The three associates of a nitrogen center with an *outside* vacancy (**H1(a)** center, **H1(b)** center, and **H1(c)** center) are *metastable* and are *precursors* of the more stable centers **NV** (9.6b), **H3** (9.7b), and **H4** (9.8b). Absorption of the metastable centers is observed in the infrared (see Table A.1.1). This book provides complete assignments for the

H1 lines, especially the **H1**(b,d,g)-DAP30 and **H1**(c,e,f)-DAP31 donor–acceptor transitions (see Sect. 10.1).

In irradiated diamonds with *isolated nitrogen* (C center), the reactions (9.6a,b) occur after annealing above 600 °C, and the **NV** is formed. The NV center is relatively stable up to temperatures of 1,400–1,700 °C.

$$V_1(\text{distant}) + (N_1{}^{\circ} = C \cdot \text{ce.}) \rightarrow (V_1C_4 - N_1{}^{\circ} = \text{H1(a)ce.}) + 2\,\text{eV}, \quad (9.6a)$$

$$(V_1C_4 - N_1{}^{\circ} = \text{H1(a)center}) \rightarrow (V_1N_1C_3{}^{\circ} = \text{NV} \cdot \text{ce.}) + 5\,\text{eV}. \quad (9.6b)$$

In irradiated type IaA diamonds (nitrogen pairs are dominant), the **H3** center is formed by the reactions (9.7a,b)

$$V_1(\text{distant}) + (N_2 = A \cdot \text{ce.}) \rightarrow (V_1 - N_1 - N_1 = \text{H1(b, d, g)ce.}) + 1\,\text{eV} \quad (9.7a)$$

$$V_1 - N_1 - N_1 \rightarrow (N_1 - V_1 - N_1 = V_1N_2{}^{\circ} = \text{H3 center}) + 4\,\text{eV}. \quad (9.7b)$$

In irradiated type IaB diamonds (with dominant B centers), the **H4** center is formed by the reactions (9.8a,b)

$$V_1(\text{distant}) + (V_1N_4 = B \cdot \text{ce.}) \rightarrow (V_1N_4 - V_1 = \text{H1(c, e, f)ce.}) + 6\,\text{eV} \quad (9.8a)$$

$$V_1N_4 - V_1 \rightarrow (N_3V_1V_1N_1 = N_3V_2N_1{}^{\circ} = \text{H4 center}) + 5\,\text{eV}. \quad (9.8b)$$

The precursor centers **H1**(a) in (9.6a), **H1**(b,d,g) in (9.7a), and **H1**(c,e,f) in (9.8a) are metastable and can be observed in absorption (see Tables 8.1.3.1 and 8.1.3.4). Theoretical calculations for the activation energy give 2.4 eV for the reactions of (9.6a), (9.7a), and (9.8a) (vacancy migration) [Mai94a]. The final reordering (site exchange between the vacancy and a carbon or a nitrogen atom) in the reactions of (9.6b), (9.7b), and (9.8b) requires a higher (estimated) activation energy (> 3 eV) with corresponding higher annealing temperatures ($T > 800$ °C) [Mai94a].

9.2 Boron in Diamond

Boron is the second abundant impurity (after nitrogen) in natural diamond. In synthetic diamonds, boron can be introduced as a dopand. In the periodic system, the elements boron and nitrogen are neighbors of carbon, and both can be easily incorporated into the diamond lattice.

Isolated substitutional boron has its ground state 0.37 eV above the valence band. This shallow acceptor center is responsible for the p-type semiconductor character of type IIb diamonds (where the concentration of compensating donors is lower than the boron concentration).

There are six types of boron centers in diamond. The majority (3–5) contains intrinsic defects, which dominantly determine the properties (see Chap. 8):

1. Isolated substitutional boron: The $B_1{}^{\circ}$ acceptor (see Table 9.2.1).
2. Standard DAP of boron with nitrogen = DAP1 (see Table 9.1.5) or with phosphorous = DAP2 (see Table 9.2.2).
3. Associate with a lattice vacancy (see Table 8.1.7.1).
4. Associates of B_1, B_2, or B_3 with the $\langle 1\ 0\ 0\rangle$ $(C_2)_i$ split interstitial (see Table 8.2.3).
5. Associates of B_1 with the $\langle 1\ 0\ 0\rangle$ (N_1C_1) split interstitial in three charge states (see Tables 8.2.4).
6. Boron-bound exciton lines (see Table 9.2.6).

9.2.1 *Isolated Substitutional Boron*

The absorption spectrum of B_1° was explained by a calculation, using effective mass theory [Cro67b]. The ground state is 0.372 eV above the valence band, and the transitions to the excited states occur at 0.2680–0.3646 eV (see Table 9.2.1). The three most intense transitions (at 0.3042, 0.34710, and 0.36273 eV) have sidebands with one to three phonons of energy 161 meV. This LVM frequency is close that of the fundamental lattice phonon (165 meV), indicating a slightly lower force constant for the B–C, compared to C–C stretch vibration.

An interesting phenomenon is observed in the photoconductivity spectrum (at 120 K). The so-called photothermal ionization process (photon + thermal phonon) explains the photoconductivity peaks well below the ionization threshold (these peaks coincide with the absorption peaks) [Col68].

9.2.2 *Standard DAP of Boron with Nitrogen (DAP1, 9, 18, 19, 20, 67, 70), Phosphorous (DAP2), or Oxygen (DAP17)*

The nitrogen containing centers (DAP1, 9, 18, 19, 20, 67, 70) are already listed and described in Table 9.1.5, Sect. 9.1.5. Other data are collected in Table 9.2.2.

In samples which contain boron and phosphorous (as evidenced by their respective bound excitons), 23 lines in the range 4.711–5.275 eV are observed, arising from the $B_1{}^{\circ} + P_1{}^{\circ}$ – DAP2 transitions [Ste99a,b]. The large number of resolved DAP lines has probably to do with the fact that both impurities have shallow states in the bandgap: $E_D(P) = E_c - 0.620$ eV, and $E_A(B) = E_v + 0.370$ eV, with a corresponding large Bohr radius, leading to a better overlap than for the deep donor nitrogen ($E_D(N) = 3.230$ eV).

Table 9.2.1 Absorption and photoconductivity lines from the substitutional single boron in diamond (after [Smi62, Cro67b, Wal79, Zai01])

Line[a]	Observed	(meV)	Assigned[b]	I–II[c]	Name/comment
LVM	NA, HA, MA	159.8	LVM	–	B–C stretch
B′	PC[e]	268.0	I/II → 1	–	1s(Γ_7/Γ_8) → 2s (forbidden); 2 SBs
B(a)	**NA**	**304.2**	**I/II → 2**	–	**1s(Γ_7/Γ_8) → 2p (Γ_6/Γ_8)**
B(b)	NA	335.6	II → 3	–	1s(Γ_7) → 2p (Γ_8)
B(c)	NA	337.3	I → 3	(1.7)	1s(Γ_8) → 2p (Γ_8)
B(d)[f]	*NA*	*340.4*	–	–	–
B(e)	NA	341.48	II → 4	–	1s(Γ_7) → 2p (Γ_8)
B(f)[f]	*NA*	*342.10*	–	–	–
B(g)	NA	343.53	I → 4	2.05	1s(Γ_8) → 2p (Γ_8)
B(h)[f]	*NA*	*345.1*	–	–	–
B(i)[f]	*NA*	*346.37*	–	–	–
B(j)	**NA, HA, LA**	**347.10**	**II → 5**	–	**1s(Γ_7) → 2p (Γ_6)**
B(k)	NA	349.13	I → 5	2.03	1s(Γ_8) → 2p (Γ_6)
B(l)	NA	352.4	II → 6	–	1s → 2p
B(m)	NA	354.56	I → 6	(2.2)	1s → 2p
B(n)	NA	355.79	II → 7	–	1s → 2p
B(o)	NA	357.89	I → 7	2.10	1s → 2p
B(p)	**NA, HA, LA, MA**	**362.73**	**II → 8**	–	**1s → 2p**
B(q)	NA	364.6	I → 8	(1.9)	1s → 2p
E_A	NA, HA, LA, MA, PC	372	I/II → VB		Ioniztion energy
B′–s1	PC	430	B′ + 0.162	–	SB (1 × 162 meV)[d]
B(a)-s1	NA, HA, LA	466	B(a) + 0.162	–	SB (1 × 162 meV)
B(j)-s1	NA, HA, LA	508	B(j) + 0.161	–	SB (1 × 161 meV)
B(p)-s1	NA, HA, LA	527	B(p) + 0.164	–	SB (1 × 164 meV)
B′-s2	PC	591	B′ + 324	–	SB (2 × 162 meV)
B(a)-s2	NA, HA, LA	625	B(a) + 0.321	–	SB (2 × 161 meV)
B(j)-s2	NA, HA, LA	670	B(j) + 0.323	–	SB (2 × 161 meV)
B(p)-s2	PC	686	B(p) + 0.323	–	SB (2 × 162 meV)
B(a)-s3	PC	790	B(a) + 0.486	–	SB (3 × 162 meV)
B(j)-s3	NA, HA, LA	830	B(j) + 0.483	–	SB (3 × 161 meV)

(continued)

Table 9.2.1 (continued)

Line[a]	Observed	(meV)	Assigned[b]	I–II[c]	Name/comment
B(p)-s3	PC	850	B(p) + 0.487	–	SB (3 × 162 meV)
B(a)-s4	PC	953	B(a) + 0.649	–	SB (4 × 162 meV)
B(p)-s4	PC	1,015	B(p) + 0.652	–	SB (4 × 163 meV)
B(p)-s5	PC	1,180	B(p) + 0.817	–	SB (5 × 163 meV)

[a]Bold = intense line;
[b]I, II = 1s ground state (2.06 meV splitting); 2–8 = excited 2p states
[c]Difference, arising from 2.06 meV gr. st. splitting;
[d]LVM frequency is 160 meV in the ground state (see line 1) and 162 meV in the excited state (see sidebands)
[e]Indirect observed;
[f]Additional lines from 19% ^{10}B or disturbed boron

Table 9.2.2 DAP of boron with phosphorous [Ste99a, b] or with oxygen [Rua93a]

Defect	Observed	L (eV)	Lines (range) (eV)	*Standard* DAP	D (eV)	Comment
$^*B_1{}^\circ + P_1{}^\circ$	HL, LL	4.500	4.711–5.275	2a–w	+1.72	Shells: 3, 5, 9, 11, 12, 14, 15, 16, 17, 18, 19, 20, 22, 24, 25, 26, 30, 31, 32,37, 39, 50/$E_D + E_A$ = 0.990; E_D(P) = 0.620 from E_A = 0.370 eV
$^*B_1{}^\circ\ ^*O_1{}^\circ$	LL	2.534	2.676–2.725	17a–g	+1.29	Shells: 34, 44, 48, 50, 58, 62/$E_D + E_A$ = 2.956; E_D(O) = 2.586 from E_A = 0.370 eV

For boron with nitrogen (six centers), see Table 9.1.5
DAP: L = limiting energy (no ZPL for *standard* DAP), D = dielectric factor

9.2.3 Associate of Boron with a Lattice Vacancy

There are two transitions with ZPLs at 2.395 eV ($^*V_1B_1(a)^\circ$) and 3.120 eV ($^*V_1B_1(b)^\circ$) (see Table 8.1.7.1) The sideband DAPs are seen as structure on broad bands. The ZPL at 2.395 eV is weak in X-ray luminescence but is intense in photoluminescence. The observed QLVM has $f = 50$ and $w = 25$ meV, which can be tentatively assigned to $^{11}B_1{}^{12}C_3$ with calculated values $f = 50.4$ and $w = 27.0$ meV (the natural abundance of the boron isotopes is $^{11}B/^{10}B = 81/19\%$).

9.2.4 Associates of B_1, B_2, or B_3 with the $\langle 1\,0\,0 \rangle$ $(C_2)_i$ Split Self-interstitial

These centers are named 2BD (type **IIb d**amaged) and are listed in Table 8.2.3. The lines from $^*(C_2)_iB_2{}^\circ$ – DAP97 transitions are weak, and no other 2BD–DAP transitions are observed.

9.2.5 Associates of B_1 with the $\langle 1\,0\,0 \rangle$ Split $(N_1C_1)_i$ Interstitial in Three Charge States

A short discussion of these centers is given in Sect. 8.2.4. In as-grown CVD synthetic diamonds, boron (identified by its bound exciton) can be present by intentional doping or by unintentional transfer from the boron-doped silicon substrate [Ste96a]. In the cathodoluminescence (CL) of such samples, the lines of $V_1N_1{}^\circ$ at 2.154 eV, $^*(C_2)_iN_1^-$ at 3.188 eV, and $^*(C_2)_iN_1{}^\circ$ (= 5RL) at 4.582 eV are observed, together with a group of nine lines at 2.417–2.792 eV [Ste96a]. These latter lines are typical DAP transitions, which can be tentatively assigned to $^*(N_1C_1)_iB_1{}^\circ$ with the ZPL at 2.792 eV, and the sideband DAP99 (see Table 8.2.4).

In similar CVD samples, CL, lines of $^*(C_2)_iN_1^-$ at 2.807 eV, and $^*(N_1C_1)_i{}^\circ$ at 3.188 eV and lines from DAP1 (= $^*N_1{}^\circ + B_1{}^\circ$; 2.067–2.668 eV) are observed, together with a group of nine lines at 2.725–3.092 eV [Col89c, Dis94a]. This group of lines is tentatively assigned to $^*(N_1C_1)_iB_1{}^+$ with the ZPL at 3.092 eV, and the sideband DAP100 (Table 8.2.4).

In a boron-doped hot filament grown CVD sample, the $(N_1C_1)_iB_1$ center is probably observed in three charge states: Weak ZPLs at 2.792 eV (from $^*(N_1C_1)_iB_1{}^\circ$) and 3.092 eV (from $^*(N_1C_1)_iB_1{}^+$), and an intense group of eight lines at 2.330–2.992 eV, which can be tentatively assigned to $^*(N_1C_1)_iB_1{}^-$ with the ZPL at 2.992 eV, and the sideband DAP101 (Table 8.2.4).

9.2.6 Boron-Bound Exciton Lines

Cathodoluminescence lines from the boron bound exciton are listed in Table 9.2e. The spectrum is typical for a shallow defect center in diamond. The ionization energy (E_i) is 370 meV, and the binding energy (E_b) is 51 meV, in excellent agreement with Haynes' rule, which predicts $E_i/E_b = 7$ [Ste99a].

Table 9.2.6 Cathodoluminescence lines from boron-bound exciton (BE) transitions in diamond (after Figs. **5**.152, **5**.169, **7**.157 [Dea65b, Ste97b, Ruf98, Zai01])

Line [Dea65]	Assignment	Relat. int.	Observed	(eV)	Diff. (meV)[a]	Name/comment (SB difference in meV)[b]
D_0'	BE-B(b)	1	NL, HL, LL	5.3671	–	No phonon BE; weak SBs at 5.3652(−1.9 meV), 5.3693(+2.2), 5.2709(+3.8)[b]
D_0	BE-B(a)	2	NL, HL, LL	5.3558	−11[c]	No phonon BE; weak SBs at 5.3539 (−1.9 meV), 5.3575(+1.7), 5.3590(+3.2)[b]
D_1'	B(b)-1Ph	5	NL	5.227	−140	TO phonon
D_1	B(a)-1Ph	9	NL, HL, LLML	5.215	−141	TO phonon
D_1''	B(a)-1Ph	1	NL, HL, LL, ML	5.193	−163	LA/LO phonon
D_2'	B(b)-2Ph	6	NL,HL,LL	5.060	−307	165(LO) + 142(TO)
D_2	B(a)-2Ph	5	NL, HL, LL, ML	5.048	−308	165(LO) + 143(TO)
D_2''	B(a)-2Ph	1	NL	5.023	−333	(2×) 162(LO)
D_3'	B(b)-3Ph	3	NL	4.903	−464	(2×) 162(LO) +140(TO)
D_3	B(a)-3Ph	2	NL	4.890	−466	(2×) 163(LO) +140(TO)
D_4	B(a)-4Ph	2	NL	4.755	−601	(2×) 161(LO) +(2×) 140(TO)

[a]Lattice phonons which couple to the free exciton are 87(TA), 141(TO), and 163(LA/LO) meV
[b]Splitting of −1.9 meV from split acceptor ground state (see Table 9.2.1), and of +3.3 meV from initial bound exciton state [Ste97b, Ruf98]
[c]Splitting of −11 meV from spin–orbit splitting of the valence band

It may be mentioned that similar bound exciton spectra are observed for phosphorous ($E_i = 620$, $E_b = 88$ meV) [Ste99b] and for lithium ($E_b = 130$ meV) [Zai97a] (see Tables 5.3.12.1–5.3.12.2).

9.2.7 Discussion

The blue color of natural diamonds (all of type IIb) is caused by boron. It has been reported that the so-called Cape yellow diamonds exhibit bright blue luminescence when excited in the ultraviolet [Col82b].

In boron-doped synthetic diamonds, the ultraviolet absorption from daylight causes a bright blue color, and this has been used for some commercial applications.

9.3 Hydrogen in Diamond

Hydrogen in diamond is of special interest because of the following:

1. Synthetic CVD diamond is usually grown in a hydrogen-rich atmosphere.
2. Passivation of the surface has been observed [Dav94a].
3. A possible passivation of donors and acceptors has been discussed [Dav94b].
4. Several spectral lines from hydrogen centers have been observed in different classes of diamond.

In most natural type Ia diamonds, absorption lines from C–H vibrations of the $^*(V_1H_1)^\circ$ center are observed, with two sharp fundamental lines and seven sidelines (see Sect. 9.3.1 and Table 9.3.1).

In the grain boundaries of polycrystalline (heteroepitaxial or homoepitaxial) CVD diamond, 18 absorption lines from C–H bend vibrations (Table 9.3.2) and 11 lines from C–H stretch vibrations (Table 9.3.3) are observed. It is remarkable that these frequencies are essentially identical with those from C–H vibrations in (amorphous) diamond-like carbon (DLC). However, the line width for the hydrogen lines in the grain boundaries and in DLC is considerably larger (0.4–15%), compared to 0.1% for the $^*(V_1H_1)^\circ$ center in the diamond crystal (see Tables 9.3.1–9.3.3).

In ion-implanted diamond, a luminescence line is observed at 2.272 eV for protons (Table 5.3.7.3) and at 2.274 eV for deuterons (Table 7.1) [Gip83]. The frequency shift is within experimental error, and is no proof for hydrogen involvement.

Several other lines in diamond, which in the literature are ascribed to hydrogen, are marked by "H?" in Tables 2.1.1.1–7.4.

9.3.1 Hydrogen Lines in Natural Diamond

Natural C–H lines. In Table 9.3.1, two main absorption lines at 174.2 and 385.2 meV together with seven sidelines are listed. These lines have been observed in bulk natural diamond (mainly of type Ia) [Cha59, Woo83], [Dav84b], [Fri91a] and also in natural diamond "coat" [Run71]. Confirmation of the line assignment to the fundamental bend and stretch vibrations of C–H came from a careful analysis of the sidelines. Especially, the weak line at 0.3841 eV from the 1.1% abundant ^{13}C isotope ruled out an N–H vibration [Woo83]. The six other sidebands are overtones or combinations of the fundamental vibrations [Dav84, Fri91]. Remarkable are the anharmonicities (including level mixing and repulsion) of up to 2.3%. All C–H lines in Table 9.3.1 originate from the same center, as confirmed by a correlation factor 0.97–0.99 for the intensities of the five dominant lines, tested on seven samples [Dav84].

From the experimental observations and the data listed in Table 9.3.1, the following conclusions can be drawn:

Table 9.3.1 C–H and N–H vibrational lines from three centers in *natural* diamond, $*(V_1H_1)^\circ$, $*(V_1H_1)N_1{}^\circ$, and $*(V_1H_1)N_4{}^\circ$, and one center in *non-natural* (modified or CVD) diamond, $*(V_1H_1)N_3{}^\circ$

Line	Energy (meV)	Frequ. (cm^{-1})	Width (%)	Relat. int. (%)[a]	Assignment	Anharm. shift (%)	Fig. [Zai01]	References
Natural								
$*(V_1H_1)^\circ$								
NA0174a	174.2	1,405	0.11	20	C–H bend = b	–	3.8	[Dav84b]
NA0174a-s1	345.4	2,786	0.09	3	2b	−0.9	3.7	[Dav84b]
i13-NA0174b	384.1	3,098	0.09	1.4	^{13}C–H (s)	–	–	[Dav84b]
NA0174b[b]	385.2	3,107	0.09	100	C–H str. = s	–	3.7	[Woo83]
NA0174a-s2	516.7	4,168	0.11	0.6	3b	−1.1	3.5	[Dav84b]
NA0174b-s1	557.8	4,499	0.11	5	1s + 1b	−0.3	3.5	[Dav84b]
NA0174a-s3	688.7	5,555	–	0.04	4b	−1.2	3.5	[Fri91a]
NA0174b-s2	729.0	5,880	–	0.06	1s + 2b	−0.6	3.5	[Fri91a]
NA0174b-s3	752.5	6,070	–	0.4	2s	−2.3	3.5	[Fri91a]
$*(V_1H_1)N_1{}^\circ$								
NA0182a	182.2	1,470	–	–	N–H bend = b	–	3.13	[Che94b]
NA0182a–s1	362.0	2,920	–	–	2b	−0.7	3.26	[Fri91a]
NA0182b	401.2	3,236	–	–	N–H str. = s		3.7	[Fer96]
NA0182b-s1	583.2	4,704	–	–	1s + 1b	−0.1	3.5	[Fri91a]
$*(V_1H_1)N_4{}^\circ$								
NA0192a	191.8	1,547	–	–	N–H bend	–	3.7	[Fer96]
NA0192b	423.8?	3,418	–	–	N–H stretch	–	3.7	[Fer96]?
Non-natural								
$*(V_1H_1)N_3{}^\circ$								
MA0198a	198.4	1,600	–	–	N–H bend	–	3.13	[Che94b]
LA/MA0198b	390.5	3,150	–	–	N–H stretch	–	3.13	[Che94b, Mil95]

[a]For the relative intensities, the strongest line was fixed at 100%
[b]The 385 meV line is the ZPL of DAP41a–e (see Table 2.1.1.1)

1. The center has rotational symmetry, because there is only one (doubly degenerate) bending vibration.
2. The hydrogen is bonded to only one carbon, because the relative intensity of the ^{13}C–H line at 384.1 meV is 1.4%, which is much closer to the value 1.1% (for bonding to one carbon) than to a value of 2.2% (for bonding to two carbons). The observed isotope shift is –0.29%, which is close to the theoretical shift (0.30%, calculated from (11.1) with $\alpha = 1$).
3. The observation of certain overtones (e.g., 2× bend at 345.4 meV) unambiguously rules out inversion symmetry for the center (from group theory) [Dav84b]).
4. The C–H bond must be of a rather unusual nature, because the frequency is 6.6% higher than for a normal sp^3CH_1 bond (Table 9.3.3), and because the relatively high intensities and large anharmonic shifts of the overtone or combination vibrations are rather unexpected.

The best agreement with the above observations is obtained by a configuration of a "hydrogen-terminated carbon atom at some form of lattice imperfection" [Dav94b]. It can now be taken as granted that this lattice imperfection is a carbon vacancy, filled with one hydrogen and denoted by $^*(V_1H_1)$. This interpretation is strongly supported by the similarity with the newly identified $^*(V_4H_4)$ center in homoepitaxial CVD diamond (see Sects. 9.3.6 and 10.1). The C–H stretch frequency for the $^*(V_4H_4)$ center at 387.3 meV [Fuc95b] is 0.5% higher than for the $^*(V_1H_1)$ center. Both stretch vibrations have very similar sidebands from coupling to DAP transitions [Dis03] (see $^*(V_4H_4)$b-DAP40 and $^*(V_1H_1)$b-DAP41 in Table 10.1).

It should be mentioned that the bond center position (between two carbon atoms) is evidently preferred by the (extremely short-lived) "anomalous" muonium [Est87]. For the long-lived hydrogen center in Table 9.3.1, the bond center position can be ruled out by the observations (2) and (3) above.

Natural N–H lines. In Table 9.3.1, three centers with N–H bend and stretch vibrations are listed, two ($^*(V_1H_1)N_1{}^\circ$ and $^*(V_1H_1)N_4{}^\circ$) in natural and one ($^*(V_1H_1)N_3{}^\circ$) in non-natural diamond. The intensity of these lines is not directly correlated with those of the $^*(V_1H_1)^\circ$ center. The three centers are observed in diamonds with a high concentration of hydrogen and nitrogen, and can therefore be assigned to N–H vibrations, i.e., the hydrogen is in a slightly different environment. Note that the N–H vibrational stretch energies are higher than the respective energies for C–H vibrations. The difference is 4.5% for $^*(V_1H_1)N_1{}^\circ$, 1.8% for $^*(V_1H_1)N_3{}^\circ$, and 10% for $^*(V_1H_1)N_4{}^\circ$.

9.3.2 Hydrogen Bend Lines in Polycrystalline CVD Diamond

The hydrogen content in polycrystalline CVD diamond is relatively high, reaching up to 20 at. % H in heteroepitaxial films [Dis93] and up to 3.3 at. % H in homoepitaxial films [Jan92]. It was shown by electron probe microanalysis that the hydrogen is predominantly located in the *diamond grain boundaries* (DGB), which consist of disordered amorphous carbon [Fal93]. Independently, this conclusion was drawn from the similarity between the C–H vibrational absorption lines in polycrystalline CVD diamond and in amorphous DLC [Dis93] (see also Tables 9.3.1–9.3.4). The lines in Tables 9.3.1–9.3.4 are positively identified as C–H vibrations by the observed isotope shift for $^{1}H/^{2}D$ and $^{12}C/^{13}C$ substitution [Dis87, Dis92, Fuc95b].

In Table 9.3.2, the 18 infrared absorption lines from C–H bend vibrations are listed. These relatively broad lines are observed in the range 80–190 meV (680–1,500 cm^{-1}). In some samples, there is overlap with the defect-induced one-phonon absorption of crystalline diamond in the range 70–170 meV (550–1,350 cm^{-1}), and of C–C vibrational absorption lines in the range 100–270 meV (840–2,180 cm^{-1}). Because of these overlaps and of the broadness, the C–H bend lines are less well documented than the C–H stretch lines. There exists a qualitative intensity correlation between bend and stretch vibrations for the eight different C–H configurations listed in Tables 9.3.1–9.3.4.

9.3.3 *Hydrogen Stretch Lines in Polycrystalline CVD Diamond*

In Table 9.3.3, the 11 infrared absorption lines from C–H stretch vibrations are listed. The spectra consist of partly overlapping lines in the range 350–410 meV (2,820–3,330 cm^{-1}). After early reports [Wil89, Bi_90, Jan91], deconvolution of the overlapping lines [Dis87, Dis92, Dis93, Joh94] allowed a quantitative analysis and the assignment to the configurations given in Tables 9.3.3 and 9.3.4.

Of special interest is the line at 349.9 meV (2,822 cm^{-1}) [Dis93], which is completely absent in DLC [Dis87] (see Table 9.3.3). On the contrary, a single sharp (0.3% width) C–H stretch line at 350.9 meV (2,830 cm^{-1}) was observed for a hydrogen monolayer on a diamond (111) cleavage plane [Chi92]. Therefore, a common configuration for both lines can be assumed, i.e., a single hydrogen atom, bonded to a tetrahedral lattice carbon, denoted by "lattice sp^3CH_1." The intensity of this C–H stretch line correlates with that of the relatively sharp C–H bend line at 155.2 meV (1,252 cm^{-1}) in Table 9.3.2 [Jan92].

This new "lattice sp^3CH_1" line, together with the ten lines from the seven traditional C–H configurations (*aliphatic* sp^3CH_3, sp^3CH_2, sp^3CH_1, *olefinic* sp^2CH_2, sp^2CH_1, *aromatic* sp^2CH_1, and *acetylenic* sp^1CH_1), has been observed in CVD diamond (see Tables 9.3.1–9.3.4). However, the acetylenic line from sp^1CH_1 at 412 meV (3,323 cm^{-1}) is absent in most samples, except in homoepitaxial CVD diamond from microwave deposition [Dis92, Fuc95b].

9.3.4 *Abundances of C–H Configurations*

From the absorption intensities, the abundances of the different C–H configurations and the total hydrogen content can be derived. In Table 9.3.4, the percentages of

Table 9.3.2 C–H bend vibrational absorption lines, observed in the diamond grain boundaries (DGB) of polycrystalline (heteroepitaxial or homoepitaxial) CVD diamond; in natural brown diamond (NBD), in natural gray diamond (NGD), in amorphous diamond-like carbon (DLC), and in polymer-like amorphous carbon (PAC) (see also Table 6.1.1.1)

Line	Assessment		Diamond: DGB, NBD, NGD				DLC	PAC	
	Configuration[c]	Sym.[d]	Energy (meV)	Frequ. (cm^{-1})	Width[b] (%)	Refs.	Energy (meV)	Energy (meV)	Refs.
	sp^3 CH_3	$C_{3v}:A_2$	(34.1)	(275)	Torsion[e]	(predicted)	(34.1)	(34.1)	[Dis87]
RA0085a1	aliph. sp^3 CH_2	$C_{2v}:B_1$	85.17	687	1–7	[Rei98]	86.8	86.8	[Dis87]
RA0087a	acetyl. sp^1 CH_1	C_{∞}:E	–	–	–	-	–	86.8	[Dis87]
RA0088a1	olef. sp^2 CH_2	$C_{2v}:B_1$	88.02	710	13–15	[Fri91a]	––	86.8	[Dis87]
RA0095a1	arom. sp^2 CH_1	$C_S:B$	95.50	770	12–14	[Fri91a]	93.6	93.6	[Dis87]
RA0108a1	olef. sp^2 CH_1	C_S:B	107.9	870	2–7	[Fri91a]	104.1	104.1	[Dis87]
RA0088a2	olef. sp^2 CH_2	$C_{2v}:A_2$	114.1	920	4–6	[Ghe92b]	112.8	112.8	[Dis87]
RA0085a2	aliph. sp^3 CH_2	$C_{2v}:B_2$	125.2	1,010	2–6	[Fri91a]	127.7	127.7	[Dis87]
RA0133a1	aliph. sp^3 CH_3	$C_{3v}:E$	133.3	1,075	3–5	[Ghe92b]	–	133.3	[Dis87]
RA0088a3	olef. sp^2 CH_2	$C_{2v}:B_2$	137.6	1,110	2–6	[Jan92]	–	137.6	[Dis87]

RA0085a3	aliph. sp^3 CH_2	C_{2v} : A_2	145.1	1,170	1–5	[Rei98]	145.1	146.3	[Dis87]
LA0155a	lattice sp^3 CH_1?	C_{3v} : E	155.2	1,252	1–2	[Jan92]	(not	obs.)	–
RA0108a2	olef. sp^2 CH_1	C_S : A	159.9	1,290	5–7	[Jan91]	159.9	158.7	[Dis87]
RA0133a2	aliph. sp^3 CH_3	C_{3v} : A_1	164.9	1,330	2–4	[Jan91]	–	164.3	[Dis87]
RA0171a	aliph. sp^3 CH_1	C_{3v} : E	171.1	1,380	4–6	[Che94b]	169.8	–	[Dis87]
RA0095a2	arom. sp^2 CH_1	C_S : A	177.3	1,430	1–2	[Fri91a]	177.9	179.1	[Dis87]
RA0085a4	aliph. sp^3 CH_2	C_{2v} : A_1	178.5	1,440	4–7	[Jan91]	178.5	179.8	[Dis87]
RA0088a4	olef. sp^2 CH_2	C_{2v} : A_1	181.6	1,465	2–4	[Jan91]	–	179.8	[Dis87]
RA0133a3	aliph. sp^3 CH_3	C_{3v} : E	185.3	1,495	1–3	[Fri91a]	–	184.7	[Dis87]

[a] Assignment from [Dis85, Dis87]
[b] Width (sample dependent: range)
[c] Configuration: *aliph.* aliphatic, *acetyl* acetylenic, *olef* olefinic, *arom* aromatic
[d] *Sym.* Symmetry
[e] Torsion: IR-inactive

Table 9.3.3 **C–H stretch** vibrational absorption lines (LA, MA, and RA), observed in diamond grain boundaries (DGB) of polycrystalline (heteroepitaxial or homoepitaxial) CVD diamond, in natural brown diamond (NBD), in natural gray diamond (NGD), in amorphous diamond-like carbon (DLC), and in PAC. Four MA lines are observed from hydrogen-covered diamond crystal surface (HCS)

Line	Assignment		Diamond: DGB, NBD, NGD, HCS					DLC		PAC	
	Configuration	Sym.	Energy (meV)	Fr. (cm^{-1})	Wi. (%)	R. i. (%)	Ref.	Energy (meV)	Wi. (%)	Energy (meV)	Refs.
LA0155b	Lattice sp^3 CH_1?	C_{3v} : A_1	349.9	2,822	0.4–1.6	1–24	[Dis93]	absent		–	absent
MA0155b	HCS, see Table 5.1.2.2	C_{3v} : A_1	350.2	2,825	–	–	[And94]	–	–	–	–
RA0085b1	aliph. sp^3 CH_2	C_{2v} : A_1	353.3	2,850	0.7–2.1	5–38	[Dis93]	353.3	2.7	353.3	[Dis87]
MA0085b1	HCS, see Table 5.1.2.2	C_{2v} : A_1	354.2	2,857	–	–	[Chi92]	–	–	–	–
RA0133b1	aliph. sp^3 CH_3	C_{3v} : A_1	356.4	2,875	1.0–2.9	1–20	[Dis93]	–	–	356.4	[Dis87]
MA0133b1	HCS, see Table 5.1.2.2	C_{3v} : A_1	358.9	2,895	–	–	[And94]	–	–	–	–

RA0171b	aliph. sp^3 CH_1	C_{3v} : A_1	361.3	2,914	0.5–1.8	1–35	[Dis93]	362.0	3.0	–	[Dis87]
RA0085b2	aliph. sp^3 CH_2	C_{2v} : B_1	361.6	2,917	0.9–2.1	5–38	[Dis93]	362.0	3.0	362.0	[Dis87]
RA0133b2	aliph. sp^3 CH_3	C_{3v} : E	367.8	2,967	1.2–2.9	2–30	[Dis93]	–	–	365.1	[Dis87]
MA0133b2	HCS, see Table 5.1.2.2	C_{3v} : E	366.4	2,955	–	–	[And94]	–	–	–	–
RA0088b1	olef. sp^2 CH_2	C_{2v} : A_1	368.5	2,972	0.5–1.2	0–10	[Dis93]	–	–	368.2	[Dis87]
RA0108b	olef. sp^2 CH_1	C_s : A	371.9	3,000	0.9–2.3	0–8	[Dis93]	371.9	2.6	371.9	[Dis87]
RA0088b2	olef. sp^2 CH_2	C_{2v} : B_1	375.0	3,025	0.8–1.2	0–10	[Dis93]	–	–	375.0	[Dis87]
RA0095b	arom. sp^2 CH_1	C_s : A	378.1	3,050	1.1–2.3	0–11	[Dis93]	377.5	2.2	379.4	[Dis87]
RA0087b	acetyl. sp^1 CH_1	C_∞:A	412.0	3,323	0.5–0.7	0–11	[Dis92]	409.1	1.3	409.1	[Dis87]
NA0087b[1]	acetyl. sp^1 CH_1	C_∞:A	412.0	3,323	0.2	–	[Woo83]	–	–	–	–

Fr. frequency, *Wi.* width (sample dependent), *R.i.* relative intensity (sample dependent), *Sym.* symmetry. Configuration: *aliph.* aliphatic, *acetyl.* acetylenic, *olef.* olefinic, *arom.* aromatic,

[1] After natural irradiation (see Table 2.1.2.2)

the eight configurations are listed for 19 different types of samples. The results are clearly sample dependent. Without going into details, it is clear that several interesting conclusions may be drawn from the data, and these can possibly contribute to a better understanding of the growth mechanism and texture development [Wil94, Mül96]. It may also give more insight into the structure of the grain boundaries [Fal93]. A brief discussion on the sample dependence is given below.

No. 1, 2: Increasing the hydrocarbon content in the feed gas of hot-filament deposited heteroepitaxial CVD diamond from 1% (HF-1%-CH) to 5% (HF-5%-CH) leads to a 50 times higher hydrogen content (due to smaller grains) [Dis93].

No. 3–6: Different substrates were used in a simultaneous microwave CVD deposition experiment. The label MW-Si denotes heteroepitaxial CVD on silicon, and MW-100, MW-110, and MW-111 denote homoepitaxial CVD on (1 0 0), (1 1 0), and (1 1 1) diamond substrates [Dis92].

No. 7–10: A set of flame-grown homoepitaxial CVD samples was deposited on (1 0 0), (1 1 0), and (1 1 1) diamond substrates. The deposition temperature was 930–1,020 °C, except for the high temperature sample (FL-111-HT) with 1,120 °C [Jan92].

No. 11, 12: The influence of nitrogen addition in the feed gas was studied in microwave CVD deposition with 0% and 1% nitrogen gas [Dis92].

No. 13, 14: The application of a negative DC bias (0 and 250 V) during microwave CVD deposition leads to a remarkable increase in sp^2 hybridization from 1% to 11% [Joh94].

No. 15, 16: After heat treatment at 1,475 °C of a hot-filament deposited CVD film, considerable modification of the C–H configurations was observed [Dis92].

No. 17–19: For comparison, some results from DLC (as-grown and heat treated for 4 h at 600 °C) [Dis83] and from polymer-like amorphous carbon (PAC) [Dis87] are included in Table 9.3.4. Remarkable is the high sp^2 percentage in as-grown samples (30–45%) and the complete transformation into aromatic sp^2 (from 8% to 98%) after the above-mentioned heat treatment.

The *average* C–H hybridization for the 16 CVD diamond samples in Table 9.1.4 is given below line 16: $sp^3 : sp^2 : sp^1 = 88 : 11 : 1\%$. The dominant configuration is sp^3CH_2 (average 47%). Exceptions with a different dominant configuration are sp^3CH_3 in (HF-1%-CH, MW-0-Volt) or sp^3CH_1 in (FL-100-diam., HF-H-1475).

The results in Table 9.3.4 for the hybridization of the C–H bonds are also representative for the hybridization of the C–C bonds in the amorphous carbon network (Table 9.3.5). This has been confirmed by electron energy loss spectroscopy for DLC [Fin84] and also by Raman spectroscopy for polycrystalline (CVD) diamond [Wag89] and amorphous DLC [Dis85, Wag89].

9.3.5 Frequencies of C–C Vibrations from H-Bonded Carbon

The atoms from the carbon network which are bonded to hydrogen show characteristic infrared-active C–C vibrations. These lines (see Table 9.3.6) are observed in natural diamond coat (NDC), DLC, and PAC. Discrimination against the C–H

Table 9.3.4 Abundances of C–H configurations in DGB, in polycrystalline (heteroepitaxial or homoepitaxial) CVD diamond (Nos. 1–16), in amorphous diamond-like carbon (DLC, Nos. 17, 18), and in PAC (No. 19), as obtained from deconvoluted C–H stretch lines

Nos.	Sample	Film thickness (μm)	Hydrogen (at%)	aliph. sp^3	aliph. sp^3	aliph. sp^3	latt. sp^3	total sp^3 (%)	olef. sp^2	olef. sp^2	arom. sp^2	Total sp^2	acetyl. sp^1	Refs.
				CH_3 (%)	CH_2 (%)	CH_1 (%)	CH_1 (%)		CH_2 (%)	CH_1 (%)	CH_1 (%)		CH_1 (%)	
1	HF-1%-CH	95	0.4	49	11	28	9	91	5	1	3	9	0	[Dis93]
2	HF-5%-CH	25	20.0	13	69	1	8	91	5	3	1	9	0	[Dis93]
3	MW-Si	50	1.0	5	76	1	11	93	2	4	1	7	0	[Dis92]
4	MW-100-diam.	198	0.2	19	31	26	5	79	2	5	3	10	11	[Dis92]
5	MW-110-diam.	180	0.5	10	66	3	12	90	3	4	2	9	1	[Dis92]
6	MW-111-diam.	231	0.7	5	73	2	10	90	3	3	1	7	3	[Dis92]
7	FL-100-diam.	~17	0.4	22	10	34	1	67	19	8	6	33	0	[Jan92]
8	FL-110-diam.	~17	3.0	22	62	6	6	94	4	1	1	6	0	[Jan92]

(continued)

Table 9.3.4 (continued)

Nos.	Sample	Film thickness (μm)	Hydrogen (at%)	aliph. sp^3 CH_3 (%)	aliph. sp^3 CH_2 (%)	aliph. sp^3 CH_1 (%)	latt. sp^3 CH_1 (%)	total sp^3 (%)	olef. sp^2 CH_2 (%)	olef. sp^2 CH_1 (%)	arom. sp^2 CH_1 (%)	Total sp^2	acetyl. sp^1 CH_1 (%)	Refs.
9	FL-111-diam.	∼ 5	1.5	23	55	2	8	88	8	3	1	12	0	[Jan92]
10	FL-111-HT-1,020 °C	∼ 4	0.2	21	27	15	24	87	9	4	0	13	0	[Jan92]
11	MW-0%-N_2	11	0.9	15	74	3	2	94	2	2	2	6	0	[Dis92]
12	MW-1%-N_2	12	1.9	3	48	19	16	86	10	4	0	14	0	[Dis92]
13	MW-0-Volt	ATR	–	32	28	27	12	99	0	0	1	1	0	[Joh94]
14	MW-250-Volt	ATR	–	22	38	22	7	89	0	0	11	11	0	[Joh94]
15	HF-as-grown	65	2.0	22	60	4	3	89	5	3	3	11	0	[Dis92]
16	HF-HT-475 °C	65	2.0	25	15	34	17	91	2	2	5	9	0	[Dis92]
	Average (CVD)	80	2.5	19	47	14	8	88	5	3	3	11	1	–
17	DLC-as-grown	2	21.0	0	40	28	–	68	0	22	8	30	2	[Dis83]
18	DLC-HT-600 °C	2	9.0	0	1	1	–	2	0	0	98	98	0	[Dis83]
19	PAC	2	22.0	25	28	0	–	53	32	8	5	45	2	[Dis87]

The CVD samples were prepared by microwave plasma (MW), hot filament (HF), or flame (FL)

Latt. lattice, *aliph.* aliphatic, *olef.* olefinic, *arom.* aromatic, *acetyl.* acetylenic, *diam.* diamond substrate, *ATR* attenuated total reflection, *HT* annealed at 475 °C (No. 16), 600 °C (No. 18), 1,020 °C (No. 10)

Table 9.3.5 Absorption lines from C–C bend and stretch vibrations from H-bonded carbon in diamond-related materials (RA): natural diamond coat (NDC), amorphous diamond-like carbon (DLC), and polymer-like amorphous carbon (PAC (see also Table 11.1)

Line	Assignment	NDC:H [Ang65] Energy (meV)	DLC:H [Dis87] Energy (meV)	DLC:D [Dis87] i2-shift (%)	PAC:H [Dis87] Energy (meV)	PAC:D [Dis87] i2-shift (%)
	arom./olef. (sp^2) ***CC bend***					
	aliph. (sp^3) CC ***bend***					
RA0104	C=C (aromatic sp^2)	104	104.1	0	104.1	0
RA0110	C–C (sp^3)	–	109.7	0	109.7	0
RA0120	C=C (olefinic sp^2)	121	120.3	−1.0	120.3	−1.0
	(sp^3) C–C ***stretch***					
RA0144	C–C (sp^3)	–	143.8	0	143.8	0
	(mixed sp^2/sp^3) *C–C* ***stretch***					
RA0154	C–C (mixed sp^2/sp^3)	–	154.4	0	154.4	0
RA0161	C–C (mixed sp^2/sp^3)	–	161.2	0	157.5	+1.2
RA0188	C–C (mixed sp^2/sp^3)	–	187.8	0	187.8	0
	(sp^2) C=C ***stretch***					
RA0196	C=C (aromatic sp^2)	–	195.9	−1.3	195.9	−1.3
RA0201	C=C (olefinic sp^2)	–	200.8	−1.2	198.4	−1.2
	(sp^1) C=C ***stretch***					
RA0270	C=C (acetylenic sp^1)	–	–	–	270.3	–
Raman	Lattice C–C (sp^3)	–	165	–	–	–

bonding lines (in the same spectral range) was achieved by a careful analysis of the spectra from deuterated samples [Dis87].

9.3.6 Additional Hydrogen Lines in Homoepitaxial CVD Diamond

In (1 0 0) epitaxial CVD diamond, two centers with correlated intensities are observed [Fuc95a],b. These lines are here tentatively assigned to $*(V_4H_4)^{\circ}$.

The ZPL at 165 meV (DAP39) is typical for a C–C bond vibration. The ZPL at 412 meV (DAP40) has very small ^{2}H (−0.5%) and ^{13}C (−0.2%) isotope shifts. Therefore, it could be a nonvibrational transition (see Table 8.1.7.2).

9.4 Silicon in Diamond

9.4.1 *Five Silicon Centers in Diamond*

Optical silicon centers in diamond seem to be of limited interest, because only few respective spectra have been published (with the exception of the *2Si 1.68 center). Accordingly, Tables 8.1.6, 9.4.1, and 9.4.2 contain no more than five silicon centers. Three silicon centers occur in natural diamond (*2Si 1.68 = $(V_3Si_2)^+$, *Si 2.05* = $(V_3Si_2)^\circ$, and *Si 2.52* = $(V_3Si_2)^-$), four centers in CVD diamond, which are usually grown on Si substrates (*2Si 1.68 = *$(V_3Si_2)^+$, *Si 2.05 = *$(V_3Si_2)^\circ$, *1Si 2.65 = *$(V_1Si_1)^\circ$, and *1Si 2.99 = * $(V_1Si_1)^-$), and two centers are observed after Si ion implantation (*2Si 1.68 = *$(V_3Si_2)^+$ and *Si 2.52 = *$(V_3Si_2)^-$).

9.4.2 *Resolved ZPL Isotope Shifts from Natural Silicon*

The natural abundance of silicon isotopes is given by 92.17% (^{28}Si), 4.71% (^{29}Si), and 3.12% (^{30}Si). For two equivalent silicon atoms, the relative abundances for the five Si_2 combinations are 84.95% ($^{56}Si_2$), 8.68% ($^{57}Si_2$), 5.97% ($^{58}Si_2$), 0.29% ($^{59}Si_2$), and 0.10% ($^{60}Si_2$).

From high-quality samples, high-resolution spectra in the range 1.681–1.684 eV have been obtained, showing 12 lines [Cla95]. Experimental photoluminescence and absorption results for the *2Si 1.68 =* $(V_3Si_2)^+$ center are given in Table 9.4.2. The agreement between calculated and observed data is excellent, both for frequencies and intensities. The isotope shift for a mass increase of one unit is −0.35 meV. From the relative intensities of the isotope shifted lines in luminescence and absorption, it is clear that two equivalent silicon isotopes are involved. Theoretically, a five-line pattern is expected, but only a three-line pattern has intensities above the detection limit of ca. 0.5% (see Table 9.4.2). It has been proposed that the three-line pattern arises from a single silicon [Cla95]; however, in this case, the lines named a57–d57 and a58–d58 in Table 9.4.2 should have roughly half the intensity of the observed values.

An elegant proof for the five-line pattern could come from a sample where the isotopes 59Si2 and 60Si2 are enriched, allowing the observation of the lines at 1.6807–1.6812 (see Table 9.4.2).

Table 9.4.1 Optical centers from silicon in diamond; for (*$(V_3Si_2)^+$; see also Table 9.4.2)

Defect	Observed	ZPL/*L* (eV)	SB-/*sta*.-DAP	*D* (eV)	Name/comment
Si–C interface	LA				99 meV/Si–C vibration
Si–C interface	LA				103 meV/Si–C vibration
*$Si_1{}^{\circ}$+?	LL	$L = 1.474$	*sta. 8a–g*	+1.33	*Si 1.56/DAP lines 1.556–1.637 eV; QLVM: −46 (calc. 1Si + 2C : 47.1 meV)
*$V_1Si_1{}^{-}$	NL, LL, ML	2.991	113a–m	−1.33	*1Si 2.99/DAP lines 2.246–2.874 eV; Si_1 from QLVM(−74 meV, calc. 74.5) NL = Si rich nat. brown diamond (NBD) LL = Si substrate, ML = Si implanted; strain-induced line splittings and shifts
*$V_1Si_1{}^{\circ}$	LE	2.646	109a–h (weak)	+1.49	*1Si 2.65/QLVM: +74, $w = 65$ meV/Si_1 from QLVM/broad band at 2.86 eV
*$V_1Si_1{}^{\circ}$	LL	2.646	110a–g (weak)	−1.49	*1Si 2.65/QLVM: −74, $w = 65$ meV/broad band at 2.45 eV
*$(V_3Si_2)^-$	NE, NL Si impl. ME, ML	2.523	75a–j	−1.25	*2Si 2.52/QLVM: −26, −44 meV, Si_2 from QLVM (calc. 2Si = 44.9 meV)/PLE of *2Si 1.68/decay time 1.0–1.8 µs
*$(V_3Si_2)^{\circ}$	NE, NL	2.052	65a–e	−1.39	*2Si 2.05/QLVM: −36, −45; LVM: −70 meV, Si_2 from QLVM (calc. 2Si = 44.9 meV)/PLE of *2Si 1.68/decay time 150–350 ns
*$(V_3Si_2)^+$	HA, LA Si impl. MA	1.681	53a–n	1.60	*2Si 1.68/QLVM: +43; LVM: +67 meV, Si_2 from QLVM (calc. 2Si = 44.9 meV)/photochromic
*$(V_3Si_2)^+$	HL, LL Si impl. ML	1.681	54a–g	−1.60	*2Si 1.68/QLVM: −36, −45; LVM:−65 meV/gr. st. spl. = 0.2; exc. st. spl. = 1.1 meV/Si_2 from shifts of two equivalent Si isotopes (see Table 9.4.2)/⟨1 1 0⟩ oriented dipole, decay time from 2.4 ns to 105 µs/photochromic

sta-DAP/SB-DAP = *standard*/sideband DAP with ZPL = *L* = limiting energy and *D* = dielectric factor; gr. st. spl. = ground state splitting; exc. st. spl. = excited state splitting

Table 9.4.2 Calculated and observed isotope shifts of the *2Si 1.68 =* $(V_3Si_2)^+$ center in diamond from natural silicon isotopes (92% ^{28}Si, 5% ^{29}Si, and 3% ^{30}Si) in photoluminescence and absorption ($T = 10\,K$)

Line[a]	Calculated for Si_2^{b-e}			Observed [Cla95, Ste95]		
	Frequ. (eV)	Lumin. intens. (%)	Absorpt. intens. (%)	Frequ. (eV)	Lumin. intens. (%)	Absorpt. intens. (%)
a60	1.68070	0.03	0.01	–	–	–
b60	1.68090	0.04	0.04	–	–	–
a59	1.68105	0.09	0.04	–	–	–
b59	1.68125	0.13	0.12	–	–	–
a58	**1.68140**	**1.8**	**0.8**	1.6814	1.4(−0.4)	1.0(+0.2)
b58	**1.68160**	**2.7**	**2.4**	1.6816	2.9(+0.2)	2.3(−0.1)
a57	**1.68175**	**2.7**	**1.2**	1.6817	2.4(−0.3)	1.5(+0.3)
c60	1.68180	0.02	0.03	–	–	–
b57	**1.68195**	**3.9**	**3.6**	1.6819	4.8(+0.9)	3.8(+0.2)
d60	1.68200	0.01	0.02	–	–	–
a56	**1.68210**	**26.3**	**10.9**	1.6821	26.7(+0.4)	11.9(+1.0)
c59	1.68215	0.05	0.08	–	–	–
b56	**1.68230**	**38.2**	**35.8**	1.6823	38.2(±0)	34.8(−1.0)
d59	1.68235	0.02	0.06	–	–	–
c58	**1.68250**	**1.0**	**1.6**	1.6825	1.0(±0)	1.5(−0.1)
d58	**1.68270**	**0.5**	**1.1**	1.6827	0.5(±0)	1.0(−0.1)
c57	**1.68285**	**1.4**	**2.3**	1.6828	1.0(−0.4)	2.3(±0)
d57	**1.68305**	**0.7**	**1.6**	1.6831	0.5(−0.2)	1.8(+0.2)
c56	**1.68320**	**13.6**	**22.9**	1.6832	13.4(−0.2)	23.0(+0.1)
d56	**1.68340**	**7.0**	**15.3**	1.6834	6.7(−0.3)	15.3(±0)
Average difference					±0.2	±0.2

[a]Two ground states (I, II) and two excited states (III, IV) with transitions II–III (a), I–III (b), II–IV (c), and II–III (d). The label indicates the transition and the combined mass ($^{56}Si_2$–$^{60}Si_2$) of the ^{28}Si, ^{29}Si, and ^{30}Si isotopes, with abundances 84.95% $^{56}Si_2$,: 8.68%:$^{57}Si_2$, 5.97% $^{58}Si_2$, 0.29% $^{59}Si_2$, 0.10% $^{60}Si_2$ (see Sect. 9.4.1.1)

[b]Frequencies: The splittings in the ground and excited states are 0.2 and 1.1 meV, respectively. The isotope shift is −0.35 meV (= −0.021%) per increasing mass unit (see Table 7.4)

[c]Intensities for luminescence: The observed ($T = 10\,K$) relative intensities a: b: c: d = 31%: 45%: 16%: 8% are used

[d]Intensities for absorption: The observed ($T = 10\,K$) relative intensities a: b: c: d = 14%: 42%: 27%: 18% are used

[e]A single silicon atom can be excluded, because the calculated intensities for a30: b30: a29: b29: a28: b28: c30: d30: c29: d29: c28: d28 are (in luminescence) 0.9%: 1.4%: 1.6%: 2.2%: 28.5%: 41.4%: 0.5%: 0.2%: 0.8%: 0.4%: 14.7%: 7.4% (average difference ±1.0), and (in absorption) 0.4%: 1.3%: 0.7%: 2.1%: 12.9%: 38.6%: 0.8%: 0.5%: 1.4%: 0.9%: 24.8%: 16.6% (average difference ±1.2). The average difference is five times larger than for the two-silicon model; in addition, there is a general misbalance: all calculated intensities for a single ^{28}Si are larger than observed, while those for ^{29}Si and ^{30}Si are all too small

9.4.3 *Effects of Annealing*

Annealing experiments are reported for the *2Si 1.68 $=^{*}(V_3Si_2)^+$ center. This is a very temperature stable center. A single crystal HTHP diamond was irradiated with 2 MeV electrons and subsequently annealed for 0.5 h at temperatures in the range 400–2,500 °C [Cla95]. Up to 600 °C, the intensity was unchanged. An order of magnitude increase occurs at 800 °C, followed by a fourfold decline at 1,200 °C. With further heating, the intensity gains a factor three with a maximum at 2,200 °C, followed by a considerable decline at 2,500 °C. The interpretation of this annealing behavior is difficult, because three charge states of the $^{*}(V_3Si_2)$ center exist, and the intensities of the lines from the other charge states were not monitored in the experiment.

9.4.4 *Time-Resolved Spectroscopy*

The luminescence decay times for three of the five silicon centers in diamond have been determined (see Table 9.4.1) [Kho94]. The *1Si 2.99 center has a medium fast decay time (63 ns), while the *2Si 2.52 center is slow (1.0–1.8 μs). The *2Si 1.68 center has two decay times: one fast (2.4 ns) and one slow (105 μs). The latter is explained by energy transfer from a reservoir. With the present knowledge, the PLE line at 2.523 (*2Si 2.52 center) is a probable candidate for non-radiative decay. It should be mentioned that in time-resolved measurements, the 2.523-eV line splits up in at least six ZPLs in the range 2.523–2.572 eV, where the relative intensities depend strongly on the type of sample [Kho94].

9.4.5 *Local Vibrational Modes*

For two centers (*2 Si 2.05 and *Si 1.68), LVM frequencies in the range 65–70 meV are observed (see Table 9.4.1). These are typical C–C bond vibrations near a vacancy (see Table 8.8).

9.4.6 *Quasi-Local Vibrational Modes*

For all five silicon centers, QLVM frequencies in the range 20–74 meV are observed, and they are listed in Tables 9.4.1. There is very good agreement with the corresponding calculated values (see Table 11.2).

9.4.7 *DAP Transition at Nickel Centers*

Weak DAP transitions of the sideband type are observed for all five vacancy-silicon associate centers (see Table 9.4.1). The variation in the dielectric factor *D* for multiple charge states is discussed in Sect. 10.6.

The DAP lines of the *Si 1.56 eV center are of the standard type, indicating a center with an unknown donor (if silicon is the acceptor).

9.4.8 *Considerations for Structure Assignments*

$^*(V_3Si_2)^{\pm}$. This proposed structure (two interstitial silicon in a triple vacancy) for the 1.681-eV center is almost certain. It is supported by four independent results:

1. The isotope shifts from natural silicon isotopes indicate *two* equivalent silicon atoms [Cla95] (see Sect. 9.4.2).
2. From polarized luminescence results, a ⟨1 1 0⟩ oriented dipole was determined;
3. The QLVM sideband indicates *two* silicon atoms (see Table 11.2).
4. Further support comes from the linear splitting into two components under ⟨1 0 0⟩ uniaxial stress [Ste94]. This is expected for a ⟨1 1 0⟩ oriented dipole.

The positive charge state is derived from the dielectric factor D of DAP53/54 (see Sects. 9.4.7 and 10.5).

$^*(V_3Si_2)^-$ and **$^*(V_3Si_2)^{\circ}$**. These centers at 2.523 and 2.052 eV provide photoluminescence excitation of the 1.681 eV $^*(V_3Si_2)^+$ center, indicating three charge states of the same structure. The DAP results (different D factors) exclude excited states of the 1.681-eV center, but provide a hint for the charge state. The QLVM sideband at 44–45 meV arises from *two* silicon atoms. Of special interest is the additional QLVM line at 26 meV from a Si_2C_8 cluster, which is expected for a (V_3Si_2) configuration, where two interstitial silicon atoms in the triple vacancy are surrounded by $(3 + 2 + 3 = 8)$ carbon atoms (see Table 11.2).

$^*V_1Si_1{}^-$ and **$^*V_1Si_1{}^{\circ}$**. The QLVM frequency of 74 meV clearly indicates a center with *one* silicon atom (see Table 11.2). From the DAP results (different D factors), the two charge states are derived (see Section and Table 10.5).

It is remarkable that in all five centers, the silicon is associated with one or three vacancies. Evidently, the large atomic radius of silicon (119 nm, compared to 76 nm for carbon, see Table 9.7.1) rules out a *substitutional* silicon center.

9.5 Nickel in Diamond

9.5.1 *Twenty Nickel Centers in Diamond*

At present, there are 20 centers, which are assigned to nickel in diamond (see Tables 9.5.1.1–9.5.1.2). Seven nickel centers occur in natural diamond, 17 centers in synthetic HPHT diamonds, which are grown from a nickel containing melt, and one center in CVD diamond. They are listed in Table 8.1.4 ($6Ni_1$, Ni_2, and Ni_3 associates with a single vacancy), in Table 8.1.5 (seven interstitial Ni_1 associates with a double vacancy), in Table 8.1.6 (two interstitial Ni_2 associates with a triple vacancy), and in Tables 9.5.1.1–9.5.1.2 (the substitutional Ni_1 center and two nickel

Table 9.5.1.1 Summary of nickel centers in diamond (see also Tables 9.5.6 (anneal), 11.1 (LVM), 11.2 (QLVM))

Nos.	ZPL (eV)	DAP lines (range) (eV)	Broad peak Abs/ Lum. (eV)	EPR	DAP[a]				Name	Structure	Table	N conc.[b]	Ni impl. + ann. (lum.)	Comment
					sta.	SB (A/L)	L (eV)	D (eV)						
1	–	1.212–1.383	1.40/1.40	NE4?	16	–	1.036	1.45	*__Ni-S8__	*$\mathbf{(V_2Ni_1)^-}$	8.1.5	lo.	–	Also res. DAP (1.22 eV)
2	1.404	–	1.55/1.25	NIRIM2	–	25/26	ZPL	1.33	1Ni 1.40	*$\mathbf{V_1Ni_1{}^+}$	8.1.4	lo.	(hi.)	1Ni from isotope splitting
3	–	1.563–2.320	1.80/1.80	–	60	–	1.460	1.53	*__Ni-S5__	*$\mathbf{(V_2 Ni_1)N_{x+y+z}{}^\circ}$	8.1.5	hi.	–	PLE of *S8
4a	1.660	–	–	–	–	–	–	–	*Ni 1.66a	*$\mathbf{V_1Ni_2{}^\circ}$	8.1.4	hi.	(lo.)	No absorption
5		1.678–1.821	–/1.70	–	4	–	–	1,66	*__Ni-S9__	*$\mathbf{(V_2Ni_1) N_{x+y+z}{}^+}$	8.1.5	hi.	–	
6a	1.693	–	1.85/–	–	–	–	–	–	*Ni 1.69a	*$\mathbf{V_1Ni_2{}^+}$	8.1.4	hi.	(lo)	Line (b) at 1.940 eV
7a	1.704	–	1.82/–	–	–	83/84	ZPL	1.34	*Ni 1.70a	*$\mathbf{V_1Ni_1{}^\circ}$	8.1.4	lo.	(lo.)	–
8	–	1.808–2.620	2.19/2.19	S = 1 ce.	11	–	1.724	1.60	*__1Ni-S7__	*$\mathbf{(V_2Ni_1)N_{x+y+z}{}^\circ}$	8.1.5	hi.	–	1Ni from isotope splitting

(continued)

Table 9.5.1.1 (continued)

Nos.	ZPL (eV)	DAP lines (range) (eV)	Broad peak Abs/ Lum. (eV)	EPR	DAP[a] sta.	SB (A/L)	L (eV)	D (eV)	Name	Structure	Table	N conc.[b]	Ni impl. + ann. (lum.)	Comment
9	1.883	–	–	–	–	107/8	ZPL	1.35	*Ni 1.88a	**$*(V_3Ni_2)^\circ$**	8.1.6	hi.	–	Lines (b, c) at 1.906/ 1.913 eV
6c	1.991	–	2.16/1.85	–	–	–/111	ZPL	1.34	*Ni 1.69c	**$*V_1Ni_2{}^+$**	8.1.4	hi.	lo.	
10a	2.071	–	–	–	–	–/87	ZPL	1.34	*Ni 2.07a	**$*V_1Ni_2{}^-$**	8.1.4	hi.	hi.	No absorption
11	2.157	–	–	–	–	–	–	–	*Ni 2.16	**?**	9.5.1.2	hi.	hi.	Table 9.5.3
12	2.267	–	–	–	–	–	–	–	*Ni 2.27	**?**	9.5.1.2	hi.	–	Table 9.5.3–
10b	2.298	–	2.48/	–	–	27/28	ZPL	1.34	*Ni 2.07b	**$*V_1Ni_2{}^-$**	8.1.4	hi.	hi.	PLE of *Ni 2.07a
13	2.370	–	–/2.20	–	–	91/92	ZPL	1.34	*Ni 2.37	**$*V_1Ni_3{}^\circ$**	8.1.4	hi.	lo.	"A" line, Ni_3 from QLVM
14	–	2.496–2.890	2.85/2.85	NE1	58	–	2.383	1.63	**Ni-S3**	**$(V_2Ni_1)N_{2+0+0}{}^+$**	8.1.5	hi.	lo.	–

7b	2.401	–	2.63/–	–	–	29/–	ZPL	1.34	*Ni 1.70b	**$V_1Ni_1^{\circ}$**	8.1.4	lo.	lo.	PLE of *Ni 1.70a
15	–	2.515–3.341	2.79/2.79	NE2	57	–	2.422	1.63	**Ni-S2**	**$(V_2Ni_1)N_{2+0+1}^{+}$**	8.1.6	High	hi.	"B–E" lines
4b	2.427	–	–	–	–	–	–	–	*N 1.66b	**$^{*}V_1Ni_2^{\circ}$**	8.1.4	hi.	lo.	PLE of *Ni 1.66a
16	2.510	–	–	W8	–	–	–	–	Ni 2.51	**Ni_1^{-}**	9.5.1.2	hi.	lo.	T_d symmetry
17a	2.562	–	−/2.35	–	–	105/6	ZPL	1.55	*2Ni 2.56a	**$^{*}(V_3Ni_2)^{+}$**	8.1.6	lo.	hi.	2Ni from isotope splitting
18	–	2.750–3.530	3.00/300	NE8?	61	–	2.637	1.63	***Ni-S6**	**$^{*}(V_2Ni_1)N_{x+y+z}^{+}$**	8.1.5	hi.	–	PLE of *S5, *S8
17c	3.065	–	3.54/–	–	–	–	–	–	*2Ni 2.56c	**$^{*}(V_3Ni_2)^{+}$**	8.1.6	lo.	(hi.)	Lines (b, d) at 2.588, 3.076 eV
19	–	3.333–3.820	3.30/3.30	NE3	59	–	3.191	1.53	**Ni-*S4**	**$(V_2Ni_1)N_{1+2+0}^{\circ}$**	8.1.8	hi.	–	PLE of S2, *S8

[a] L = limiting energy, D = dielectric factor

[b] N conc.: lo. = ca. 5–10 ppm, hi. = ca. 250 ppm. References [Law93a, Kup99, Col00, Zai01]

Table 9.5.1.2 Additional data on nickel centers in diamond

Nos.	ZPL (eV)	Structure	Observed	Name	Comment
16	2.510	$^*Ni_1^-$	HA, MA, HL, ML *EPR:W8*	*Ni 2.51(a)	T_d symmetry/3 components at 4 K: 2.509, 2.510, 2.511 eV/correlates with W8 EPR center of Ni_1^- ($3d^7$, $S = 3/2$)
16	2.523	$^*Ni_1^-$	HA,MA, HL,ML	*Ni 2.51(b)	Line (b) is three times weaker than line (a)
20	2.436	$^*(C_2)_i\ Ni_1^{\circ}$	ML	*Ni 2.44	Resonance/SB DAP14(2.066–2.271 eV), resonance lines at 2.156 ($V_1N_1^{\circ}$) and 2.427 eV ($V_1Ni_2^+$(b)); QLVM: −75 meV
11	2.157	Ni+?	ML	*Ni 2.16	Ni implanted +1, 400 °C annealing; QLVM: −40 meV (calc. 38.9 meV for 1Ni + 1C)
12	2.267	Ni+?	HA,MA, HL,ML	*Ni 2.27	QLVM: +49 meV (calc. 49.7 meV for 4C)

centers with unknown structure). It is remarkable that at least 15 of the 20 nickel centers are nickel-vacancy associates, which is probably due to the large ionic radius of nickel.

For three centers, the involvement of nickel is evident from resolved splitting due to nickel isotopes (Sect. 9.5.2). Direct evidence for nickel involvement is obtained for ten centers, which are produced by nickel implantation (Sect. 9.5.3). Further arguments for the assignment to nickel come from the observation of related EPR centers (Sect. 9.5.4).

Only for three nickel centers, the structure is confirmed by EPR (S2, S3, and *S4; see Tables 8.1.5 and 9.5.1.1). For 13 additional centers, the proposed structure (marked by a preceding *) is based on information from characteristic LVM frequencies (Sect. 9.5.7 and Table 9.5.7), from QLVM frequencies (Sect. 9.5.8 and Table 9.5.8), and from DAP transition analysis (Sect. 9.5.9).

In Table 9.5.1.2 additional data for the nickel centers Nos. 11, 12, 16, and 20 are given. For two nickel centers (nos. 11, 12) the structure is unknown. The substitutional single nickel center (no. 16 $=^* Ni_1^-$) correlates with the W8 EPR center. The associate of nickel with the carbon split self interstitial (no. 20 = $^*(C_2)_iNi_1^{\circ}$) is not contained in Tables 9.5.1.1 and 9.5.6.

9.5.2 Resolved Isotope Shifts from Natural Nickel

The natural abundance for *one* nickel atom (odd mass in *italics*) is given by 67.77% (^{58}Ni), 26.16% (^{60}Ni), *1.25% (^{61}Ni)*, 3.66% (^{62}Ni), and 1.16% (^{64}Ni).

For *two* equivalent nickel atoms, the relative abundances for the 11 Ni_2 combinations (odd masses in *italics*) are 45.93% ($^{116}Ni_2$), 35.46% ($^{118}Ni_2$), *1.15% ($^{119}Ni_2$)*, 11.80% ($^{120}Ni_2$), *0.65% ($^{121}Ni_2$)*, 3.50% ($^{122}Ni_2$), *0.09% ($^{123}Ni_2$)*, 0.74% ($^{124}Ni_2$), *0.03% ($^{125}Ni_2$)*, 0.08% ($^{126}Ni_2$), and 0.01% ($^{128}Ni_2$).

Experimental results for three nickel centers are given below and in Table 7.5. The isotope splitting for zero phonon lines is small (0.16–1.60 meV) and is resolved only in favorable cases. Surprisingly large (8–13 meV) is the splitting for the five DAP transitions of *S7 (see Table 7.5).

(2) *1Ni 1.40 center = **$*V_1N_1{}^+$**: Four resolved isotope-shifted lines from a single nickel atom are observed, with a separation of +0.16 meV (decreasing mass) for the even isotopes. A separation of +0.22 meV is estimated for the not well-resolved isotope *^{61}Ni* [Dav89]. For decreasing energy, the relative intensities (*uneven isotope in italics*) are 72%: 21%: *1.4%*: 4.2%: 1.4%. This is in good agreement with the natural Ni_1 isotope abundance.

(8) *Ni-S7 = **$*(V_2Ni_1)N_{x+y+z}{}^+$ at 1.80–2.62 eV**: Nickel isotope shifts with three resolved components are observed on the five DAP11 transitions with $s = 9$, 14, 22, 64, and 162. The separation varies between +8 and +12 meV for increasing mass (^{58}Ni, ^{60}Ni, and ^{62}Ni). The relative intensities agree with the natural abundances of a single nickel atom, as demonstrated by simulated spectra [Iak00g].

(17a) *2Ni 2.56 center = **$*(V_3Ni_2)^+$(a)**: Five groups of isotope-shifted luminescence lines are observed with a separation of +1.6 meV (for decreasing mass), denoted by I (2.558 eV)–V (2.564 eV). The exact separation is 1.60 ± 0.05 meV for I–II–III, 1.56 ± 0.05 meV for III–IV, and 1.49 ± 0.05 meV for IV–V. Each group consists of three components (denoted a,b,c) with a separation of 0.37 meV for a–b, and 0.30 meV for b–c [Col83b,Col90c,Col97]. From the temperature dependence of the spectra, it is evident that the 1.49–1.60 meV splitting I–V (isotope shift) occurs in the excited state, and the 0.30–0.37 meV splitting a–c occurs in the ground state. The relative intensities of the components a:b:c at the temperature 16 K are (I) 1%: 3%: 0%, (II) 16%: 21%: 0%, (III) 4%: 12%: 0%, (IV) 0%: 8%: 4%, and (V) 0%: 0.2%: 1%.

Unfortunately, the relative intensities of I–V do not allow a straightforward analysis for three reasons:

1. The luminescence decay time varies drastically from 10.5 ns at 77 K over 0.8 ms at 20 K to 140 μs at 2.3 K, indicating interaction with a reservoir;
2. Thermal depopulation of the higher states in the excited state starts below 88 K (begin of resolved structure) and is very pronounced down to 1.8 K;
3. The relative intensities of some lines are not reproducible (the spectrum at 4.2 K is very different from the spectra at other temperatures). This is a hint for partial absorption inside the crystal (see below).

To eliminate the influence of depopulation, Boltzmann statistics can be applied. For the present analysis, depopulation-corrected intensities have been calculated. The result for the 16 K spectrum is 0.12: 3.5: 4.6: 12.4: 4.3 (for calibration, see below).

A more serious difficulty is the loss of intensity by self-absorption. This loss is different for the groups I–V. Since the loss of group II is the smallest, this group can be used for a pseudo-calibration (0% loss assumed), i.e., the intensity of group II is given the theoretical value of 3.5% (see above). By comparison with the theoretical intensities, the relative absorption losses (at 16 K) can be obtained, and are 80% (I), 60% (III), 65% (IV), and 90% (V). Depth-dependant luminescence spectra are desirable.

In spite of this unsatisfactory situation, the involvement of two equivalent nickel atoms can be concluded from the *five* equidistant groups I–V (a single nickel atom can account only for *four* equidistant groups). In addition, there is great similarity to the well-understood corresponding silicon center $^*(V_3Si_2)^+$ at 1.681 eV (see Sect. 9.4). This silicon center has a 0.22-meV splitting in the ground state and +0.35-meV isotope splitting (for decreasing mass) in two excited states (with 1.1 meV separation). For the corresponding nickel center, a similar second excited state $(^*(V_3Ni_2)^\circ$(b) at 2.588 eV) is observed, with a spectral separation of 24 meV and an activation energy of 23.8 eV [Col83b].

In absorption and photoluminescence excitation, two further excited states $(^*(V_3Ni_2)^+$(c,d)) are observed at 3.065 and 3.076 (see Table 8.1.6). The weak structure on the 3.065 eV line may arise from isotope splitting.

9.5.3 *Nickel Implantation and Resulting Centers*

A natural type IIa diamond (containing nitrogen) has been implanted with 340 keV Ni^+ ions at a dose of $10^{14}\,cm^{-2}$, and subsequently annealed at 1,400 °C [Zai00a]. The published CL spectrum is restricted to the energy range 1.95–2.70 eV. The following ten nickel centers are observed (intense lines are underlined): *Ni 1.69(c) at 1.991 eV; *Ni 2.1(a) from $^*V_1Ni_2^-$ at 2.071 and (b) at 2.298 eV; *Ni 2.16 at 2.157 eV; *Ni 2.37 from $^*V_1Ni_3^\circ$ at 2.370 eV; *Ni 1.70(b) from $^*V_1Ni_1^\circ$ at 2.401 eV; *Ni 1.66(b) from $^*V_1Ni_2^+$ at 2.427 eV; *Ni 2.51 from Ni_1^- at 2.510 eV; Ni-S2 from $(V_2Ni_1)N_{2+0+1}{}^+$ with lines B at 2.537, C at 2.597, and D at 2.623 eV; *2Ni 2.56 from $^*(V_3Ni_2)^+$ at 2.562 eV; and Ni-S3 from $(V_2Ni_1)N_{2+0+0}{}^+$ at 2.653 eV.

Additional lines (not related to nickel) are observed from $V_1N_1^\circ$ (NV center) at 2.154 eV, from $V_1N_2^\circ$ (H3 center) at 2.464 eV, and from $V_2N_4^\circ$ (H4 center) at 2.497 eV. Also observed are 14 lines from DAP1 transitions (nitrogen + boron pair) in the range 2.070–2.649 eV.

In a previous experiment (nickel implantation and 1,400 °C annealing), it had been found that the *1Ni 1.40 center (from $^*V_1Ni_1^+$) and the *2Ni 2.56 center (see above) arise from nickel [Gip83].

9.5.4 Electron Paramagnetic Resonance and Zeeman Effect of Nickel Centers

The most important EPR centers are *NE1*, *NE2*, and *NE3* with their optical counterparts S3, S2, and *S4. Note, that the *NE3* EPR center is in the positive state, while the optical counterpart *S4 is in the neutral charge state. Evidently the two charge states coexist. The complicated structure of these centers was derived from the analysis of the hyperfine interaction [Nad93]. The interstitial nickel atom resides midway between two adjacent lattice vacancies. Different numbers of nitrogen atoms are found in the first, second, and third shell around the nickel atom. In the unrelaxed lattice, the distance to these sites (relative to the diamond lattice bond r_0) are $r_1 = 1.26\ r_0$, $r_2 = 1.82\ r_0$, and $r_3 = 2.06\ r_0$. The average hyperfine interaction is $A_1 = 15.5$, $A_2 = 5.6$, and $A_3 = 2.6$ Gauss. This is in good agreement with the ratio $1/r^3$ of the three shells, which is 15.5: 5.3: 3.4. Counting the number of nitrogen atoms in the three shells leads to the notation $S2 = (V_2Ni_1)N_{2+0+1}$, $S3 = (V_2Ni_1)N_{2+0+0}$, and $^*S4 = (V_2Ni_1)N_{1+2+0}$ (see also Sects. 8.1.5 and 9.5.3).

The **W8** EPR center arises from Ni_1^- ($3d^7$) with an effective spin of 3/2 and is assigned to the optical 2.510 eV center (substitutional Ni_1 with T_d symmetry) [Col00].

The **NIRIM 2** EPR center has been assigned to the optical 1.40 eV center [Iso90a].

The 1Ni*S7 center at 1.808–2.620 eV (observed only in CVD samples) is closely related to the **S = 1** EPR center [Iak00a].

Four EPR Ni centers without a confirmed optical counterpart are **NE4**, **NE5**, **NE8**, and **NIRIM 1** [Col00].

9.5.5 Influence of Nitrogen Concentration

It is a remarkable property of the nickel centers that 14 of them occur (more or less exclusively) in diamond with *high* nitrogen concentration, and only four centers in diamond with *low* nitrogen concentration. The latter are $^*(V_2Ni_1)^-$, $^*V_1Ni_1^+$, $^*V_1Ni_1^\circ$, and $^*(V_3Ni_2)^-$ (see Tables 9.5.1.1 and 9.5.6).

In the present structure assignment, only the seven centers S2–*S7, *S9 contain nitrogen. The centers $V_1Ni_2^-$, $V_1Ni_2^\circ$, $V_1Ni_2^+$, and $V_1Ni_3^\circ$ might contain nitrogen. Experiments with the search for ^{15}N isotope shifts are desirable.

9.5.6 Effects of Annealing of Nickel Centers

The annealing effects and thermal stability of nickel centers in diamond are summarized in Table 9.5.6. Fourteen nickel centers are observed in as-grown HPHT diamonds (nos. 3–5, 14 are missing in low temperature-grown HPHT diamonds

Table 9.5.6 Observed annealing effects and thermal stability of nickel centers in diamond (see also Table 9.5.1.1)

No	ZPL (eV)	DAP lines (range) (eV)	Name	Structure	N conc.[a]	Before anneal: Natural	Before anneal: HPHT	Anneals out at 1,400–1,800 °C	Anneals out at 2,000–2,500 °C	Text in [Zai01] page	Comment temperatures in °C
1	–	1.212–1.383	***S8**	*$(V_2Ni_1)^-$	lo.	–	HA, HL	–	*>1,800*	134	Photochromic at < 1.7 eV
2	1.404	–	1Ni 1.40	*$V_1Ni_1{}^+$	lo.(hi.)	NA, NL	HA, HL	*>1,800*	*>2,500*	140	C_{3v} symmetry, photochromic at > 1.7 eV
3	–	1.563—2.320	***S5**	*$(V_2Ni_1)\,N_{x+y+z}{}^\circ$	hi.	NA, NL	HA, HL	*>1,800*	*>1,800*	148	Strengthened at $T > 1{,}500$
4a	1.660	–	*Ni 1.66a	*$V_1Ni_2{}^\circ$	hi.	–	–	–	**< 2,200**	156	ML appears after $T > 1{,}500$; no abs.
5		1.678–1.821	***S9**	*$(V_2Ni_1)\,N_{x+y}+z^+$	hi.	–	–	–	–	183	ML appears after N irrad. + $T = 950$
6a	1.693	–	*Ni 1.69a	*$V_1Ni_2{}^+$	hi.	–	HA, HL	**< 1,800**	*<1,800*	182	Strengthened at $1{,}500 < T < 1{,}800$
7a	1.704	–	*Ni 1.70a	*$V_1Ni_1{}^\circ$	lo.	–	HA, HL	**< 1,800**	*<1,800*	183	

8	–	1.808—2.620	*S7	*$(\mathbf{V_2Ni_1})\mathbf{N_{x+y+z}}^+$	hi.	–	–	*>1,800*	?	–	Only in CVD samples [Iak00a]
9	1.883	–	*Ni 1.88a	*$(\mathbf{V_3Ni_2})^\circ$	hi.(lo.)	–	HA, HL	< **1,800**	–	192	Transforms into *$V_1Ni_2^+$
6c	1.991	–	*Ni 1.69c	*$\mathbf{V_1Ni_2}^+$	hi.	–	HA, HL	*>1,800*	< **2,200**	197	Strengthened at $T > 1,500$; line (b) at 1.940
10a	2.071	–	*Ni 2.07a	*$\mathbf{V_1Ni_2}^-$	hi.	–	–	*>1,800*	< **2,200**	209	Appears after $T > 1,500$; No absorption
11	2.157	–	*Ni 2.16	**?**	hi.	–	–	–	*>1,800*	226	Appears after $T > 1,700$
12	2.267	–	*Ni 2.27	**?**	hi.	–	HA, HL	–	*>1,800*	234	Strengthened at $T > 1,500$
10b	2.298	–	*Ni 2.07b	*$\mathbf{V_1Ni_2}^-$	hi.	–	–	*>1,800*	< **2,200**	236	Appears after $T > 1,500$
13	2.370	–	*Ni 2.37	*$\mathbf{V_1Ni_3}^\circ$	hi.	NA, NL	–	–	*>1,800*	248	"A" line
14	–	2.496—2.890	**S3**	$(\mathbf{V_2Ni_1})\mathbf{N_{2+0+0}}^+$	hi.	NA, NL	HA, HL	*>1,800*	*>1,800*	277	Inactive in CL

(continued)

Table 9.5.6 (continued)

No	ZPL (eV)	DAP lines (range) (eV)	Name	Structure	N conc.[a]	Before anneal		Anneals out at 1,400–1,800 °C	Anneals out at 2,000–2,500 °C	Text in [Zai01] page	Comment temperatures in °C
						Natural	HPHT				
7b	2.401	–	*Ni 1.70b	$\mathbf{V_1Ni_1}^{\circ}$	lo.	–	HA	–	< **2,200**	255	PLE of *Ni 1.70a
15	–	2.515–3.341	**S2**	$\mathbf{(V_2Ni_1)N_{2+0+1}}^{+}$	hi.	NA, NL	HA, HL	*>1,800*	*>1,800*	287	"B–E" lines, inactive in CL
4b	2.427	–	*N 1.66b	$^{*}\mathbf{V_1Ni_2}^{\circ}$	hi.	–	HA, HL	*>1,800*	< **2,200**	256	PLE of *Ni 1.66a, strengthened at T >1,500
16	2.510	–	*Ni 2.51	$^{*}\mathbf{Ni_1}^{-}$	hi.(lo.)	–	HA, HL	< **1,600**	–	283	
17a	2.562	–	*2Ni 2.56a	$^{*}\mathbf{(V_3Ni_2)}^{+}$	lo.	NA, NL	HL	*>1,800*	*>1,800*	290	Line (b) at 2.588 eV
18	–	2.750–3.530	***S6**	$^{*}\mathbf{(V_2Ni_1)N_{x+y+z}}^{+}$	hi.	–	–	*>1,800*	*>1,800*	303	Appears after T >2,200, PLE of *S5
17c	3.064	–	*2Ni 2.56c	$^{*}\mathbf{(V_3Ni_2)}^{+}$	lo.	NA/NE	HA/HE	*>1,800*	*>1,800*	324	PLE of 2Ni 2.56a,b, line (d) at 3.076 eV
19	–	3.333–3.820	***S4**	$^{*}\mathbf{(V_2Ni_1)N_{1+2+0}}^{\circ}$	hi.	NA/NE	HA/HE	–	*>1,800*	334	PLE of S2, *S8

[a]N concentration: lo. = ca. 5–10 ppm, hi. = ca. 250 ppm [Law93a, Kup99, Col00, Zai01]

[Law93a]), three centers (nos. 9, 10, 17; all with high nitrogen concentration) appear after annealing at 1,500–2,500 °C, and one center (no. 7) exists only in CVD diamond.

In diamonds with high nitrogen concentration, substantial changes in the nickel spectra occur in the temperature range 1,500–1,900 °C. This is the temperature regime where single substitutional nitrogen is transformed into aggregated structures (see Sect. 9.1.6.1). In parallel with the nickel spectra, the infrared absorption of nitrogen centers was monitored with the following result: *as grown*: 100 ppm N_I (C-center)+6 ppm N_2 (A center); *after 1,600 °C anneal*: 37 ppm N_I + 82 ppm N_2; *after 1,800 °C anneal*: 4 ppm N_I + 115 ppm N_2 [Law93a]. These infrared spectra also show intense absorption from the $N_I{}^+$ (X center, lines at 165, 130, and 118 meV (or 1,332; 1,046; and 950 cm^{-1}); see Table 9.1.2), which is preferentially observed in nickel-containing diamonds, and anneals out at 1,900 °C. This absorption is proportional to the concentration of compensated boron [Zai01].

In contrast to the nitrogen-rich samples, the nickel spectra in diamonds with low nitrogen concentration show almost no annealing effects, except for a general weakening of the signals at 1,800 °C [Law93a].

The influence of nitrogen aggregation on the annealing behavior of nickel centers in diamond with high nitrogen concentration is quite evident. After the nitrogen aggregation, a shift in the Fermi level toward the valence band is expected, with a corresponding change in charge state of the nickel centers. This effect cannot be further discussed because the charge state for most of the nickel centers is not yet confirmed.

Alternatively, a change in structure can occur when vacancies in nickel-vacancy associates recombine with migrating carbon interstitials (see (9.2b) and (9.3b)). A direct proof for this annealing effect (at $T < 1,800$ °C) is given by the quantitative transformation of $^*(V_3Ni_2)^\circ$ at 1.883 eV into $^*V_1Ni_2{}^+$ at 1.693 eV [Law93a].

A further possibility is the partial dissociation of nickel associates. Of interest is also the appearance of the DAP1 absorption transitions. Some ten lines (1.938–2.648 eV) arise from this center, which is free of nickel, and has been assigned to the $N_I{}^\circ + B_I{}^\circ$ pair (see Sect. 9.1.5 [Dis94a],b). In the samples with high nitrogen concentration, these lines appear after 1,500 °C annealing. After a 1,800–2,200 °C annealing, only the close pairs in shells 1–3 are present, while all distant pairs have disappeared (by dissociation) [Law93a, Kup99].

The thermal stability of the nickel centers has been investigated in the temperature range 1,600–1,900 [Law93a] and 2,200–2,500 °C [Kup99]. The results are listed in Table 9.5.6. All seven nickel centers, which occur in natural diamond, are stable at least up to 2,200 °C.

9.5.7 Local Vibrational Modes

LVM frequencies for nickel centers in diamond are listed in Tables 2.1.1.1–5.4.4. There is good agreement between the predicted and observed LVM frequencies, which supports the present structure assignments (see Table 9.5.7).

Table 9.5.7 Predicted and observed LVM frequencies of nickel centers in diamond

		Observed LVM frequencies (meV)				
		8.1.4	8.1.4	8.1.4	8.1.4	8.1.6
	Table Defect	V_1Ni_1	V_1Ni_2	V_1Ni_3	$(V_2Ni_1)N_n$	(V_3Ni_2)
Type of vibration	Predicted[a]					
$Ni(sp^3)$ – –$C/N(sp^3)$ bend	27–45[b]	–	–	–	–	–
$C/N(sp^3)$ – –$C(sp^3)$ bend	45–75[c]	61	49–61 $(1x–7x)$[d]	49	–	–
$Ni(sp^3)$ – –$Ni(sp^3)$ stretch	45–81[e]	–	–	–	–	61–65
$Ni(sp^3)$ – –$C/N(sp^3)$ stretch	58–106[b]	74–80	64–66	70	63–99	–
$C/N(sp^3)$ – –$C(sp^3)$ stretch	97–177[c]	100–127	–	–	102–144	–

[a]The predicted values are obtained from data in Table 8.8 by multiplication (for the heavier mass of ^{59}Ni) with the appropriate factor
[b]Factor 0.60
[c]Factor 1.0
[d]Number of harmonics
[e]Factor 0.46

9.5.8 Quasi-Local Vibrational Modes

Valuable information on the structure of nickel centers is obtained from the QLVM sidebands, as shown in Table 9.5.8. This table is subdivided in QLVM frequencies of 3 Ni (23 meV), 2 Ni (25–30 meV), and 1 Ni (29–46 meV).

9.5.9 DAP Transition at Nickel Centers

With three exceptions (nos. 10, 11, and 15 in Tables 9.5.1.1 and 9.5.1.2), all nickel centers are associates with one, two, or three lattice vacancies. In principle, DAP transitions from all these centers should exist, and indeed for only two (nos. 4 and 5), the DAP lines are not yet observed. A special case is that of *standard* DAPs of the centers S2 to *S9, where the interstitial nickel resides midway between two vacancies. The nickel d electrons form bonds with the six equidistant carbon neighbors, thus avoiding dangling bonds, which would produce a *sideband* DAP. For the centers S2 to *S9 the dielectric factor D depends on the charge state, and is 1.60–1.66 eV for $S2^+$, $S3^+$, $*S6^+$, $*S7^+$, $*S9^+$, is 1.53 eV for $*S4°$, $*S5°$, and

Table 9.5.8 Calculated and observed QLVM frequencies of Ni centers

Atoms	Calculated[a] (meV)		Observed (meV)		ZPL/*or* L^b	Center	Tables
	Frequ.	Width	Frequ.	Width	(eV)		
3 Ni	23.3	5.8	−23	10	2.370	$^*V_1Ni_3{}^{\circ}$	8.1.4
2 Ni + 2C	26.2	7.3	−25	8	2.562	$^*(V_3Ni_2)^{\circ}$(a)	8.1.6
			+26	7	3.064	$^*(V_3Ni_2)^{\circ}$(b) = 2.562(b)	8.1.6
			−26	10	1.883	$^*(V_3Ni_2)^+$	8.1.6
2 Ni	29.0	9.0	+30	11	2.298	$^*V_1Ni_2{}^{\circ}$(b) = 2.071(b)	8.1.4
			+28	12	2.427	$^*V_1Ni_2{}^+$(b) = 1.660(b)	8.1.4
1 Ni + 2N + 2C	30.0	9.5	+29		*2.422*	**S2** = $(V_2Ni_1)N_{2+0+1}{}^+$	8.1.5
			+30		*2.383*	**S3** = $(V_2Ni_1)N_{2+0+0}{}^+$	8.1.5
1 Ni + 1N + 3C	30.3	9.7	−31		*3.191*	***S4** = $(V_2Ni_1)N_{1+2+0}{}^+$	8.1.5
			−33		*1.460*	***S5** = $(V_2Ni_1)N_{1+x+y}{}^{\circ}$	8.1.5
1 Ni + 3C	32.8	11.4	−32	12	2.071	$^*V_1Ni_2{}^{\circ}$(a)	8.1.4
1 Ni + 2N	34.5	12.6	+34		*2.422*	**S2** = $(V_2Ni_1)N_{2+0+1}{}^+$	8.1.5
1 Ni + 2C	35.5	13.3	+34	11	2.401	$^*V_1Ni_1{}^{\circ}$(b)= 1.704(b)	8.1.4
			−36	15	2.562	$^*(V_3Ni_2)^{\circ}$(a)	8.1.6
1 Ni + 1C	38.9	16.1	−40	18	2.157	$^*Ni_1$ + X	9.5.1.2
1 Ni	43.6	20.2	+42		*1.460*	***S5** = $(V_2Ni_1)N_{1+x+y}{}^{\circ}$	8.1.5
			+42		*2.637*	***S6** = $(V_2Ni_1)N_{2+x+y}{}^+$	8.1.5
			−44	20	1.704	$^*V_1Ni_1{}^{\circ}$(a)	8.1.4
			+45	25	1.404	$^*V_1Ni_1{}^+$	8.1.4
			−46		*2.422*	**S2** = $(V_2Ni_1)N_{2+0+1}{}^+$	8.1.5

[a] See Table A.3
[b] L = limiting energy for standard DAPs (no ZPL)

is 1.45 eV for $^*S8^-$ (see Sect. 10.5). Since the vacancy is the dominant defect in nickel-vacancy associates, the corresponding DAP spectra are already discussed in Sect. 8.3.2.

A decisive help in the analysis of DAP16 ($^*S8 =^* (V_2Ni_1)^-$) is the simultaneous appearance of resonance DAP lines from this center in some samples [Law93c]. The 12 lines peak at 1.22 eV and allow the determination of the parameters L and D for this DAP, which in most samples produces only an unstructured broad band, peaking at 1.40 eV.

9.5.10 The S2 to *S9 Family (DAP57–61, 11, 16, 4) with the (V_2Ni_1) Nucleus and 0–3 Nitrogen Ligands

The almost 100 lines of the S2 to *S9 family have never been really explained until this book was written. The observation of coinciding energies for absorption and luminescence is very intriguing, and an explanation by zero phonon lines was generally assumed [Zai01]. The present DAP interpretation (see Tables 8.1.9, 9.5.10 and 10.1) can satisfactorily explain the line positions and the coincidence of absorption and luminescence. The latter is characteristic for *standard* DAPs (see Sect. 10.1).

By the pioneering work of [Yel92a],b and [Nad93], the structure of three optical centers was determined from their EPR counterpart. These are **S2** with EPR center NE2, **S3** with EPR center NE1, and ***S4** (sometimes named "523 nm" center) with EPR center NE3. (It turned later out that the 523 nm = 2.370 eV line "A" is not part of the *S4/NE3 center, but is the ZPL transition of DAP91/92 = $V_1N_3{}^\circ$ with a very effective energy transfer from *S4 to $V_1N_3{}^\circ$). For this book, the names *S5 to *S9 are introduced for centers with very similar structure (see Tables 8.1.9 and 9.5.6).

The S2 to *S9 family provides interesting insight into the mechanism of DAP transitions. From Table 9.5.10, it is evident that both *intra-center* DAP pairs (shell numbers 1 and 2) and distant pairs are observed. For *S8 (with zero nitrogen) only distant pairs are observed. This observation indicates that in the *intra-center* pairs (shell numbers 1 and 2 of S2, *S5, *S6, and *S7), the nitrogen ligands act as donor, and the nickel nucleus acts as acceptor.

9.5.11 Discussion of 20 Individual Nickel Centers

In the past, only four structure assignments have been made: S2, S3, *S4, and *$Ni_1{}^-$. The present structure assignments for 12 additional centers are based on a combination of arguments.

The QLVM frequencies indicate the number of Nickel atoms in the center (see Sect. 9.5.8 and Table 9.5.8).

A further hint is obtained from the LVM frequencies (see Tables 8.1.4, 8.2.3, and 9.5.7). Characteristic for the V_1Ni_2 centers is the bonding vibrations with frequencies of 49–57 meV with up to six harmonics.

The 20 nickel centers are discussed in order in Tables 9.5.1.1, 9.5.1.2, and 9.5.6:

1. ***(V_1Ni_1)$^-$** = Ni-S8-DAP16a–l (1.212–1.383 eV): This is the only center of the S2 to *S9 family with *no* nitrogen ligands. It is observed in diamonds with low nitrogen concentration (e.g., grown with nitrogen getter). The DAP16 transitions are of the *resonance* type (see Chap. 10) with neighboring shell numbers: 42, 43, 44, and 47, 48, 49, 50, 51. Possibly, the EPR counterpart is the paramagnetic NE4 center.
2. ***$V_1Ni_1{}^+$** = 1Ni 1.40 at 1.401 and 1.404 eV with DAP25a–k (1.485–2.070 eV) in absorption, and DAP26a–k (1.000–1.333 eV) in luminescence.

Table 9.5.10 Survey of the DAP analysis of the S2 to *S9 family

Shell no.[a]	S2	S3	S4	S5	S6	S7	S8	S9	Comment	Name	Dielectric factor	Limiting energy, L (eV)	Structure
228	a	a	–	–	–	a	–	–		**S2**	+1.63	2.422	(V_2Ni_1) $N_{2+0+1}{}^{+}$
162	b	b	–	a	a	b	–	–	S2: line **B**	DAP57			
98	c	–	–	–	–	–	–	–		**S3**	+1.63	2.383	(V_2Ni_1) $N_{2+0+0}{}^{+}$
86	–	–	a	b	–	–	–	–		DAP58			
72	d	–	–	–	–	–	–	a		***S4**	+1.53	3.191	(V_2Ni_1) $N_{1+2+0}{}^{\circ}$
64	e	–	–	–	–	c	–	b	S2: line C	DAP59			
58	–	–	–	c	–	–	–	d	S9: c = 62	***S5**	+1.53	1.460	$^*(V_2Ni_1)$ $N_{x+y+z}{}^{\circ}$
50	f	c	b	d	–	–	b	g	S8: a = 51	DAP60			
48	g	–	–	–	–	–	d	–	S8: c = 49	***S6**	+1.63	2.637	$^*(V_2Ni_1)$ $N_{x+y+z}{}^{+}$
44	h	d	–	–	–	–	f	–	S8: e = 47	DAP61			
42	–	e	–	–	–	–	h	h	S8: g = 43	***S7**	+1.60	1.724	$^*(V_2Ni_1)$ $N_{x+y+z}{}^{+}$
38	i	–	–	–	–	–	–	–		DAP11			
34	j	f	c	–	–	d	–	–		***S8**	+1.45	1.036	$^*(V_2Ni_1)^{-}$
32	k	–	–	–	–	–	–	–		DAP16			
30	l	–	–	e	–	–	–	–		***S9**	+1.66	1.509	$^*(V_2Ni_1)$ $N_{x+y+z}{}^{+}$

(continued)

Table 9.5.10 (continued)

Shell no.[a]	S2	S3	S4	S5	S6	S7	S8	S9	Comment	Name	Dielectric factor	Limiting energy, L (eV)	Structure
26	m	g	d	–	–	–	i	–					
22	n	–	e	–	b	e	–	i					
18	–	–	f	–	–	–	j	–					
16	–	–	–	–	c	f	–	–					
14	–	–	g	–	d	g	k	–					
13	–	–	–	–	–	–	l	–					
12	–	h	h	–	e	–	–	–					
10	–	i	i	–	–	–	–	–					
9	–	–	–	–	f	h	–	–					
8	–	j	j	f	g	–	–	–					
7	–	–	–	g	h	–	–	–					
6	–	–	k	h	i	i	–	–					
5	–	–	–	–	j	j	–	–					
4	–	–	l	i	k	k	–	–					
3	–	–	–	–	–	l	–	–					
2	o	–	–	j	l	m	–	–					
1	p	–	–	k	m	n	–	–					

[a] High abundance shells in bold type (see Table A.2)

The involvement of a single nickel atom is evident from the resolved isotope shifts (see Table 7.5 and Sect. 9.5.2). The center is a typical vacancy associate (see Table 8.1.4). The charge state is derived from the EPR counterpart NIRIM2.

3. **$^*(V_2Ni_1)N_{x+y+z}{}^{\circ}$** $=^*$ Ni-**S5**-DAP60a–k (1.563–2.320 eV). This is a typical member of the S2 to *S9 family (see Sect. 9.5.3 and Table 9.5.6). For the number of nitrogen atoms in the first shell, see Sect. 9.5.2.9. The neutral charge state is assigned according to Sect. 10.5.
4. **(a,b) $^*V_1Ni_2{}^{\circ}$** $=^*$ Ni 1.66a,b at (a) 1.660 and (b) 2.427 eV. This center is absent in as-grown synthetic or natural diamonds, but appears after heat treatment ($T >$ 1,500 °C). The Ni_2 structure is derived from the QLVM frequency (28 meV) (see Table 9.5.8). The charge state is proposed in a comparison with the other V_1Ni_2 centers, i.e., the $^*V_1Ni_2{}^+$ center (6) at 1.693, 1.940, and 1.991 eV, and the $^*V_1Ni_2{}^-$ center (10) at 2.071 and 2.298 eV. There is a general trend for decreasing charge state with increasing energy.
5. **$^*(V_2Ni_1)N_{x+y+z}{}^+$** $=^*$ Ni-S9-DAP4a–I (1.678–1.821 eV). This is a typical member of the S2 to *S9 family (see Sect. 9.5.10 and Table 9.5.10). This center is absent in as-grown synthetic or natural diamonds, but appears after nitrogen implantation and 950 °C annealing.
6. **(a–c) $^*V_1Ni_2{}^+$** $=$ Ni 1.69a–c at (a) 1.693, (b) 1.940, and (c) 1.991 eV. Broad peaks are observed in absorption from (a) at 1.85 eV and from (b,c) at 2.16 eV. In luminescence, the broad peak is at 1.85 eV from (b,c)-DAP11a–f (1.708–1.898 eV). The expected resolved DAP transitions from the other lines are not yet observed. For the charge state, see line (4) at 1.660 eV.
7. **(a,b) $^*V_1Ni_1{}^{\circ}$** $=$ Ni 1.70a,b at (a) 1.704 and (b) 2.401 eV. Line (a) with DAP83a–f (1.816–2.070 eV) in absorption, and with DAP84a–c (1.568–1.627 eV) in luminescence. Line (b) with DAP29a–f (2.713–2.965 eV) in absorption.
8. **$^*(V_2Ni_1)N_{x+y+z}{}^+$** $=^*$ Ni-S7-DAP11a–n (1.808–2.620 eV). The EPR counterpart is the paramagnetic $S = 1$ center [Iak00a]. The single nickel atom follows from the isotope shifts (see Table 7.5 and Sect. 9.5.2). For the number of nitrogen atoms in the first shell, see Sect. 9.5.2.9 and Table 9.5.6.
9. **(a–c) $^*(V_3Ni_2)^{\circ}$** $=^*$ Ni 1.88a–c at (a) 1.883, (b) 1.906, and (c) 1.913 eV. Line (a) with DAP107a–g (1.958–2.300 eV) in absorption, and DAP108a,b (1.779, 1.808 eV) in luminescence. The structure is derived from the similarity with the better established $^*(V_3Si_2)^{\circ}$ center at 1.681 eV (see Sect. 8.1.6 and Table 8.1.6). The involvement of Ni_2 is derived from the LVM frequency (61 meV) (see Table 9.5.7). The neutral charge state follows from Tables 10.4 and 10.5.
10. **(a, b) $^*V_1Ni_2{}^-$** $=^*$ Ni 2.07 at (8a) 2.071 and (b) 2.298 eV. Line (a) with DAP87a–n (1.709–1.962 eV) in luminescence. Line (b) with DAP27a–j (2.3909–2.568 eV) in absorption and DAP28a–I (1.905–2.089 eV) in luminescence. The involvement of Ni_2 follows from the QLVM frequency (30 meV) (see Table 9.5.8). For the charge state, see line (4) at 1.660 eV.
11. **Ni + X** $=^*$ Ni 2.16 at 2.157 eV. This center with an unknown structure is observed after nickel implantation and 1,440 °C annealing [Zai00a]. The QLVM frequency (40 meV) indicates a (1 Ni + 1 C) vibration (see Table 9.5.8).

12. **Ni+X** =* Ni 2.27 at 2.267 eV. For this center with an unknown structure, many references exist in the literature (see [Zai01]). The line is assigned to nickel because of the close relation to the established nickel centers **S2** and **S3**.
13. ***V_1Ni_3°** = Ni 2.37 at 2.370 eV (named "A" line) with DAP91a–g (2.460–2.641 eV) in absorption and DAP92a–f (2.170–2.280 eV) in luminescence. The involvement of Ni_3 is derived from the QLVM frequency (23 meV) (see Table 9.5.8). The neutral charge state follows from the dielectric factor D (1.34 eV), which is expected for 3Ni with 84 total electrons (see Table 10.4).
14. **$(V_2Ni_1)N_{2+0+0}{}^+$** = Ni-S3-DAP58a–i (2.496–2.890 eV). The complex structure and charge state of this center are established, because the EPR counterpart NE1 has been carefully analyzed [Yel92a],b, [Nad93] (see Sects. 9.5.4 and 9.5.10).
15. **$(V_2Ni_1)N_{2+0+1}{}^+$** = Ni-S2-DAP57a–n (2.515–3.341 eV). Similar to line (14) = S3, the structure is established by analysis of the EPR counterpart NE2.
16. **$Ni_1{}^-$** = Ni 2.51 at 2.510 eV. The structure and charge state are derived from the EPR counterpart WE8 (see Sect. 9.5.4).
17. **(a–d) *$(V_3Ni_2)^+$** = 2Ni 2.56 at (a) 2.562, (b) 2.588, (c) 3.065, and (d) 3.076 eV. Line (a) with DAP105a–d (2.780–3.221 eV) in absorption and DAP106a–f (2.383–2.466 eV) in luminescence. The incorporation of two equivalent nickel atoms is evident from the isotope line shifts (see Table 7.5 and Sect. 9.5.2). For the positive charge state, see line (9) and Table 10.5.
18. ***$(V_2Ni_1)N^+_{x+y+z}$** =* Ni-S6-DAP61a–m (2.750–3.530 eV). This center is absent in as-grown synthetic or natural diamonds, but appears after heat treatment ($T > 2{,}200$ °C). The involvement of a single nickel atom follows from the QLVM frequency (42 meV) (see Table 9.5.8). For the positive charge state, see Sect. 10.5. A possible EPR counterpart is the NE8 center.
19. **$(V_2Ni_1)N_{1+2+0}$°** = Ni-*S4-DAP59a–l (3.333–3.820 eV). This center is sometimes named "NE3 center" after the EPR counterpart NE3, from which the structure is established. In this book, the name ***S4** is introduced. Note the different charge states: Positive (EPR) and neutral (optics).
20. ***$(C_2)_iNi_1$°** =* Ni 2.44 at 2.436 eV with DAP14a–f (2.066–2.271 eV). The structure *$(C_2)_iNi_1$° is assigned to the 2.436 eV -DAP14 center (Table 9.5.1.2) with the following arguments: 1.) The sample of Fig. 7.90 in [Zai98] contains nickel (see the line at 2.427 eV from *V_1Ni_2°(b)= no. 4(b) above). 2.) The sample with the Ni2.44 eV center contains $(C_2)_i$ (see the line at 3.188 eV from *$(C_2)_iN_1^-$) 3.) The dielectric factor ($D = -1.22$) is in the expected range for associates with $(C_2)_i$ (see Table 8.2.3). 4.) The *Ni2.44 eV center anneals out at 1300°C, which is typical for associates with $(C_2)_i$ (see Sect. 8.3.1). 5.) The ZPL (2.436 eV) of the ∗Ni2.44-DAP14 lines coincides with a single line from the so-called 509 nm center (with unknown structure); however, the 509 nm center has quite different properties: It appears (as irradiated) after ion implantation, and is stable up to 1400°C, while the DAP14 lines disappear at 1300°C. This center is absent in as-grown synthetic or natural diamonds, but is created after carbon ion implantation and 1000°C annealing [Zai92a].

9.6 Cobalt in Diamond

In the synthesis of HTHP diamond, transition metals are normally employed as solvent catalysts. Nickel, cobalt, and iron are often used, occasionally also manganese or chromium. While nickel forms at least 18 optical centers in diamond (see Sect. 9.5), its direct neighbor in the periodic system behaves surprisingly differently.

The concentration of cobalt in HTHP diamond, which is grown from pure cobalt, is an order of magnitude less than the nickel concentration in comparable HTHP diamond, grown from a pure nickel melt. The first cobalt-related optical lines were reported in 1991 from HTHP diamond, grown from a pure cobalt melt at high temperature [Col91c], [Fie92]. As with the nickel centers, the cobalt centers are found only in the (1 1 1) growth sectors, which dominate in the high temperature-grown octahedral diamonds. With the present knowledge (see Table 9.6), six cobalt centers were observed: *Co1.47, *Co2.59, *Co2.89, *Co1.85, *Co-S5 with nine lines at 2.135–2.820 eV, and *Co 1.99. For three centers, a DAP analysis was possible: Standard *Co-S5 DAP104a–i, sideband *Co 1.85 DAP102a–f, and sideband *Co 1.99 DAP112a–f.

A very detailed study of cobalt-related centers appeared in 1996, including measurements on annealing effects (1, 500–1, 800 °C), radiative decay times (20 ns–159 μs), and temperature dependence (4–150 K) [Law96].

As shown in Table 9.6, all known cobalt lines can be assigned to six cobalt centers, which are all associates with one, two, or three lattice vacancies. Only one center (named *Co-S5, in analogy to the nickel *S5 center) contains nitrogen in addition.

After cobalt ion implantation and > 600 °C annealing, spectral lines from two cobalt centers (*Co-S5 and *Co2.59) were observed [Zai79a, Zai01].

9.6.1 *The *Co 1.47 Center at 1.472 eV from* $^{*}V_1Co_1{}^{+}$

9.6.2 *The *Co 2.59 Center at 2.590 eV from* $^{*}V_1Co_2{}^{-}$

The luminescence of this center is fast (decay time ca. 80 ns), and is more intense in cathodoluminescence than in photoluminescence. No detailed spectra is shown in [Law96]; therefore, no results on phonon sidebands or DAP transitions can be given. The corresponding lines of $^{*}V_1Ni_2{}^{-}$ occur at 2.071 and 2.298 eV (see Table 8.3a).

9.6.3 *The *Co 2.89 Center at 2.889 eV from* $^{*}V_1Co_2{}^{\circ}$

The luminescence of this center is fast. This center appears after annealing at $T >$ 1,500 °C.

Table 9.6 Cobalt centers in diamond

Defect[a]	Observed	ZPL (eV)	DAP lines (eV)	DAP[b]	L (eV)	D (eV)	Name/comment
$^*V_1Co_1{}^+$	HL	1.472	–	–	–	–	*Co 1.47/relatively broad (20 meV); in regions of low nitrogen concentration
$^*V_1Co_2{}^-$	ML (1,500 °C) Co impl. (600 °C)	2.590	–	–	–	–	*Co 2.59/medium fast luminescence (ca. 80 ns)
$^*V_1Co_2{}^\circ$	ML (20 h, 1800 °C)	2.889	–	–	–	–	*Co 2.89/fast lum./appears after annealing at $T > 1,500\,°C$
$^*V_1Co_2{}^+$	MA (24h, 1600 °C)	1.852	1.992–2.187	SB 102a–f (weak)	ZPL	+1.30	*Co 1.85 a/very weak absorption/LVM: $(4x) + 44$ meV
$^*(V_2Co_1)N_{1+x+y}{}^\circ$ =*Co-S5	HL Co impl. ML	–	2.135–2.820	*sta. 104a–i*	1.948	+1.53	***Co-S5**/slow lum. (109 μs)/additional lines from $L' = 1.952$, 1.960, and 1.966 eV (from exc. st. splitting of 4, 12, and 18 meV)/QLVM: −38; LVM: −100 meV
$^*(V_3Co_2)^\circ$	HL (as-grown, anneals out at $T >$ 1,500°)	1.989	1.772–1.914	SB 112a–f (weak)	ZPL	−1.37	*Co 1.99/fast lum. (< 20 ns)/additional lines at 1.984, 1.991 eV (the latter from 2 meV exc. st. spl.)/QLVM: $(2x) - 25$; LVM: −56 meV

[a] N_{a+b+c} indicates the number of nitrogen atoms in the first (a), second (b), or third shell (c)
[b] *Standard DAP104* and sideband DAPs 102 and 112 with L = limiting energy and D = dielectric factor

9.6.4 The *Co 1.85 Center at 1.852 eV from *$V_1Co_2^+$

This center appears after annealing at $T>1{,}500\,°C$. Simultaneously, the *Co 1.99 $(V_3Co_2)°$ center anneals out, probably by trapping two carbon interstitials, which are created during V_1N_3 (N3 center) formation, and start to migrate at this temperature (see Sect. 9.1.6.1).

The 1.852 line is not yet observed in luminescence (1.693 eV luminescence for $V_1Ni_2^+$(a), see Table 8.1.4), but the *Co 2.89 line of *$V_1Co_2°$ is clearly present in the luminescence spectrum with 0 μs delay, together with the V_1N_3 (N3 center) line at 2.985 eV.

9.6.5 The *Co-S5 Center in the Range 2.135–2.820 eV from *$(V_2Co_1)N_{x+y+z}°$

The spectroscopic lines of this center (luminescence only for the cobalt center, but both luminescence and absorption for the nickel *S5 center) arise from the standard DAP104a–h transitions. There is no ZPL, and the relative intensities are sample dependent, with intense lines at **2.135** eV ($s=50$), **2.207** ($s=26$), **2.277** ($s=16$), and **2.367** ($s=10$), and weak lines at 2.244 eV ($s=26$), 2.330 ($s=12$), 2.682 ($s=3$), and 2.820 ($s=1$). The radiative decay time for the slow luminescence has been determined for four lines [Law96]: 63 μs for $s=10$ (donor–acceptor separation $=6.3\ r_0$), 109 μs for $s=16$ ($8.0\ r_0$), 159 μs for $s=26$ ($10.2\ r_0$), and 235 μs for $s=50$ ($14.1\ r_0$). The decay time increases approximately with the square of the distance, and this is a direct confirmation for the DAP interpretation of the *Co-S5 lines.

The dielectric factor of +1.53 eV indicates association with a single nitrogen in the first shell (in analogy to the *$(V_2Ni_1)N_{x+y+z}°$ = *S5 center in the range 1.563–2.320 eV and $D=+1.53$ eV, see Table 8.3b). The weaker sidebands with separation of 4, 12, and 18 meV arise from DAP transitions with a slightly higher limiting energy L': 1.952, 1.960, and 1.966 eV. The temperature dependence indicates a corresponding splitting in the excited state. Like in Sect. 9.6.3, an alternate interpretation is a possible depopulation by energy transfer between slightly inequivalent centers.

9.6.6 The *Co 1.99 Center at 1.989 eV from *$(V_3Co_2)^+$

The *$(V_3Co_2)°$ center is observed in as-grown samples, and anneals out at $T>1{,}500\,°C$. The ZPL is split into three components (1.9835, 1.9886, and 1.9906 eV). The difference (2 meV) between the latter two lines is ascribed to an excited state splitting. However, the Boltzmann analysis yields besides the correct activation energy (2.47 meV) an additive constant of 0.29 to the ratio of the intensities. Therefore, an alternate interpretation is a possible depopulation by energy transfer between slightly inequivalent centers.

The corresponding $^*(V_3Ni_2)^+$ center has three lines at 1.883, 1.905, and 1.913 eV (see Table 8.1.6).

9.6.7 Discussion of Spectral Data from Cobalt Centers in Diamond

Cobalt has only one stable isotope of mass 59, and no frequency shifts from other isotopes can occur. Possibly, a paramagnetic center from cobalt has been observed [Col00].

In comparison with the corresponding nickel centers, the cobalt lines have the following:

Higher ZPL (10%) energies for *Co 1.85 and *Co 1.99 (*Ni 1.69 a and *Ni 1.88 a).
Higher ZPL (20%) energies for *Co 2.59 (*Ni 2.07 a).
Higher L (30%) (limiting energy) for *Co-S5 (L = 1.948) (*Ni-S5, L = 1.460 eV).

In contrast to the nickel centers, the influence of the nitrogen concentration on the cobalt lines is unspecific. In samples with low nitrogen concentration and in samples with 90% Fe dilution (10% Co), most Co lines are absent or very weak.

The LVM and QLVM frequencies are very similar for cobalt and nickel centers (see Tables 11.1 and 11.2).

The observed LVM frequencies are as follows:

(4x) + 44 meV for $^*(V_1Co_2)^\circ$(a) from Co(sp$_3$)–C(sp$_3$) bend.
(2x) – 60 meV for $^*(V_3Co_2)^+$ from Co(sp$_3$)–C(sp$_3$) stretch.
(1x) ca. –100 meV for $^*(V_2Co_1)N_{1+x+y}{}^\circ$ from Co–C/N(sp$_3$) stretch.

QLVM frequencies are observed at

(2x) – 24 meV for $^*(V_3Co_2)^+$ from 2Co + 4C (calculated 24.1 meV).
(1x) – 44 meV for $^*(V_2Co_1)N_{1+x+y}{}^\circ$ from 1Co (calculated 43.6 meV).

In theoretical cluster calculations on the transition metals V, Mn, Fe, Co, Ni, and Cu, their solubility in diamond and their long-range elastic strain have been investigated [Joh02]. Solvation enthalpy calculations for the impurity sites "substitutional," "interstitial (no vacancy)," and "interstitial with divacancy" lead to the conclusion that nickel is the most likely transition metal impurity in diamond, and cobalt and copper following next. The lighter metals V, Mn, and Fe have unfavorable conditions for incorporation into diamond. These preliminary calculations are in good agreement with the experimental findings [Joh02].

9.7 Other Impurities in Diamond

Optical centers from 25 impurities (plus purely intrinsic centers from lattice vacancies or carbon interstitials) are observed in diamond. Only six impurities (H (see Sect. 9.3), B (see Sect. 9.2), N (see Sect. 9.1), Si (see Sect. 9.4), Ni (see Sect. 9.5), and Al (see NA2530 and NL2693a–e)) are found in natural diamond. The remaining 19 impurities are introduced by doping or ion implantation.

The 25 impurities can be divided into 15 main group impurities (Sect. 9.7.1 and Table 9.7.1) and ten transition metal impurities (Sect. 9.7.2 and Table 9.7.2). Six important impurities in diamond were treated in the preceding sections (Sects. 9.1–9.6). For 19 other impurities, optical centers are observed in diamond. Within Tables 9.7.1 and 9.7.2, the impurities are listed according to the main groups 1–8A and the transition metal groups 1–8B.

9.7.1 Main Group Impurities: H, Li, B, Al, In, Tl, Si, N, P, As, Sb, O, He, Ne, and Xe in Diamond

Out of the 36 main group elements, 15 impurities form optical centers in diamond (see Table 9.7.1). From group 1A, the light elements hydrogen (see Sect. 9.3) and lithium (also with bound excitons, see Sect. 9.7.6) are observed. Elements from group 2A are totally missing. From group 3A, the light elements boron (direct neighbor of carbon, see Sect. 9.2) and aluminum, together with the heavy elements indium (doping) and thallium (implanted), are optically active. For group 4A, only the light impurity silicon (see Sect. 9.4) is observed. In group 5A, the important nitrogen (direct neighbor of carbon, see Sect. 9.1), the light element phosphorous, and the medium heavy elements arsenic (doping) and antimony (implanted) give rise to optical centers. Only the light element oxygen (doping) from group 6A is optically active. Finally, the noble gas elements from group 8A, helium, neon, and xenon (all implanted), give rise to numerous lines.

9.7.2 Transition Metal Impurities: Ag, Zn, Ti, Zr, Ta, Cr, W, Fe, Co, and Ni in Diamond

Neglecting the lanthanides (element nos. 58–71) and actinides (nos. 90–103), there are ten elements (out of 30) in groups 1B–8B (see Table 9.7.2) with optical centers in diamond. From group 1B, *silver* (implanted) is observed. From theory [Joh02], copper (group 1B) is expected to be a stable impurity in diamond. In group 2B, *zinc* (implanted) is observed. No elements are found from group 3B. The two elements *titanium* (implanted) and *zirconium* (doping in the melt) are observed from group 4B. From group 5B, the heavy element *tantalum* (from hot filament

Table 9.7.1 Optical centers from main group impurities in diamond

Atom	Main group	Element no. = Z	Radius (nm)[a]	Electron configuration[b]	Observed lines and parameters (eV)[c–e] Comment	Defect
H	1 A	1	(140)	$1s^1$	***11 centers, see Sect. 9.3***, DAP39, 40	
Li	1 A	3	77	He + $2s^1$	(Lum)* 1.407–<u>1.881</u> = SB-DAP68a–g, $D = -1.58$; (Lum) BE: 4,757–5,280	$*V_1Li_1$
B	3 A	5	(86)	He + $2s^22p^1$	***16 centers, see Sect. 9.2***; DAP1, 2, 17–20, 67, 70, 73, 74, 97, 99–101, (Lum) BE: 4,755–5,367	
Al	3 A	13	141	Ne + $3s^23p^1$	(A) 2.530, (Lum)* 2.693–<u>2.974</u> = SB-DAP66a–e, $D = -1.52$, (A,B) 2.990	$*V_1Al_1$
In	3 A	49	144	Kr + $4d^{10}5s^25p^1$	(A) 1.722	$*V_1In_1{}^\circ$?
Tl	3 A	81	199	Xe + $4f^{14}5d^{10}6s^26p^1$	(Lum)* 1.682–<u>2.019</u> = SB-DAP94a–f, $D = -1.29$; implanted + HT	$*V_1Tl_1{}^\circ$
C	4 A	6	76	He + $2s^22p^2$	***9 impurity-free centers (vacancies, interstitials), see Chap. 8.***	
Si	4 A	14	119	Ne + $3s^23p^2$	***5 centers, see Sect. 9.4***, DAP8, 53, 54, 65, 75, 109, 110, and 113	
N	5 A	7	70	He + $2s^22p^3$	***42 centers, see Sect. 9.1***, DAP1, 4, 9, 18–20, 45, 67	
P	5 A	15	(110)	Ne + $3s^23p^3$	(Lum, B) 1.900, (Lum, B) 2.066, (A) 2.637, (Lu,B) 4.450, (Lum)* <u>4.711–5.275</u> = sta.-DAP2a–w, $D = +1.72$; $L = 4.500$; (Lum) BE: 5.020–5.333	$P_1{}^\circ + B_1{}^\circ$,
As	5 A	33	120	Ar + $3d^{10}4s^24p^3$	(A) <u>0.113</u>–0.578 = SB-DAP115a–f, $D = +1.32$, * PC threshold = 0.524 eV	$*V_1As_1{}^\circ$

Sb	5 A	51	134	Kr + $4d^{10}5s^25p^3$	(Lum) 2.705	
O	6 A	8	60	He + $2s^22p^4$	(Lum) 2.676–<u>2.725</u> = sta.-DAP17a–f, $D = +1.29$	$^*O_1{}^\circ + B_1{}^\circ$
He	8 A	2	(90)	$1s^2$	(Lum)*2.049–<u>2.212</u> = SB-DAP95a–f, $D = -1.57$, (Lum) 2.317, (Lum) 2.415	$^*V_1He_1$
Ne	8 A	10	150	He + $2s^22p^6$	(Lum) 1.723, (Lum) 1.731, (Lum) 1.881, (Lum) 1.884, (Lum.) 2.393	$^*(V_2Ne_1)^\circ$?
Xe	8 A	54	233	Kr + $4d^{10}5s^25p^6$	(Lum) *1.235–<u>1.528</u> = SB-DAP42a–j, $D = -1.33$	$^*(V_2Xe_1)^\circ$

[a] [Fin56], Fig. 1, with (interpolated) values

[b] [Fin56], Table 11, complete shells: He = $1s^2$; Ne = He + $2p^6$, Ar = Ne + $3s^23p^6$, Kr = Ar + $3d^{10}4s^24p^6$, Xe = Kr + $4d^{10}5s^25p^6$

[c] Observed: (A) = absorption, (Lum) = luminescence, (B) = broad, BE = bound exciton

[d] DAP: sta. = standard, SB = sideband, D = dielectric factor, L = limiting energy, underlined = ZPL

[e] Quasi-local vibrational modes (QLVM), see Table 8.7

Table 9.7.2 Optical centers from transition metal (TM) impurities in diamond

Atom	TM group	Element no. = Z	Radius (nm)[a]	Electron configuration[b]	Observed lines and parameters[c,d] (eV)	Name/ defect
Ag	1 B	47	144	Kr + $4d^{10}5s^1$	(Lum)2.842–<u>3.111</u> = SB-DAP85a–f, $D = -1.31$, line (b) at 3.117	$^*(V_2Ag_1)^\circ$
Zn	2 B	30	134	Ar + $3d^{10}4s^2$	(Lum) 2.117—2<u>.394</u> = SB-DAP114a–f, $D = -1.42$	$^*(V_2Zn_1)^\circ$
Ti	4 B	22	150	Ar + $3d^2 4s^2$	(Lum) 1.249, (Lum) <u>1.277</u>, (Lum) 2.017, (Lum) 2.507	$^*(V_2Ti_1)^\circ$
Zr	4 B	40	163	Kr + $4d^2 5s^2$	(Lum) <u>1.443–1.565</u> = sta.-DAP5a–j, $D = +1.39$, $L = 1.364$	$^*(V_2Zr_1)^\circ$
Ta	5 B	73	141	Xe + $4f^{14}5d^3 6s^2$	(Lum) <u>1.774–1.817</u> = sta.-DAP7a–e, $D = +1.30$, $L = 1.685$	$^*(V_2Ta_1)^\circ$
Cr	6 B	24	126	Ar + $3d^5 4s^1$	(Lum) 1.672, (Lum, B) 2.360	$^*V_1Cr_1^\circ$
W	6 B	74	138	Xe + $4f^{14}5d^4 6s^2$	(Lum) 1.723–1.759 = sta. (res.)-DAP6a–f, $D = +1.24$, LE = 1.608	$^*(V_2W_1)^\circ$
Fe	8 (B)	26	129	Ar + $3d^6 4s^2$	(Lum) 2.557	

Co	8 (B)	27	128	Ar + $3d^7 4s^2$	***6 centers, see Sect. 9.6***, SB-DAP102a–f, $D = +1.30$, ZPL = 1.852	*Co 1.85 =* $V_1Co_2{}^+$
					sta.-DAP104a–h, $D = +1.53$, $L = 1.948$	*Co-S5=* (V_2Co_1) $N_{1+x+y}{}^\circ$
					SB-DAP112a–f, $D = -1.37$, ZPL = 1.989	*Co 1.99 =* $(V_3Co_2)^\circ$
Ni	8 (B)	28	127	Ar + $3d^8 4s^2$	***18 centers, see Sect. 9.5***, SB-DAP83/84, $D = \pm 1.34$, ZPL = 1.704	*Ni 1.70a =* $V_1Ni_1{}^\circ$
					sta.-DAP60a–g, $D = +1.53$, $L = 1.460$	*Ni-S5=* (V_2Ni_1) $N_{1+x+y}{}^\circ$
					SB-DAP107/108, $D = \pm 1.34$, ZPL = 1.883	*Ni 1.88 =* $(V_3Ni_2)^\circ$

[a] [Fin56], Fig. 1
[b] [Fin56], Table 11, complete shells: Ar = Ne + $3s^2 3p^6$, Kr = Ar + $3d^{10} 4s^2 4p^6$, Xe = Kr + $4d^{10} 5s^2 5p^6$
[c] Observed: (A) = absorption, (Lum) = luminescence, (B) = broad
[d] DAP: sta. = standard, res. = resonance, SB = sideband, D = dielectric factor, L = limiting energy

CVD deposition) is found. In group 6B, the elements *chromium* (implanted) and *tungsten* (from hot filament CVD deposition) are observed. No elements are found from group 7B. The light elements of group 8B are important for the catalytic melt of HPHT synthesis, and all three (with probability Ni > Co > Fe [Joh02]) are incorporated in diamond. Optical lines from *iron*, *cobalt* (see Sect. 9.6), and *nickel* (see Sect. 9.5) are observed. For many of the 18 Ni centers, the HPHT melt is not the only source of nickel, because these centers are also found in natural or ion-implanted diamond.

9.7.3 Effects of Annealing

After ion implantation, annealing at $T > 1{,}000\,°\mathrm{C}$ is usually necessary. The interesting annealing effects of Ni centers are listed in Table 9.5.6.

9.7.4 Local Vibrational Modes

The observed LVM frequencies of H, B, N, Tl, Si, Ag, and Ni are discussed in Sect. 11.1.

9.7.5 Quasi-Local Vibrational Modes

The QLVM frequencies and line widths are discussed in Sect. 11.2. In Table 11.2 observed QLVM data are compared with calculated values. The agreement is very good.

9.7.6 DAP Transitions

For most of the 26 impurities, resolved DAP transitions are observed (see Tables 9.7.1 and 9.7.2). More DAP discussion is given in Sect. 9.7.7.

9.7.7 Bound Excitons from Li, B and P

In diamond, bound exciton lines from Li, B, and P are observed (see Table 9.2.6). In the literature, the boron [Ste97a] and phosphorous [Ste99a] bound excitons have been discussed.

9.7.8 Considerations for Structure Assignments

In columns 4 and 5 of Tables 9.7.1 and 9.7.2, the atomic radius and the electron configuration are given. The impurity X can form an associate V_1X_1 with the single vacancy, or (V_2Y_1) with the double vacancy (Y in an interstitial position). The atomic radius is an important hint for the stable configuration. Interestingly, Co and Ni with $r = 127$–128 nm occur in both configurations. The large Xe ($r = 233$ nm) requires the double vacancy. The configuration assignment is further supported by the QLVM frequency, which includes C_3 for V_1X_1 and C_8 for (V_2Y_1) (see Table 11.2).

The electron configuration is related to the type of DAP transitions (see Chap. 10). If the impurity in the double vacancy has a *partly* filled 3d, 4d, or 5d shell (like for Zr, Ta, W, Co, and Ni), the DAP transitions are of the *standard* type. In the case of a complete 3d or 4d shell (Ag, Zn, and Xe), the DAP transitions are of the *sideband* type. For the V_1X_n configurations ($n = 1, 2, 3, 4$), the DAP transitions are always of the *sideband* type.

Chapter 10
Donor–Acceptor Pair Transitions in Diamond

10.1 Resolved Lines in Absorption and Luminescence

Luminescence lines from donor–acceptor pair (DAP) recombination are known since 1963 [Hop63], and have been treated in detail [Dea73]. An electron is transferred from the ionized acceptor to the ionized donor, and energy is released by photon emission:

$$A^- + D^+ \rightarrow A^\circ + D^\circ + h\nu. \qquad (10.1)$$

The energy of the donor–acceptor transition in shell number s is given in (10.2) [Wil70, Dea73].

$$E(s) = E_g - E_A - E_D + D\ C(s) - E_{corr} + E_{mp}(\text{vector}). \qquad (10.2)$$

Here s is the shell number, E_g is the band-gap energy, E_A and E_D are the binding energies of the acceptor and donor, $D(=e^2/\varepsilon)$ is the dielectric factor (1.64 eV for pure diamond [Dea65a]), $C(s)$ is the Coulomb factor (see (10.3)), E_{corr} is a correction term, which vanishes for $s > 7$ (see near shell correction, below), and E_{mp}(vec) is a small term for *multipole interaction*, depending on the *lattice vector*.

In the limit of infinite donor–acceptor separation, (10.2) is reduced to (10.3) and the "limiting" energy L is introduced:

$$L = E(s = \infty) = E_g - E_A - E_D. \qquad (10.3)$$

With the donor and acceptor on substitutional lattice sites, the pair separation r_{DA} (and consequently the Coulomb factor) can assume only discrete values:

$$C(s) = r_b/r_{DA} = [0.75/(s - x)]^{1/2}. \qquad (10.4)$$

The bond length r_b is 0.1545 nm in diamond, and $x = 0.25$ for *odd* shell numbers while $x = 0$ for *even* shell numbers s [Wil70, Dea73].

DOI 10.1007/978-3-642-22215-3_10,

The alternating values for x, 0.25 (odd s), and 0 (even s) create a characteristic fingerprint for the line spacing in a DAP series. The ratio $E(s)/E(s+1)$ alternates between values that are larger (s odd) or smaller (s even) than the average spacing. This is observed for DAP1 in diamond [Dis94a].

A further fingerprint is the small splitting (due to E_{mp}(vector)), which can be observed for shells with two or more different lattice vectors. Such a splitting (for $s = 7$; $\langle 5, 1, 1\rangle$ and $\langle 3, 3, 3\rangle$) is indeed observed for DAP1 (9 meV [Law93a, Law93b]) and for DAP2 (6 meV [Ste99a, Ste99b]).

A third fingerprint is the occurrence of "empty" shells, where no lattice vector exists (e.g., $s = 28,\ 60$; see Sect. A.2). In diamond, only one example is known ($s = 28$ for DAP2).

Originally, the DAP transitions were observed only in *luminescence* and termed donor acceptor pair recombination (DAPR) [Dea73]. In diamond, the DAP transitions are also observed in *absorption*, which is the reverse process of DAP luminescence:

$$A^{\circ} + D^{\circ} + h\nu \rightarrow A^{-} + D^{+}. \tag{10.5}$$

It is expected that the DAP absorption can be accompanied by photoconductivity. Examples are the photoconductivity spectra observed for $^{*}V_1{}^{+}$-DAP32 (named GR2-8, [Far72]) and $^{*}V_1N_4{}^{-}$-DAP63 (named N9, [Den67].

Calculated values for $C(s)$ in the range 1.0 ($s = 1$) to 0.057 ($s = 228$), together with the lattice vectors, degeneracy, and experimental abundance, are listed in Table A.2.

An overview of the DAP results in diamond is given in Table 10.1. The DAP numbering is in chronological order. The pairs DAP1 [Dis94a, Dis94b] and DAP2 [Ste99a, Ste99b] have already been published. With the new spectra from DAP3 to DAP116, an assignment for over 1,000 spectral lines is presented for the first time [Dis03]. This can be regarded as a breakthrough in diamond spectroscopy.

Note that in Table 10.1, separate DAP numbers are given to sideband DAP spectra of the same defect, if both absorption (E(DAP) > ZPL) and luminescence (E(DAP) < ZPL) are observed (e.g., DAP23/24). In Table 10.1, weak and seldom observed lines are also included for completeness and because of a possible interest to theorists.

Of special interest are those well-known lines, which have a name in the literature, and for which—after many speculations—a convincing interpretation is finally found. An outstanding example is the group of 11 resolved lines GR2 to GR8a–e, which for many years have been a "***basic puzzle***" [Sto92]. Other examples are the TR14–TR18 lines (see DAP35), the H9 to H18 lines (see DAP37, DAP50, and DAP52), the S2a–f lines (see DAP57), and the S3a–e lines (see DAP58).

In general, the proof for the DAP interpretation is given by the close agreement between observed and calculated line positions (see Tables 2.1.2.3–5.3.12.2).

Additional support came from uniaxial stress experiments, where the *defect orientation* is in agreement with the DAP lattice vector (see DAP18, [Moh82b]).

Table 10.1 Resolved lines of donor–acceptor pairs (DAP) in diamond

Name	Label	$L(s=\infty)$ (eV)	$E_D + E_A$ = 5.49–L	D(eV)	Impur./Defect	Type	Shells/comment/References[a]
DAP1	HA,MA/ NL,LL, ML2070a-m	1.893	3.597	+1.43	$^*N_1{}^\circ + B_1{}^\circ$	Sta	1, 2, 3, 4, 5, 6, 7a,b, 8, 10, 13, 16, 34, 50/see Table 9.1.5
DAP2	HL,LL4709a-w	4.500	0.990	+1.72	$P_1{}^\circ + B_1{}^\circ$	Sta	3, 5, 7a,b, 9, 11, 12, 14, 15, 16, 17, 18, 19, 20, 22, 24, 25, 26, 30, 31, 32, 37, 39, 50/see Table 9.2.2
DAP3	NA0386a-i	0.212	L = ZPL	+1.24	$^*(N_1C_1)_i^\circ$	Res	26, 27, 29, 30, 34, 36, 38/see Table 8.2.4
DAP4	MA,ML1678a-i	1.509	3.981	+1.66	$^*(V_2Ni_1)N_x^+$	Sta	22, 42, 50, 52, 54, 58, 62, 64, 72/***S9**
DAP5	HL1443a-j	1.364	4.126	+1.39	$^*(V_2Zr_1)^\circ$	Sta	36, 42, 50, 58, 72, 86, 106, 130, 162, 228/**Zr** from melt/Table 8.1.7.2
DAP6	LL1723a-f	1.608	3.882	+1.24	$^*(V_2W_1)^\circ$	Res	50, 54, 58, 64, 72, 86/**W** from hot filament/see Table 8.1.7.2
DAP7	LL1774a-g	1.667	3.804	+1.23	$^*(V_2Ta_1)^\circ$	Sta	50, 58, 72, 86, 98, 106, 162/**Ta** from hot filament/see Table 8.1.7.2
DAP8	LL1474a-g	1.474	4.106	+1.33	$^*Si_1{}^\circ$+?	Sta	50, 54, 58, 86, 106, 130, 198/see Table 9.4.1
DAP9	NA,LA/ NL1743a-t	1.641	3.849	+1.53	$^*N_1{}^+ + B_1{}^\circ$	Sta	7, 9, 10, 12, 13, 14, 16, 18, 20, 22, 24, 26, 34, 36, 50, 58, 74, 98, 106, 162/see Table 9.1.5
DAP10	ML2012a-l	1.856	3.634	+1.66	?	Sta	1, 2, 3, 4, 11, 14, 16, 20, 22, 50, 82, 86
DAP11	LE,LL1808a-n	1.724	3.766	+1.60	$^*(V_2Ni_1)N_x{}^+$	Sta	1, 2, 3, 4, 5, 6, 9, 14, 16, 22, 34, 64, 162, 228/named ***S7**, Table 9.6.10
DAP12	ML2226a-f	2.140	3.350	+1.55	?	Sta	20, 22, 34, 50, 118, 228
DAP13	ML2179a-l	2.466	3.024	−1.59	$^*(V_1He_1)^\circ$(b)	SB	23, 26, 30, 34, 42, 50, 52, 58, 78, 130, 162, 228/**He** implanted
DAP14	ML2066a-f	2.436	L = ZPL	−1.22	$^*(C_2)_iNi_1{}^\circ$	SB	8, 10, 18, 20, 26, 42/see Table 8.2.3
$V_2{}^\circ$-DAP15	MA2543a-l	2.419	L = ZPL	+1.63	$V_2{}^\circ$	SB	8, 16, 34, 50, 54, 58, 68, 74, 86, 98, 114, 130/ZPL = **TH5**/Table 8.1.1

(continued)

Table 10.1 (continued)

Name	Label	$L(s=\infty)$ (eV)	$E_D + E_A$ = 5.49–L	D(eV)	Impur./Defect	Type	Shells/comment/References[a]
S8-DAP16	HA1212a-l	1.036	4.455	+1.45	$*(V_2Ni_1)^-$	Res	13, 14, 18, 26, 42, 43, 44, 47, 48, 49, 50, 51/named ***S8**/Table 8.1.5
DAP17	LL2676a-g	2.534	2.956	+1.29	$*O_1{}^\circ + B_1{}^\circ$	Res	26, 34, 44, 48, 50, 58, 62/see Table 9.1.5
DAP18	NL2649a-g	2.334	3.156	+1.54	$*N_2{}^{2+} + B_1{}^\circ$	Sta	1, 2, 4, 6, 10, 12, 14, 18/named **Yellow band**/see Table 9.1.5
DAP19	NL2095a-j	1.890	3.600	+1.30	$*N_2{}^\circ + B_1{}^-$	Sta	1, 2, 3, 4, 6, 10, 12, 16, 20, 32/see Table 9.1.5
DAP20	NL1792a-g	1.407	4.083	+1.38	$*N_2{}^+ + B_1{}^\circ$	Sta	2, 4, 5, 6, 7, 8, 10/see Table 9.1.5
DAP21	NE5521a-l	5.411	0.095	+1.32	?	Sta	6, 8, 10, 12, 14, 16, 20, 24, 26, 30, 50, 98/PLE of A bands
(3.4eV)-DAP22	NL2652a-l	3.414	L = ZPL	−1.36	$*V_1N_3{}^+$	SB	1, 2, 4, 5, 8, 10, 12, 18, 22, 34, 50, 98
V_1N_3(c)-DAP23	NA,NE,LA, MA3075a-p	2.985	L = ZPL	±1.33	$V_1N_3{}^\circ$	SB	1, 2, 3, 4, 8, 11, 12, 14, 22, 34, 42, 50, 58, 64, 162/ZPL = **N3(c)**; lines m,o named **N4, N5**;/N3(a,b) see DAP55 and DAP56
V_1N_3(c)-DAP24	NL,HL,LL, ML2440a-q	2.985	L = ZPL	−1.33	$V_1N_3{}^\circ$	SB	4, 8, 9, 12, 14, 22, 42, 50, 54, 58, 62, 64, 74, 86, 106, 130, 162/ZPL = **N3(c)**; lines b–h, o, q named **N3-(D8-D2)** [Dea65]
1Ni(1.40)-DAP25	HA1485a-k	1.403	L = ZPL	+1.33	$*V_1Ni_1{}^+$	SB	2, 4, 6, 8, 10, 14, 22, 50, 64, 130, 198
1Ni(1.40)-DAP26	HL1004a-i	1.403	L = ZPL	−1.33	$*V_1Ni_1{}^+$	SB	8, 12, 18, 26, 50, 54, 86, 130, 198
Ni(2.07)b-DAP27	ME2390a-j	2.298	L = ZPL	+1.34	$*V_1Ni_2{}^-$	SB	18, 20, 22, 26, 30, 34, 44, 58, 106, 162/Ni(2.07)a at 2.071 eV
Ni(2.07)b-DAP28	ML1905a-m	2.298	L = ZPL	−1.34	$*V_1Ni_2{}^-$	SB	9, 11, 12, 14, 16, 18, 20, 26, 30, 34, 50, 98, 162
Ni(1.70)b-DAP29	ME2630a-f	2.401	L = ZPL	+1.34	$*V_1Ni_1{}^\circ$	SB	4, 7, 9, 10, 14, 26/Ni(1.70)a at 1.704 eV, see DAP83
$V_1C_4N_2$-DAP30	MA0331a-d	0.154	L = ZPL	+1.45	$*V_1C_4N_2{}^\circ$	SB	8, 10, 18, 50, named **H1b,d,g/precursor** of **H3** =* $\mathbf{V_1N_2{}^\circ}$

i13-DAP30	*I13-MA0331a,c*	*0.152*	*L = ZPL*	*+1.46*	$*V_1C_4N_2{}^{\circ}$	*SB*	*10, 50*
$V_1N_4V_1$-DAP31	MA0362a-e	0.147	L = ZPL	+1.45	$*V_1N_4V_1{}^{\circ}$	SB	6, 10, 24, 34, 36, named **H1c,e,f/precursor** of **H4** = $*\mathbf{N_3V_2N_1{}^{\circ}}$
i13-DAP31	*i13-MA0362b,d*	*0.141*	*L = ZPL*	*+1.47*	$*V_1N_4V_1{}^{\circ}$	*SB*	*10, 34*
$V_1{}^{+}$-DAP32	MA,ME2881a-m	2.781	L = ZPL	+1.31	$*V_1{}^{+}$	SB	16, 18, 25, 26, 27, 32, 34, 40, 42, 50, 86, 118, 130/very weak ZPL/lines a–k = **GR2–GR8(a–e)** = PLE of **GR1**/see Tables 8.1.1
i13-DAP32	*i13-MA2881a-h*	*2.788*	*L = ZPL*	*+1.31*	$*V_1{}^{+}$	*SB*	*26, 27, 32, 34, 40, 42, 118,130*
$V_1{}^{\circ}$(b)-DAP33	MA1771a-k	1.673	L = ZPL	+1.20	$V_1{}^{\circ}$	SB	12, 16, 40, 44, 48, 50, 54, 58, 62, 78, 118/ZPL = **GR1(b)**
i13-DAP33	*I13-MA1771a-h*	*1.676*	*L = ZPL*	*+1.20*	$V_1{}^{\circ}$	*SB*	*42, 48, 50, 58, 62, 78, 118*
$V_1{}^{\circ}$(b)-DAP34	ML1508a-g	1.673	L = ZPL	−1.20	$V_1{}^{\circ}$	SB	40, 44, 50, 58, 62, 66, 78, 118/L = **GR1(b)**/see Table 8.1.1
$(C_2)_i{}^{+}$(b)-DAP35	MA2759a-n	2.638	L = ZPL	+1.30	$*(C_2)_i{}^{+}$(a)	SB	12, 14, 16, 18, 26, 30, 32, 40, 42, 50, 54, 64, 72, 86/(a) DAP86 ZPL = **TR12** = ⟨100⟩ split self interstitial/see Table 8.2.1
$(C_2)_i{}^{+}$(b)-DAP36	ML2129a-n	2.638	L = ZPL	−1.30	$*(C_2)_i{}^{+}$(a)	SB	5, 12, 14, 16, 18, 24, 26, 30, 40, 42, 54, 64, 72, 86/ZPL = **TR12**
$N_3V_2N_1{}^{\circ}$(b)-DAP37	MA2652a-n	2.498	L = ZPL	+1.60	$N_3V_2N_1{}^{\circ}$	SB	1, 2, 3, 4, 6, 7, 8, 9, 10, 11, 16, 34, 54, 86/ZPL = **H4b/**lines e, g, i, j, l–n = **H6-H12/**H4a at 2.417 eV/see Table 8.1.3.4

(continued)

Table 10.1 (continued)

Name	Label	$L(s=\infty)$ (eV)	$E_D + E_A = 5.49–L$	D(eV)	Impur./Defect	Type	Shells/comment/References[a]
$N_3V_2N_1{}^{\circ}$(b)-DAP38	ML2276a-g	2.498	L = ZPL	−1.60	$N_3V_2N_1{}^{\circ}$	SB	34, 54, 72, 86, 98, 106, 130/ZPL = **H4b**/see Table 8.1.3.4
$(V_4H_4)^{\circ}$-DAP39	LA0165g	0.165	L = ZPL	+1.31	*$(V_4H_4)^{\circ}$	SB	26 (*lines 18–228,* see *i13-DAP39*)/L = **CC** stretch; see Table 11.1
i13-$(V_4H_4)^{\circ}$-DAP39	*i13-LA0165a-j*	*0.159*	*L = ZPL*	*+1.36*	*$(V_4H_4)^{\circ}$*	*SB*	*18, 26, 29, 30, 40, 42, 46, 118, 162, 228/i13 shift: −3.8%*
$(V_4H_4)^{+}$-DAP40	LA0599a-g	0.412	L = ZPL	+1.43	*$(V_4H_4)^{+}$	SB	4, 6, 7, 8, 20, 36, 42/not CC or CH vibration (*see shifts below*)
i2-$(V_4H_4)^{+}$-DAP40	*i2-LA0599a-g*	*0.410*	*L = ZPL*	*+1.41*	*$(V_4H_4)^{+}$*	*SB*	*4, 6, 7, 8, 20, 36, 42/i2 shift: −0.5%; Table 7.1*
i13-$(V_4H_4)^{+}$-DAP40	*i13-LA0599a-g*	*0.411*	*L = ZPL*	*+1.44*	*$(V_4H_4)^{+}$*	*SB*	*4, 6, 7, 8, 20, 36, 42/i13 shift: −0.2%; Table 7.2*
(V_1H_1)b-DAP41	NA0882a-e	0.385	L = ZPL	+1.54	*(V_1H_1)b	SB	3, 4, 5, 6, 7/L = **natural CH** stretch; see Table 9.3.1
(V_2Xe_1)-DAP42	ML1235a-j	1.528	L = ZPL	−1.33	*$(V_2Xe_1)^{\circ}$	SB	16, 17, 22, 23, 48, 50, 64, 78, 130, 198/**Xe** implanted
$V_1N_2{}^{+}$(b)-DAP43	NL,ML2100a-g	2.463	L = ZPL	−1.33	*$V_1N_2{}^{+}$(b)	SB	10, 12, 20, 34, 58, 86, 162/named **S1(b)** center; S1(a) at 2.429 eV (20× weaker)
DAP44	*Deleted*	–	–	–	–	–	–
$V_1C_4N_1$-DAP45	MA1834a-i	1.702	L = ZPL	+1.53	*$V_1C_4N_1{}^{\circ}$	SB	4, 8, 12, 18, 26, 34, 40, 42, 106/precursor of $V_1N_1{}^{\circ}$
$V_1N_1{}^{\circ}$-DAP46	MA2244a-f	2.154	L = ZPL	+1.39	$V_1N_1{}^{\circ}$	SB	10, 16, 24, 30, 50, 162/named **NV** or **575-nm** center
$V_1N_1{}^{\circ}$-DAP47	ML1862a-i	2.154	L = ZPL	−1.39	$V_1N_1{}^{\circ}$	SB	24, 30, 42, 50, 64, 86, 130, 162, 198

i15-$V_1N_1^{\circ}$-DAP47	*ML1989d-i*	*2.151*	*L = ZPL*	*−1.33*	*$V_1N_1^{\circ}$*	*SB*	*50, 64, 86, 130, 162, 198/i15 shift: −0.1%; Table 7.3*
$V_1N_1^-$-DAP48	MA2019a-j	1.943	L = ZPL	+1.24	$V_1N_1^-$	SB	8, 12, 14, 22, 24, 42, 50, 64, 74, 198/named **638-nm** center
$V_1N_1^-$-DAP49	LL,ML1564a-o	1.943	L = ZPL	−1.24	$V_1N_1^-$	SB	8, 10, 12, 14, 16, 20, 24, 30, 42, 50, 64, 74, 98, 162, 198
$V_1N_2^{\circ}$(a)-DAP50	MA2617a-d	2.464	L = ZPL	+1.34	$V_1N_2^{\circ}$	SB	14, 22, 34, 58/L = **H3a**; H3b see DAP52
$V_1N_2^{\circ}$(a)-DAP51	ML2237a-f	2.464	L = ZPL	−1.34	$V_1N_2^{\circ}$	SB	26, 36, 50, 54, 58, 64/L = **H3a**
$V_1N_2^{\circ}$(b)-DAP52	MA3461a-h	3.361	L = ZPL	+1.34	$V_1N_2^{\circ}$	SB	2, 4, 30, 34, 42, 50, 64, 78, 136/L = **H3b** (named **H 13**); lines a–f = **H15a,b, H16a,b, H17,H18a,b**
$(V_3Si_2)^+$-DAP53	LA,LE1785a-m	1.682	L = ZPL	+1.60	*$(V_3Si_2)^+$	SB	4, 6, 8, 10, 12, 42, 50, 64, 72, 86, 106, 118, 162
$(V_3Si_2)^+$-DAP54	LL1487a-g	1.682	L = ZPL	−1.60	*$(V_3Si_2)^+$	SB	50, 72, 78, 82, 86, 118, 162
V_1N_3(a)-DAP55	NL2116a-e	2.297	L = ZPL	−1.33	$V_1N_3(a)^{\circ}$	SB	16, 22, 42, 86, 162/ZPL named **N3B**=higher ground state of N3(c)=DAP24
V_1N_3(b)-DAP56	NL2500a-d	2.680	L = ZPL	−1.33	$V_1N_3(b)^{\circ}$	SB	16, 22, 42, 86/ZPL named **N3A**=higher ground state of N3(c)
S2-DAP57	NA,HA,MA /ME,NE/ NL,HL,ML 2515a-m	2.422	3.068	+1.63	$(V_2Ni_1)\ N_{2+0+1}{}^+$	Sta	1, 30, 32, 38, 44, 48, 50, 64, 72, 98, 162, 228/named **S2**; lines b, e, f, h = **B, C, D, E**/SBs at 32 and 44 meV/**NE2** (EPR)
S3-DAP58	NA,HA,MA /HE,ME/ NL,HL,ML 2477a-j	2.383	3.107	+1.63	$(V_2Ni_1)\ N_{2+0+0}{}^+$	Sta	8, 10, 12, 26, 34, 42, 44, 50, 162, 228/named **S3**; SBs at 30 and 41 meV/**NE1** (EPR)

(continued)

Table 10.1 (continued)

Name	Label	$L(s=\infty)$ (eV)	E_D+ $E_A=5.49$–L	D(eV)	Impur./Defect	Type	Shells/comment/References[a]
S4-DAP59	NA,HA,MA /NE,HE ME 3333a-l	3.191	2.299	+1.53	(V_2Ni_1) $N_{1+2+0}{}^{\circ}$	Sta	4, 6, 8, 10, 12, 14, 18, 22, 26, 34, 50, 86/named ***S4**; SBs at 31 and 48 meV/**NE3** (EPR)
S5-DAP60	NA,HA,MA /ME/NL, HL,LL,ML 1563a-k	1.460	4.030	+1.53	*(V_2Ni_1) $N_{x+y+z}{}^{\circ}$	Sta	1, 2, 4, 6, 7, 8, 30, 50, 58, 86, 162/named ***S5**; SBs at 33 and 44 meV
S6-DAP61	ME 2750a-m	2.637	2.853	+1.63	*(V_2Ni_1) $N_{x+y+z}{}^{+}$	Sta	1, 2, 4, 5, 6, 7, 8, 9, 12, 14, 16, 22, 162/named ***S6**; SB at (3×) 42 meV
DAP62	ML 1425a-e	2.101	L = ZPL	−1.42	?	SB	3, 6, 12, 34, 198
N9(a)-DAP63	NA,NE 5351a-g	5.252	L = ZPL	+1.43	*$V_1N_4{}^{-}$	SB	20, 26, 50, 64, 78, 106, 198/similar DAP from **N9(b)** at 5.262 eV
N9(a)-DAP64	NL 4810a-i	5.252	L = ZPL	−1.43	*$V_1N_4{}^{-}$	SB	8, 18, 20, 26, 34, 58, 64, 78, 198/ZPL = $\mathbf{X_0}$; lines a,b,h,i = $\mathbf{X_1}$–$\mathbf{X_3}$, $\mathbf{Y_1}$
(V_3Si_2)-DAP65	NL 1844a-e	2.052	L = ZPL	−1.39	*$(V_3Si_2)^{\circ}$	SB	34, 64, 78, 130, 228/SB at −45 meV
V_1Al_1-DAP66	NL 2681a-i	2.974	L = ZPL	−1.52	* $V_1Al_1{}^{\circ}$(a)?	SB	22, 30, 34, 42, 50, 86, 106, 162, 228/SB at −38 meV
DAP67	HA 3862a-i	3.684	1.806	+1.43	*$N_2{}^{+} + B_1{}^{\circ}$	Sta	4, 6, 10, 12, 14, 26, 34, 50/correlates exactly with A-center ($N_2{}^{\circ}$)
V_1Li_1-DAP68	ML 1407a-d	1.881	L = ZPL	−1.58	*$V_1Li_1{}^{\circ}$	SB	8, 20, 30, 64/**Li** implanted/SB at −36 meV
$(N_1C_1)_i{}^{-}$-DAP69	MA 4060a-l	3.988	L = ZPL	+1.26	*$(N_1C_1)_i{}^{-}$	SB	4, 5, 6, 7, 8, 10, 12, 22, 32, 50, 86, 228/named **R11**(EPR)
DAP70	ML 1696a-q	1.567	3.923	+1.52	*$N_1{}^{2+} + B_1{}^{\circ}$	Sta	1, 2, 3, 10, 12, 13, 14, 15, 16, 18, 20, 24, 26, 30, 34, 50, 106
$(C_2)_i{}^{\circ}$-DAP71	MA 1825a-c	1.685	L = ZPL	+1.30	*$(C_2)_i{}^{\circ}$	SB	12, 18, 64/⟨100⟩ split self interstitial, named **R2** (EPR)

$^*(C_2)_iN_1^-$-DAP72	HL,LL,ML 2673a-o	3.188	L = ZPL	−1.26	$^*(C_2)_iN_1^-$	SB	1, 2, 3, 4, 6, 8, 10, 12, 22, 26/named **3.188-eV** center
i13-DAP72	*i13-HL 2844a-p*	*3.191*	*L = ZPL*	*−1.22*	$^*(C_2)_i N_I^-$	*SB*	*9, 10, 14, 16, 26, 30, 36, 38,42, 44, 48, 50, 58, 72, 78, 86, 130, 228*
$V_1B_1{}^\circ$(a)-DAP73	NL 1750a-f	2.395	L = ZPL	−1.30	$^*V_1B_1{}^\circ$(a)	SB	2, 4, 8, 12, 34, 50/type IIb, structure on ***A**$_{\mathbf{semic}}$**(b)** band
$V_1B_1{}^\circ$(b)-DAP74	NL 2476a-f	3.120	L = ZPL	−1.30	$^*V_1B_1{}^\circ$(b)	SB	2, 4, 8, 22, 34, 50/type IIb, structure on ***A**$_{\mathbf{semic}}$**(d)** band
$(V_3Si_2)^-$-DAP75	NL,ML2041a-j	2.523	L = ZPL	−1.25	$^*(V_3Si_2)^-$	SB	6, 8, 12, 20 34, 42, 50, 98, 162, 228/SBs at −31, −44 meV
$V_1N_3{}^-$-DAP76	NL,ML2164a-m	2.526	L = ZPL	−1.19	$^*V_1N_3{}^-$	SB	8, 10, 12, 14, 22, 36, 48, 50, 64, 98, 162, 228/SB at −46 meV
$V_1N_2{}^-$-DAP77	HA,MA1333a-j	1.256	L = ZPL	+1.42	$^*V_1N_2{}^-$	SB	1, 3, 6, 8, 10, 20, 30, 64, 86, 106/named **H2**
$V_1N_2{}^-$-DAP78	HL,ML1020a-d	1.256	L = ZPL	−1.42	$^*V_1N_2{}^-$	SB	26, 48, 64, 106/named **H2**
$(C_2)_iN_1{}^\circ$-DAP79	MA4653a-j	4.582	L = ZPL	+1.26	$^*(C_2)_iN_1{}^\circ$	SB	2, 3, 4, 6, 10, 14, 58, 78, 162, 228/named **5RL**/SBs (2×) 202 meV
$(C_2)_iN_1{}^\circ$-DAP80	ML4204a-f	4.582	L = ZPL	−1.26	$^*(C_2)_iN_1^\circ$	SB	8, 10, 14, 58, 78, 118/named **5RL**/SBs (4x) 237 meV
i13-DAP80	*ML4226a-e*	*4.590*	*L = ZPL*	*−1.22*	$^*(C_2)_i N_I{}^\circ$	*SB*	*8, 10, 14, 58, 118/named **5RL/SBs** (4x) 227 meV*
$V_1N_4(C_2)_i$-DAP81	NA,HA4721a-g	4.567	L = ZPL	+1.67	$^*V_1N_4(C_2)_i{}^\circ$	SB	18, 22, 30, 34, 50, 64, 86/named **F**center, see Table 8.1.3.6
$V_1N_4{}^\circ$(a)–DAP82	NA,MA4307a-e	4.191	L = ZPL	+1.56	$\mathbf{V_1N_4}{}^\circ$	SB	30, 34, 42, 72, 130/named **B**-center, see Table 8.1.3.5
$V_1Ni_1^\circ$(a)-DAP83	HA, MA1816a-f	1.704	L = ZPL	+1.34	$^*V_1Ni_1{}^\circ$	SB	10, 14, 18, 26, 50, 106; line (b) at 2.401 = DAP29
$V_1Ni_1{}^\circ$ (a)-DAP84	HL, ML1568a-c	1.704	L = ZPL	−1.34	$^*V_1Ni_1{}^\circ$	SB	72, 198, 228

(continued)

Table 10.1 (continued)

Name	Label	$L(s=\infty)$ (eV)	$E_D + E_A$ = 5.49–L	D(eV)	Impur./Defect	Type	Shells/comment/References[a]
$(V_2Ag_1)^{\circ}$(a)-DAP85	ML2842a-f	3.111	L = ZPL	−1.31	$^*(V_2Ag_1)^{\circ}$(a)	SB	18, 32, 42, 54, 72, 198/SB (2×) 22 meV/line (b) at 3.117 eV
$(C_2)_i^-$-DAP86	ML2138a-i	2.462	L = ZPL	−1.31	$^*(C_2)_i$–-DAP86	SB	12, 18, 26, 30, 34, 48, 58, 86, 118/ = **3H**(Table 8.2.1)
$V_1Ni_2^-$(a)-DAP87	ML1709a-n	2.071	L = ZPL	−1.34	$^*V_1Ni_2^-$(a)	SB	10, 11, 12, 14, 16, 20, 22, 26, 30, 34, 38, 50, 72, 106/see Table 8.1.8
$(C_2)_iN_2^-$-DAP88	MA2442a-f	2.367	L = ZPL	+1.25	$^*(C_2)_iN_2^-$	SB	22, 50, 64, 86, 130, 198/see Table 8.2.3
$(C_2)_iN_2^-$-DAP89	ML2075a-h	2.367	L = ZPL	−1.25	$^*(C_2)_iN_2^-$	SB	14, 18, 26, 32, 54, 106, 130, 198/see Table 8.2.3
$(C_2)_iN_2^+$-DAP90	MA2676a-c	2.535	L = ZPL	+1.25	$^*(C_2)_iN_2^+$	SB	26, 42, 58/see Table 8.2.3
$V_1N_3^{\circ}$-DAP91	NA,MA2460a-g	2.370	L = ZPL	+1.34	$^*V_1Ni_3^{\circ}$	SB	18, 20, 24, 26, 118, 130, 162/**A line**/see Table 8.1.8
$V_1N_3^{\circ}$-DAP92	NL,ML2170a-f	2.370	L = ZPL	−1.34	$^*V_1Ni_3^{\circ}$	SB	34, 50, 58, 72, 106, 162/**A line**/see Table 8.1.8
V_1^--DAP93	MA3290a-f	3.150	L = ZPL	+1.10	$^*V_1^-$	SB	1, 2, 3, 4, 6, 48/named **R10 = ND1**/see Table 8.1.1
$(V_2Tl_1)^{\circ}$-DAP94	ML1680a-h	2.019	L = ZPL	−1.29	$^*(V_2Tl_1)^{\circ}$	SB	11, 12, 16, 18, 24, 48, 54, 64/**Tl** implanted/see Table 8.1.12
$V_1He_1^{\circ}$-DAP95	ML2049a-f	2.212	L = ZPL	−1.57	$^*V_1He_1^{\circ}$(a)	SB	72, 86, 98, 162, 198, 228/see Table 8.1.11
$(C_{2n})_i$-DAP96	NL0995a-f	1.526	L = ZPL	−1.30	$^*(C_{2n})_i^{\circ}$	SB	4, 8, 10, 12, 16, 18, 64/**large platelets**/see Tables 8.2.1, 8.2.2
$(C_2)_iB_1^{\circ}$-DAP97	ML4417a-f	4.777	L = ZPL	−1.25	$^*(C_2)_iB_1^{\circ}$	SB	3, 4, 9, 16, 50, 98/**2BD(F)**in irradiated type IIb/see Table 8.2.3
$(N_1C_1)_i^+$-DAP98	LL,ML2608a-i	2.807	L = ZPL	−1.25	$^*(N_1C_1)_i^+$	SB	30, 32, 38, 40, 42, 86, 130, 162, 228/see Table 8.2.4

i13-DAP98	i13-ML2618a-e	2.810	L = ZPL	−1.21	*$(N_1C_1)_i^+$	SB	30, 32, 38, 40, 42
$(N_1C_1)_iB_1{}^\circ$-DAP99	LL2417a-h	2.792	L = ZPL	−1.25	*$(N_1C_1)_iB_1{}^\circ$	SB	8, 12, 16, 26, 34, 72, 106, 130/see Table 8.2.4
$(N_1C_1)_iB_1{}^+$-DAP100	LL2715a-i	3.092	L = ZPL	−1.27	*$(N_1C_1)_iB_1{}^+$	SB	8, 10, 16, 18, 34, 44, 50, 72, 106/see Table 8.2.4
$(N_1C_1)_iB_1^-$-DAP101	LL2330a-d	2.992	L = ZPL	−1.22	*$(N_1C_1)_iB_1^-$	SB	1, 4, 7, 12/see Table 8.2.4
$V_1Co_2{}^\circ$-DAP102	MA1992a-e	1.852	L = ZPL	+1.34	*$V_1Co_2{}^\circ$	SB	12, 16, 20, 30, 42, 64/see Table 9.6
DAP103	Deleted	–	–	–	–	–	–
$(V_2Co_1)N_1{}^\circ$-DAP104	ML2135a-i	1.948	3.542	+1.53	*$(V_2Co_1)N_{1+x}{}^\circ$	Sta	1, 3, 4, 10, 12, 16, 20, 26, 50/= ***Co-S5**, *see Table* 9.6
$(V_3Ni_2)^+$-DAP105	HA,HE 2780a-d	2.588	L = ZPL	+1.55	*$(V_3Ni_2)^+$	SB	4, 8, 12, 50/see Tables 7.5, 8.1.6, 9.5.1.1
$(V_3Ni_2)^+$-DAP106	HL2383a-f	2.588	L = ZPL	−1.55	*$(V_3Ni_2)^+$	SB	42, 50, 54, 64, 86, 162/see Tables 8.3c, 9.5.1, 9.5.2
$(V_3Ni_2)\,^\circ$-DAP107	HA1958a-g	1.883	L = ZPL	+1.36	*$(V_3Ni_2)\,^\circ$	SB	8, 14, 34, 64, 78, 130, 228/see Table 8.1.6
$(V_3Ni_2)\,^\circ$-DAP108	HL1779a,b	1.883	L = ZPL	−1.36	*$(V_3Ni_2)\,^\circ$	SB	130, 228/see Table 8.1.6
$V_1Si_1{}^\circ$-DAP109	LE2785a-h	2.646	L = ZPL	+1.49	*$V_1Si_1{}^\circ$	SB	10, 14, 22, 30, 42, 50, 72, 86/see Table 9.4.2
$V_1Si_1{}^\circ$-DAP110	LL2241a-g	2.646	L = ZPL	−1.49	*$V_1Si_1{}^\circ$	SB	10, 12, 14, 22, 42, 72, 86/see Table 9.4.2
$V_1Ni_2^+$-DAP111	HL,ML1708a-f	1.991	L = ZPL	−1.34	*$V_1Ni_2^+$(c)	SB	16, 18, 26, 34, 50, 162/lines (a), (b) at 1.693 and 1.940 eV/see Table 8.1.4, 9.5.1.1
$(V_3Co_2)\,^\circ$-DAP112	HL1772a-f	1.989	L = ZPL	−1.37	*$(V_3Co_2)\,^\circ$	SB	30, 54, 86, 98, 130, 228/see Tables 8.1.6, 9.6

(continued)

Table 10.1 (continued)

Name	Label	$L(s=\infty)$ (eV)	$E_D + E_A$ = 5.49–L	D(eV)	Impur./Defect	Type	Shells/comment/References[a]
$V_1Si_1^-$-DAP113	LL2246a-m	2.991	L = ZPL	−1.33	*$V_1Si_1^-$	SB	1, 2, 4, 5, 6, 7, 8, 16, 24, 42, 64, 72, 86/see Table 9.4.2
$V_1Zn_1{}^\circ$-DAP114	ML2117a-f	2.394	L = ZPL	−1.42	*$V_1Zn_1{}^\circ$	SB	20, 36, 42, 58, 86,130/Zn implanted/see Table 8.1.7.1
$V_1As_1{}^\circ$-DAP115	HA0278a-f	0.080	L = ZPL	+1.32	*$V_1As_1{}^\circ$	SB	6, 8, 12, 22, 42, 50/As doped/see Table 8.1.7.1
DAP116	NL,ML1606a-k	1.871	3.619	−1.43	*$N_1{}^\circ$+B_1^-	Sta	16, 20, 22, 24, 26, 32, 34, 54, 58, 198, 228

L = limiting energy; E_D, E_A = ionization energy of donor (E_D) and acceptor (E_A); D = dielectric factor of Coulomb energy

Types: *Sta*=Standard, *Res*=Resonance, *SB*=Sideband

[a]References for donor–acceptor pair analysis: DAP1 [Dis94a, Dis94b], DAP2[Ste99a, b], DAP3–DAP116 [Dis03]

Also characteristic for DAP transitions is the observation of *spectral hole burning*. The latter can be readily understood by slightly different energies for specific pairs in a perturbed lattice (see DAP4, [Sil94a] and DAP45, [Sil95]).

An elegant proof for DAP transitions is given by four measured *radiative decay times* τ_{DA} with two examples: 1) Value of τ_{DA} = (63–235 μs) for the shells nos. 10, 16, 26, 50 of the *Co-S5 center (DAP104, [Law96]). The decay times are approximately proportional to the inverse square of the donor acceptor distance (point change model) (see Sect. 9.6.5). 2) For the shell nos. 1, 4, 8, 16 of the DAPI center, very short τ_{DA} values are observed (0.02–4.0 ns) [Sch95]. Here, the dependence on the pair separation d_{DA} is different and the τ_{DA}/d_{DA}^3 value is quasi constant (large Bohr radius model). For details see the discussion of DAP1 below.

Important new findings for DAPs are presented here. In diamond, the DAPs can be divided into three types: The ***"standard"*** type (old), the ***"resonance"*** type (new), and the ***"sideband"*** type (new).

Standard DAP (with 19 entries in Table 10.1). These pairs resemble the DAPs in other host crystals [Dea73] like GaP, GaAs, GaN, InP, Si, c-SiC, CdS, and ZnS. A typical standard DAP is observed in DAP2a-w.

Resonance DAP (with eight entries in Table 10.1). In some cases, the DAP lines are constraint to a narrow region (with a cutoff, see DAP3) or grouped around one or more non-DAP lines (see DAP4). This suggests that some sort of energy transfer occurs. A very good example for the resonance type is given by DAP3a-i with nine shells in the range of $s = 46$ (0.386 eV) to $s = 32$ (0.421 eV). The resonance line at 0.385 eV (natural CH stretch vibration, see Table 9.2) is also observed, even in the absence of the DAP3 lines, i.e., the intensity of the resonance line is not directly correlated with that of the DAP3 lines. It must be admitted that the resonance DAP can be *basically* of the standard type (positive Coulomb energy in luminescence, e.g., DAP4-6) or the sideband type (possibly DAP3).

Sideband DAP (85 entries in Table 10.1). Note that in the notation for sideband DAPs, the defect name or the ZPL energy precedes the DAP number. It is well known that local vibrational modes (LVM) or quasi local vibrational modes (QLVM) can couple to an electronic transition, producing vibronic sideband lines.

For the first time, a very similar coupling is reported here for donor–acceptor transitions. Completely unexpected, DAP lines with "negative" Coulomb energies (relative to the ZPL) are observed in luminescence. In absorption (or photoluminescence excitation), DAP lines with exactly the same "positive" Coulomb energies (relative to the ZPL) are observed.

This mirror image is much more perfect than those from phonon sidebands, where the frequencies and intensities are usually different for luminescence and absorption. Typical examples for perfect mirror images are **N3C** = V_1N_3(c)-DAP23/24 (ZPL at 2.985 eV) and N9a-DAP63/64 (ZPL at 5.252 eV).

It must be emphasized that for standard and resonance DAPs, the Coulomb energy is always positive, both in luminescence and in absorption (or photo luminescence excitation), i.e., the line positions are identical in luminescence and absorption (see, e.g., DAP1).

For some sideband DAPs, only one side of the mirror image is observed, because the corresponding zero phonon line occurs exclusively in absorption (e.g., to a higher lying excited state) or exclusively in luminescence (e.g., to a higher lying ground state).

It is quite remarkable that the majority of DAP spectra in Table 10.1 are of the new "sideband" type. Most of these defects contain a vacancy or a self-interstitial, suggesting an involvement of electron dangling bonds.

In most cases, the DAP-zero phonon line (ZPL) is an electronic transition. However, in a few very special cases, coupling to a vibrational line as a quasi-ZPL occurs, with typical LVM frequencies of 154 meV (C–N stretch, DAP30), 147 meV (C–N stretch, DAP31), and 385 meV (natural C–H stretch, DAP41).

In the course of the DAP analysis several remarkable aspects have been found: ***Nearest neighbor pairs, near shell corrections, high intensity shells***, and ***vibrational sidebands*** on DAP lines. These four important items are discussed below.

Nearest neighbor pairs. Reports on transitions from the nearest neighbor pair (first shell) are rare. In the literature, only two examples of nearest neighbor pairs are mentioned: GaP:O,Zn and GaP:O,Cd [Dea73]. Coulomb attraction or direct chemical bonding considerations during growth or annealing can stabilize such associates, and for the intensity of the first shell transitions exceptionally high values have been found [Dea73]. In diamond, the luminescence intensity of the first shell in DAP1 is up to 30 times higher than the expected statistical value [Dis94a, Dis94b].

Near shell corrections. It has been previously observed [Dis94a, Dis94b] that for near shells (s = 1–6), the observed Coulomb energy is less than expected from theory. Therefore, the ideal point charge model for the Coulomb energy has to be corrected. A useful set of empirical corrections has been derived by averaging all the observed deviations (for $s = 1$–6) vs. the theoretical values. Listed in Table A.2 are the following averaged results: −44% ($s = 1$), −18% ($s = 2$), −10% ($s = 3$), −5% ($s = 4$), −3% ($s = 5$), −0.3% ($s = 6$).

High intensity shells. For diamond the observed intensities within a group of DAP lines deviate significantly from the expected theoretical values, which are given by the degeneracy of the shells (see Table A.2). On the other hand, spectra with only a limited number of lines are dominated by the lines from the high intensity shells. This observation has been very helpful for the shell assignment in the analysis of many DAPs. The phenomenon of high intensity shells has been noticed before [Dis94a, Dis94b]. The observed relative intensities and the statistical results in Tables 10.1 and A.2 (last row) allow the confirmation and extension of these previous findings. A preliminary list of 18 high intensity shells is given by $s = 4$, 8, 12, 16, 22, 26, 34, 42, 50, 64, 86, 98, 106, 118, 130, 162, 198, and 228 (shown in bold type in Table A.2).

It is remarkable that for a different host (Si) with different growth conditions, a very similar result is observed in the photo luminescence excitation for the DAP system Si:P,In [Sch83]. In the spectrum from over 50 shells, there are 14 lines with higher intensity than expected from the statistical degeneracy. These are the shells $s = 4$, 8, 12, 16, 22, 26, 34, 42, 50, 58, 86, 98, 106, and 118.

No straightforward explanation for the phenomenon of high intensity shells can be given. This unusual behavior was not mentioned in the literature before, except in [Dis94a,Dis94b]. Hopefully, theoretical analysis will find a solution. One possibility is site preference during crystal growth or annealing. Also considered should be an orientation dependent wave function overlap of donor and acceptor with relatively small Bohr radius (relatively large ionization energy). The following facts may assist a future analysis:

(i) All high intensity shells have an even number.
(ii) The lattice vectors of 14 high intensity shells contain at least two equal numbers.
(iii) The lattice vectors of ten high intensity shells contain one or more times the number zero. The facts (i) and (iii) may be related, because the lattice vectors of odd shell numbers never contain a zero.
(iv) The only exception to the rules (ii) and (iii) is the shell 42 $\langle 10, 8, 2\rangle$.

Vibrational sidebands on DAP lines. It was mentioned above that in sideband DAPs, the coulomb energy couples to a ZPL similar to a vibration. In rare cases, vibrational sidebands on a DAP transition are observed, i.e., the DAP transition appears as a pseudo ZPL. Examples are DAP18 (53 meV), Ni1.4-DAP26, Ni2.1-DAP28, Ni1.7-Dap29, N_2(V)-DAP30 (64 meV), V_1N_4(V)-DAP31 (64 meV), Ni2.4-DAP43, and DAP57–DAP62. A sideband spectrum with eight lines (75–281 meV) is observed for DAP1 ($s = 1$) in the photoluminescence of a particular sample [Dis94a].

Discussion of Individual DAP Centers. In the discussion of the 116 individual DAP spectra some additional information will be given.

DAP1 (standard, 13 absorption and luminescence lines) $=^*(\mathbf{N_1^\circ} + \mathbf{B_1}^\circ)$. The very beginning of the DAP analysis in diamond was marked by the assignment of a luminescence spectrum to resolved DAP transitions [Dis94a, Dis94b]. Meanwhile, the DAP1 lines are identified in a great variety of samples, both in absorption and luminescence. These samples include natural [Dea64c], high-pressure synthetic [Law93a], low-pressure synthetic [Col93c, Kho93, Dis94a, Dis94b, Sch95], heat treated [Col85b], and implanted [Zai00a] diamonds. Even in electro-luminescence from a p-i-n diode, DAP1 lines are seen [Zai98d]. Therefore, it is likely that two frequently occurring impurities are involved, i.e., *nitrogen* and *boron* (nickel can be excluded by SIMS analysis [Dis94a, Dis94b, and unpublished]). When $E_A = 0.370$ eV (for the boron acceptor), then $E_D = 3.227$ eV (see Sect. 10.3). In an experiment with time resolved photoluminescence, the intrinsic decay times τ_{DA} for the shells # 1, 4, 8, 16 have been determined [Sch95]: #1(<0.02 ns), #4(0.25 ± 0.05 ns), #8(0.75 ± 0.01 ns), #16(4.0 ± 3.0 ns). For these values, the express in τ_{DA}/d_{DA}^3 is quasi constant: <5.4 (#1), 5.6 ± 1.1 (#4), 5.9 ± 0.01 (#8), and 1.1 ± 8.0 (#16).

DAP2 (standard, 23 luminescence lines) $=^*(\mathbf{P_1}^\circ + \mathbf{B_1}^\circ)$. The DAP2 spectrum is outstanding in several respects: (a) The largest number of DAP lines (23) is seen in the spectrum. (b) For both donor and acceptor, the chemical identity and the binding energy have been determined [Ste99b]. These are *phosphorus* ($E_D = 0.62$ eV) and *boron* ($E_A = 0.37$ eV). (c) Both donor and acceptor are relatively shallow, i.e., the Bohr radius is relatively large. This is in contrast to all other *standard* DAPs in Table 10.1, where at least one partner of the pair is deep, which follows from the fact that the sum $E_D + E_A$ is between 2.299 eV (DAP59) and 4.162 eV (DAP5). (d) The shell assignment is confirmed by the observation of the gap at the empty shell 28, and by the 6-meV splitting from multipole interaction of the lines from shell 7, due to the two nonequivalent lattice vectors 7a $\langle 3, 3, 3\rangle$ and 7b $\langle 5, 1, 1\rangle$ (see Table A.2).

DAP3 (resonance, nine absorption lines) $=^*(\mathbf{N_1C_1})_i^\circ$. For DAP3 the resonance effect is evident. The resonance line (0.3852 eV) is the well-known C–H stretch vibration of the $^*(V_1H_1)$ center in natural diamond (see Sect. 9.3). It is at the low energy cut-off limit of the spectrum (shell 38). The limiting energy ($L = 0.212$ eV) indicates a vibrational center, similar to DAP30 (0.154 eV) and DAP31 (0.147 eV).

*S9 - DAP4 $=^*$ $(\mathbf{V_2Ni_1})\mathbf{N}_{x+y+z}{}^+$ (resonance (standard), 11 absorption and luminescence lines). By comparison of the absorption and luminescence spectra of DAP4, it appears that there are two resonance lines around 1.633 eV (weak) and 1.700 eV (strong).

DAP5 $=^*$ $(\mathbf{V_2Zr_1})^\circ$ (standard, 10 luminescence lines). The DAPs 5, 6, and 7 from heavy transition metal impurities are very similar. Zirconium (DAP5) is incorporated from the melt. Tungsten (DAP6) and tantalum (DAP7) are incorporated from the hot filament during CVD growth. Typical QLVM sidebands of the intense resonance lines are observed for tungsten (24 meV) and for tantalum (25 meV).

DAP6 (resonance (standard), six luminescence lines) $=^*$ $(\mathbf{V_2W_1})^\circ$. DAP6 and DAP7 are observed in hot filament CVD diamond, where the filament metal (tungsten) is incorporated like a doping impurity.

DAP7 (standard, seven luminescence lines), $=^*$ $(\mathbf{V_2Ta_1})^\circ$. Tantalum from hot filament; see also DAP6.

DAP8 (standard, seven luminescence lines) $=^*$ $(\mathbf{Si_1}^\circ + ?)$. The characteristic structure on top of a broad band is identified as a DAP spectrum. Some additional lines are too weak for a positive identification (Table 9.4.1)

DAP9 (standard, 20 absorption and luminescence lines) $=^*$ $(\mathbf{N_1}^+ + \mathbf{B_1}^\circ)$. The structure on top of the **B-band** in boron rich natural type IIb diamond is a typical standard DAP spectrum with identical line positions in absorption and luminescence.

DAP10 (standard, 12 luminescence lines). This spectrum is probably due to an unknown intrinsic defect, because the sample is a natural type IIa diamond, implanted with carbon.

*S7-DAP11 $=^*$ $(\mathbf{V_2Ni_1})\mathbf{N}_{x+y+z}{}^+$ (standard, nine PLE and luminescence lines). The same line positions are observed in luminescence and in photoluminescence excitation. The low energy part of the DAP11 luminescence spectrum is excited at the DAP line positions in the high energy part of the spectrum. The PL spectrum of DAP62 is of special interest, because the isotope shift of ^{60}Ni (9–13 meV) and ^{62}Ni (18–26 meV) is resolved for five lines, indicating a single nickel atom in the defect (see Table 7.5). See also DAP57.

Table 10.2.1 Broad bands from unresolved lines of donor–acceptor pairs in diamond

Name of band	Label	Peak (eV) (shell)	W (meV)	$L(s=\infty)$ (eV)	$E(s=1)$ (eV)	D(eV)	Impur./defect	Type	Comment
DAP1	NB,LB2450	2.45 ($s=4$)	320	1.893	2.66	+1.43	$^*N_1^\circ + B_1^\circ$	Sta	**A, CL, EL, PL**
DAP9	NB,LB1800	1.80 ($s=74$)	410	1.641	2.47	+1.53	$^*N_1^+ + B_1^\circ$	Sta	**A, CL;** named **B-band**
DAP18	NB1910	2.91 ($s=6$)	460	2.334	3.20	+1.54	$^*N_2^+ + B_1^\circ$	Sta	**CL**
DAP19	NB2350	2.35 ($s=6$)	360	1.890	2.62	+1.30	$^*N_2^+ + B_1^-$	Sta	**CL;** named **Yellow band**
DAP21	NB5580	5.57 ($s=46$)	340	5.411	6.15	+1.32	?	Sta	**PLE** of A bands
V_1N_3(c)-DAP23	NB,LB,MB3210	3.21 ($s=28$)	400	2.985	3.76	+1.33	$V_1N_3^\circ$	SB	**A;** ZPL = **N3(c)**
Ni(1.4)-DAP25	HB1560	1.56 ($s=50$)	410	1.403	2.12	+1.33	Ni_1	SB	**A**
Ni(1.4)-DAP26	HB1240	1.24 ($s=50$)	280	1.403	0.66	−1.33	Ni_1	SB	**CL**
Ni(1.7)b-DAP29	MB2630	2.63 ($s=26$)	210	2.401	3.15	+1.34	Ni	SB	**PLE** of Ni(1.7)a at 1.704 eV
V_1^+(c)-DAP32	MB2930	3.10 ($s=16$)	300	2.781	3.52	+1.31	V_1^+	SB	**A, PLE** of V_1°(c) = **GR1(c)** at 2.781 eV
$V_1N_1^\circ$-DAP46	MB2480	2.48 ($s=16$)	370	2.154	2.93	+1.39	$V_1N_1^\circ$	SB	**A, PLE** of $V_1N_1^\circ$ = **NV** at 2.154 eV
$V_1N_1^\circ$-DAP47	MB2010	2.01 ($s=72$)	240	2.154	1.38	−1.39	$V_1N_1^\circ$	SB	**CL, PL;** named NV or 575-nm center
$V_1N_1^-$-DAP48	MB2210	2.21 ($s=16$)	440	1.943	2.64	+1.24	$V_1N_1^-$	SB	**A;** named **638-nm** center
$V_1N_1^-$-DAP49	LB,MB1750	1.75 ($s=34$)	300	1.943	1.25	−1.24	$V_1N_1^-$	SB	**PL**
DAP57	NB,HB,MB2790	2.79 ($s=14$)	390	2.422	3.34	+1.63	$(V_2Ni_1)N_3^+$	Sta	**PLE** of 2.537-eV line, named **S2**
DAP58	NB,HB,MB2850	2.85 ($s=10$)	450	2.383	3.30	+1.63	$(V_2Ni_1)N_2^+$	Sta	**PLE** of 2.496-eV line, named **S3**
DAP59	NB,HB,MB3540	3.54 ($s=14$)	320	3.191	4.05	+1.52	$^*(V_2Ni_1)N_3^\circ$	Sta	**PLE** of 2.370-eV line, named *__S4__

(continued)

Table 10.2.1 (continued)

Name of band	Label	Peak (eV) (shell)	W (meV)	$L(s=\infty)$ (eV)	$E(s=1)$ (eV)	D(eV)	Impur./defect	Type	Comment
DAP60	NB,HB,MB1810	1.81 ($s=14$)	400	1.460	2.32	+1.53	$^*(V_2Ni_1)N_x^{\circ}$	Sta	**PLE** of 1.563-eV line, named ***S5**
DAP61	MB3000	3.00 ($s=16$)	270	2.637	3.53	+1.63	$^*(V_2Ni_1)N_x^{+}$	Sta	**PLE** of 1.648-eV line, named ***S6**-band
DAP11	LB,MB2190	2.19 ($s=8$)	360	1.724	2.62	+1.60	$^*(V_2Ni_1)^{+}$	Sta	**PL, PLE** of ***S7** lines and of 1.648-eV line
(2.1 eV)-DAP65	NB1920	1.92 ($s=82$)	190	2.052	1.27	−1.39	Si_2?	SB	**PL**
3.0 eV(b)-DAP66	NB2740	2.74 ($s=32$)	300	2.974	2.12	−1.52	Al?	SB	**PL**; ZPL = 2.974-eV center
Li(1.9)-DAP68	MB1610	1.61 ($s=18$)	260	1.881	1.12	−1.35	Li	SB	**CL**; low N natural diamond, **Li** implanted
$(C_2)_i^{\circ}$-DAP71	MB1970	1.97 ($s=16$)	320	1.685	2.41	+1.30	$(C_2)_i^{\circ}$	SB	**A**; ⟨*100*⟩ split self-interstitial, named**R2** (EPR)
(3.2eV)-DAP72	HB,LB,MB2880	2.88 ($s=12$)	320	3.188	2.48	−1.27	$(NC)_i^{\circ}$	SB	**CL, PL**; named **3.188-eV** center
$^*(V_3Si_2)^{-}$-DAP75	NB/MB2310	2.31 ($s=30$)	350	2.52	1.83	−1.25	$^*(V_3Si_2)^{\circ}$	SB	**NL, ML**
$^*V_1N_3^{-}$-DAP76	NB2360	2.36(s = 40)	340	2.526	1.86	−1.19	$^*V_1N_3^{-}$	SB	**NL**
H2-DAP77	HB/MB1500	1.50($s=26$)	360	1.256	2.01	+1.35	$^*V_1N_2^{-}$	SB	**HA, MA**
H2-DAP78	HB/MB1070	1.07 ($s=48$)	250	1.256	0.50	−1.35	$^*V_1N_2^{-}$	SB	**HL, ML**
5RL-DAP79	MB4800	4.80 ($s=40$)	400	4.582	5.288	+1.26	$^*(C_2)_iN_1^{\circ}$	SB	**MA**
5RL-DAP80	MB4400	4.40 ($s=34$)	260	4.582	3.876	−1.26	$^*(C_2)_i\ N_1^{\circ}$	SB	**ML**
F(c)-DAP81	NB/HB5000	5.00 ($s=10$)	600	4.567	5.502	+1.67	$^*V_1N_4\ (C_2)_i^{\circ}$	SB	**NA, HA**
B(b)-DAP82	NB/MB4500	4.50 ($s=18$)	540	4.191	5.065	+1.56	$^*V_1N_4^{\circ}$(b)	SB	**NA, MA**
***V$_1$Ni$_1$**$^{\circ}$-DAP83	HB/MB1820	1.82 ($s=106$)	160	1.704	2.454	+1.34	$^*V_1Ni_1^{\circ}$	SB	**HA, MA**

*$(V_2Ag_1)^\circ$-DAP85	MB	2.84 ($s = 18$)	170	3.111	2.477	−1.31	*$(V_2Ag_1)^\circ$	SB	**ML**
*$(C_2)_iN_2^-$-DAP89	MB	2.20 ($s = 42$)	270	2.367	1.667	−1.25	*$(C_2)_iN_2^-$	SB	**ML**
*$V_1Tl_1^\circ$-DAP94	MB	1.65 ($s = 8$)	2.019	–1.29	*$V_1Tl_1^\circ$	SB	**ML**		
*$(C_{2n})_i^\circ$-DAP96	NB/MB	1.25 ($s = 16$)	300	1.526	0.798	−1.30	*$(C_{2n})_i^\circ$	SB	**NL, ML;** large platelets = D center (Table 8.2.1)
*$\mathbf{Co\text{-}S5}$-DAP104	MB	2.20 ($s = 28$)	1.948	2.805	+1.53	*$(V_2Co_1)N_{1+x}$	sta	**ML**	
*1Si2.65-DAP109	LB	2.86 ($s = 42$)	2.646	+1.49	*$V_1Si_1^\circ$	SB	**LE**		
*1Si2.65-DAP110	LB	2.45 ($s = 34$)	2.646	−1.49	*$V_1Si_1^\circ$	SB	**LL**		

W = FWHM; L = limiting energy; D = dielectric factor of Coulomb energy;
Types: *Sta* = Standard, *SB* = Sideband (L = ZPL)

DAP12 (standard, six luminescence lines). These DAP lines are observed in low nitrogen diamond after carbon implantation or electron irradiation with subsequent annealing above 500 °C. The lines appear and disappear simultaneously with the intrinsic defects GR1 (= $V_1°$) and TR12 (= $(C_2)_i{}^+$). Therefore an involvement of an unknown intrinsic defect in DAP12 is probable.

DAP13 (sideband, 15 luminescence lines) =* $\mathbf{V_1He_1(b)}°$. After helium implantation and 600 °C anneal, a complicated luminescence spectrum is observed with three He lines (at 2.212, 2.316, and 2.415 eV) and DAP lines in the range 2.179–2.430 eV (weak ZPL at 2.466 eV). The intensity of the DAP lines relative to the He lines is sample dependent. See also DAP95 =* $V_1He_1(a)°$.

DAP14 (resonance, nine luminescence lines) =* $\mathbf{(C_2)_iNi_1}°$. The DAP lines appear in low nitrogen diamond after carbon implantation and 1,000 °C anneal. Probable resonance lines are at 2.154 eV (= NV-center, $V_1N_1°$), and at 2.427 eV $(V_1Ni_2{}^+(b))$.

***$\mathbf{V_2}°$-DAP15 = TH5** (sideband, 12 absorption lines). This DAP spectrum is observed in heavily irradiated diamond. The 12 sideband lines reveal that the true ZPL of the **TH5**-center is a weak line at 2.419 eV (and not the proposed line at 2.543 eV). This latter value is the position of the first DAP line (shell 130).

S8-DAP16 =* $\mathbf{(V_2Ni_1)^-}$ (resonance, 12 absorption lines). In nickel containing brown HPHT synthetic diamond, these mostly overlapping lines are observed. The *S8 =* $(V_2Ni_1)^-$ center is the nucleus of the nitrogen-containing centers S2–*S7 and *S9. The probable resonance lines are at 1.219, 1.228, and 1.383 eV.

***$\mathbf{(O_1° + B_1°)}$-DAP17** (resonance (standard), six luminescence lines). In CVD diamond grown with oxygen, a set of luminescence lines is observed. These lines are superimposed on a broad band, between 2.65 and 2.75 eV.

***$\mathbf{(N_2{}^{2+} + B_1°)}$-DAP18** (standard, eight absorption, PLE and luminescence lines). The luminescence lines are superimposed on the 2.91-eV Yellow band. Intense lines have been given letters (C, D, E, F, H = DAP transitions with s = 18, 14, 10, 2, 1). In absorption and luminescence, the same line positions are observed. DAP18 is of special interest because the symmetry of some lines has been determined by uniaxial stress experiments [Moh82b]. The results confirm the DAP model and the shell assignments: A $\langle 1,1,1\rangle$ dipole is found for line H in agreement with the $\langle 1,1,1\rangle$ lattice vector of shell 1. Three $\langle 1,1,0\rangle$ dipoles are found for the lines C, D, and E, which are the shells 18, 14, and 10 with the lattice vectors $\langle 6,6,0\rangle$, $\langle 6,4,2\rangle$, and $\langle 6,2,0\rangle$. DAP18 is also a good example for vibrational sidebands on DAP lines. Up to four very intense harmonic sidebands with a vibrational frequency of 53 meV are observed for the shells 1, 4, and 50.

***$\mathbf{(N_2° + B_1{}^-)}$-DAP19** (standard, ten luminescence lines). The lines are superimposed on the 2.35-eV broad band. Intense lines are given letters (A, B, X, Y, Z = DAP lines with s = 4, 3, 12, 10, 6)) [Moh82b].

***$\mathbf{(N_2{}^+ + B_1°)}$-DAP20** (standard, seven luminescence lines). These relatively weak lines are superimposed on the 1.88-eV broad band of yellow luminescent diamonds [Moh82b].

DAP21 (standard, 12 PLE lines). This DAP spectrum of an unknown defect is observed in the photoluminescence excitation spectra of the A-bands and of the

N3(c) = V_1N_3 center. Remarkable is the high limiting energy $L = 5.310$ eV, which leads to $E_D + E_A = 0.180$ eV. This very small value indicates very shallow defects. An alternate interpretation is the involvement of the direct energy gap at 7.12 eV.

$^*V_1N_3{}^+$-DAP22 (sideband, 13 luminescence lines). A typical sideband DAP appearing as structure on the broad A_{insul}(d) band (see Table 10.2.2). There is a weak ZPL at 3.414 eV.

$V_1N_3^\circ$(c)-DAP23 (sideband, 15 absorption and PLE lines). DAP23 and DAP24 are each superimposed on a broadband and form an almost perfect mirror image on both sides of the intense ZPL at 2.985 eV, named N3(c). Some of the DAP lines may be in resonance with vibrations. DAP23 contains the well-known lines N4 (3.603 eV) and N5 (3.762 eV), = DAP transitions with $s = 3$, 1). Broad band A_{insul}(c) (see Table 10.2.2).

$V_1N_3^\circ$(c)-DAP24 (sideband, 17 luminescence lines, see also DAP23). DAP 24 contains the well-known lines D_2–D_8 [Dea65a], = DAP transitions with $s = 162$, 106, 50, 42, 22, 14, 8.

$^*V_1Ni_1{}^+$-DAP25 (sideband, 11 absorption lines). DAP25 (absorption) and DAP26 (luminescence) are observed in HPHT synthetic diamonds (with nickel catalyst). The Ni(1.4 eV) center at 1.403 eV is very prominent. This helps to identify the relatively weak DAP lines.

$^*V_1Ni_1{}^+$-DAP26 (sideband, nine luminescence lines, see also DAP25).

$^*V_1Ni_2{}^-$(b)-DAP27 (sideband, ten PLE lines). DAP27 and DAP28 are observed in nickel and nitrogen containing HPHT synthetic diamond after 1,500 °C anneal. The DAP27 PLE spectrum excites the $^*V_1Ni_2{}^-$(a) luminescence center at 2.071 eV which terminates at a higher lying ground state (see DAP87).

$^*V_1Ni_2{}^-$(a)-DAP28 (sideband, 13 luminescence lines, see also DAP27).

$^*V_1Ni_1{}^\circ$(b)-DAP29 (sideband, seven PLE lines). Effective PLE of the $^*V_1$ $Ni_1{}^\circ$(a) at 1.704 eV (see DAP83) This center (ZPL at 2.401 eV) is observed in nickel containing HPHT synthetic diamond after 1,700 °C anneal. The relatively weak DAP lines are superimposed on the broad 2.63-eV band.

$V_1C_4N_2{}^\circ$-DAP30 (sideband, four absorption lines) DAP30 and DAP31 are observed after heavy electron irradiation and annealing above 600 °C. It appears that vacancies migrate in IaA diamond to the N_2 center (named A) and in IaB diamond to the V_1N_4 center (named B). The DAP30 lines are sidebands of the dominant A-center line at 158.9 meV. The DAP31 lines are sidebands of the dominant B-center line at 145.7 meV. The lines named H1b, H1g, and H1d are DAP30 transitions with $s = 8$, 10, 50. The $^*V_1C_4N_2{}^\circ$ defect is the precursor of $^*V_1N_2$ (DAP27/28, DAP87, DAP111).

$V_1N_4V_1^\circ$-DAP31 (sideband, five absorption lines, see also DAP30). The lines named H1c, H1f, and H1e are DAP31 transitions with $s = 6$, 10, 34. The $^*V_1N_4V_1{}^\circ$ defect is the precursor of $^*N_3V_2N_1{}^\circ$ (DAP37/38).

$V_1{}^+$-DAP32 = GR2-GR8 (sideband, 13 absorption and PLE lines). This spectrum is observed in irradiated diamond. The DAP32 lines are very efficient in the PLE of the $V_1{}^\circ$(a/b) center (at 1.665/1.673 eV, named GR1). The DAP32 lines are sidebands of the weak ZPL transition at 2.781 eV and are also observed in photoconductivity [Ver74, Wal79].

Table 10.2.2 Broad luminescence "A" bands in diamond

Name of band[a]	Label	Peak (eV) (shell no.)	W (meV)	$L(s=\infty)$ (eV)	$E(s=1)$ (eV)	D (eV)	Impur./defect	Type	Comment
*A_{insul} (a)-DAP55	NB2150	2.12 ($s=42$)	320	2.30	1.55	−1.33	*$V_1N_3^{\circ}$(a)	SB	**PL, XL;** type I, IIa; also named N3(a)
*A_{insul} (b)-DAP56	NB2450	2.53 ($s=64$)	370	2.70	1.96	−1.33	*$V_1N_3^{\circ}$(b)	SB	**CL, PL, XL;** type I, IIa; also named N3(b)
*A_{insul} (c)-DAP24	NB,HB,LB2750	2.72 ($s=20$)	330	2.985	2.24	−1.33	*$V_1N_3^{\circ}$(c)	SB	**PL;** type I, IIa; also named N3(c)
*A_{insul} (d)-DAP22	NB3000	3.00 ($s=8$)	360	3.414	2.65	−1.36	*$V_1N_3^{+}$	SB	**PL;** type I, IIa; ZPL named 3.4-eV line
*A_{insul} (e)	NB3300	3.30	400	(3.74)	(3.00)		?		**CL, PL, XL;** type I, IIa
*A_{semic} (a)	NB,MB,HB2050	2.05	350	(2.20)	(1.46)		B?		**XL;** type IIb
*A_{semic} (b)-DAP73	NB,HB2200	2.20 ($s=34$)	400	2.395	1.67	−1.30	*$V_1B_1^{\circ}$(a)	SB	**CL, XL;** type IIb
*A_{semic} (c)	HB,MB2500	2.50	350	(2.80)	(2.06)		B?		**CL, XL;** type IIb
*A_{semic} (d)-DAP74	NB,HB2800	2.80 ($s=12$)	350	3.120	2.38	−1.30	*$V_1B_1^{\circ}$(b)	SB	**CL, XL;** type IIb

[a] A_{insul} in insulating diamond (type I, IIa); A_{semic} in semiconducting diamond (type IIb)

$V_1{}^\circ$(b)-DAP33 (sideband, ten absorption lines). DAP33 and DAP34 are observed as weak sideband spectra of the intense ZPL at 1.673 eV (mirror image). The $V_1{}^\circ$(a) ZPL (at 1.665 eV) is much weaker than the $V_1{}^\circ$(b) ZPL. The DAP interpretation is confirmed by the expected isotopic shifts in the spectrum of a ^{13}C diamond (see Table 7.2).

$V_1{}^\circ$(b)-DAP34 (sideband, eight luminescence lines, see also DAP33).

$(C_2)_i{}^+$(a)-DAP-35 = TR12 (sideband, 14 absorption lines). DAP35 and DAP36 are observed in irradiated diamond and are mirror image sidebands of the 2.638-eV transition from the ⟨100⟩ split self-interstitial (named **TR12**). The $(C_2)_i{}^-$-DAP86 = 3H center has its ZPL at 2.462 eV. For the neutral charge state of this defect, see $(C_2)_i{}^\circ$-DAP71.

$(C_2)_i{}^+$(a)-DAP36 (sideband, 14 luminescence lines, see also DAP35).

***$N_3V_2N_1{}^\circ$(b)-DAP37** (sideband, 14 absorption lines). DAP37 and DAP38 are observed after irradiation and annealing above 700 °C. The ZPL at 2.498 eV (named H4(b)) is ten times more intense (in absorption) than the H4(a) line at 2.417 eV. The lines named H6-H12 are DAP37 transitions.

***$N_3V_2N_1{}^\circ$(b)-DAP38** (sideband, seven luminescence lines, see also DAP37).

***$(V_4H_4)^\circ$-DAP39** (sideband, ten absorption lines in a ^{13}C sample, only one line in a ^{12}C sample, see also DAP40). The spectra are observed in homoepitaxial CVD diamond. The pseudo ZPL at 0.165 eV is a C-C stretch vibration (−3.6% ^{13}C shift).

***$(V_4H_4)^+$-DAP40** (sideband, seven absorption lines) observed in homoepitaxial CVD diamond. Valuable information is obtained from the spectra with isotopic substitution. For a sample with 50% deuterium, the $s = 20$ line of DAP40 is split in a pattern with intensities 1:4:6:4:1 (3.7-meV separation, see Table 7.1). This indicates an involvement of four equivalent H atoms. For the ZPL of DAP40 (at 0.412 eV) an almost vanishing isotope shift is observed (−0.5% for deuterium and −0.2% for ^{13}C). Therefore, it can be concluded that the transition is electronic and not vibrational. In contrast, for the pseudo ZPL of DAP39 (at 0.165 eV), the ^{13}C isotope shift is −3.8%, which is characteristic for a CC stretch vibration.

(V_1H_1)b-DAP41 (sideband, five absorption lines). Observed in hydrogen-rich natural diamond with very intense lines at 0.1742 eV ((V_1H_1)a = natural CH bend) and 0.385 eV ((C_1H_1)b = natural CH stretch, see Table 9.3.1).

***$(V_2Xe_1)^\circ$-DAP42** (sideband, ten luminescence lines). Observed in the CL spectrum of a type IIa diamond after Xe implantation and 1,400 °C anneal. The characteristic QLVM sideband of Xe appears at −29 meV (calculated −27.3 meV).

***$V_1N_2^+$(b)-DAP43** (sideband, seven luminescence lines). This intense ZPL with weak sidebands is a common feature in the PL spectra of natural diamond. The ZPL at 2.463 eV is named **S1**. The excited state of the center is split by 34 meV, causing the 20 times weaker ZPL V_1N_2(a) at 2.429 eV (**S1**(a)).

DAP44 *deleted.*

***$V_1C_4N_1{}^\circ$(a)-DAP45** (sideband, nine absorption lines). This weak spectrum is observed in nitrogen-containing diamond after high energy (>1 MeV) neutron irradiation and annealing at 800–950 °C, which is the same treatment as for DAP4. Spectral hole burning is observed for the line at 1.909 eV [Sil94a, Sil95]. Possibly, three luminescence lines from the mirror image of DAP45 (with $D = -1.53$ eV) are reported at 1.177 eV ($s = 6$), 1.525 eV ($s = 58$), and 1.602 eV ($s = 162$) [Vin88a, Vin88b, Sil95].

V_1N_1°-DAP46 (sideband, six absorption lines). DAP46 and DAP47 form a sideband mirror image on both sides of the well-known ZPL at 2.154 eV, named **NV** or **575-nm** center. This center occurs in natural diamond, in as-grown HPHT and in CVD diamond.

V_1N_1°-DAP47 (sideband, nine luminescence lines, see also DAP46)

$V_1N_1^-$(a)-**DAP48** (sideband, ten absorption lines) DAP48 and DAP49 are observed, when the NV center (see DAP46/47) is in its negative charge state with the ZPL at 1.943 eV, named **638-nm** center.

$V_1N_1^-$(a)-**DAP49** (sideband, 15 luminescence lines, see also DAP48)

V_1N_2°(a)-DAP50 (sideband, eight absorption lines) DAP50 and DAP51 are observed on both sides of the ZPL at 2.464 eV, named **H3(a)**. For H3(b) see DAP52. See also DAP77/78.

V_1N_2°(a)-DAP51 (sideband, six luminescence lines, see also DAP50)

V_1N_2°(b)-DAP52 (sideband, seven lines) is observed only in absorption. The ZPL at 3.361 eV is named **H3(b)** or **H13**, and terminates at a higher excited state of H3(a).

***$(V_3Si_2)^+$-DAP53** (sideband, 13 absorption or PLE lines). DAP53 and DAP54 are observed on both sides of the ZPL at 1.682 eV, named **Si center**. Other charge states of this defect are *$(V_3Si_2)^-$-DAP75, and *(V_3Si_2)°-DAP65. The *$(V_3Si_2)^+$ center is frequently observed in CVD films, grown on silicon substrates. It can also be created by silicon ion implantation and annealing at 1, 400 °C.

***$(V_3Si_2)^+$-DAP54** (sideband, seven luminescence lines, see also DAP53)

V_1N_3(a)-DAP55 (sideband, five luminescence lines). V_1N_3(a)-DAP55 (ZPL at 2.297 eV) and V_1N_3(b)-DAP56 (ZPL at 2.680 eV) terminate at higher lying ground states of the N3(c) center (ZPL at 2.985 eV). It is the great similarity with V_1N_3(c)-DAP23/24, which allows a DAP analysis in spite of the small number of observed lines. Broadband A_{insul}(a), see Table 10.2.2.

V_1N_3(b)-DAP56 (sideband, four luminescence lines, see also DAP55). Broadband A_{insul}(b), see Table 10.2.2.

DAP57 to **DAP61** = **S2 to *S6** and ***S7**(DAP11), ***S8**(DAP16), ***S9**(DAP4). All these spectra are produced by similar defects: one nickel atom at the center of a double vacancy, surrounded by different numbers and positions of nitrogen atoms. These eight DAP centers are of the *standard* type, i.e., no ZPL occurs. Instead, the DAP transitions were previously misinterpreted as ZPLs with individual names (B to E). Frequently observed are **S2**-DAP57 and **S3**-DAP58. (See also Sect. 8.1.5).

S2-DAP57 = **$(V_2Ni_1)N_{2+0+1}^+$** (standard, 15 absorption, PLE, and luminescence lines). Four lines are prominent and are named B to E: Line **B** (2.537 eV) from $s = 162$, line **C** (2.597 eV) from $s = 64$, line **D** (2.623 eV) from $s = 50$, and line **E** (2.634 eV) from $s = 44$. Monitor for the PLE spectrum was at $B = 2.537$ eV. The EPR analog is the **NE2** center.

S3-DAP58 $= (\mathbf{V_2Ni_1})N_{2+0+0}{}^+$ (standard, ten absorption, PLE, and luminescence lines). Monitor for the PLE spectrum was at 2.496 eV ($s = 162$ line). The EPR analog is the **NE1** center.

***S4-DAP59** $= (\mathbf{V_2Ni_1})N_{1+2+0}{}^\circ$ (standard, 12 absorption, and PLE lines). Luminescence is suppressed by energy transfer to the A line (2.370 eV, see DAP92). Monitor for the PLE spectrum was at 2.370 eV (which does not belong to the *S4 defect, but is related to *S4 by a very effective energy transfer from S4-DAP59 transitions (3.333–3.820 eV) to the A line (ZPL) of *$V_3Ni_3{}^\circ$-DAP92). The EPR analog is the **NE3** center.

***S5-DAP60** $= (\mathbf{V_2Ni_1})N_{x+y+z}{}^0$ (standard, 11 absorption, PLE, and luminescence lines). Unusual for this DAP center is the dominant appearance of the line at 1.563 eV ($s = 162$), which seems to be a cutoff ZPL in absorption, PLE, and luminescence. However, the analogy of the NE8 EPR center (ascribed to *S5) to the NE1-NE3 EPR centers of S2–S4, and the similarity of the DAP transition to the S2–S9 centers support the present structure assignment. In luminescence, the DAP 1.563 line has a QLVM sideband at 28 meV (1Ni + 1N + 3C) and unresolved sidebands in the range 1.20–1.50 eV. Monitor for the PLE spectrum was at 1.35 eV (maximum of the sidebands from the *S5-DAP60 s = 162 luminescence line).

***S6-DAP61** $= (\mathbf{V_2Ni_1})N_{x+y+z}{}^+$ (standard, 13 PLE lines). Luminescence is suppressed by energy transfer from the *S6 DAP61 lines (2.750–3.530 eV) to the *S5-DAP60 lines (1.563–2.320 eV). Monitor for the PLE spectrum was at 1.35 eV (maximum of the *S5-DAP60 luminescence band).

DAP62 (sideband, 5 luminescence lines). This defect is observed in heavily neutron irradiated diamond with annealing at $T > 200\,°C$.

*$\mathbf{V_1N_4{}^-}$**(a)-DAP63** (sideband, seven absorption, and PLE lines, named **N9(a)**). DAP63 and DAP64 form a perfect mirror image on both sides of the ZPL at 5.252 eV. The defect occurs in natural diamond and has a second line at 5.262 eV (**N9(b)**) with similar DAP sidebands.

*$\mathbf{V_1N_4{}^-}$**(a)-DAP64** (sideband, nine luminescence lines, see also DAP63)

*$\mathbf{(V_3Si_2)^\circ}$**-DAP65** (sideband, five luminescence lines). This center is observed in natural type Ia diamonds. See also DAP 53/54 = * $(V_3Si_2)^+$ and DAP75 = *$(V_3Si_2)^-$.

$\mathbf{V_1Al_1{}^\circ}$**(a)-DAP66** (sideband, nine luminescence lines) is observed in brown natural diamonds and in HTHP diamonds from a melt containing aluminum (as a nitrogen getter). In natural diamond, a time delay of 0.1–2.0 ms is necessary to suppress the otherwise dominant $V_1N_3{}^\circ$(c)-DAP24 spectrum. The two ZPLs at 2.964 eV (line (a)) and 2.974 eV (line (b)) have very similar but weak DAP sidebands. At $T = 90$ K line (b) is dominant, while at $T = 15$ K only line (a) is observed, indicating a 10-meV splitting in the excited state.

*$\mathbf{(N_2{}^+ + B_1{}^\circ)}$**-DAP67** (standard, nine absorption lines). This center is observed in natural type IaA diamonds and correlates exactly with the A-center ($N_2{}^\circ$) concentration.

*$\mathbf{V_1Li_1{}^\circ}$**-DAP68** (sideband, seven luminescence lines). This spectrum appears as weak structure on a broadband (maximum at 1.61 eV) in lithium implanted diamond

(annealed at 1, 400 °C). Also the ZPL at 1.881 eV is weak and has a QLVM sideband at −36 meV (1Li + 6C).

***$(N_1C_1)_i^-$-DAP69** (sideband, 12 absorption lines). This center is observed in as-irradiated (electrons, 2 MeV) type IIa natural or CVD diamond. It anneals out above 400 °C. The ZPL at 3.988 eV has a 36-meV QLVM sideband (1N + 5C). The corresponding EPR center is named **R11**. Two other charge states of the $(N_1C_1)_i$ defect are also observed: $(N_1C_1)_i^\circ$-DAP3 and $(N_1C_1)_i^+$ DAP98 (see Table 8.2.4).

***$(N_1^{2+} + B_1^\circ)$-DAP70** (standard, 17 luminescence lines). This group of lines is observed in boron doped CVD diamonds, as-irradiated with 300-keV electrons. The microscopic photoluminescence studies reveal considerable variation from one location to another [Ste99d]. Two spectra (a and b) for different locations show lines from $s = 1, 3, 10, 12, 14, 16, 18, 20, 24, 30, 32, 106$ in (a), and lines from $s = 2, 12, 13, 14, 20, 26, 50$ in (b). This result is an elegant proof for the DAP interpretation.

***$(C_2)_i^\circ$-DAP71** (sideband, three absorption lines). This center is observed as structure on a broadband. The defect is the ⟨100⟩ split self-interstitial in the neutral charge state, with the ZPL at 1.685 eV and a QLVM sideband at 40 meV (6C). The corresponding EPR center is named **R2**. (For the positive charge state, see DAP35).

$(C_2)_iN_1^-$-DAP72 (sideband, 22 luminescence lines). This spectrum is observed in as-grown CVD diamond, but more often (with high intensity) in as-irradiated or as-implanted (N) diamond. The sideband spectrum of the ZPL (3.188 eV) is a complicated superposition of DAP lines, QLVM, and LVM lines. There is a prominent QLVM line at 77 meV (1N + 1C) and LVM lines at 161 (C–C), 179 (N–C), and 190 (C–C) meV. In absorption, only a very weak ZPL is observed. See also DAP79/80 = $(C_2)_iN_1^\circ$.

***$V_1B_1^\circ$(a)-DAP73** (sideband, six luminescence lines). In X-irradiated type IIb diamonds, the lines appear as structure on the broad A_{semic}(b) band (see Table 10.2.2), with a weak ZPL. In as-grown boron doped HPHT diamond, the ZPL at 2.395 eV is sharp and intense (probably due to energy transfer from *$V_1B_1^\circ$(b)-DAP74, which is absent in this spectrum). There is a QLVM sideband at 50 meV (1B + 3C).

***$V_1B_1^\circ$(b)-DAP74** (sideband, six luminescence lines). In X-irradiated type IIb diamonds, the lines appear as structure on the broad A_{semic}(d) band (see Table 10.2.2), with a weak ZPL.

***$(V_3Si_2)^-$-DAP75** (sideband, ten luminescence lines). This spectrum (with the ZPL at 2.523 eV) is observed in natural diamond, or in silicon as-implanted type Ia diamond.

***$V_1N_3^-$-DAP76** (sideband, 13 luminescence lines). This center occurs in natural or irradiated diamond. A good fit with calculated phonon sidebands [Naz92] indicates a resonance of DAP and phonon sidebands.

***$V_1N_2^-$-DAP77 = H2** (sideband, ten absorption lines). For DAP77 and DAP78, the DAP lines appear as structure on the broad 1.50-eV band.(absorption) or on the 1.07-eV band (luminescence) in irradiated and annealed diamond. The H2 center is photochromic with the H3(a,b) center (*$V_1N_2^\circ$-DAP50/51 and DAP52).

***$V_1N_2^-$-DAP78 = H2** (sideband, four luminescence lines).See also DAP77.

***$(C_2)_iN_1{}^\circ$-DAP79 = 5RL** (sideband, ten absorption lines). DAP79 and DAP80 arise from a typical irradiation-induced center ("5RL" = five radiation lines). This defect occurs also in as-grown CVD diamond.

***$(C_2)_iN_1{}^\circ$-DAP80 = 5RL** (sideband, six luminescence lines). The spectrum is dominated by four phonon sidebands of the main LVM (237 meV). An interesting effect occurs in the spectrum from a CVD diamond. The phonon sidebands are in resonance with the unresolved DAP80 transitions, leading to intensities, which are much higher than that of the ZPL [Col92b]. Uniaxial stress splittings of the ZPL (4.582 eV) fully confirm tilted $\langle 100 \rangle$ orientation of the defect [Col86b].

***$V_1N_4(C_2)_i{}^\circ$-DAP81 = F(c) center** (sideband seven absorption lines). The F-center is formed during thermally induced nitrogen aggregation at $T > 1,900\,°C$ (see Sect. 9.1.6.1). It is named after its intense infrared absorption bands (see Table 8.1.3.6).

$V_1N_4{}^\circ$-DAP82 = B center (sideband, five absorption lines).The B center is one of the fundamental defects in diamond (see Tables 1.1.1 and 1.2). It occurs in natural diamond, and is formed during thermally induced nitrogen aggregation at $T > 1,900\,°C$ (see Sect. 9.1.6.1). The ZPL at 4.191 eV is a triplet with 6-meV splitting. Well known are the characteristic B center infrared absorption bands (see Table 8.1.3.5).

***$V_1Ni_1{}^\circ$(a)-DAP83** (sideband, 6 absorption lines). DAP83, DAP84, and *$V_1Ni_1{}^\circ$(b)-DAP29 are observed in nickel containing diamonds after annealing at $T > 1,700\,°C$.

***$V_1Ni_1{}^\circ$(a)-DAP84** (sideband, three luminescence lines).See also DAP83.

***$(V_2Ag_1)^\circ$(a)-DAP85** (sideband, six luminescence lines). This center is observed after silver implantation in diamond and 1, 400 °C anneal.

***$(C_2)_i{}^-$-DAP86 = 3H** (sideband, nine luminescence lines). This center is observed in as-irradiated diamond. There are two additional charge states: *$(C_2)_i^\circ$-DAP71 at 1.685 eV (R2 center), and *$(C_2)_i^+$(a)-DAP35/36 at 2.638 eV (TR12 center). The name 3H is chosen, because the ZPL at 2.462 eV is very close to the ZPL at 2.464 eV of the H3 center (see DAP50/51).

***$V_1Ni_2{}^\circ$(a)-DAP87** (sideband, 14 luminescence lines). See DAP27/28.

***$(C_2)_iN_2{}^-$-DAP88** (sideband, six absorption lines). DAP88 and DAP89 are observed in low temperature irradiated diamond and anneals out at 100–250 °C. The DAP88 center with the ZPL at 2.367 eV is photochromic with the *$(C_2)_iN_2{}^\circ$ center at 1.979 eV (no DAP lines) and with the *$(C_2)_iN_2{}^+$ center at 2.535 eV (DAP90).

***$(C_2)_iN_2{}^-$-DAP89** (sideband, seven luminescence lines).See DAP88.

***$(C_2)_iN_2{}^+$DAP90** (sideband, three absorption lines). See DAP88.

***$V_1Ni_3{}^\circ$-DAP91 = A line** (sideband, seven absorption lines). DAP91 and DAP92 with the ZPL at 2.370 eV is observed in natural diamonds and in annealed Ni containing HPHT synthetic diamonds.

***$V_1Ni_3{}^\circ$-DAP92 = A line** (sideband, six luminescence lines). See DAP91.

***$V_1{}^-$-DAP93 = R10 = ND1** (sideband, six absorption lines). The isolated vacancy in irradiated diamond occurs in three charge states: *$V_1{}^-$ with ZPL at

3.150 eV, V_1° = GR1 with ZPL at 1.672 eV (see DAP33/34), and *V_1^{+} with ZPL at 2.781 eV (see DAP32, PLE of V_1°). *V_1^{-} and V_1° are a photochromic pair.

***$(V_2Tl_1)^{\circ}$-DAP94** (sideband, eight luminescence lines). This center is observed after Tl implantation in diamond and 1,400 °C anneal.

***$V_1He_1^{\circ}$(a)-DAP95** (sideband, six luminescence lines). This center is observed after He implantation in diamond and 1,000 °C anneal.

***$(C_{2n})_i^{\circ}$-DAP96 = D center = platelets** (sideband, six luminescence lines). The formation of platelets (ZPL at 1.526 eV) by aggregation of isolated *$(C_2)_i^{\circ}$ centers (ZPL at 1.685 eV, see DAP71) is part of the thermally induced nitrogen aggregation in diamond (see Sect. 9.1.6.1). The corresponding infrared absorption bands are named D center lines (see Table 8.2.2).

***$(C_2)_iB_1^{-}$-DAP97 = 2BD(F)** (sideband, four luminescence lines). This is the dominant group of boron related lines in irradiated IIb diamond (2BD = II b damaged). Other groups are 2BD(C) and 2BD(G).

***$(N_1C_1)_i^{+}$-DAP98** (sideband, nine luminescence lines). This defect is observed in as-grown CVD diamond and in irradiated HPHT synthetic diamond (see also *$(N_1C_1)_i^{-}$-DAP69 and *$(N_1C_1)_i^{\circ}$-DAP3).

***$(N_1C_1)_iB_1^{\circ}$-DAP99** (sideband, eight luminescence lines). This defect (with the ZPL at 2.792 eV) is observed in boron contaminated CVD diamond, together with *$(C_2)_iN_1^{-}$ at 3.188 eV(see DAP72) and *$(C_2)_iN_1^{\circ}$ = 5RL at 4.582 eV (see DAP80).

***$(N_1C_1)_iB_1^{+}$-DAP100** (sideband, nine luminescence lines). This defect (with the ZPL at 3.092 eV) is observed in boron contaminated CVD diamond.

***$(N_1C_1)_iB_1^{-}$-DAP101** (sideband, four luminescence lines). This defect (with the ZPL at 2.992 eV) is observed in boron doped CVD diamond.

$V_1Co_2^{\circ}$-DAP102 (sideband, five absorption lines). This spectrum (with an unfavorable signal-to-noise ratio) is found in HPHT synthetic diamond, grown from a pure cobalt melt. Three intense sidebands are from cobalt LVM with 44 meV.

DAP103 ***deleted***

***Co-S5-DAP104 = $(V_2Co_1)N^{\circ}_{x+y+z}$** (standard, nine luminescence lines). This defect is observed in HPHT synthetic diamonds, grown from a pure cobalt melt, and annealed at 1,800 °C. It is similar to the corresponding *S5 Ni center (see DAP60). The spectrum was recorded with a 1-μs delay in order to discriminate against overlapping lines from other defects (see Sect. 9.6).

***$(V_3Ni_2)^{+}$-DAP105** (sideband, four absorption lines). DAP105 and DAP106 are observed in Ni containing HPHT synthetic diamonds or in Ni implanted and annealed (T = 1,400 °C) diamonds. The ZPL at 2.562 eV (with a split-off line at 2.588 eV from ground state splitting) is observed in luminescence (decay time 140 μs), but in absorption only the 2.588-eV ZPL is (indirectly) observed. The absorption spectrum starts at 2.780 eV (DAP line; s = 50). Four QLVM sidebands with 26 meV (2Ni + 2C) are observed for the 3.064 eV (s = 8) line. See also DAP106/107.

***$(V_3Ni_2)^{+}$-DAP106** (sideband, six luminescence lines).In high resolution, the ZPL at 2.562 eV is split into 15 lines (of which nine are observed): A quintet from

Ni_2 isotopes (see Table 7.5) with 1.56-meV separation, and each quintet line further splits into a triplet with 0.33-meV separation.

$^*(V_3Ni_2)^\circ$-DAP107 (sideband, seven absorption lines). DAP106 and DAP107 occur in as-grown HPHT synthetic diamonds or in as-grown CVD diamonds. The ZPL at 1.883 eV is accompanied by lines at 1.906 and 1.914 eV (from excited state splitting). Two LVM sidebands with 61 meV are observed.

$^*(V_3Ni_2)^\circ$-DAP108 (sideband, two luminescence lines).The luminescence lines are weak, and only two DAP lines are seen. See also DAP107.

$^*V_1Si_1{}^\circ$-DAP109 (sideband, eight PLE lines). DAP108 and DAP109 are observed in Si contaminated CVD diamonds (ZPL at 2.646 eV). The DAP lines appear as weak structure on the broad bands. The PLE of the own luminescence (DAP109) is monitored at the maximum of the band at 2.4 eV.

$^*V_1Si_1{}^\circ$-DAP110 (sideband, seven luminescence lines). See also DAP109.

$^*V_1Ni_2{}^+$(c)-DAP111 (sideband, six luminescence lines). This spectrum is observed in nickel containing HPHT synthetic diamonds after annealing at $T = 1,500\,°C$. From the three ZPL lines of $^*V_1Ni_2{}^+$(a) at 1.693 eV, $^*V_1Ni_2{}^+$(b) at 1.940 eV, and $^*V_1Ni_2{}^+$(c) at 1.991 eV, only one DAP spectrum (DAP111) is observed with relatively weak lines.

$^*(V_3Co_2)^\circ$-DAP112 (sideband, six luminescence lines). This defect is observed in HPHT synthetic diamonds, grown from a pure cobalt melt, and annealed at $1,800\,°C$. The ZPL is split into three lines at 1.984, 1.989 (main), and 1.991 eV.

$^*V_1Si_1{}^-$-DAP113 (sideband, 11 luminescence lines). The spectrum (with ZPL at 2.991 eV) was recorded with a gate delay of 50 ns. See also DAP110.

$^*V_1Zn_1{}^\circ$-DAP114 (sideband, six luminescence lines).This center occurs in zinc implanted and at $1,400\,°C$ annealed diamond. There are two QLVM sidebands with 32 meV (1Zn + 3C).

$^*V_1As_1{}^\circ$-DAP115 (sideband, six absorption lines). These infrared lines (ZPL at 0.113 eV) are observed in arsenic doped HPHT synthetic diamonds.

$^*(N_1{}^\circ + B_1{}^-)$-DAP116 (standard, 11 luminescence lines). This spectrum is observed in natural type Ia diamonds (see Table 9.1.5).

10.2 Broad Bands from Unresolved Lines

In an early publication [Dea65a], it was proposed that the band-A luminescence can be consistently interpreted by unresolved DAP transitions. A list of 50 bands from unresolved DAP transitions is given in Tables 10.2.1 and 10.2.2. Among them are also the six bands for which the DAP character is now established (see Table 10.2.2).

Discussion. Some empirical rules can be derived from Tables 10.2.1 and 10.2.2. The *halfwidths* W vary from 190 to 450 meV. This appears reasonable, because the band should be confined between the end points of the respective DAP system, i.e., between $L(s = \infty)$ and $E(s = 1)$, a spread which varies from 690 to 910 meV (full width) or 345 to 455 meV (halfwidths) for $D = 1.24$–1.63.

Luminescence A-bands. In a 15 page publication, a group of overlapping luminescence bands (named A-bands) with emission from 1.5 to 3.8 eV has been described [Dea65a]. This emission occurs in natural diamonds (insulating = type I and IIa, or semiconducting = type IIb) and has been studied by photo-, cathodo-, and X-ray excitation [Dea64c, Dea65a]. The A-bands are contrasted with other visible luminescence bands containing a pronounced zero phonon line, especially those with a ZPL at 2.154 eV (named NV, see DAP47), at 2.464 eV (named H3, see DAP51), at 2.498 eV (named H4, see DAP38), and at 2.985 eV (named N3, see DAP24) [Dea64c, Dea65a]. A further common feature of the A-bands is the photoexcitation in two PLE bands at 5.37 eV (DAP20) and 5.58 eV (DAP21).

For the present book, a tentative separation into nine individual overlapping bands was undertaken (now named $^*A_{insul}$(a–e) for *insulating* diamond and $^*A_{semic}$(a–d) for *semiconducting* diamond, see Table 10.2.2).

The five $^*A_{insul}$(a–e) bands have peaks at (a) 2.15, (b) 2.45, (c) 2.72, (d) 3.00, and (e) 3.30 eV. The $^*A_{insul}$(c) peak at 2.72 eV is dominant, and a typical intensity distribution (for sample "DS21" [see [Dea65a], Fig. 1B]) is $^*A_{insul}$(a):(b):(c):(d):(e) = 7:17:51:19:6%.

The four A_{semic}(a–d) bands have peaks at (a)2.05, (b)2.20, (c)2.50, and (d)2.80 eV. A fifth band at 3.30 eV is probably identical with the band A_{insul}(e) for insulating diamond. At *80 K*, either the overlapping A_{semic}(a) and A_{semic}(b) bands (unresolved peak at 2.1 eV) or the A_{semic}(d) band at 2.80 eV dominate, depending on differences in the concentrations of acceptors and donors($N_A - N_D$). For small ($N_A - N_D$) the 2.1-eV band, dominates and for large ($N_A - N_D$) the 2.8-eV band [Dea65a, Fig. 5]. The sample temperature is also important: At *295 K* the A_{semic}(c) band at 2.50 eV is dominant [Dea65a, Fig. 9].

On six bands (A_{insul}(a, b, c, d) and A_{semic}(b, d) a structure is superimposed, which arises from resolved DAP transitions (see Tables 10.1 and 10.2.2). These six DAPs are of the sideband type with a broad and very weak zero phonon line. Of special interest is the A_{insul}(c) band with the peak at 2.72 eV and the ZPL at 2.985 eV. This ZPL (named N3(c)) can be sharp and intense, depending on the type of excitation and on the crystalline quality of the sample. The variation in the peak intensities (ratio ZPL: band) is demonstrated in three spectra. This ratio is 2:1 for the sample "Sierra Leone" [Dea65a, Fig. 11], is 1:1 for sample "DS21" [Dea64c, Fig. 1b], and is 1:20 for sample "K82" [Dea64c, Fig. 1a]. This latter value is typical also for the bands A_{insul}(d) and A_{semic}(b,d).

Six of the A-bands show DAP structure. The underlying bands arise definitely from unresolved DAP transitions (in agreement with the interpretation in [Dea65a]). Considering the similarity of the remaining three bands with the other A-bands, it can be concluded that probably all broad A-bands arise from unresolved DAP transitions. Using interpolation between the six analyzed and the three presumed DAP systems, a set of predicted values for the end points $L(s=\infty)$ and $E(s=1)$ are given in parenthesis in Table 10.2.2.

Except for the A_{insul}(c) band at 2.75 eV, which arises from the $V_1N_3{}^\circ$-DAP23/24 defect (ZPL = V_1N_3(c) at 2.985 eV), the atomic nature of A-band defects is not yet established. With high probability the bands A_{insul}(a) and A_{insul}(b) arise from the higher lying ground states of the N3(c) = $V_1N_3°$ defect: N3(a)-DAP55 (ZPL at 2.297 eV) and N3(b)-DAP56 (ZPL at 2.680 eV). If structure could be observed on these weak bands, this would be the final proof for the assignment.

For the A_{semic}(a–d)-bands from the semiconducting (type IIb) diamonds, the involvement of boron is almost certain, and structures for DAP73 and DAP74 are proposed (see Table 10.2.2).

In *insulating HTHP synthetic* diamond, a "green" band extending from 1.4 to 3.0 eV and peaking at 2.1 eV is reported [Dea65a]. This band is evidently quite different from the "blue" A-bands in *insulating natural* diamond, because the luminescence decay time for the *slow "green" band* at 80 K is 25 s, while only 10 ms are measured for the *fast "blue" band.* In [Dea65a] it is mentioned that this difference provides a useful way to distinguish between natural and synthetic diamonds (see also Chap. 14).

10.3 Depth of Donors (N, O, P) from Analysis of DAPs with B Acceptor

For *standard* DAPs, the depth of the donor level (E_D) can be determined, if the depth of the acceptor level (E_A) is known (see (10.6a) and Table 10.3).

Fortunately, the ionization energy of the isolated boron acceptor is known ($E_A = 0.370$ eV, see (10.6b) and Table 9.2.1).

$$E_D = E_g - L - E_A, \tag{10.6a}$$

$$E_D = 5.490 - L - 0.370\text{eV}. \tag{10.6b}$$

From (10.8b), the depth of the nitrogen centers $N_1°$ = C center (3.227 eV), $*N_1^+$ = E center (3.479 eV), $N_2°$ = A center (3.713 eV), $*N_2^+$ (2.625 eV), and $*N_2^{2+}$ (2.786 eV) have been determined (see Table 10.3). These values are discussed in Sect. 9.1.5.

The donor depths of oxygen and phosphorous are also given in Table 10.3.

10.4 Dependence of the Dielectric Factor *D* on the Number of Electrons

For neutral defect centers there is a pronounced dependence of the dielectric factor D on the total number of electrons involved (see Table 10.4). The number of electrons for each neutral atom is identical to the element number Z (see Tables 7.9a, b). The variation of D is from 1.20 eV (for 168 electrons) to 1.63 eV (for 18 electrons). An approximate linear relationship can be expressed by the empirical equation (10.7).

Table 10.3 Depth of nitrogen, oxygen, or phosphorous donor centers from analysis of standard DAPs with a boron acceptor

Iso-lated donor	Name	Composition of corresponding DAP	*stand.* DAP[a]	Range (eV)	D^a (eV)	L^a (eV)	$E_D + E_A^a$(eV)	E_D^b (eV)	Table
N_1°	**C**	$^*N_1^\circ + B_1^\circ$	1	2.02–2.67	+1.43	1.893	3.597	3.227	9.1.5
$^*N_1^+$	**E=X**	$^*N_1^+ + B_1^\circ$	9	1.74–2.18	+1.53	1.641	3.849	3.479	9.1.5
N_2°	**A**	$^*N_2^\circ + B_1^\circ$	20	1.79–2.10	+1.38	1.407	4.083	3.713	9.1.5
$^*N_2^+$	–	$^*N_2^+ + B_1^\circ$	67	3.86–4.28	+1.43	3.684	1.806	1.436	9.1.5
$^*N_2^{2+}$	–	$^*N_2^{2+} + B_1^\circ$	18	2.65–3.20	+1.54	2.334	3.156	2.786	9.1.5
$^*O_1^\circ$	–	$O_1^\circ + B_1^\circ$	17	2.68–2.72	+1.29	2.534	2.956	2.586[c]	9.2.2
P_1°	–	$P_1^\circ + B_1^\circ$	2	4.17–5.28	+1.72	4.500	0.990	0.620[c]	9.2.2

[a] Standard DAP: D dielectric factor, L limiting energy, $E_D + E_A = E_g - L$, E_g =5.490 eV
[b] Obtained from the preceding column after inserting E_A =0.370 eV for the boron acceptor
[c] In agreement with E_D =0.640 eV from the bound exciton analysis [Ste99a, Ste99b]

Table 10.4 Dependence of the dielectric factor D of neutral DAP centers on the total number of electrons involved

D (eV)	Impurity	Group	Total electrons	Atomic radius (nm)[a]	DAP	Type	Defect	Table
±1.20	(20 C)	1 A	120	(76)	33/34	SB	$^*V_1^\circ$ = **GR1**, distorted	8.1.1
+1.23	1 Ta + 6 C	5 B	109	141	7 a-f	*sta.*	$^*(V_2Ta_1)^\circ$	8.1.7.2, 9.7.2
+1.24	1 W + 6 C	6 B	110	138	6 a-f	*sta.*	$^*(V_2W_1)^\circ$	8.1.7.2, 9.7.2
+1.29	1 O +10C +1 B + 3C	6/3 A	91	60/(86)	19 a-h	*sta.*	$^*O_1^\circ$ *distorted*+ $^*B_1^\circ$ *distorted*	9.7.1
−1.29	1 Tl + 2 C	3 A	93	199	94 a-h	SB	$^*(V_2Tl_1)^\circ$, *distorted*	8.1.7.2, 9.7.1
−1.31	1 Ag + 6 C	1 B	83	144	85 a-f	SB	$^*(V_2Ag_1)^\circ$	8.1.7.2, 9.7.2
+1.32	1 As + 9 C	5 A	87	120	150 a-f	SB	$^*V_1As_1^\circ$, *distorted*	9.7.1
−1.33	1 Xe + 3 C	8 A	72	233	42 a-j	SB	$^*(V_2Xe_1)^\circ$, *distorted*	9.7.1
±1.34	2 N + 10 C	5 A	74	70	50/51	SB	$^*V_1N_2^\circ$ = **H3**, *distorted*	8.1.3.1
±1.34	1 Ni + 7 C	8 (B)	70	127	27/28/87	SB	$^*V_1Ni_1$(a,b)	8.1.4
±1.36	2 Ni + 2 C	8 (B)	68	127	107/108	SB	$^*(V_3Ni_2)^\circ$	8.1.6
−1.37	2 Co + 2 C	8 (B)	66	128	112 a-f	SB	$^*(V_3Co_2)^\circ$	8.1.6
+1.38	2N + 5C + 1B + 3 C	5/3 A	67	70/(86)	20 a-f	*sta.*	$^*N_2^\circ$ *distorted* + B_1° *distorted*	9.1.5
±1.39	1 N + 10 C	5 A	67	70	46/47	SB	$V_1N_1^\circ$ = **NV**	8.1.3.1
−1.39	2Si + 6C	4 A	64	119	65 a-e	SB	$^*(V_3Si_2)^\circ$	8.1.6
+1.39	1Zr + 3C	4 B	58	163	5 a-j	*sta.*	$^*(V_2Zr_1)^\circ$	8.1.7.2
+1.43	4 H + 8 C	1 A	52	(140)	40 a-g	SB	$^*(V_4H_4)^\circ$	8.1.7.2
−1.42	1Zn + 3C	2 B	48	134	114 a-f	SB	$^*V_1Zn_1^\circ$	8.1.7.2
+1.43	1N + 3C + 1B + 3 C	5/3 A	48	70/(86)	1 a-m	*sta.*	$^*N_1^\circ$ *distorted*+$^*B_1^\circ$ *distorted*	9.1.5

(continued)

Table 10.4 (continued)

D (eV)	Impurity	Group	Total electrons	Atomic radius (nm)[a]	DAP	Type	Defect	Table
+1.43	1 B + 6 C	3 A	41	(86)	67	SB	$^*V_1B_1^\circ$	8.1.7.1
±1.49	1 Si + 3 C	4 A	32	119	109/110	SB	$^*V_1Si_1^\circ$	8.1.4
−1.52	1 Al + 3 C	3 A	31	141	66 a-e	SB	$^*V_1Al_1^\circ$	8.1.7.1
+1.54	1 H + 4 C	1 A	25	(140)	41 a-e	SB	$^*(V_1H_1)^\circ$	8.1.7.1
+1.56	4 N + 0 C	5 A	28	70	82 a-e	SB	$V_1N_4^\circ = B(b)$	8.1.3.5
−1.57	1He + 3C	8 A	20	(90)	95 a-f	SB	$^*V_1He_1^\circ$	8.1.7.1
−1.58	1Li + 3C	1 A	21	77	68 a-g	SB	$^*V_1Li_1^\circ$	8.1.7.1
+1.60	1 N + 2 C	5 A	19	70	37 a-l	SB	$^*N_3V_2N_1^\circ = \mathbf{H4}$	8.1.3.4
+1.63	3 C	4 A	18	76	15 a-l	SB	$^*V_2^\circ = \mathbf{TH5}$	8.1.1
1.64	(2 C)	4 A	(12)	76	–	–	Pure diamond crystal	9.7.1

[a][Fin56], Fig. 1, with (interpolated) values, SB-DAP=sideband DAP/*sta. DAP=standard DAP* with *D* = dielectric factor

Table 10.5 Dependence of the dielectric factor D of DAP centers on the charge state of the defect (for number of electrons see Tables 9.7.1 and 9.7.2)

D^a (eV)	Contributing atoms	Total electrons	DAP	Type	ZPL/L (eV)	Defect	Tables
+1.10	$(20\ C)^-$	121	93 a–f	SB	3.150	$*V_1^- =$ **ND1**	8.1.1
±1.20	**$(20\ C)^\circ$**	**120**	33/34	SB	1.673	$*V_1^\circ =$ **GR1**	8.1.1
+1.31	$(20\ C)^+$	119	32 a–l	SB	2.781	$*V_1^+ =$ **GR2–8**	8.1.1
±1.34	**$(2\ N + 12\ C)^\circ$**	**86**	50/51	SB	2.464	$*V_1N_2^\circ =$ **H3**	8.3.1
±1.42	$(2\ N + 12\ C)^+$	85	77/78	SB	1.256	$*V_1N_2^+ =$ **H2**	8.3.1
+1.30	$(2N + 1B + 8C)^-$	68	19 a–j	*sta.*	1.890	$*N_2^\circ + B_1^-$	9.1.5
+1.38	**$(2N + 1B + 8C)^\circ$**	**67**	20 a–f	*sta.*	1.407	$*N_2^\circ + B_1^\circ$	9.1.5
+1.43	$(2N + 1B + 8C)^+$	66	67 a–i	*sta.*	3.684	$*N_2^+ + B_1^\circ$	9.1.5
+1.54	$(2N + 1B + 8C)^{2+}$	65	18 a–g	*sta.*	2.334	$*N_2^{2+} + B_1^\circ$	9.1.5
±1.36	$(2Ni + 2C)^\circ$	68	107/108	SB	1.883	$*(V_3Ni_2)^\circ$	8.1.6
±1.55	$(2Ni + 0C)^+$	55	105/106	SB	2.562	$*(V_3Ni_2)^+$	8.1.6
−1.25	$(2Si + 6C)^-$	65	75 a–i	SB	2.523	$*(V_3Si_2)^-$	8.1.6
−1.39	**$(2Si + 6C)^\circ$**	**64**	65 a–e	SB	2.052	$*(V_3Si_2)^\circ$	8.1.6
±1.60	$(2Si + 2C)^+$	39	53/54	SB	1.681	$*(V_3Si_2)^+$	8.1.6
±1.24	$(1\ N + 9\ C)^-$	62	48/49	SB	1.943	$*V_1N_1^-$	8.1.3.1
±1.39	**$(1\ N + 9\ C)^\circ$**	**61**	46/47	SB	2.154	$V_1N_1^\circ =$ **NV**	8.1.3.1
+1.31	$(4\ H + 8\ C)^-$	53	39 a–j	SB	0.165	$*(V_4H_4)^-$	8.1.7.2
+1.43	**$(4\ H + 8\ C)^\circ$**	**52**	40 a–g	SB	0.412	$*(V_4H_4)^\circ$	8.1.7.2
+1.43	**$(1N + 1B + 6C)^\circ$**	**48**	1 a–m	*sta.*	1.893	$*N_1^\circ + B_1^\circ$	9.1.5
+1.53	$(1N + 1B + 6C)^+$	49	9 a–v	*sta.*	1.641	$*N_1^+ + B_1^\circ$	9.1.5

(continued)

Table 10.5 (continued)

D^a (eV)	Contributing atoms	Total electrons	DAP	Type	ZPL/L (eV)	Defect	Tables
±1.30	$(1\,B + 6\,C)^-$	42	73/74	SB	2.395	$^*V_1B_1^-$	8.1.7.1
+1.43	**$(1\,B + 6\,C)^\circ$**	**41**	67 a–i	SB	3.684	$^*V_1B_1^\circ$	8.1.7.1
+1.33	$(1\,Si + 3\,C)^-$	33	113 a–i	SB	2.991	$^*V_1Si_1^-$	8.1.4
±1.49	**$(1\,Si + 3\,C)^\circ$**	**32**	109/110	SB	2.646	$^*V_1Si_1^\circ$	8.1.4
±1.45	$(1\,Ni + 0\,C)^-$	29	16a-1	Sta	1.036	$^*(V_2Ni_1)^- =^* S8$	8.1.5
±1.53	$(1\,$**Ni**$ + $**0**$\,C)^\circ$	28	59/60	Sta	–	$^*(V_2Ni_1)^0 =^* S4,^* S5$	8.1.5
±1.63	$(1\,Ni + 0\,C)^+$	27	4/11/57/58/61	Sta	–	$^*(V_2Ni_1)^+ =$ $S2, S3, ^*S6, ^*S7, ^*S9$	8.1.5
±1.43	$(4\,N + 0\,C)^-$	29	63/64	SB	5.252	$^*V_1N_4^- =$ **N9**	8.1.3.3
+1.56	**$(4\,N + 0\,C)^\circ$**	**28**	82 a–e	SB	4.191	$V_1N_4^\circ =$ **B(b)**	8.1.3.3
+1.67	$(4\,N + 0\,C)^+$	27	81 a–g	SB	4.567	$^*V_1N_4^+ =$ **F**	8.1.3.3

[a]The sign of D is always positive for *standard* DAPs. For sideband DAPs the sign is negative in luminescence, and positive in absorption and PLE

$$D\ (\text{eV}) = 1.64 - 0.0039\ (\text{total number of electrons}). \tag{10.7}$$

It must be emphasized that Table 10.4 and (10.7) are based on assumptions concerning the number of ligands involved. Nevertheless, a guideline is presented, which allows to predict the dielectric factor D for the DAP analysis of a *known impurity*. If an unquestionable value for the dielectric factor D *is known*, conclusions on the defect structure are possible.

Exceptions are expected for unusual DAP centers.

1. For ($P_1{}^\circ + B_1{}^\circ$)-DAP2 the electrons are highly delocalized (two shallow impurities). The unquestionable D value is 1.72 eV, while (10.7) predicts $D = 1.41$ eV.
2. For the $\langle 1,0,0 \rangle$ split shells interstitials there are four different DAP centers with D factors 1.30 and 1.31 eV (see Table 8.2.1)
3. Associates of the $\langle 1,0,0 \rangle$ split shells interstitials with nitrogen, boron or nickel have D factors of 1.22 to 1.25 eV (see Table 8.2.3)
4. The $\langle 1,0,0 \rangle$ $(N_1C_1)_i$ interstitials and boron associates have D factors of 1.20 to 1.27 eV (see Table 8.2.4)
5. Evidently, the different structures and charge states for the exception 1–4 do not follow the relationships of (10.7) and (10.8a, b)

10.5 Dependence of the Dielectric Factor *D* on the Charge State

There are several centers with DAP transitions, which exist in two or more different charge states. In Table 10.5, the corresponding D values are listed. There is a quite evident change of the D value with the change of the charge state. When the charge state changes from singly negative to neutral to singly positive, the D value is increased for each step by **0.11 ± 0.05 eV** (10.8a, b). The physical reason for this relationship is not known. However, this regularity helps in structure assignments to defects with DAP transitions.

$$D(\text{neutral defect}) = D(\text{negative defect}) + 0.11\,\text{eV}, \tag{10.8a}$$

$$D(\text{positive defect}) = D(\text{neutral defect}) + 0.11\ \text{eV}. \tag{10.8b}$$

Chapter 11
Vibrational Frequencies of Defect Centers in Diamond

In many spectra, the zero phonon line (ZPL) is accompanied by sidebands which arise from a coupling to localized vibrational modes at the defect center (see Table 5.10 in [Zai01]). These sidebands occur on the high energy side in absorption and on the low energy side in luminescence spectra. In some cases, the respective frequencies are also observed as infrared absorption bands.

For light impurity atoms ($m < 12$), true local vibrational modes (LVM) are observed (Sect. 11.1). Exceptions are weakly bonded heavier atoms like silicon or nickel. In general, the LVM frequency in absorption is a few percent higher than in luminescence. Examples are the 5RL center (+2%, see Table 8.2.3) or the TR12 center (+8%, see Table 8.2.1).

For heavy impurity atoms ($m > 25$), quasilocal vibrational modes (QLVM), which have been recently discussed [Zai00a], can be observed (Sect. 11.2). The QLVM frequencies are below the one-phonon lattice vibrations (86–165 meV). Exception are multiple light atoms in a cluster like N_1C_1 from $^*(N_1C_1)_i$, or C_3 from distorted $N_1{}^\circ = C$ center (see Tables 11.2 and A.3).

Lattice phonons (1LP) are not observed as sidebands of normal optical transitions. However, the 1LP frequencies LA = 87 meV, TA = 141 meV, and LO = 163 meV occur in sidebands from free or bound excitons (Sect. 11.3).

11.1 Local Vibrational Modes

The usual equation for the LVM frequency is derived from a vibrating diatomic molecule and is given by [Tho72]

$$f = \left[k\left(\frac{1}{m} + \frac{1}{\alpha M}\right)\right]^{1/2}, \tag{11.1}$$

B. Dischler, *Handbook of Spectral Lines in Diamond*,
DOI 10.1007/978-3-642-22215-3_11, © Springer-Verlag Berlin Heidelberg 2012

Table 11.1 LVM frequencies for C–C or N–C vibrations at intrinsic defects in diamond and comparison with C–C vibrations in diamond related materials (Table 9.3.5)

Observed LVM frequencies (meV)								
Tables	8.1.1	8.1.3.1–8.1.3.5	8.1.4.1–8.1.7.2	8.2.1	8.2.1	8.2.3	8.2.4	9.3.5[a]
Defect	V_1	$V_m N_n$	$V_m X_n$	$(C_2)_i$	$(C_{2n})_I$ platel.	$(C_2)_i\, N_n(C_2)_i\, B_n$	$(N_1 C_1)_i$	$C\text{–}CH_n$
Type of vibration								
C/N(sp^3)–C(sp^3) bend	64–80	41–75	45–70	–	–	–	–	
C/N(sp^3)–C/N(sp^3) stretch	–	97–177	128	–	–	–	–	104–144
C(sp^2) = C(sp^2) out of plane bend	–	–	–	65–80	41	90	105–108	–
C/N(sp^2)–C/N(sp^3) mixed stretch	–	–	–	120–199	150–173	160–195	120–191	154–188
C/N(sp^2) = C/N(sp^2) stretch	–	–	–	209–218	177	202–245	199–238	196–198
C(sp^1)≡C(sp^1) stretch	–	–	–	–	–	–	–	270

[a]In Table 9.3.5 frequencies of C–C bond and stretch vibrations from H bonded carbon in diamond-related materials are listed
DLC amorphous *d*iamond-*l*ike *c*arbon, *PAC* *p*olymer-like *a*morphous *c*arbon, *NDC* *n*atural *d*iamond *c*oat

Table 11.2 Observed and calculated QLVM data (Ni centers see Table 9.5.9)

Defect	Observed				Calculated (see Table A.3)				Comment
	Tables	ZPL/L (eV)	Frequ. (meV)	Width (meV)	Atoms	Total mass	Frequ. (meV)	Width(meV)	
$*(V_2Tl_1)^{\circ}$	8.1.7.2	2.017	21	6	$^{204}Tl_1$	204	21.5	4.9	[Zai00a]
$*(V_2Ag_1)^{\circ}$	8.1.7.2	3.111	22	7	$^{108}Ag_1\,^{12}C_8$	204	21.5	4.9	[Zai00a][a]
$*V_1Ni_3{}^{\circ}$	8.1.4.1	2.370	23	7	$^{59}Ni_3$	177	23.3	5.8	
$*(V_2W_1)^{\circ}$	8.1.7.2	1.735	24	6	$^{184}W_1$	184	22.7	5.5	[Zai00a]
$*(V_2Ta_1)^{\circ}$	8.1.7.2	1.774	24	6	$^{181}Ta_1$	181	22.9	5.6	[Zai00a]
$*(V_3Co_2)^{\circ}$	8.1.6	1.989	25	8	$^{59}Co_2\,^{12}C_2$	142	26.1	7.3	[Zai00a]
$*(V_3Si_2)^{-}$	8.1.6	2.523	26	20	$^{28}Si_2\,^{12}C_8$	152	25.2	6.7	
$*(V_2Xe_1)^{\circ}$	8.1.7.2	1.528	29	8	$^{131}Xe_1$	131	27.3	7.9	
$*V_1Zn_1$	8.1.7.1	2.393	30	12	$^{65}Zn_1\,^{12}C_3$	101	31.6	10.6	[Zai00a][a]
$*V_1Cr_1$	8.1.7.1	1.673	31	12	$^{52}Cr_1\,^{12}C_3$	88	34.2	12.4	[Zai00a][a]
$*(V_2Ti_1)^{\circ}$	8.1.7.2	1.249	35[b]	15[b]	$^{48}Ti_1\,^{12}C_3$	84	35.1	13.1	[Zai00a][a]
$*V_1Li_1{}^{\circ}$	8.1.7.1	1.881	36	15	$^{7}Li_1\,^{12}C_6$	79	36.4	14.1	
$*V_1He_1(a)^{\circ}$	8.1.7.1	2.212	38	17	$^{4}He_1\,^{12}C_6$	76	37.3	14.7	
$*(C_2)_iN_2{}^{-}$	8.2.3	2.367	38	15	$^{14}N_2\,^{12}C_4$	76	37.3	14.7	
$*(C_2)_iN_2{}^{-/\circ}$	8.2.3	1.979	37	15	$^{14}N_2\,^{12}C_4$	76	37.3	14.7	
$*(N_1C_1)_i{}^{-}$	8.2.4	3.988	38	16	$^{14}N_1\,^{12}C_5$	74	37.9	15.2	
$*(C_2)_i{}^{\circ}$	8.2.1	1.685	38–40	20	$^{12}C_6$	72	38.5	15.7	**R2center**
$*V_1Al_1{}^{\circ}$	8.1.7.1	2.974	42	20	$^{27}Al_1\,^{12}C_3$	63	41.7	18.5	
$*(V_3Si_2)^{-}$	8.1.6	2.523	44	40	$^{28}Si_2$	56	44.9	21.4	[Zai00a]
$*(V_3Si_2)^{\circ}$	8.1.6	2.052	45	23	$^{28}Si_2$	56	44.9	21.4	[Zai00a]
$*(V_3Si_2)^{+}$	8.1.6	1.681	43	25	$^{28}Si_2$	56	44.9	21.4	[Zai00a]

(continued)

Table 11.2 (continued)

Defect	Observed				Calculated (see Table A.3)				Comment
	Tables	ZPL/L (eV)	Frequ. (meV)	Width (meV)	Atoms	Total mass	Frequ. (meV)	Width(meV)	
$*(C_2)_iN_2^-$	8.2.3	2.367	48	30	$^{14}N_2\,^{12}C_2$	52	47.1	23.6	
N_2°	9.1.3	4.470	60	40	$^{14}N_1\,^{12}C_2$	38	58.5	36.3	**A center**
$*(C_2)_iN_1^\circ$	8.1.2	4.582	58	40	$^{14}N_1\,^{12}C_2$	38	58.5	36.3	**5RL center**
$*(C_2)_iN_1^+$	8.1.2	4.676	58	40	$^{14}N_1\,^{12}C_2$	38	58.5	36.3	
$*(N_1C_1)_i^\circ$	8.2.4	3.188	58	40	$^{14}N_1\,^{12}C_2$	38	58.5	36.3	[Zai00a]
N_1°	9.1.1	4.059	61	40	$^{12}C_3$	36	60.9	39.3	**C center**
$*V_1N_1^-$	8.1.3.1	1.943	63	40	$^{12}C_3$	36	60.9	39.3	
$*V_1Si_1^-$	9.4.1.1	2.991	74	60	$^{28}Si_1$	28	74.5	59.0	
$*V_1Si_1^\circ$	9.4.1.1	2.646	74	65	$^{28}Si_1$	28	74.5	59.0	
$*V_2N_2^+$	8.1.3.4	2.086	75	60	$^{14}N_2$	28	74.5	59.0	
$*(C_2)_iN_2^-$	8.2.3	2.367	75	60	$^{14}N_2$	28	74.5	59.0	
$*(C_2)_iN_2^+$	8.2.3	2.535	73	65	$^{14}N_2$	28	74.5	59.0	
$*(N_1C_1)_i^-$	8.2.4	3.988	79–80	70	$^{14}N_1\,^{12}C_1$	26	79.7	67.4	**R11 center**
$*(N_1C_1)_i^\circ$	8.2.4	3.188	79–80	70	$^{14}N_1\,^{12}C_1$	26	79.7	67.4	**3.188 eV**

[a] In [Zai00a] rather improbable pair assignments are proposed, i.e., Ag_2, Zn_2, Cr_2, and Ti_2

[b] In [Zai00a] too small values of the observed frequency and line width for $*(V_2Ti_1)^\circ$ are quoted, neglecting the coincidence of the broad QLVM line with a sharp LVM line

where k is the force constant (in units of the frequency squared), m is the impurity mass, M is the ligand mass ($M > m$), and α is a fitting parameter ($\alpha = 1$ for the diatomic molecule, $\alpha = 1.22$ for Si:B, $\alpha = 1.46$ for Si:C, and $\alpha = 5.98$ for GaP:C(Ga) [Tho72]). The values for α are obtained from the known isotope shifts. For diamond, the observed isotope shifts indicate a very large α value, i.e., the approximations of (11.2a, b) are valid.

Some characteristic LVM frequencies are listed in Tables 11.1 and 9.5.8 and are discussed below. For the calculation of the force constant k, it was assumed that $1/\alpha M$ in (11.1) is negligible, which allows the use of the approximations

$$f = \left(\frac{k}{m}\right)^{1/2}, \quad (11.2a)$$

$$k = f^2 m. \quad (11.2b)$$

Hydrogen ($m = 1$): The C–H stretch frequencies (from Table 9.3.1) are in the range 350–368 meV for sp^3 bonding, 368–378 meV for sp^2 bonding, and 412 meV for sp^1 bonding. The corresponding average force constants [in units 10^4 $(meV)^2$] are 13, 14, and 17, respectively.
Boron ($m = 11$): The boron LVM frequency (see Table 9.2.1) is 160–162 meV with a resulting force constant of 29×10^4 $(meV)^2$.
Carbon ($m = 12$): The C–C stretch frequencies (see Table 11.1) are 97–177 meV for sp^3 bonding. For the split self-interstitial centers, frequencies of 120–199 meV for mixed sp^2/sp^3 bonding and 177–245 meV for sp^2 bonding are observed. For completeness, the 270-meV frequency for sp^1 bonding (in polymer-like amorphous carbon) is included. This leads to average force constants (in units10^4 $(meV)^2$) of 22, 31, 53, and 88, for sp^3, mixed sp^2/sp^3, sp^2, and sp^1, respectively. The force constant for the lattice Raman frequency (165.2 meV) is 32.7×10^4 $(meV)^2$.

Note, that there are generally reduced vibrational frequencies in platelets ($(C_{2n})_i$, in Table 8.2.1, compared to the isolated self-interstitials ($(C_2)_i$, in Table 8.2.1), indicating a lengthening of the bonds. The reduction is −43% for the $(sp^2) = (sp^2)$ out of plane bend, −2% for the mixed (sp^2)–(sp^3) stretch at the surface of the platelets (150–165 meV), +6% for the mixed (sp^2)–(sp^3) stretch at the rim of the platelets (168–172 meV), and −18% for the $(sp^2) = (sp^2)$ stretch.
Nitrogen ($m = 14$): The frequencies of N–C vibrations are very similar to those of the C–C vibrations (see Table 11.1).
Silicon ($m = 28$): The LVM frequencies for the $^*(V_3Si_2)^{o/+}$ centers are 65–70 meV. The corresponding force constant is 13×10^4 $(meV)^2$. This low value illustrates the weak bond between the two interstitial silicon atoms (each between two vacancies).
Nickel ($m = 58.7$): Evidently, for the Ni–Ni or Ni–C bonds, the force constants are very similar to those of the C–C bonds (see Table 11.1).
Silver ($m = 108$): The 67-meV frequency of the $^*(V_2Ag_1)^\circ$ center (see Table 8.4) corresponds to a force constant of 48×10^4 $(meV)^2$.
Thallium ($m = 204$): A force constant of 41×10^4 $(meV)^2$ is derived from the 45-meV frequency of the $(V_1Tl_1)^\circ$ center (see Table 8.1.7.2).

11.2 Quasilocal Vibrational Modes

It has been shown that the theory for QLVM [Bro62] can explain the frequencies and line width of heavy impurities (^{28}Si, ^{48}Ti, ^{52}Cr, ^{59}Ni, ^{59}Co, ^{64}Zn, ^{108}Ag, ^{181}Ti, ^{184}W, and ^{204}Tl) [Zai00a]. The theoretical expressions for the frequency f and the line width w of a acoustic QLVM vibration are:

$$f = f_D \left[\frac{M_C}{3} (nM_I - M_C) \right]^{1/2}, \qquad (11.3)$$

$$w = f_D \frac{(\pi M_C/6)}{(nM_I - M_C)}, \qquad (11.4)$$

where $f_D = 150$ meV is the Debye frequency of diamond, $M_C = 12$ is the host atom mass, and n is the number of impurity atoms M_I. After insertion of f_D and introducing the reduced mass $M_R = (nM_I - M_C)$, the (11.3) and (11.4) become:

$$f\,(\text{meV}) = \left[\frac{10^6}{(11.25 M_R)} \right]^{1/2}, \qquad (11.5)$$

$$w\,(\text{meV}) = \left[\frac{10^4}{(10.6 M_R)} \right]^{1/2}. \qquad (11.6)$$

Equations (11.5) and (11.6) are used to calculate the QLVM data in Table A.3.

In Table 11.2 observed QLVM data are compared with calculated values. The observed frequencies are in the range 21–80 meV, corresponding to total masses of 204 (Tl_1) – 26 (N_1C_1). The agreement with the calculated values is very good (deviation < 2 meV). It should be pointed out that of the 34 vibrating units in Table 11.2, only eight are single impurity atoms, while 26 are clusters with up to eight carbon atoms.

11.3 Lattice Phonons in Free and Bound Exciton Sidebands

Lattice phonons at $k = 0.76$ in the $\langle 0\ 0\ 1 \rangle$ direction (minimum of the conduction band) have values of 87 meV (TA), 141 meV (TO), and 163 meV (LO). In sidebands of the free exciton, the 141-meV phonon is dominant and is observed up to the fourth harmonic [Dea65b]. The 87-meV phonon is weaker (up to the third harmonic), and the 163-meV phonon is very weak (no harmonics). The bound exciton spectra of lithium, boron (see Table 9.2.6), and phosphorous show a similar coupling to lattice phonons.

Chapter 12
Modification of Diamond by Irradiation and Heat

More optical centers are observed in modified diamond than in natural diamond. An example is given by the 75 centers of intrinsic defects and associates in Chap. 8, where only 24 occur in natural diamond.

An important method in creating additional defect centers in diamond is the irradiation with X-rays, high energy electrons, or fast neutrons (Sect. 12.1).

Ion implantation is used to introduce specific impurities with simultaneous radiation damage (Sect. 12.2).

With heat treatment, the migration of intrinsic defects and of impurities is activated. Both aggregation and dissociation can occur, i.e., new defect centers are created or annealed out (Sect. 12.3).

12.1 High Energy Irradiation by X-rays, Electrons, or Neutrons

The basic radiation damage is the displacement of a carbon atom from its lattice site, leaving behind one vacancy and one self-interstitial. Also some lattice disorder occurs. The three types of irradiation differ by their penetration depth and by the resulting lattice disorder.

12.2 Ion Implantation

With ion implantation and subsequent annealing at 800–1,400 °C, optically active centers of He, Li, Ti, Cr, Zn, Ag, Xe, and Tl have been successfully produced (see Tables 8.1.7.1 and 8.1.7.2). All these centers are associated with one or two vacancies. Nitrogen ion implantation is very efficient in producing nitrogen containing optical centers in diamond. One example is the $^{*}(C_2)_i N_1^-$ center at

B. Dischler, *Handbook of Spectral Lines in Diamond*,
DOI 10.1007/978-3-642-22215-3_12, © Springer-Verlag Berlin Heidelberg 2012

3.188 eV (see Table 8.2.3). With carbon ion implantation the just mentioned center can also be obtained. For many ions the implantation did not produce optical centers (Be, F, Na, S, Cu, Ga, Y, Nb, Mo, Pd, Cd, Sb, and Au) [Zai00a].

12.3 Heat Treatment

Heat treatment is a powerful tool to remove lattice disorder after irradiation or ion implantation. The self-interstitial $(C_2)_i{}^\circ$ with a migration activation energy of 1.7 eV anneals out at temperatures of 420–700 °C. At slightly higher temperatures, i.e., 500–1,000 °C the single vacancy anneals out, having a migration activation energy of 2.4 eV [Mai94b].

At higher temperatures, a protecting pressure is necessary during anneal to avoid graphitization. This is usually accomplished by using the same apparatus as for HPHT diamond growth.

The heat (1,500–1,900 °C) induced nitrogen aggregation ($N_1{}^\circ$ (C center) to $N_2{}^\circ$ (A center) to $V_1N_4{}^\circ$ (B center)) is described in Sect. 9.1.6.1 for *nonirradiated* samples. In the presence of vacancies (*irradiated* samples), association with nitrogen centers starts above 600 °C and centers are formed, which also occur naturally: NV (= $V_1N_1{}^\circ$) from $N_1{}^\circ$ (C center), H3 (= $V_1N_2{}^\circ$) from $N_2{}^\circ$ (A center), and H4 (= $V_2N_4{}^\circ$) from $V_1N_4{}^\circ$ (B center) (see Sect. 9.1.6.2).

For nickel centers in nitrogen-rich samples, interesting annealing effects occur in the range 1,500–1,900 °C (see Sect. 9.5.6 and Table 9.5.6). The quantitative transformation of $^*(V_3Ni_2)^\circ$ at 1.883 eV into $^*V_1Ni_2{}^+$ at 1.693 eV (at $T <$ 1,800 °C) has been documented [Law93a].

In a forthcoming book [Zai10] the optical changes, which result from HPHT treatment of diamond, are described in detail.

Chapter 13
Isotopic Line Shifts in Diamond

Isotopic line shifts from $^{2}H/^{1}H$, $^{13}C/^{12}C$, and $^{15}N/^{14}N$ substitution are collected in Tables 7.1, 7.2, and 7.3, respectively. The effects of ^{13}C or ^{15}N substitution have been discussed in the literature [Col87a], [Col87b,Col88c,Col90b,Dav94b,Dav99a, Dav01a]. The shifts from the natural isotopes ^{60}Ni, ^{61}Ni, ^{62}Ni, and ^{64}Ni are given in Table 7.5. Calculated and observed shifts and intensities from the natural isotopes ^{28}Si, ^{29}Si, and ^{30}Si are shown in Tables 7.4 and 9.4.2.

13.1 Isotopic Line Shifts at Zero Phonon Lines

^{2}Hydrogen = D: The isotopic line shifts in zero phonon lines (ZPL) are expected to be small. For $^{2}H = D$ substitution (Table 7.1), an isotope shift of –0.5% is observed for the 0.412 eV ZPL= $(V_4D_4)^{\circ}$–DAP40. Remarkable is the splitting of the DAP40c (shell = 20) line into five equidistant components (intensities 1:4:6:4:1) for 50% D substitution with a total shift of –2.2%. The involvement of four equivalent hydrogen atoms is thus established.

^{13}Carbon: In ^{13}C substituted samples (Table 7.2), isotope shifts of –0.3% to +0.3% are found. In 17 cases, the shift is *negative*, in two cases there is no shift, and in 38 cases the shift is *positive*. In ^{13}C diamond, the bond length is decreased by 0.015%, and a compression of the lattice usually increases the zero phonon energy [Dav99a].

Nickel: Line splittings arising from the five natural isotopes of nickel (68% ^{58}Ni, 26% ^{60}Ni, 1% ^{61}Ni, 4% ^{62}Ni, and 1% ^{64}Ni) are observed for three centers (see Table 7.5). A very small shift of –0.006% per increasing mass unit is observed for the *1Ni 1.40 center (single nickel atom). Only the three most abundant isotopes are seen in the DAP62 transitions for the shells 9, 14, 22, 64, and 162 $(^{*}(V_2Ni_1)N_{x+y+z}$ = *S7 center). The average shift is +0.28% per increasing mass unit. The splitting into five equidistant line groups reveals the involvement of two

DOI 10.1007/978-3-642-22215-3_13, © Springer-Verlag Berlin Heidelberg 2012

equivalent nickel atoms in the *2Ni 2.56 center. The shift is +0.03% per increasing mass unit.

Silicon: The three natural isotopes of silicon (92% ^{28}Si, 5% ^{29}Si, and 3% ^{30}Si) give rise to splittings from two equivalent silicon atoms in the spectra (absorption and luminescence) of the *2Si 1.68 = *$(V_3Si_2)^+$ center (see Tables 7.4 and 9.4.2). The average shift is –0.02% per increasing mass unit. A splitting into five lines from the combined masses 56, 57, 58, 59, and 60 is expected. However, the calculated intensities for the masses 59 and 60 are below 0.13%, which is too weak for observation. The almost perfect fit of the observed and calculated intensities (for two equivalent silicon atoms) rules out the possible involvement of a single silicon atom (see Table 9.4.2).

13.2 Isotopic Frequency Shifts at Local Vibrational Modes

^{2}Hydrogen=D: The observed shifts for vibrational C–H lines after ^{2}H=D substitution are close to the theoretical value of –29.3% (see Table 7.1). Some smaller shifts (–21.7% to –22.8%) are observed especially for sp^2 and sp^1 hybridization.

^{13}Carbon: After ^{13}C substitution, the observed frequency shifts for C–C vibrations (including the lattice phonons 1LP(q), 2LP(c, e, o, p), and 3LP(a, b)) are close to the theoretical value (–3.9%). Observed shifts for centers containing N atoms are in the range –1.6% to –2.9%. This indicates a partial localization of the vibrating energy at the carbon atom in the C–N vibration. Examples are: 60% N for the 0.212-eV line of *$(N_1C_1)_i{}^+$ and 30% N for the C(d′) line at 0.140 eV (see Tables 7.2 and 9.1.1).

^{15}Nitrogen: The full theoretical shift for N–N vibration is –3.4%. Vibrational line shifts in the range –1.3 to –1.9% are observed for ^{15}N substitution. This implies a partial localization of the vibrating energy at the carbon atom in the N–C vibration. Examples are 70% C for the C(d′) line at 0.140 eV and of 40% C for the 0.212 eV line of *$(N_1C_1)_i{}^+$. An interesting result is the –2% shift for the *$(N_2)_i{}^\circ$ center (0.230 eV line), which indicates a 40% C participation in the central N–N vibration.

Partial ^{15}N substitution revealed a single nitrogen atom (two lines for mixed isotopes) in the *$(N_1C_1)_i{}^\circ$ center (H1a line at 0.180 eV). Two equivalent nitrogen atoms (three lines for mixed isotopes) were found for the* $(N_2)_i{}^\circ$ center (0.230 eV line) (see Table 7.3).

13.3 Isotopic Frequency Shifts at Quasilocal Vibrational Modes

^{13}Carbon: The isotope shift for quasilocal vibrational modes (QLVM) frequencies is larger than for local vibrational modes (LVM) vibrations, and it depends on the number of atoms involved (see Table A.3). For ^{13}C substitution the theoretical **QLVM** shift is –7.4% for 2C (79.7/86.1 meV), –5.7% for 3C (57.4/60.9 meV),

−5.2% for 4C (47.1/49.7 meV), and −4.7% for 5C (41.0/43.0 meV). Experimental shifts are observed for two sidebands of the 3.188 eV = $^{*}(C_2)_iN_1^-$ center: −3.6 (calculated −3.4 for 1N+1C) meV at 3.116 eV, and −3.4 (calculated −3.8 for 1N+2C) meV at 3.136 eV (see Table 7.2).

Chapter 14
Spectroscopic Discrimination Between Natural and Nonnatural Diamond

Nonnatural diamonds are, of course, all diamonds from HPHT or CVD synthesis. Of increasing interest are diamonds, which were originally natural, but have been artificially changed in their properties.

The first question is: Are there spectroscopic properties, which are observed exclusively either in natural or in nonnatural diamond?

The answer for synthetic diamond is: Yes. Now, the second question concerns the spectroscopic method: Preferable is a simple apparatus, which can finally be developed into a handheld version. The apparatus should also accept raw or multifacetted diamonds, in contrast to the polished parallel plates, generally used in spectrometers.

Indeed, the method named "diamond view" perfectly meets these requirements (see Sect. 14.1).

14.1 Existing Methods

Researchers (Paul Spear et al.) at the discrimination laboratory of De Beers in London have presented the "diamond view" method [Pen00]. A bright blue luminescence is observed (after excitation with hard ultraviolet light) in synthetic diamonds. Very characteristic is a phosphorescence for 3–5 s after the UV light is switched off. In natural diamonds the phosphorescence is totally absent, and the blue luminescence is also absent or very weak [Shi92].

No satisfactory physical explanation for the phenomenon can be given at present. Initially, it was proposed, the difference should arise from the different growth planes of diamond. These are dominant octahedral in natural and dominant cubic in synthetic diamonds. This explanation turned out to be insufficient, after purely octahedral synthetic diamonds were grown [Pen00]. The phenomenon originates rather from the significantly different growth history. The natural diamonds have a growth time of many years and experience annealing conditions during millions of years. In contrast, the synthetic diamonds are grown within several days.

B. Dischler, *Handbook of Spectral Lines in Diamond*,

DOI 10.1007/978-3-642-22215-3_14,

14.2 Alternate Methods

(a) Synthetics: For the future is may be necessary to develop alternate methods, replacing the very efficient "diamond view" method, which is the best spectroscopic choice at present, but can possibly be ruled out by novel synthesis methods.

One existing alternate method is the search for growth features, which are characteristic for synthetic diamonds. Here again, these features could be avoided in the future with refined technology.

It is promising to look for spectral features, which occur *exclusively* in *natural* diamonds. The (V_1H_1) center can serve as an example. It is a common defect in natural diamonds, but is absent in synthetic diamonds. There are two *sharp* absorption lines arising from C–H stretch (0.3852 eV) and bend (0.1742 eV) vibrations (see Table 9.3.1). Using a tunable laser, the absorption ratio measured at the line and in the neighborhood should provide high sensitivity.

In Table 14.2, additional lines are listed, which after present knowledge should be exclusive to natural diamond. Broad bands are omitted, because their identification is more difficult than for sharp lines.

(b) Fancy Colors: The artificial coloring of diamond by radiation damage and annealing has been discussed in detail in [Col97]. More recent results on the recognition of HPHT treatments of diamond will be published in a forthcoming book [Zai10].

Table 14.2 Lines (probably) exclusively observed in natural diamond

Label	Energy (meV)	Defect	Name	Comment
NA0058	57.65	?		Fig. 3.7, [Fer96]
NA0174a	174.2	$^*(V_1H_1)^o$		Natural C–H bend
NA0182a	182.2	$^*(V_1N_1H_1)^o$		Natural N–H bend
NA0174b	385.2	$^*(V_1H_1)^o$		Natural C–H stretch
NA0182b	401.2	$^*(V_1N_1H_1)^o$		Natural N–H stretch
NL1264	1.264	?		
NL1328	1.328	?		
NL1360	1.360	?		
NA1500	1.500	?	**N1**	
NL1559	1.559	?		
NL1946	1.946	?		NBD
NL2001	2.001	?		
NL2024	2.024	?		
NL2082	2.082	?		NBD
NL2088	2.088	?		NBD
NL2099	2.099	?		
NL2134	2.134	?		
NL2157	2.157	?		NBD, related to NL2082?
NL2167	2.167	?		NBD, related to NL2082?
NL2175	2.175	?		NBD, related to NL2082?

(continued)

Table 14.2 (continued)

Label	Energy (eV)	Defect	Name	Comment
NA2202	2.202	?		NBD
NL2205	2.205	?		NBD
NL2391	2.391	?		NBD
NA2596	2.596	?	**N2**	
NA2615	2.615	?		NBD
NA2767	2.767	?		NGD
NA2792	2.792	?		NGD
NA3593	3.593	?		
NA3877	3.877	?		
NA4191	4.191	?	**N8**	
NA4468	4.468	?		
NA4646	4.646	?		
NL4986	4.986	?	**N** line	
NA4990	4.990	?	**N10**	
NL4999	4.999	?	**M** line	
NL5037	5.037	?	**L** line	
NL5092	5.092	?	**K** line	

? unknown defect

Chapter 15
Conclusions and Outlook

15.1 Conclusions

One basic idea for this book was to split up the huge number of spectral lines into 27 tables (see Chaps. 2–7), based on the two criteria, (a) the type of sample and (b) the method of observation. A further aim was the interpretation, i.e., assignment of the lines to specific defects, and (hopefully) identification of the corresponding structure.

Structure assignments: This handbook is a breakthrough in the understanding of the large number of spectral lines in diamond. Data on more than 2,000 lines and bands are presented in 200 tables, including many unpublished results.

With a novel organization scheme, the search for a specific line is greatly simplified as a benefit for researchers and students.

In order to meet the interest in the understanding of the spectra, structure assignments for 80% of the lines are given, of which 15% only were published before. The majority of the structures for the 300 centers is explained in most cases for the first time.

Examples for successful structure assignments are: The infrared centers D = platelets =* $(C_{2n})_i^\circ$, E =* N_1^+, F =* $V_1N_4(C_2)_i^\circ$, and the visible or ultraviolet centers TR12-TR17 =* $(C_2)_i^-$, and N9 = $V_1N_4^-$.

Donor–acceptor pair (DAP) transitions: A considerable progress is achieved by the identification of 97 donor–acceptor pairs (116 DAP entries in Table 10.1), which (with two exceptions: DAP1 [Dis94a, Dis94b] and DAP2 [Ste99a, Ste99b]) are published for the first time. The group of lines GR2–GR8, which for many years was a "basic puzzle," is now explained by DAP32. The DAP results are of great help when discussing the defect structures.

Sideband type DAPs: These occur exclusively with intrinsic defects (vacancies, interstitials, and their associates). If a sideband DAP is found, the respective center must contain an intrinsic defect. Conversely, if the ZPL of a center with an intrinsic

B. Dischler, *Handbook of Spectral Lines in Diamond*,
DOI 10.1007/978-3-642-22215-3_15, © Springer-Verlag Berlin Heidelberg 2012

defect is found, one can look for DAP transitions, which can be weak enough to be overlooked (see Table 10.1).

Depth of donor centers: In Table 10.3, the depth of five deep nitrogen centers (E_D between 1.44 and 3.48 eV) are listed, together with 2.59 eV for oxygen and 0.62 eV for the shallow phosphorous donor.

Dependence of the dielectric factor D on the total number of electrons: For neutral DAP centers, there is a close relationship between the dielectric factor D and the total number of electrons. This allows to determine the number of ligand carbon atoms involved (see Table 10.4).

Dependence of the dielectric factor D on the charge state: In Table 10.5 an empirical law is demonstrated. Going from the negative charge state over neutral to the positive charge state, the dielectric factor is increased by 0.05–0.16 eV with an average of 0.11 eV. One surprise was the positive charge state of the GR2-8 center ($V_1{}^+$). This explains the intense photoconductive response of GR2-8.

Transition metals: Until recently, only six nickel centers were described [Col97b]. In Tables 9.5.1.1–9.5.1.2 the present knowledge on 20 nickel centers with eight standard DAPs (S2-*S9), and eight sideband DAPs, and four centers without DAP lines are listed.

Surprisingly, cobalt centers in diamond were unknown until a pioneering article appeared [Law96]. In Table 9.6 all observed lines are assigned to six cobalt centers, including three sets of DAP transitions.

Of the other solvent metals, only lines from chromium and iron are found (see Table 9.7.2). Using doping or ion implantation, optical centers of Ti, Zn, Zr, Ag, Ta, and W have been produced (see Table 9.7.2).

15.2 Outlook

Continued research is necessary for a better understanding of those optical centers in diamond, which have only a proposed assignment and structure (marked by a preceding *). Also, there are still 20% of the lines, which are not yet assigned.

Spectra: A better signal-to-noise ratio is desirable for many DAP transitions. Often, these sidelines are weak and were of no interest before the present DAP analysis.

Isotopes: One powerful method in defect structure assignment is isotopic line shift analysis: (^{2}H, ^{13}C, ^{15}N, ^{29}Si, ^{30}Si, ^{60}Ni, ^{62}Ni, see Tables 7.1–7.5). Especially, *partial* isotope substitution can reveal the number of atoms involved (e.g., 3 lines for $^*(N_2)_i{}^\circ$ (see Table 7.3) or 5 lines for $^*(V_4D_4)^\circ$ (see Table 7.1).

Photoconductivity spectra: In Chap. 10 of [Zai01], 59 PC peaks or thresholds are listed. Until now, only in few cases a correlation with optical centers is given. The defects with DAP transitions are good candidates for PC measurements. Two excellent examples, where optical absorption and PC spectra coincide, are $^*V_1{}^+$ – DAP32 = GR2-8, and $^*V_1N_4{}^-$ – DAP63 = N9(a).

Electron spin resonance: More ESR results are desirable. Many defects have several charge states (see Table 10.5) and are probable candidates for ESR measurements.

Correlations: It is important to measure for one particular sample the optical lines in different spectral regions. Until the pioneering work with IR and UV correlation appeared [Naz87], the well known A, B, and C centers were regarded as isolated IR centers, ignoring their UV lines. It is expected, that, more such correlations can be established. An elegant proof for the correlation of two charge states of the same defect is the quantitative photochromic interconversion between two spectra (e.g., $V_1{}^\circ$ = GR1 and $V_1{}^-$ = R10 = ND1). Many more defects exist in several charge states (see Table 10.5).

Theory: The theoretical treatment of defects in diamond should be extended. Theorists can hopefully provide a comprehensive treatment for the new field of DAP transitions, including the phenomenon of high intensity shells, and the near shell correction.

Nanodiamond: These new materials (nanowires, nanoparticles) are of high interest for applications. A comprehensive description of the *optical* properties is desirable.

Appendix A

A.1 Historical, Existing, and New Names of Centers in Diamond

It is not surprising that with the large number of spectral lines in diamond a considerable number of names appeared in the literature.

A.1.1 Historical Names

The 53 *historical* names in Table A.1.1 were introduced in a systematic way by [Cla56a, Cla56c]: Letters A–F for the infrared centers; names N1–N10 for lines in *natural* diamond; GR1–GR8 for lines after *general radiation*; radiation lines R9–R11, radiation lines TR12–TR17 in *type II* diamonds (now also in type I); and lines H1–H18 in *heat treated* diamonds (usually after irradiation).

A.1.2 Existing Names

The 66 *existing* names as mentioned in [Zai01] are given in Table A.1.2.

A.1.3 New Names

The 72 *new* names in Table A.1.3 are introduced in this book.

B. Dischler, *Handbook of Spectral Lines in Diamond*,

DOI 10.1007/978-3-642-22215-3,

Table A.1.1 Historical names of 17 defect centers or lines in diamond [Cla56a, Cla56c],[Wal79]

Name	Alternate name	Defect/impurity	Energy (eV)	Observed	Comment	Tables
A center	. . .	$N_2{}^{\circ}$	0.0600–0.1589	NA/HA/MA	Typical IR center	9.1.1.3
A center (a)	N6	$N_2{}^{\circ}$	3.928	NA/HA/MA	SB at +156 (N7′ = 4.084 eV)	9.1.1.3
B center	. . .	$V_1N_4{}^{\circ}$	0.0967–0.1651	NA/MA	Typical IR center	8.1.3.5
B center	DAP82a–d	$V_1N_4{}^{\circ}$	4.191–4.438	NA/MA	Split ZPL: 4.184, 4.191, 4.197	8.1.3.5
C center	. . .	$N_1{}^{\circ}$	0.1360–0.1666	NA/HA/LA	Typical IR center	9.1.1.1
C center	. . .	$N_1{}^{\circ}$	4.059	NA/HA/LA	SBs at +61 meV	9.1.1.1
D center	Platelets, B′	$*(C_{2n})_i{}^{\circ}$	0.0409–0.1773	NA/HA/MA	Typical IR center	8.2.2
D center	Platelets, B′	$*(C_{2n})_i{}^{\circ}$	1.528	NL/HL/ML	ZPL of D(α) band at 1.25	8.2.2
E center	X center	$*N_1{}^{+}$	0.1178–0.1651	NA/HA	Typical IR center	9.1.1.2
F center	. . .	$*V_1N_4(C_2)_i{}^{\circ}$	0.1210–0.1959	NA/MA	Typical IR center	8.1.3.6
F center (a, b)	. . .	$*V_1N_4(C_2)_i{}^{\circ}$	2.145, 2.721	NA/HA/NL	. . .	8.1.3.6
F center (c)	DAP81a–f	$*V_1N_4(C_2)_i{}^{\circ}$	4.567–4.906	NA/HA	. . .	8.1.3.6
N1	. . .	?	1.500	NA	**N** = "natural"; SBs at +60, 140 meV	2.1.1.1–2.1.8
N2	. . .	?	2.596	NA	SBs at +89, +154, +241 meV	2.1.1.1–2.1.8
N3(c)	DAP23a–o	$V_1N_3{}^{\circ}$	2.985–3.762	NA/NE/HA/LA/MA	N3(a, b) see Table A.1.2	8.1.3.2
N3(c)	DAP24a–q	$V_1N_3{}^{\circ}$	2.440–2.985	NL/HL/LL/ML	. . .	8.1.3.2
N4	DAP23, s = 3	$V_1N_3{}^{\circ}$	3.603	NA/NE/HA/LA/MA	. . .	8.1.3.2
N5(a)	DAP23, s = 1	$V_1N_3{}^{\circ}$	3.762	NA/NE/HA/LA/MA	. . .	8.1.3.2
N5(b)	DAP52, s = 4	$V_1N_2{}^{\circ}$	3.901	NA/MA	. . .	8.1.3.1
N6	A center (a)	$N_2{}^{\circ}$	3.928	NA/HA/MA	. . .	9.1.1.3
N7	DAP52, s = 2	$V_1N_2{}^{\circ}$	4.045	NA/MA	Resonant with +117 meV SB of N6	8.1.3.1
N8	. . .	?	4.190	NA	. . .	2.1.1.1–2.1.8
N9(a)	DAP63a–g	$V_1N_4{}^{-}$	5.252–5.527	NA/NE/MA	. . .	8.1.3.3

N9(a)	DAP64a–i	$V_1N_4^-$	4.810–$\underline{5.252}$	NL/ML	...	8.1.3.3
N9(b)	...	$V_1N_4^-$	5.262	NA/MA/NL/ML	Same DAP SBs as N9(a)	8.1.3.3
N10(a, b)	...	?	4.990, 5.165	NA/NE	...	2.1.1.1–2.1.8
GR1	NV°, DAP33z, a–j	$^*V_1{}^\circ$	$\underline{1.673}$–1.980	MA	**GR** = "general radiation"	8.1.1
GR1	NV°, DAP34a–h, z	$^*V_1{}^\circ$	1.508–$\underline{1.673}$	ML	...	8.1.1
GR2–GR8	DAP32, ZPL	$^*V_1{}^+$	2.781	MA	...	8.1.1
GR2–GR8	DAP32a–k	$^*V_1{}^+$	2.881–3.065	MA/ME	...	8.1.1
R9	...	?	3.040	MA	**R** = "radiation"	5.1.1.1–5.1.13
R10	ND1, DAP93z, a–d	$^*V_1{}^{2-}$	$\underline{3.150}$–3.890	MA	...	8.1.1
R11	DAP69z, a–l	$^*(N_1C_1)_i{}^-$	$\underline{3.988}$–4.488	MA	...	8.2.4
TR12	DAP35, ZPL	$^*(C_2)_i{}^+$	2.638	MA	**TR** = "type two irradiated" (low N)	8.2.1
TR12	DAP36, ZPL	$^*(C_2)_i{}^+$	2.638	ML	...	8.2.1
TR13	...	$^*(C_2)_i{}^+$	2.670	MA/ME	To higher excited state (+32 meV)	8.2.1
TR14–TR16	DAP35c–j	$^*(C_2)_i{}^+$	2.777–2.812	MA/ME	...	5.1.1.1–5.1.13
TR17	...	$^*(C_2)_i{}^+$	2.826	MA/ME	SB (+188 meV) of TR12	5.1.1.1–5.1.13
H1a	...	$^*V_1C_4N_1^\circ$	0.1798	MA	**H** = "irradiated and heated", LVM	8.1.3.1
H1 d, g, b	DAP30a–d	$^*V_1C_4N_2{}^\circ$	0.3308–0.6125	MA	H1g = 0.5508 eV	8.1.3.1
H1 e, f, c	DAP31a–e	$^*V_1N_4V_1{}^\circ$	0.3611–0.6404	MA	H1f = 0.5450	8.1.3.1

(continued)

Table A.1.1 (continued)

Name	Alternate name	Defect/impurity	Energy (eV)	Observed	Comment	Tables
H2	DAP77z, a–k	$*V_1N_2^-$	1.256–2.020	HA/MA	...	8.1.3.1
H2	DAP78a–c, z	$*V_1N_2^-$	1.015–1.256	HL/ML	...	8.1.3.1
H3a	DAP50z, a–d	$*V_1N_2^\circ$	2.464–2.777	NA/HA/LA/MA	H3b = H13	8.1.3.1
H3a	DAP51a–f, z	$*V_1N_2^\circ$	2.237–2.464	NL/HL/LL/ML	H3b = H13	8.1.3.1
H4a	...	$*N_3V_2N_1^\circ$	2.417	ML	H4b–81 meV (LVM?)	8.1.3.4
H4b	DAP37, ZPL	$*N_3V_2N_1^\circ$	2.498	MA	...	8.1.3.4
H4b	DAP38, ZPL	$*N_3V_2N_1^\circ$	2.498	ML	....	8.1.3.4
H5	(DAP37a?)		2.652 ?	MA	Not in [Zai01]	5.1.1.1–5.1.13
H6–H12	DAP37a–m	$*N_3V_2N_1^\circ$	2.498–3.343	MA	...	8.1.3.4
H13	DAP52, ZPL	$*V_1N_2^\circ$	3.361	NA/HA/LA/MA	...	8.1.3.1
H14	...	$*V_1N_2^\circ$	3.403	MA	related to H3	5.1.1.1–5.1.13
H15–H18	DAP52a–h	$*V_1N_2^\circ$	3.361–4.042	NA/HA/LA/MA	...	8.1.3.1

Table A.1.2 Existing names [Zai01] of 66 defect centers or lines in diamond (not contained in Table A.1.1)

Name	Alternate name	Defect/impurity	Energy (eV)	Observed	Tables	Comment	[Zai01] page
A band	AI(a–e), insulat. AS(a–d), semic.	See Table 10.2.1, 10.2.2, and A.1.3	2.00–3.30	NB/HB/LB/MB	10.2.2	A_{insul}(a–e) 2.15–3.30 eV; A_{semic}(a–d): 2.00–2.80 eV	309
AH3	–	?	2.463	ML	5.3.1–5.3.12.2	Related to H3?	
A line	DAP19g, Yellow	$^{*}N_2{}^{\circ}+B_1{}^{-}$	2.403	NL	9.1.5	ZPL at 1.890 eV; NBD	255
amber ce. (a,b)	–	?	0.520, 0.770	NA	2.1.1.1–2.1.8	In type Ib, SBs 83, 89 meV	133
amber ce. (c,d)	–	?	2.20, 3.30	NB	2.4.1–2.4.3	In type Ib	228, 334
A_s band	–	?	2.480	MB	5.4.1–5.4.4		190
B band	DAP19, Yellow	$^{*}N_2{}^{\circ}+B_1{}^{-}$	1.743–2.270	NA/NL	9.1.5	L = 1.641	
B line	DAP19h, Yellow	$^{*}N_2{}^{\circ}+B_1{}^{-}$	2.478	NL	9.1.5	ZPL at 1.890 eV; NBD	276
Boron acceptor	–	$B_1{}^{\circ}$	0.304–1.180	NA/HA/LA/MA	9.2.1		125
β-band	Ni-S3(β2)	See Table A.1.3	3.600	HB	2.4.1–2.4.3	PLE of Ni-S3	337
B_s band	–	?	2.140	MB	5.4.1–5.4.4		
C band	–	?	1.330	NB (lum.)	2.4.1–2.4.3	ZPL at 1.673 eV, 11(×)-53 meV	
C band	–	?	1.800	NB (abs.)	2.4.1–2.4.3	ZPL at 1.673 eV	
C-line	DAP18a	$^{*}N_2{}^{+}+B_1{}^{\circ}$	2.649	NA	9.1.5	ZPL at 2.334 eV; NBD	298
D-line	DAP18b	$^{*}N_2{}^{+}+B_1{}^{\circ}$	2.699	NA	9.1.5	ZPL at 2.334 eV; NBD	301

(continued)

Table A.1.2 (continued)

Name	Alternate name	Defect/impurity	Energy (eV)	Observed	Tables	Comment	[Zai01] page
D′ center	–	B, N?	0.123–0.160	HA	3.1.1.1–3.1.1.2	Lines a, b (0.132 eV), c	
E-line	DAP18d	$*N_2{}^+ + B_1{}^\circ$	2.748	NA	2.1.1.1–2.1.8	ZPL at 2.334 eV; NBD	302
F-line	DAP18g	$*N_2{}^+ + B_1{}^\circ$	3.147	NA, NL	2.1.1.1–2.1.8	ZPL at 2.334 eV; NBD,NGD	Fig. 7.132
Green band	–	?	2.10–2.73	HB/LB/MB	3.4.2/4.4.2/5.4.2	Possibly three components	240
H-line	DAP18h	$*N_2{}^+ + B_1{}^\circ$	3.204	NA,NL	9.1.5	ZPL at 2.334 eV; NBD,NGD	332
K,L,M,N-line	–	?	4.986–5.092	NL	2.3.11.1	SBs with 141–152 meV	348–350
L-band	5RL	$*(C_2)_iN_i^\circ$	4.00	MB	5.4.4	Also in p–n diode (electro-lum.)	339
M1	–	?	2.400	MA	5.1.7.2	SB: +21 meV	254
M2	–	?	2.445	MA	5.1.8.1	SB: +21 meV	258
N-band	–	?	3.10	MB	5.4.4	also in CVD, as-grown or T = 600°C	325
ND1	R10	See Table A.1.1	3.150	MA	8.1.1	ND1 named by [Dye65b]	327
NE3	S2–“A”, DAP91/92	$*V_1Ni_3{}^+$	2.370	NA/MA/ NL/ML	2.1.4/5.1.7.1/ 2.3.6.2/5.3.7.5	Not ESR NE3 center	248
NG	–	$V_1Ne_1?$	1.723, 1.731	ML	5.3.2.4	Ne implanted, SB 25 meV	186

NV^-	638 nm center DAP48/49	$*V_1N_1^-$	1.943	NA/HA/LA/MA/NL/HL/LL/ML	8.1.3.1	QLVM at +63 meV QLVM at −63 meV	197
NV°	575 nm center DAP46/47	$*V_1N_1^\circ$	2.154	NA/HA/LA/MA/NL/HL/LL/ML	8.1.3.1	QLVM at +90 meV QLVM at −48 meV	218
N3B	N3(a), DAP55a–e	$V_1N_3^\circ$	2.297	NL/HL/ML	8.1.3.2	–	238
N3A	N3(b), DAP56a–d	$V_1N_3^\circ$	2.680	NL/HL/ML	8.1.3.2	–	301
Raman freq.	–	–	0.1652	MA	5.1.1.1–5.1.13	–	50, 90
Si–C interface	–	–	0.0993	LA	4.1.1.1–4.1.4	CVD on Si substrate	67
S_0	–	?	5.258	HL	3.3.1–3.3.7.2	–	354
S1	DAP43a–g, z	$V_1N_2^+$	2.100–<u>2.463</u>	NL/ML	8.1.3.1	QLVM:–48, LVM:–67 meV	257, 272
S2	sta.-DAP57a–m	$(V_2Ni_1)N_{2+0+1}$	2.515–3.341	NA/HA/MA/ME/NL/HL/ML	8.1.5	L = 2.422;B, C, D, E lines at 2.537, 2.597, 2.623, 2.634 eV	287
S3	sta. DAP58a–h	$(V_2Ni_1)N_{2+0+0}$	2.477–2.890	NA/HA/MA/ME/NL/HL/ML	8.1.5	L = 2.383	277

(continued)

Table A.1.2 (continued)

Name	Alternate name	Defect/impurity	Energy (eV)	Observed	Tables	Comment	[Zai01] page
TH5	DAP15a	*$V_2{}^\circ$	2.543	MA	8.1.1	ZPL at 2.419 eV	289
T1	NV°, 575 nm	See NV°	2.154	NA/HA/LA/MA/NL/HL/LL/ML	8.1.3.1	QLVM at +90 meV QLVM at −48 meV	218
Water (a)	O–H bend	H_2O	0.2033	RA	6.1.1.1–6.1.3	Inclusion or physisorbed	30
Water (b)	O–H stretch	H_2O	0.2033	RA	6.1.1.1–6.1.3	Inclusion or physisorbed	30
W1–W5	DAP6a–f	*$(V_2W_1)^\circ$	1.723–1.759	LL	9.7.2	W from hot filament	186
Yellow band	DAP19a–j	*$N_2{}^\circ+B_1{}^-$	2.35	NB	9.1.5	ZPL at 1.890 eV NBD	Fig. 7.132
2BD(C)	–	*$(C_2)_iB_3{}^\circ$	4.803	ML	8.2.3	2BD = "II b damaged" LVM: (3×) −263 meV	346
2BD(F)	DAP97a–d	*$(C_2)_iB_1{}^\circ$	4.777	ML	8.2.3	LVM: (3×) −210 meV	345
2BD(G)	–	*$(C_2)_iB_2{}^\circ$	4.698	ML	8.2.3	LVM: (3×) −220 meV	345
3H	–	*$(C_2)_i{}^-$	2.462	MA/ME/ML	8.2.1	Very close to H3a (2.464 eV)	260
5RL	DAP79–j DAP80a–e	*$(C_2)_iN_1{}^\circ$	4.582	LA/MA/LL/ML	8.2.3	5RL = "5 radiation lines" = ZPL + 4 sharp LVM-SBs: −237 meV	342

1.40 eV	DAP25/26	$^*V_1Ni_1{}^+$	1.401, 1.404	HA/HL/MA/ML	3.1.3/3.3.2.1/5.1.3.1/5.3.2.1/8.1.4	Isotope shifts from a single Ni	140
1.68 eV	3H-DAP71z, a–c	$^*(C_2)_i{}^-$	<u>1.685</u>–2.010	MA	8.2.1	6.2 meV gr. st. spl.	181
1.91 eV	DAP45c	$^*V_1C_4N_1{}^\circ$	1.909	MA	8.1.3.1	ZPL at 1.702 eV	195
1.98 eV		$^*(C_2)_iN_2{}^\circ$	1.979	MA	8.2.3	4.1 meV gr. st. spl.; SBs: + 37, 67 meV	204
2.05 eV	DAP65a–e	$^*(V_3Si_2)^\circ$	2.053	NL	8.1.6	QLVM 45 meV	208
2.07 eV	DAP87a–n	$^*V_1Ni_2{}^-$(a)	2.071	ML	8.1.3.1	LVM: (3×) −55 meV	209
2.56 eV	484 nm, DAP105/106	$^*(V_3Ni_2{}^+)$(a)	2.562 = 484 nm	NA/HA/MA/HL/LL/ML	8.1.3.1	4 ZPL(a–d), 2Ni from isotopes (Table 7.5)	290
2.65 eV	DAP109/110	$^*V_1Si_1{}^\circ$	2.646	LE/LL	9.4.1	1Si 2.65; QLVM: ±74 meV	300
2.68 eV	DAP17z, a–f	$^*O_1{}^\circ+B_1{}^\circ$	<u>2.676</u>–2.725	LL	9.2.2	Oxygen + boron doped CVD	301
2.97 eV	DAP66	$^*V_1Al_1{}^\circ$(a,b)	2.964, 2.974	NL/HL	8.1.7.2	Al doped HPHT; exc. st. spl. 10 meV	317

(continued)

Table A.1.2 (continued)

Name	Alternate name	Defect/impurity	Energy (eV)	Observed	Tables	Comment	[Zai01] page
3.10 eV	2.56 eV(c)	$*(V_3Ni_2{}^+)(c)$	3.064	NA/HA/HE/MA	8.1.6	4 ZPL(a–d), 2Ni from isotopes (Table 7.5)	324
3.188 eV	DAP72a–n	$*(C_2)_iN_1{}^-$	3.188	(LA/MA)/LL/ML	8.2.3	Weak absorption, many SBs and IR lines	328
389 nm	DAP72a–n	$*(C_2)_iN_1{}^-$	3.118 = 388.9 nm	(LA/MA)/LL/ML	8.2.3	Weak absorption, many SBs and IR lines	328
2.562 = 484 nm	2.56 eV, DAP105/106	$*V_3Ni_2{}^+(a)$	484 nm	NA/HA/MA/ HL/LL/ML	8.1.6	4 ZPL(a–d), 2Ni from isotopes (Table 7.5)	290
540 nm	DAP27/28	$*V_1Ni_2{}^-$	2.298 = 539.4 nm	ME/ML	8.1.4		236
575 nm	NV°	$*V_1N_1{}^\circ$	2.154 = 575.5 nm	NA/HA/LA/MA/ NL/HL/LL/ML	8.1.3.1	QLVM at +90 meV QLVM at −48 meV	218
595 nm	DAP45g	$*V_1C_4N_1{}^\circ$	2.086 = 594.4 nm	MA	8.1.3.1	ZPL at 1.702 eV	213
638 nm	NV⁻	$*V_1N_1{}^-$	1.943 = 638.1 nm	NA/HA/LA/MA/ NL/HL/LL/ML	8.1.3.1	QLVM at +63 meV QLVM at −63 meV	197

Table A.1.3 New names of 84 defect centers or lines in diamond

Name	Alter-nate name	Defect/impurity	Energy (eV)	Observed	Tables	Comment	[Zai01] [Zai98] page or Figure
*FE(a)	A_0	Free exciton	5.423	(NL, indirect)	2.3.11.3	Upper valence band, forbidden NPL	354
*FE(b)	A_0’	Free exciton	5.430	(NL, indirect)	2.3.11.3	Lower valence band, 7 meV spin–orbit split	354
*FE(a)-1TA	A_1	SB, free exciton	5.322	NL	2.3.11.2	Upper valence band, −87 meV(=TA-phonon)	355, Fig. 5.152
*FE(b)-1TA	A_1’	SB, free exciton	5.329	NL	2.3.11.2	Lower valence band, −87 meV(=TA-phonon)	357, Fig. 5.152
*FE(a)-1TO	B_1	SB, free exciton	5.268	NL	2.3.11.2	Upper valence band, −141 meV(=TO-phonon)	354, Fig. 5.152
*FE(b)-1TO	B_1’	SB, free exciton	5.275	NL	2.3.11.2	Lower valence band, −141 meV(=TO-phonon)	355, Fig. 5.152
*FE(a)-1LO	C_1	SB, free exciton	5.246	NL	2.3.11.2	Upper valence band, −163 meV(=LO-phonon)	353, Fig. 5.152
*FE(b)-1LO	C_1’	SB, free exciton	5.252	NL	2.3.11.2	Lower valence band, −163 meV(=LO-phonon)	354, Fig. 5.152
*BE-B(a)	D_0	B, Table 9.2.6	5.356	NL/HL/LL/ML	2.3.11.2/3.3.7.2/4.3.9/5.3.12.2	Upper valence band, no phonon line (NPL)	357, Fig. 5.159

(continued)

Table A.1.3 (continued)

Name	Alter-nate name	Defect/impurity	Energy (eV)	Observed	Tables	Comment	[Zai01] [Zai98] page or Figure
*BE-B(b)	D_0'	B, Table 9.2.6	5.367	NL/HL/LL/ML	2.3.11.2/3.3.7.2/ 4.3.9/5.3.12.2	Lower valence band, no phonon line (NPL)	357, Fig. 5.152
*BE-Li	G_0	Li	5.283	LL	4.3.9	No phonon line (NPL)	Fig. 7.158
*BE-Li-1LO	G_1	Li, SB	5.120	LL	4.3.9	NPL −163 meV(=LO phonon)	Fig. 7.158
*BE-Li-2LO	G_2	Li, SB	4.955	LL	4.3.9	NPL −328 meV(=2× LO phonon)	Fig. 7.158
*BE-Li-4TA	X	Li, SB	4.832	LL	4.3.9	NPL −348 meV(=4× TA phonon)	Fig. 5.151
*BE-Li-3TO	X	Li, SB	4.757	LL	4.3.9	NPL −423 meV(=3× TO phonon)	Fig. 5.151
*BE-Li-4LO	X	Li, SB	4.617	LL	4.3.9	NPL −663 meV(=4× LO phonon)	Fig. 5.151
*BE-P(a)	F_0	P	5.316	LL/ML	4.3.9/5.3.12.2	Upper valence band, no phonon line (NPL)	Fig. 5.155a
*BE-P(b)	F_0'	P	5.333	LL/ML	4.3.9/5.3.12.2	Lower valence band, no phonon line (NPL)	Fig. 5.155a
*BE-P(b)-1TA	F_1(b)'	P, SB	5.245	LL	4.3.9	NPL −88 meV(=TA phonon)	Fig. 5.155a
*BE-P(a)-1TO	F_1(a)	P, SB	5.170	LL/ML	4.3.9/5.3.12.2	NPL −146 meV(=TO phonon)	351
*BE-P(b)-1TO	F_1(b)'	P, SB	5.190	LL/ML	4.3.9/5.3.12.2	NPL −143 meV(=TO phonon)	Fig. 5.155a

*BE-P(a)-2TO	$F_2(a)$	P, SB	5.020	LL	4.3.9	NPL −296 meV(=2× TO phonon)	Fig. 5.155a
*BE-Y	E_0	?	5.145	NL	2.3.11.2	No phonon line (NPL)	Fig. 5.155b
*BE-Y-1LO	E_1	?	5.005	NL	2.3.11.1	NPL −140 meV(=TO phonon)	Fig. 5.155b
*1Ni-S2	DAP57	$(V_2Ni_1)N_{2+0+1}{}^+$	2.515–3.341	See Table A.1.2	8.1.5	L = 2.422 eV	287
*1Ni-S2(α2)		$(V_2Ni_1)N_{2+0+1}{}^+$	2.79	MB	5.4.3	PLE of S2	Fig. 7.104
*1Ni-S2(β2)		$(V_2Ni_1)N_{2+0+1}{}^+$	3.54	MB	5.4.4	PLE of S2	Fig. 7.104
*1Ni-S2(γ2)		$(V_2Ni_1)N_{2+0+1}{}^+$	4.68	MB	5.4.4	PLE of S2	Fig. 7.104
*1Ni-S3	DAP58	$(V_2Ni_1)N_{2+0+0}{}^+$	2.477–2.890	See Table A.1.2	8.1.5	L = 2.383 eV	277
*1Ni-S3(α2)		$(V_2Ni_1)N_{2+0+0}{}^+$	2.85	MB	5.4.3	PLE of S3	Fig. 7.104
*1Ni-S3(β2)		$(V_2Ni_1)N_{2+0+0}{}^+$	3.60	MB	5.4.4	PLE of S3	Fig. 7.104
*1Ni-S3(γ2)		$(V_2Ni_1)N_{2+0+0}{}^+$	3.97	MB	5.4.4	PLE of S3	Fig. 7.104

(continued)

Table A.1.3 (continued)

Name	Alternate name	Defect/impurity	Energy (eV)	Observed	Tables	Comment	[Zai01] [Zai98] page or Figure
*Ni-S4	DAP59	$*(V_2Ni_1)N_{1+2+0}{}^{\circ}$	3.333–3.820	NA/NE/HA/ HE/MA/ME	2.16/2.22/3.16/3.21/ 5.1.11.1–5.1.11.2/ 5.2.6	L = 3.191 eV	Fig. 7.148
*Ni-S4(α2)		$*(V_2Ni_1)N_{1+2+0}{}^{\circ}$	2.72	NB/HB/MB	2.4.1–2.4.3/3.4.1–3.4.3/ 5.4.1–5.4.4	PLE of S4	Fig. 7.104
*Ni-S4(β2)		$*(V_2Ni_1)N_{1+2+0}{}^{\circ}$	3.31	NB/HB/MB	2.4.1–2.4.3/3.4.1–3.4.3/ 5.4.1–5.4.4	PLE of S4	Fig. 7.104
*Ni-S5	DAP60	$*(V_2Ni_1)N_{x+y+z}{}^{\circ}$	1.563–2.320	NA/HA/MA/ MENL/HL/ LL/ML	2.31/3.13/5.1.3.1–5.1.7.1/ 5.2.1–5.2.2/2.3.2/3.3.2.1/ 4.3.1.1/5.3.2.2–5.3.2.3	L = 1.460 eV	
*Ni-S5(α2)		$*(V_2Ni_1)N_{x+y+z}{}^{\circ}$	1.81	MB	5.4.2	PLE of S5	Fig. 5.19
*Ni-S5(β2)		$*(V_2Ni_1)N_{x+y+z}{}^{\circ}$	2.20	MB	5.4.2	PLE of S5	Fig. 5.19
*Ni-S5(γ2)		$*(V_2Ni_1)N_{x+y+z}{}^{\circ}$	3.00	MB	2.4.1–2.4.3/3.4.1–3.4.3/ 5.4.1–5.4.4	PLE of S5	Fig. 5.19
*Ni-S6	DAP61	$*(V_2Ni_1)N_{x+y+z}{}^{+}$	2.750–3.530	ME	4.2.1,4.2.2/5.2.1–5.2.6	L = 2.637 eV; PLE of S5	Fig. 5.26

*Ni-S6(α2)		$*(V_2Ni_1)N_{x+y+z}{}^+$	3.00	MB	5.4.1–5.4.4	PLE of S5	Fig. 5.26
*Ni-S6(β2)		$*(V_2Ni_1)N_{x+y+z}{}^+$	3.30	MB	5.4.1–5.4.4	PLE of S5	Fig. 5.26
*1Ni-S7	DAP11	$*(V_2Ni_1)N_{x+y+z}{}^+$	1.802–2.620	LE/LL/ME	4.2.1,4.2.2/ 4.3.1.1–4.3.9/ 5.2.1–5.2.6	L = 1.724 eV; PLE of S5	Fig. 5.26
*1Ni-S7(α2)		$*(V_2Ni_1)N_{x+y+z}{}^+$	2.19	MB	5.4.1–5.4.4	PLE of S5	Fig. 5.26
*1Ni-S7(β2)		$*(V_2Ni_1)N_{x+y+z}{}^+$	2.35	MB	5.4.1–5.4.4	PLE of S5	Fig. 5.26
*Ni-S8	DAP16	$*(V_2Ni_1)^-$	1.212–1.383	HA	3.1.1.1–3.1.6	L = 1.036 eV	Fig. 5.7
*Ni-S8(α2)		$*(V_2Ni_1)^-$	1.40	HB	3.4.1–3.4.3	Absorption; named Ni1.4 eV band	140
*Ni-S9	DAP4a–i	$*(V_2Ni_1)N_{x+y+z}{}^+$	1.678–1.821	ML	5.3.2.3–5.3.3	L = 1.509 eV	Fig. 5.15
*1Ni-1.40	DAP25/26	$*V_1Ni_1{}^+$	1.404	NA/HA/MA	2.1.3/3.1.3/ 5.1.3.1/8.1.4	Ni_1 from isotopes	140
*Ni-1.66a	–	$*V_1Ni_2{}^\circ$	1.660	ML	5.3.2.3	–	156
*Ni-1.66b	DAP29z, a–f	$*V_1Ni_2{}^\circ$	2.401	ME	5.2.2	Broad band at 2.63 eV	254
*Ni-1.69a	–	$*V_1Ni_2{}^+$	1.693	HA/MA/ML	3.1.3/5.1.3.2/ 5.3.2.3	–	182
*Ni-1.69b	–	$*V_1Ni_2{}^+$	1.940	HA/MA/ HL/ML	3.1.4/5.1.5.2/ 3.3.3/5.3.4.1	–	197

(continued)

Table A.1.3 (continued)

Name	Alternate name	Defect/impurity	Energy (eV)	Observed	Tables	Comment	[Zai01] [Zai98] page or Figure
*Ni-1.69c	DAP111a–f	*$V_1Ni_2^+$	1.991	HA/HL/ML	3.1.4/5.1.5.2/ 3.3.3/5.3.4.2	–	204
*Ni-1.70a	DAP83/84	*$V_1Ni_1^\circ$(a)	1.704	HA/MA/ML	3.1.3/3.3.2.2/ 5.3.2.4	–	183
*Ni-1.70b	DAP29a–f	*$V_1Ni_1^\circ$(b)	2.401	HA/ME	3.1.3/3.3.2.2/ 5.3.2.4	–	254
*Ni-1.88a	DAP107/108	*$(V_3Ni_2)^\circ$	1.883	HA/LA/HL	3.1.4/4.1.4.7/ 3.3.2/8.1.6	–	192, Fig.7.77
*Ni-1.88b	–	*$(V_3Ni_2)^\circ$	1.905	HA	3.1.4	–	192, Fig.7.77
*Ni-1.88c	–	*$(V_3Ni_2)^\circ$	1.913	HA	3.1.4	–	192, Fig.7.77
*Ni-2.07a	DAP87a–n	*$V_1Ni_2^-$	2.071	ML	5.3.5		209, Fig. 5.69a
*Ni-2.07b		*$V_1Ni_2^-$	2.298	ME	5.2.2		236, Fig. 5.69b
*Ni-2.37	DAP91/92	*$V_1Ni_3^+$	2.370	NA/MA/NL/ML	2.1.4/5.1.7.1/ 2.3.6.2/5.3.7.5	Formerly named S2-A-line	247, Fig. 5.97
*Ni-2.51a		*Ni_1^-	2.510	HA/HL	3.1.5/3.3.5		283, Fig. 7.77
*Ni-2.51b		*Ni_1^-	2.523	HA	3.1.5		Fig. 7.77
*2Ni-2.56a	DAP105a–c DAP106a–f	*$(V_3Ni_2)^+$	2.562	HA/HL/ML	3.1.5/3.3.5/ 5.3.8.3/8.1.6	See also Table A.1.2 (2.56 eV/484 nm)	290

*2Ni-2.56b		*$(V_3Ni_2)^+$	2.588	HA/HL/ML	3.1.5/3.3.6/8.1.6		290
*2Ni-2.56c	3.10 eV	*$(V_3Ni_2)^+$	3.064	HA/HE	3.1.5/3.2.1/8.1.6	See also Table A.1.2 (3.10 eV)	324
*2Ni-2.56d	3.10 eV	*$(V_3Ni_2)^+$	3.076	HA/HE	3.1.5/3.2.1/8.1.6		324
*Co-1.47		*$V_1Co_1{}^+$	1.472	HL	3.3.2.1, 9.6		143
*Co-1.85a	102z, a–f	*$V_1Co_2{}^+$	1.852	MA	5.1.4, 9.6		192, Fig. 5.52
*Co-**S5**	sta. 104a–i, z	*$(V_2CO_1)N_{x+y+z}{}^{\circ}$	2.135–2.820	ML	5.3.10.1, 96	L = 1.948 eV; exc. st. splittings 4, 12, 18 meV; slow luminescence (63–235 μs)	235, Fig. 5.72
*Co-1.99	DAP112a-f	*$(V_3Co_2)^{\circ}$	1.989	HL	3.3.3, 9.6	Fast luminescence (<20 ns)	204, Fig. 5.62
*Co-2.59		*$V_1Co_2{}^-$	2.590	ML	5.3.9.1, 9.6	Medium fast luminescence (ca. 80 ms)	295
*Co-2.89		*$V_1Co_2{}^{\circ}$	2.889	HL	3.3.6, 9.6	Fast luminescence	315
*Si-1.56	sta. DAP8a-g	*$Si1^{\circ}+$?	1.556–1.637	LL	4.3.1.1, 9.4.1	L = 1.474 eV	156, Fig. 7.67

(continued)

Table A.1.3 (continued)

Name	Alternate name	Defect/impurity	Energy (eV)	Observed	Tables	Comment	[Zai01] [Zai98] page or Figure
*2Si-1.68	DAP53a-m	$*(V_3Si_2)^+$	1.682	HA/LA/MA	3.1.5/3.3.5	Si_2 from isotopes; named **Si**-center	174, Fig. 5.48
*2Si-1.68	DAP54a-g	$*(V_3Si_2)^+$	1.682	HL/LL/ML	3.3.2.2/4.3.1.1/ 5.3.2.3/9.4.1	Si_2 from isotopes; photochromic	174, Fig. 5.44
*2Si-2.05	DAP65a-e	$*(V_3Si_2)^\circ$	2.052	NL	2.3.4, 9.4.1	Si_2 from QLVM(–45 meV, calc. 44.9)	207, Fig. 7.85a
*2Si-2.52	DAP75a–j	$*(V_3Si_2)^-$	2.523	NL/ML	2.3.7/5.3.8.2, 9.4.1	In NBD After Si implantation	285, Fig. 7.118 285, Fig. 5.120a
*1Si-2.65	DAP109a–h	$*V_1Si_1{}^\circ$	2.646	LE	4.2.2, 9.4.1	Si_1 from QLVM(+74 meV, calc. 74.5)	300, Fig. 5.128
*1Si-2.65	DAP110a–g	$*V_1Si_1{}^\circ$	2.646	LL	4.3.6, 9.4.1	Si_1 from QLVM(–74 meV, calc. 74.5)	300, Fig. 5.128
*1Si-2.99	DAP113a–m	$*V_1Si_1{}^-$	2.991 DAP lines: 2.246–2.874 eV	LL	4.3.7.2, 9.4.1	Si_1 from QLVM(–74 meV, calc. 74.5) NL = Si rich natural brown diamond(NBD) LL = Si substrate, ML = Si implanted; strain induced line splittings and shifts	245, Fig. 7.114 305, Fig. 7.120 323, Fig. 7.135

A.2 Calculated Coulomb Factors for Donor–Acceptor Pairs

In Table A.2, calculated Coulomb factors of donor–acceptor pairs for the shells 1–228 are listed [see (10.3)]. The 18 **high intensity** shells are printed in bold, and the "near shell correction" is performed for shells 1–6 (see Sect. 10.1). For $s > 52$ only even shells and for $s > 106$, only observed shells are listed. A similar table is given in [Dea73].

The *degeneracy* of the shells is determined by the lattice vector as follows:

⟨even1, 0, 0⟩	6
⟨even1, even1, 0⟩	12
⟨even1, even2, 0⟩	24
⟨even1, even1, even1⟩	8
⟨even1, even1, even2⟩	24
⟨even1, even2, even3⟩	48
⟨odd1, odd1, odd1⟩	4
⟨odd1, odd1, odd2⟩	12
⟨odd1, odd2, odd3⟩	24

The number of empty shells (s_{empty}) can be calculated by a formula given in [Wil70]:

$$s_{\text{empty}} = 4m(8n + 7); \quad m = 1, 2, 3; \; n = 0, 1, 2$$

Numbers of empty shells are: 28, 60, 92, 112, 124, 156, 188, 220, 240, etc.

Table A.2 Calculated Coulomb factor C(s) of donor–acceptor pairs (DAP) in diamond. The last column gives the experimental abundance (high abundance shells in bold type)

Shell Numbers	Lattice vector	Degeneracy	Coulomb-factor[a]	Exper. abund.	Shell No.	Lattice vector	Degeneracy	Coulomb-factor	Exper. abund.
1	<1, 1, 1>	4	0.5600	18	20	<8, 4, 0>	24	0.1936	23
			(*1.0000*)						
2	<2, 2, 0>	12	0.5020	20	21a	<9, 1, 1>	12	0.1901	0
			(*0.6124*)		21b	<7, 5, 3>	24		
3	<3, 1, 1>	12	0.4700	13			36		
			(*0.5222*)		**22**	<6, 6, 4>	**24**	**0.1846**	**30**
4	**<4,0,0>**	**6**	0.4110	**33**					
			(*0.4330*)		23	<9, 3, 1>	24	0.1816	1
5	<3, 3, 1>	12	0.3850	11					
			(*0.3974*)		24	<8, 4, 4>	24	0.1768	15
6	<4, 2, 2>	24	0.3530	23					
			(*0.3336*)		25a	<9, 3, 3>	12	0.1741	2
7a	<5, 1, 1>	12	0.3333	15	25b	<7, 7, 1>	12		
7b	<3, 3, 3>	4			25c	<7, 5, 5>	12		
		16					36		
8	**<4,4,0>**	**12**	**0.3062**	**35**	**26a**	**<10,2,0>**	**24**	**0.1698**	**34**
					26b	<8, 6, 2>	**48**		
9	<5, 3, 1>	24	0.2928	9			**72**		
					27a	<9, 5, 1>	24	0.1674	1
10	<6, 2, 0>	24	0.2739	36	27b	<7, 7, 3>	12		
							36		
11	<5, 3, 3>	12	0.2641	7	*(28)*	*empty*	*0*	*(gap)*	–

12	**<4,4,4>**	**8**	**0.2500**	**39**	29	<9, 5, 3>	24	0.1615	0
13a	<7, 1, 1>	12	0.2425	4	30	<10, 4, 2>	48	0.1581	24
13b	<5, 5, 1>	12							
		24			31a	<11, 1, 1>	12	0.1562	1
14	<6, 4, 2>	48	0.2315	30	31b	<7, 7, 5>	12		
							24		
15a	<7, 3, 1>	24	0.2255	2	32	<8, 8, 0>	12	0.1531	10
15b	<5, 5, 3>	12							
		36			33a	<11, 3, 1>	24	0.1513	1
16	<8, 0, 0>	**6**	**0.2165**	**30**	33b	<9, 7, 1>	24		
					33c	<9, 5, 5>	12		
17	<7, 3, 3>	12	0.2116	2			60		
					34a	<10, 6, 0>	**24**	**0.1485**	**40**
18a	<8, 2, 2>	24	0.2041	28	**34b**	<8, 6, 6>	**24**		
18b	<6, 6, 0>	12					48		
		36			35a	<11, 3, 3>	12	0.1469	0
19a	<7, 5, 1>	24	0.2000	1	35b	<9, 7, 3>	24		
19b	<5, 5, 5>	4					36		
		28							
36a	<12, 0, 0>	6	0.1443	10	*even*	*shells only*			
36b	<8, 8, 4>	24							
		30			52	<12, 8, 0>	24	0.1201	2
37a	<11, 5, 1>	24	0.1429	2					
37b	<7, 7, 7>	4			54a	<14, 4, 2>	48	0.1179	17
		28			54b	<12, 6, 6>	24		
38a	<12, 2, 2>	24	0.1405	3	54c	<10, 10, 4>	24		
38b	<10, 6, 4>	48					96		
		72			56	<12, 8, 4>	48	0.1157	0

(continued)

Table A.2 (continued)

Shell Numbers	Lattice vector	Degeneracy	Coulomb-factor[a]	Exper. abund.
39a	<11, 5, 3>	24	0.1391	1
39b	<9, 7, 5>	24		
		48		
40	<12, 4, 0>	24	0.1369	6
41	<9, 9, 1>	12	0.1357	0
42	**<10,8,2>**	**48**	**0.1336**	**32**
43a	<13, 1, 1>	12	0.1325	1
43b	<11, 7, 1>	24		
43c	<11, 5, 5>	12		
43d	<9, 9, 3>	12		
		60		
44	<12, 4, 4>	24	0.1306	10
45a	<13, 3, 1>	24	0.1295	1
45b	<11, 7, 3>	24		
45c	<9, 7, 7>	12		
		60		
46	<12, 6, 2>	48	0.1277	1
47a	<13, 3, 3>	12	0.1267	1
47b	<9, 9, 5>	12		
		24		
48	<8, 8, 8>	8	0.1250	12

Shell No.	Lattice vector	Degeneracy	Coulomb-factor	Exper. abund.
58	<14, 6, 0>	24	0.1137	27
(60)	***empty***	*0*	*(gap)*	–
62a	<14, 6, 4>	48	0.1100	4
62b	<12, 10, 2>	48		
		96		
64	**<16,0,0>**	**6**	**0.1082**	**33**
66a	<16, 2, 2>	24	0.1066	1
66b	<14, 8, 2>	48		
66c	<10, 10, 8>	24		
		96		
68a	<16, 4, 0>	24	0.1050	1
68b	<12, 8, 8>	24		
		48		
70	<12, 10, 6>	48	0.1035	0
72a	<16, 4, 4>	24	0.1021	20
72b	<12, 12, 0>	12		
		36		
74a	<16, 6, 2>	48	0.1007	6
74b	<14, 10, 0>	24		
74c	<14, 8, 6>	48		
		120		
76	<12, 12, 4>	24	0.09934	0

49a	<13, 5, 1>	24	0.1240	1					
49b	<11, 7, 5>	24			78	<14, 10, 4>	48	0.09806	13
		48							
50a	**<14,2,0>**	**24**	**0.1225**	**60**	80	<16, 8, 0>	24	0.09682	0
50b	**<10,10,0>**	**12**							
50c	**<10,8,6>**	**48**			82a	<18, 2, 0>	24	0.09564	2
		84			82b	<16, 6, 6>	24		
51a	<13, 5, 3>	24	0.1216	1			48		
51b	<11, 9, 1>	24							
		48							
84	<16, 8, 4>	48	0.09449	0	*observed*	*shells only*			
86a	**<18,4,2>**	**48**	**0.09339**	**36**	114a	<16, 14, 2>	48	0.08111	1
86b	<14, 12, 2>	**48**			114b	<16, 10, 10>	24		
86c	<12, 10, 10>	**24**			114c	<14, 14, 8>	24		
		120					96		
88	<12, 12, 4>	24	0.09232	0	**118a**	**<20,6,6>**	**24**	**0.07972**	**8**
					118b	**<18,12,2>**	**48**		
90a	<18, 6, 0>	24	0.09129	0			72		
90b	<16, 10, 2>	48			**130a**	<22, 6, 0>	**24**	**0.07596**	**20**
90c	<14, 10, 8>	48			**130b**	<18, 14, 4>	**24**		
		120					48		
(92)	empty	0	(gap)	–	136a	<20, 12, 0>	24	0.07426	1
					136b	<16, 12, 12>	24		
94a	<18, 6, 4>	48	0.08932	0			48		
94b	<14, 12, 6>	48			**162a**	<24, 6, 6>	**24**	**0.06804**	**25**
		96			**162b**	<22, 10, 8>	**48**		
96	<16, 8, 8>	24	0.08839	0	**162c**	<16, 14, 14>	**24**		
							96		

(continued)

Table A.2 (continued)

Shell Numbers	Lattice vector	Degeneracy	Coulomb-factor[a]	Exper. abund.	Shell No.	Lattice vector	Degeneracy	Coulomb-factor	Exper. abund.
98a	<18, 8, 2>	**48**	**0.08748**	**12**	**198a**	<28, 2, 2>	**24**	**0.06155**	**14**
98b	<16, 10, 6>	**48**			**198b**	<26, 10, 4>	**48**		
98c	**<14,14,0>**	**12**			**198c**	<20, 14, 14>	**24**		
		108					96		
100a	<20, 0, 0>	6	0.08660	0	**228a**	<28, 8, 8>	**24**	**0.05735**	**17**
100b	<16, 12, 0>	24			**228b**	<20, 16, 16>	**24**		
		30					48		
102a	<20, 2, 2>	24	0.08575	0	_other_	_shells_			
102b	<14, 14, 4>	24							
		48			300	–	–	0.05000	0
104a	<20, 4, 0>	24	0.08492	0					
104b	<16, 12, 4>	48			400	–	–	0.04330	0
		72							
106a	**<18,10,0>**	**24**	**0.08412**	**20**	500	–	–	0.03873	0
106b	**<18,8,6>**	**48**							
		72			∞			0	0

[a]For shells 1–6, the values are given with near shell correction (*uncorrected* in parenthesis)

A.3 Calculated Frequencies and Linewidths for Quasilocal Vibrational Modes

Calculated frequencies and linewidths for quasilocal vibrations in diamond are listed in Table A.3 [see (11.5) and (11.6)]. Observed examples are indicated, and for AM > 80 only data for observed examples are given.

Table A.3 Calculated frequencies and linewidths of quasilocal vibrational modes (QLVM) in diamond

AM	*RM*	*f* (meV)	*w* (meV)	Examples (see also Table 11.2)	AM	*RM*	*f* (meV)	*w* (meV)	Examples, (see also Table 11.2)
24	*12*	86.1	78.6	$^{12}C_2$	67	*55*	40.2	17.2	
25	*13*	82.7	72.6		68	*56*	39.8	16.8	$^{28}Si_2+^{12}C_1$, $^{14}N_4+^{12}C_1$
26	*14*	79.7	67.4	$^{13}C_2$ (−7.4%), $^{14}N_1+^{12}C_1$	69	*57*	39.5	16.6	
27	*15*	77.0	62.9	$^{27}Al_1$, $^{14}N_1+^{13}C_1$(−3.4%)	70	*58*	39.1	16.3	$^{14}N_5$
28	*16*	74.5	59.0	$^{14}N_2$, $^{28}Si_1$	71	*59*	38.8	16.0	
29	*17*	72.3	55.5		72	*60*	38.5	15.7	$^{12}C_6$
30	*18*	70.3	52.4	$^{15}N_2$	73	*61*	38.2	15.5	
31	*19*	68.4	49.7	$^{31}P_1$	74	*62*	37.9	15.2	$^{12}C_5+^{14}N_1$
32	*20*	66.7	47.2		75	*63*	37.6	15.0	
33	*21*	65.1	44.9		76	*64*	37.3	14.7	$^{4}He_1+^{12}C_6$, $^{12}C_4+^{14}N_2$
34	*22*	63.6	42.9		77	*65*	37.0	14.5	
35	*23*	62.2	41.0	$^{11}B_1+^{12}C_2$	78	*66*	36.7	14.3	$^{13}C_6$, $^{12}C_3+^{14}N_3$
36	*24*	60.9	39.3	$^{12}C_3$	79	*67*	36.4	14.1	
37	*25*	59.6	37.7	$^{14}N_1+^{11}B_1+^{12}C_1$	80	*68*	36.2	13.9	$^{28}Si_2+^{12}C_2$
38	*26*	58.5	36.3	$^{14}N_1+^{12}C_2$	84	*72*	35.1	13.1	$^{12}C_7$, $^{48}Ti_1+^{12}C_3$
39	*27*	57.4	34.9	$^{13}C_3$(−5.7%)	90	*78*	33.8	12.1	
40	*28*	56.3	33.7	$^{14}N_2+^{12}C_1$, $^{14}N_1+^{13}C_2$	91	*79*	33.5	11.9	$^{13}C_7$(−4.6%),
41	*29*	55.3	32.5		96	*84*	32.5	11.2	$^{12}C_8$
42	*30*	54.4	31.4	$^{14}N_3$	101	*89*	31.6	10.6	$^{65.4}Zn_1+^{12}C_3$
43	*31*	53.5	30.4		108	*96*	30.4	9.8	$^{12}C_9$
44	*32*	52.7	29.5	$^{28}Si_1+^{16}O_1$	117	*105*	29.1	9.0	$^{58.7}Ni_2$
45	*33*	51.9	28.6	$^{15}N_3$(−4.6%)	118	*106*	29.0	8.9	$^{58.9}Co_2$

46	*34*	51.1	27.7		120	*108*	28.7	8.7	$^{12}C_{10}$
47	*35*	50.4	27.0	$^{12}C_3+^{11}B_1$	131	*119*	27.3	7.9	$^{131}Xe_1$
48	*36*	49.7	26.2	$^{12}C_4$, $^{48}Ti_1$	144	*132*	25.9	7.1	$^{12}C_{12}$
49	*37*	49.0	25.5		156	*144*	24.8	6.6	$^{13}C_{12}$
50	*38*	48.4	24.8	$^{12}C_3+^{14}N_1$	168	*156*	23.9	6.0	$^{14}C_{12}$
51	*39*	47.7	24.2		176	*164*	23.3	5.8	$^{58.7}Ni_3$
52	*40*	47.1	23.6	$^{12}C_2+^{14}N_2$, $^{12}C_2+^{28}Si_1$	181	*169*	22.9	5.6	$^{181}Ta_1$
53	*41*	46.6	23.0		184	*172*	22.7	5.5	$^{184}W_1$
54	*42*	46.0	22.5	$^{12}C_1+^{14}N_3$, $^{27}Al_2$	192	*180*	22.2	5.2	$^{12}C_{16}$
55	*43*	45.5	21.9		204	*192*	21.5	4.9	$^{204}Tl_1$
56	*44*	44.9	21.4	$^{14}N_4$, $^{28}Si_2$	216	*204*	20.9	4.6	$^{108}Ag_2$
57	*45*	44.4	21.0		272	*260*	18.5	3.6	
58	*46*	44.0	20.5		312	*300*	17.2	3.1	
59	*47*	43.5	20.1	$^{58.7}Ni_1$, $^{58.9}Co_1$	412	*400*	14.9	2.4	
60	*48*	43.0	19.7	$^{12}C_5$	612	*600*	12.2	1.6	
61	*49*	42.6	19.3		912	*900*	9.9	1.0	
62	*50*	42.2	18.9	$^{12}C_4+^{14}N_1$					
63	*51*	41.7	18.5	$^{27}Al_1+^{12}C_3$					
64	*52*	41.3	18.1	$^{12}C_3+^{14}N_2$					
65	*53*	41.0	17.8	$^{13}C_5(-4.7\%)$, $^{65.4}Zn_1$					
66	*54*	40.6	17.5	$^{12}C_2+^{14}N_3$					

Frequency(f) $=[10^6/(11.25\,RM)]^{1/2}$, width($w$) $=10^4/(10.6\,RM)$, RM = AM-12, *AM* actual mass, *RM = AM-12 = reduced mass (12 = mass of carbon)* [Zai00][Bro62]

References

[All95] Allers L, Collins A T (1995): J. Appl. Phys. **77**, 3879

[All98a] Allers L, Collins A T, Hiscock J (1998a): Diamond and Related Materials **7**, 228

[All98b] Allers L, Mainwood A (1998b): Diamond and Related Materials **7**, 261

[And94] Ando T, Aizawa T, Kaino M, Sato Y, Anzai T, Yamornoto H, Wada A, Domen K, Hirose C (1994): In *Advances in New Diamond Science and Technology*, Saito S, Fujirnory N, Fukunaga O, Kamo M, Kobashi K, Yoshikawa M (eds.). MYU, Tokyo, p. 461

[And97] Anderson F G, Dallas T, Lal S, Gangopadhyay S, Holtz M (1997): Solid. State Commun. **102**, 867

[Ang65] Angress J F, Smith S D (1965): Phil. Mag. **12**, 415

[Ant92] Anthony T R, Banholzer W F (1992): Diamond and Related Materials **1**, 717

[Bac93] Bachmann P K, Leers D, Wiechert D U (1993): Diamond and Related Materials **2**, 683

[Bad97a] Badzian A, Roy R, Mistry P, Turchan M C (1997a): In *The Physics of Diamond*, Course CXXXV, Proc. Int. School of Physics, IOS Press, p. 195

[Bau93] Bauer R, Osvet A, Sildos I, Bogner U (1993): J. of Luminescence **56**, 57

[Ber93] Bergman L, Stoner B R, Turner K F, Glass J T, Nemanich R J (1993): J. Appl. Phys. **73**, 3951

[Bet94] Bettiol A A, Jamieson D N, Prawer S, Allen M G (1994): Nucl. Instrum. and Meth. B **85**, 775

[Bi_90] Bi X X, Eklund P C, Zhang J G, Rao A M, Perry T A, Beetz C P (1990): J. Materials Res. **5**, 811

[Bie67] Bienemann-Kuespert E, Brennecke E, Flachsbart I, Pietsch-Wilke G, Stiess P, Wagner J (1967): *In* Kirschstein G, Koschel D, Kugler H K (eds) *Gmelins Handbuch der Anorganischen Chemie*, Verlag Chemie, p 217

[Bok86] Bokii G B, Bezrakov G N, Kluev Ju A, Naletov A M, Nepsha V I (1986): *Natural and Synthetic Diamonds*, Nauka, Moscow (in Russian)

[Bos89] Bosshart G (1989), J. Gemmology **21**, 351

[Bre95] Breuer S J, Briddon P R (1995): Phys. Rev. B **51**, 6984

[Bro62] Brout R, Visscher W (1962): Phys. Rev. Lett. **9**, 54

[Bro95] Brown S W, Rand S C, (1995): J. Appl. Phys. **78**, 4069

[Buh94] Buhaenko D, Beer C, Ellis P, Walker L, Stoner B, Tessmer G, Fauber R (1994): In *Advances in New Diamond Science and Technology*. Saito S, Fujimory N, Fukunaga O, Kamo M, Kobashi K, Yoshikawa M (eds.). MYU, Tokyo, p. 311

[Bun55] Bundy F P, Hall H T, Strong H M, Wentorf R H (1955): Nature **176**, 51

[But76] Butuzov V P, Laptev V A, Presnov V A, Rotner Yu M (1976): Dokl. AN SSSR, Crystallography **226**, 328

[Cel91] Celii F G, White D, Purdes A J (1991): J. Appl. Phys. **70**, 5636

B. Dischler, *Handbook of Spectral Lines in Diamond*,
DOI 10.1007/978-3-642-22215-3,

[Cha59] Charette J J (1959): Physica **25**, 1303
[Che82] Cheljushkin A G, Laptev V A, Reznik B I, et al. (1982): In *Proc. IV. All-Union Symp. on Luminescent Detectors of X-Rays*. Irkutsk, USSR, p. 121 (in Russian)
[Che94b] Chen Chia-Fu, Chen Sheng-Hsiung, (1994b): In: Saito S, Fujimori N, Fukunaga O, Kamo M, Kobashi K, Yoshikawa M (eds) *Advances in New Diamond Science and Technology*. MYU, Tokyo, p. 709
[Chi92] Chin R P, Huang J Y, Shen Y R, Chuang T J, Seki H, Buck M (1992): Phys. Rev. B **45**, 1522
[Chr77] Chrenko R M, Tuft R M, Strong H M (1977): Nature **270**, 141
[Cla56a] Clark C D, Ditchbum R W, Dyer H B (1956a): Proc. R. Soc. London A **234**, 363
[Cla56c] Clark C D, Ditchbum R W, Dyer H B (1956c): Proc. R. Soc. London A **237**, 75
[Cla65] Clark C D (1965): *In* Berman R (ed.) *The Physical Properties of Diamond*, Clarendon Oxford, p. 295–324
[Cla70] Clark C D, Norris C A (1970): J. Phys. C **3**, 651
[Cla84b] Clark C D, Davey S T (1984b): J. Phys. C **17**, 1127
[Cla94] Clark C D, Dickerson C D (1994): In *Thin Film Diamond*. Lettington A, Steeds J W (eds.). Chapman & Hall, London p. 83–90
[Cla95] Clark C D, Kanda H, Kiflawi I, Sittas G (1995): Phys. Rev. B **51**, 16681
[Col68] Collins A T, Lightowlers E C (1968): Phys. Rev. 171, 843
[Col79a] Collins A T (1979a): Inst. Phys. Conf Ser. No. **46**, 327
[Col79b] Collins A T, Lightowlers E C (1979b): In *The Properties of Diamond*. Field J E (ed.). Academic Press, Oxford, p. 79–105
[Col79c] Collins A T, Raflque S (1979c): Proc. R. Soc. London A **367**, 81
[Col82b] Collins A T (1982b): J. Phys. D **15**, 1431
[Col82d] Collins A T, Mohammed K (1982d): J. Phys. C **15**, 147
[Col82e] Collins A T, Woods G S (1982e): Phil. Mag. B **45**, 385
[Col82f] Collins A T, Woods G S (1982f): Phil. Mag. **46**, 77
[Col83a] Collins A T (1983a): J. Phys. C **16**, 6691
[Col83b] Collins A T, Thomaz M F, Jorge M I B (1983b): J. Phys. C **16**, 5417
[Col83c] Collins A T, Spear P M (1983c): J. Phys. C **16**, 963
[Col85b] Collins A T, Stanley M (1985b): J. Phys. D **18**, 2537
[Col86b] Collins A T, Spear P M (1986b): J. Phys. C **19**, 6845
[Col87a] Collins A T (1987a): J. Phys. C **20**, 2027
[Col87b] Collins A T, Stanley M, Woods G S (1987b): J. Phys. C **20**, 969
[Col87c] Collins A T, Woods G S (1987c): J. Phys. C **20**, L797
[Col88a] Collins A T, Szechi J, Tavender S (1988a): J. Phys. C **21**, L161
[Col88c] Collins A T, Davies G, Kanda H, Woods G S (1988c): J. Phys. C **21**, 1363
[Col89a] Collins A T (1989): J. Phys. Cond. Matter **1**, 439
[Col89b] Collins A T, Lawson S C (1989b): J. Phys. Condens. Matter. **1**, 6929
[Col89c] Collins A T, Lawson S C (1989c): Phil. Mag. Letters **60**, 117
[Col89d] Collins A T, Kamo M, Sato Y (1989d):J. Phys. D **22**, 1402
[Col89e] Collins A T, Kamo M, Sato Y (1989e): J. Phys. Condensed Matter **l**, 4029
[Col90b] Collins A T (1990b): In *Science and Technology of New Diamond*. Saito S, Fukunaga O, Yoshikawa M (eds.). KTK Scientific Publishers/Terra Scientific Publishing Company, Tokyo, p. 273
[Col90c] Collins A T, Kamo M, Sato Y (1990c): J. Materials Sci. **5**, 2507
[Col90d] Collins A T, Lawson S C, Davies G, Kanda H (1990d): Phys. Rev. Letters **65**, 891
[Col90e] Collins A T, Kanda H, Burns R C (1990e): Phil. Mag. B **61**, 797
[Col90h] Collins A T (1990): In *Science and Technology of New Diamond*, Saito S, Fukunaga O, Yoshikawa M (eds.), KTK Scientific Publishers/Terra scientific Publishing Company, p.273–278
[Col91b] Collins A T (1991b): In *Proc. 2nd Int. Conf. On Diamond, Diamond-like and Related Coatings*. Nice, France
[Col92b] Collins A T (1992b): Diamond and Related Materials **1**, 457

[Col93b] Collins A T (1993b): Physica B **185**, 284

[Col93c] Collins A T (1993c): In *Proc. of the IEE Colloquium on Diamond in Electronics and Optics*. Digest -No. 1993/204, London, UK. p. 2/1

[Col93d] Collins A T, Woad P J (1993d): In Proc. Diamond Conference, Bristol, UK, p. 151

[Col93e] Collins A T, Woad P J, Woods G S, Kanda H (1993e): Diamond and Related Materials **2**, 136

[Col94c] Collins A T, Lightowlers E C, Higgs V, Allers L, Sharp S J (1994c): In *Advances in New Diamond Science and Technology*, Saito S, Fujimory N, Fukunaga O, Kamo M, Kobashi K, Yoshikawa M (eds.). MYU, Tokyo, p. 307

[Col97] Collins A T (1997): In Paoletti A, Tucciarone A (eds) *The Physics of Diamond*. Course 135, Proc. Int. School of Physics, IOS Press, Amsterdam, p. 273–354

[Col00] Collins A T (2000): Diamond and Related Materials **9**, 417

[Col02] Collins A T, Ly C-H (2002): J. Phys. Condensed Matter **14**, L-467

[Cre96] Cremades A, Piqueras J, Solis J (1996): J. Appl. Phys. **79**, 8118

[Cro67b] Crowther P A, Dean P J, Sherman W F (1967b): Phys. Rev. **154**, 772

[Cus52] Custers J F H (1952): Physica **18**, 489

[Dav76a] Davies G (1976a): J. Phys. C **9**, L 537

[Dav76b] Davies G Nazare M N, Hamer M F (1976b): Proc. R. Soc. London A **351**, 245

[Dav77c] Davies G (1977c): In *Chemistry and Physics of Carbon*, Marcel Dekker, New York, Chap. **13**, 1–143

[Dav79b] Davies G (1979b): J. Phys. C **12**, 2551

[Dav80c] Davies G, Nazare M H (1980c): J. Phys. C **13**, 4127

[Dav81b] Davies G, Foy C, O'Donnell K (1981b): J. Phys. C **14**, 4153

[Dav84b] Davies G, Collins A T, Spear P (1984b): Solid State Commun. **49**, 433

[Dav89] Davies G, Neves A J, Nazare M N (1989): Europhysics Letters **9**, 47

[Dav94a] Davies G (ed) (1994a): *Properties and Growth of Diamond*. INSPEC; The Institution of Electrical Engineers, London

[Dav94b] Davies G (1994b): *In*: [Dav94a], p 121–125

[Dav94d] Davies G (1994d): *In*: [Dav94a], p 135–138

[Dav99a] Davies G (1999a): Physica B **273–274**, 15

[Dav99b] Davies G, Collins A T (1999b): In *Proc. Diamond Conference 1999*, Oxford, UK, p. 16.1

[Dav00] Davies G, Smith H, Kanda H, (2000): Phys. Rev. B **62**, 1528

[Dav01a] Davies G (2001a): *In* [Naz01a], Chap. **A 7.3**, p. 193

[Dav01b] Davies G, Campbell B, Mainwood A, Newton M, Watkins M, Kanda H, Anthony T R (2001): phys. stat. sol. (a) **186**, 187

[Dea64c] Dean P J, Male J C (1964c): J. Phys. Chem. Solids **25**, 1369

[Dea65a] Dean P J (1965a): Phys. Rev. A **139**, 588

[Dea65b] Dean P J, Lightowlers E C, Wight D R (1965b): Phys. Rev. A 140, 352

[Dea73] Dean P J (1973), *Inter-Impurity Recombinations in Semiconductors*, In McCaldin J O, Somorjai G (eds.) *Progress in Solid State Chemistry*, Pergamon, Oxford, p 1–126

[Den67] Denham P, Lightowlers E C, Dean P J (1967): Phys. Rev. **161**, 762

[Den92] Deneuville A, Papadopoulo A C, Gheeraert E, Gonon P (1992), Le vide, les couches minces, suppl. 261,p.277

[Den93] Deneuville A, Gonon P, Gheeraert E, Collins A T, Khong Y L (1993):Diamond and Related Materials **2**, 737

[DeS77] deSa E S, Davies G (1977): Proc. R. Soc. London A **357**, 231

[Dis83] Dischler B, Bubenzer A, Koidl P (1983): Solid State Commun. **48**, 105

[Dis85] Dischler B, Sah R E, Koidl P, Fluhr W, Wokaun A (1985): *In*: Timmermans C J (ed) *Proc. 7th Internat. Symposium on Plasma Chemistry*, Eindhoven

[Dis87] Dischler B (1987): In Koidl P, Oelhafen P (eds) *Amorphous Hydrogenated Carbon Films*, Proc. European Mat. Res. Soc. **17**, p 189

[Dis92] Dischler B, Wild C, Müller-Sebert W, Biebl H, Koidl P (1992): *Infrared Analysis of CVD Diamond* (unpublished)

[Dis93] Dischler B, Wild C, Müller-Sebert W, Koidl P (1993): Physica B **185**, 217

[Dis94a] Dischler B, Rothemund W, Maier K, Wild C, Biebl H, Koidl P (1994a): Diamond Relat. Mater. **3**, 825

[Dis94b] Dischler B, Rothemund W, Wild C, Locher R, Biebl H, Koidl P (1994b): Phys. Rev. B **49**, 1685

[Dis98] Dischler B, Wild C (eds.) (1998): *Low-Pressure Synthetic Diamond*, Springer-Verlag, Berlin, Heidelberg

[Dis03] Dischler B (2003): *Interpretation of spectral line groups in diamond by donor-acceptor pair transitions, named DAP3-DAP116* (unpublished 2003 to 2010)

[Dol95] Dollinger G, Bergmaier A, Frey C M, Roesler M, Verhoeven H (1995): Diamond and Related Materials **4**, 591

[Dye65b] Dyer H B, du Preez L (1965b): J. Chem. Phys. **42**, 1898

[Enc94] Enckevort W J P van (1994): In *Synthetic Diamond.- Emerging CVD Science and Technology*. Spear K E, Dismukes J P (eds.). John Wiley & Sons Inc., New York, p. 307–353

[Ert95] Ertz R, Dötter W, Jung K, Ehrhardt H (1995): Diamond and Related Materials **4**, 469

[Est87] Estle T L, Estreicher S, Marynick D S (1987): Phys. Rev. Lett. **58**, 1547

[Eva79] Evans T (1979): In *The Properties of Diamond*. Field J E (ed.). Academic Press, London, p. 403

[Fal93] Fallon P J, Brown L M (1993): Diamond Relat. Mater. **2**, 1004

[Far69] Farrer R G (1969): Solid State Comm. **7**, 685

[Far72] Farrer R G, Venneulen L A (1972): J. Phys. C **5**, 2762

[Far74] Farrer R G (1974): PhD thesis, University of Witwatersrand, Johannesburg; spectra reproduced in [Wal79], Fig. 9

[Fen93] Feng T, Schwartz B D (1993): J. Appl. Phys. **73**, 1415

[Fer96] Ferrer N, Nogues-Carulla J M (1996): Diamond Relat. Mater. **5**, 598

[Fie92] Field J. E. (ed.). (1992): *The Properties of Natural and Synthetic Diamond*. Academic Press, London

[Fil92] Filipp A R, Tkachev V V, Varichenko V S, Zaitsev A M, Chelyadinskii A R, Kluev Yu A (1992): Diamond Relat. Mater. **1**, 271

[Fin56] Finkelnburg W (1956): *Einführung in die Atomphysik*. Springer-Verlag, Berlin, Heidelberg

[Fin84] Fink J, Müller-Heinzerling T, Pflüger J, Scheerer B, Dischler B, Koidl P, Bubenzer A, Sah R E (1984) Phys. Rev. B **30**, 4713

[Fre90] Freitas J A Jr., Butler J E, Bishop S G, Carrington W A, Strom U (1990): In *Diamond, Silicon Carbide and Related Wide Bang Gap Semiconductors*. Glass J T, Messier R, Fujimori N (eds.). MRS Symp. Proc. **162**, p. 237

[Fre93a] Freitas J A, Strom U, Collins A T (1993a): Diamond and Related Materials **2**, 87

[Fre93b] Freitas J A, Klein P B, Collins A T (1993b): Electronic Letters **29**, 1727

[Fre94a] Freitas J A, Klein P B, Collins A T (1994a): Appl. Phys. Letters **64**, 2136

[Fre94c] Freitas J A, Doverspike K, Klein P B, Khong Y L, Collins A T (1994c): Diamond and Related Materials **3**, 821

[Fri91a] Fritsch E, Scarrat K, Collins A T (1991): In: *New Diamond Science and Technology*. Mater. Res. Soc. Conf. Proc., p. 671

[Fuc95a] Fuchs F, Wild C, Schwarz K, Mueller-Sebert W, Koidl P (1995a): Appl. Phys. Letters 66, 177

[Fuc95b] Fuchs F, Wild C, Schwarz K, Koidl P (1995): Diamond Relat. Mater. **4**, 652

[Fuc96] Fuchs F, private communication

[Ger71] Gerasimenko N N, Lezhejko L V (1971): Synthetic Diamonds, **3**, 19 (in Russian)

[Ghe92b] Gheeraert E, Deneuville A (1992b): Diamond Relat. Mater. **1**, 584

[Gip81] Gippius A A, Vavilov V S, Zaitsev A M (1981): Inst. Phys. Conf. Ser. No. 59, Chapter 5, p. 253

[Gip82a] Gippius A A, Zaitsev A M, Vavilov V S (1982a): Sov. Phys. Semicond. **16**, 256

[Gip83] Gippius A A, Vavilov V S, Zaitsev A M, Zhacupbekov B S (1983): Physica B **116**, 187

[Gip92] Gippius A A (1992): Materials Sci. Forum **83–87**, 1219

[Gip93b] Gippius A A, Collins A T (1993b): Solid State Commun. **88**, 637

[Gip95b] Gippius A A, Kazarian S A (1995b): In *Proc. Diamond Conference*, Bristol

[Gom60a] Gomon G O (1960a): Optics and Spectroscopy **81**, 275 (in Russian)

[Gor98] Gorokhovsky A, Zaitsev A M, (1998): *"Photoluminescence Studies of Xe-related Optical Center in Diamond"*, to be published

[Gos99] Goss J P, Jones R, Latham C D, Briddon P R, Oeberg S (1999): In *Proc. Diamond Conference 1999*, Oxford, UK, p. 9.1

[Gos01a] Goss J P, Jones R, Shaw T D, Rayson M J, Briddon P R (2001): phys. stat. sol. (a) **186**, 215

[Gra94] Graham R J, Buseck P R (1994): Phil. Mag. B **70**, 1177

[Gri99] Grill A (1999): Diamond and Related Materials **8**, 428

[Har96] Harris S J, Weiner A M, Prawer S, Nugent K (1996): J. Appl. Phys. **80**, 2187

[Hay97] Hayashi K, Watanabe H, Yamanaka S, Sekiguchi T, Okushi H, Kajimura K (1997a): Mat. Res. Soc. Symp. Proc. **442**, 699

[Hei97] Heiderhoff R. (1997): *Analyse der elektronischen Eigenschaften von Diamant mittels rastermikroskopischer Methoden.* PhD thesis, University of Wuppertal (in German)

[Hop63] Hopfield J J, Thomas D G, Gershenzon M (1963): Phys. Rev. Lett. **10**, 162

[Iak99d] Iakoubovskii K, Adriaenssens G J (1999d): Characterization of the broad green band luminescence in CVD and synthetic Ib diamond. (personal communication), appeared in (2000) Diamond Relat. Mater. 9, 1017

[Iak00a] Iakoubovskii K, Adriaenssens G J (2000a): Phys. Rev. B **61**, 10174

[Iak00d] Iakoubovskii K, Adriaenssens G J (2000d): Diamond Relat. Mater. **9**, 1017

[Iak00g] Iakoubovskii K, Stesmans A, Nouwen B, Adriaenssens G J (2000g): Phys. Rev. B **62**, 16587

[Iak01] Iakoubovskii K, Adriaenssens G J (2001): J. Phys. Condensed Matter **13**, 6015

[Ily71a] Ilyin V E, Sobolev E V, Yureva O P (1971a): Sov. Phys. Solid State **12**, 1721

[Ily71b] Ilyin V E, Yurjeva 0 P, Sobolev E V (1971b): J. Appl. Spectroscopy **14**, 158

[Iso90a] Isoya J, Kanda H, Uchida Y (1990a): Phys. Rev. B **42**, 9843

[Jan91] Janssen G, Vollenberg W, Giling J, van Enckevort W J P, Schaminee J J D, Seal M (1991): Surf. Coatings Technol. **47**, 113

[Jan92] Janssen G, van Enckevort W J P, Vollenberg W, Giling L E (1992): Diamond Relat. Mater. **1**, 789

[Joh94] John P, Milne D K, Drummond I C, Jubber M G, Wilson J I B, Savage J A (1994): Diamond Relat. Mater. **3**, 486

[Joh02] Johnston K, Mainwood A (2002): Diamond and Related Materials **11**, 631

[Jon92] Jones R, Briddon P R, Oeberg R (1992): Phil. Mag. Letters **66**, 67

[Jor83] Jorge M I B, Pereira M E, Thomaz M F, Davies G, Collins A T (1983): Portugal Phys. **14**, 195

[Jou96] Joubert D P, Li L, Lowther J E (1996): Solid State Commun. **100**, 561

[Kam82] Kamo M, Tsutsumi M, Sato Y, Setaka N (1982): 43rd Japan Appl. Phys. Soc. Fall Meeting

[Kan98] Kanda H, Watanabe K (1998): *Distribution of Nickel Related Luminescence Centers in HPHT Diamond.* Proc. ICNDST-6, Pretoria, 1998

[Kan99] Kanda H, Watanabe K (1999): *Cobalt-Related Photoluminescence in Relation to Nitrogen Aggregation of high Pressure Synthetic Diamond.* In: *Proc. Int. Conf. ADC/FCT '99*, Yoshikawa M, Koga Y, Tzeng Y, Klages C P, Miyoshi K (eds.). Tsukuba, Japan

[Kho93] Khong Y L, Collins A T (1993): Diamond and Related Materials **2**, 1

[Kho94] Khong Y L, Collins A T, Allers L (1994): Diamond and Related Materials **3**, 1023

[Kif99] Kiflawi I, Davies G, Fisher D, Kanda H (1999): Diamond and Related Materials **8**, 1576

[Kif00] Kiflawi I, Bruley (2000): Diamond and Related Materials **9**, 87

[Kle92] Klein C A, Hartnett T M, Robinson C J (1992): Phys. Rev. B **45**, 12854

[Kli75b] Klimenkova N T, Rotner Yu M, Presnov V A (1975b): Synthetic Diamonds, No. **3**, p. 11 (in Russian)

[Kli75c] Klimenkova N T, Prokopchuk E O, Laptev V A, Rotner Yu M (1975c): Optics and Spectroscopy **39**, 333 (in Russian)

[Kop86] Koppitz J, Schirmer O F, Seal M (1986): J. Phys. C **19**, 1123

[Kup99] Kupriyanov l A Gusev V A, Borzdov Yu M, Kalinin A A, Palyanov Yu N (1999): Diamond and Related Materials, **8**, 1301

[Lan68] Lanoo M, Stoneham A M (1968): J. Phys. Chem. Solids **29**, 1987

[Law92b] Lawson S C, Davies G, Collins A T, Mainwood A (1992b): J. Phys. Condens. Matter **4**, 3439

[Law93a] Lawson S C, Kanda H (1993a): Diarnond and Related Materials **2**, 130

[Law93b] Lawson S C, Kanda H (1993b): J. Appl. Phys. **73**, 3967

[Law93c] Lawson S C, Kanda H, Sekita M (1993c): Phil. Mag. **68**, 39

[Law95] Lawson S C, Kanda H, Kiyota H, Tsutsumi T, Kawarada H (1995): J. Appl. Phys. **77**, 1729

[Law96] Lawson S C, Kanda H, Watanabe K, Kiflawi I, Sato Y, Collins A T (1996): J. Appl. Phys. **79**, 4348

[Law98] Lawson S C, Fisher D, Hunt D C, Newton M E (1998): J. Phys. Condens. Matter **10**, 6171

[Lev79] Levinson J, Halperin A (1979): J. of Luminescence **18/19**, 749

[Mai78] Mainwood A (1978): J. Phys. C **11**, L 449

[Mai94a] Mainwood A (1994a): Phys. Rev. **49**, 7934

[Mai94b] Mainwood A, Collins A T, Woad P (1994b): Materials Sei. Forum **143/147**, 29

[Mai97] Mainwood A, Stoneham A M (1997): J. Phys., Condens. Matter **9**, 2453

[Mai01] Mainwood A (2001): in [Naz01] Chap. A **7.2**, p. 188

[Man95] Manfredotti C, Wang F, Polesello P, Vittone E, Fizzotti F (1995): Appl. Phys. Letters **67**, 3376

[Mat81] Matsumoto S, Sekata N (1981): 8th Japan Carbon Soc. Fall Meeting

[Maz80] Mazzaschi J, Brabant J C, Brousseau M, Viollot F (1980): Revue Phys. Appl. **15**, 4 (in French)

[McC94a] McCauley T. S., Vohra Y. K. (1994): In *Advances in New diamond Science and Technology*, Saito S, Fujimori N, Fukunaga O, Kamo M, Kobashi K, Yoshikawa M (eds.), MY, Tokyo, p. 371

[McC95] McCauley T. S., Vohra Y. K. (1995): In *Applications of Diamond Films and Related Materials*. 3rd Int. Conf., Feldman A, Tzeng Y, Yarbrough W A, Yoshikawa M, Murakawa M (eds.). p. 377

[Mel96] Melnikov A A, Zaitsev A M, Schreck M (1996): *CL Characterization of CVD Diamond Films*. Internal report, Fern-Uni Hagen, Dept. of Electronics (unpublished)

[Mel98] Melnikov A A, Denisenko A V, Zaitsev A M, Shulenkov A, Varichenko V S, Philipp A R, Dravin V A, Kanda H, Fahrner W R (1998): J. Appl. Phys. **84**, 6127

[Mey98] Meyer H O A, Seal M (1998):*Natural Diamond. In*: Prelas M A, Popovici G, Bigelow L K (eds): *Handbook of Industrial Diamonds and Diamond Films*, Marcel Dekker, New York, Chap. 10, pp 481–526

[Mil95] Miller J B, Brown D W (1995): Diamond and Related. Materials 4, 435

[Mit90] Mita Y, Nisida Y, Suito K, Onodera A, Yazu S (1990): J. Phys. Condens. Matter **2**, 8567

[Moh82b] Mohamrned K, Davies G, Collins A T (1982b): J. Phys. **C15**, 2789

[Mor92a] Mori Y, Eimori N, Kozuka H, Yokata Y, Moon J, Ma J S, Ito T, Hiraki A (1992a): Appl. Phys. Letters **60**, 47

[Mül96] Müller-Sebert W, Wörner E, Fuchs F, Wild C, Koidl P (1996): Appl. Phys. Lett. **68**, 59

[Nad93] Nadolinny V A, Yelisseyev A P (1993): Diamond and Related Materials **3**, 17

[Nad99] Nadolinny V A, Yelisseyev A P, Baker J M, Newton M E, Twitchen D J, Lawson S C, Yuryeva O P, Feigelson B N (1999): J. Phys. Condens. Matter **11**, 7357

[Nai98] Naidoo S R, Prins J F (1998): *"Electroluminescence from Electron Injection Junctions Created by Carbon and Phosphorous Ion Implantation"*, Int. Conf. ICNDST-6, Pretoria, South Africa

[Naz85a] Nazare M H, Thomas M F, Jorge M I B (1985a) Solid State Comm. **55**, 577

[Naz87] Nazare M H, Neves A J (1987): J. Phys. C **20**, 2713

[Naz91b] Nazare M H, Neves A J, Davies G (1991b): Phys. Rev. B **43**, 14196

[Naz92] Nazare M H, Woods G S, Assuncao M C (1992): Materials Sci. Engin. B **11**, 341

[Naz93c] Nazare M H, Rino L M (1993c): Materials Sci. Engin. B **21**, 329

[Naz94a] Nazare M H, Lopes J C, Kanda H (1994a): Mat. Res. Symp. Proc. **339**, 625

[Naz01a] Nazare M H, Neves A J (eds.) (2001): *Properties, Growth and Applications of Diamond*, IEE-INSPEC (London), Datareview **26**

[Ned72] Nedzvetskii D S, Gaisin V A (1972): Solid State Physics **14**, 2946

[New01a] Newton M E (2001): In [Naz01a], Ch A5.4, p. 136

[New01b] Newton M E, Baker J M, Twitchen D J (2001): In [Naz01a], Ch A7.4, p. 204

[New01c] Newton M E, Baker J M, Twitchen D J (2001): *EPR measurement on V_1 and V_2*, In [Naz01a], Ch A7.4, p. 204

[Nij97] Nijenhuis J te, Olsthoorn S M, van Enckevort W J P, Giling L J (1997): J. Appl. Phys. **82**, 419

[Nis89] Nishida Y, Mita Y, Mori K, Okuda S, Sato S, Yazu S, Nakagawa M, Okada M (1989): Materials Sci. Forum **38–41**, 561

[Nov68] Novikov N N (ed. in chief) (1968): *Synthetic Superhard Materials: Vol. 1 Synthesis of Superhard Materials*, Kiev, Naukova Dumka

[Orl73a] Orlov Yu L (1973a):*Mineralogy of Diamond*. Izd. Acad. of Sci. of the USSR, Moscow (in Russian)

[Osv97] Osvet A, Yelisseyev A P, Feigelson B N, Mironova N A, Sildos I (1997): Radiation Effects and Defects in Solids **146**, 339

[Pen00] Pennink A (2000): *"Die Diamantenmacher"*, German version of a BBC film on HPHT synthetic diamonds, broadcast on German N3 TV in November 2000

[Per86] Pereira M E, Jorge M I B, Thomaz M F (1986): J. Phys. C **19**, 1009

[Per88] Pereira E, Santos L (1988): J. Luminescence **40/41**, 139

[Per93] Pereira E, Santos L (1993): Physica B **185**, 222

[Per94b] Pereira E, Santos L (1994b): Diamond and Related Materials **4**, 26

[Per94c] Pereira E (1994c): *Tabulation of luminescence decay times in diamond* in [Dav94a], p. 228–232

[Pha65] Phaal C (1965): Phil. Mag. 11, 369

[Phi92] Phillips R, Wei J, Tzeng Y (1992): Thin Solid Films **212**, 30

[Pop96] Popovici G, Melnikov A A, Varichenko V S, Khasawinah S, Sung T, Prelas M A, Denisenko A V, Penina N M, Martinovich V A, Drozdova E N, Zaitsev A M, Farner W R, Farmer J R, White H, Chamberlain J (1996): Diamond and Related Materials **5**, 761

[Pre98] Prelas M A, Popovici G, Bigelow L K (eds): *Handbook of Industrial Diamonds and Diamond Films*, Marcel Dekker, New York

[Pri94] Prins J F (1994a): Diamond and Related Materials **3**, 922

[Rei98] Reinitz I M, Fritsch E, Shigley J E (1998): Diamond Relat. Mater. **7**, 313

[Rob34] Robertson R, Fox J J, Martin A E (1934): Phil. Trans. R. Soc. London A **232**, p. 463–535

[Rob93] Robins L H, Farabaugh E N, Feldmann A (1993): Phys. Rev. B **48**, 14167

[Ron99] Ronning C, Hofsaess H (1999): Diamond and Related Materials **8**, 1623

[Rot83] Rotner S M, Rotner Yu M., Krishchuk G V, Khrakovskaya E M, Stepanova N S, Laptev VA (1983): Sov. Phys. Semicond. **17**, 128

[Rua91b] Ruan J, Choyke W J, Partlow W D (1991b): J. Appl. Phys. **69**, 6632

[Rua92b] Ruan J, Kobashi K, Choyke W J (1992b): Appl. Phys. Letters **60**, 3138

[Rua93a] Ruan J, Choyke W J, Kobashi K (1993a): Appl. Phys. Letters **62**, 1379

[Rua93b] Ruan J, Choyke W J (1993b): American Ceramic Soc. Bulletin **72**, 139

[Ruf98] Ruf T, Cardona M, Stemschulte H, Wahl S, Thonke K, Sauer R, Pavone P, Anthony T R (1998): Solid State Communication **105**, 311

[Run71] Runciman W A, Carter T (1971): Solid State Commun. **9**, 315

[Ruo91a] Ruoff A L, Vohra Y K, Desgreniers S (1991a): In *New Diamond Science and Technology*. Mat. Res. Soc. Conf. Proc., p. 645

[San89] Sandhu G S, Swanson M L, Chu W K (1989): Appl. Phys. Letters **55**, 1397

[San90] Santos L, Pereira E (1990): In *Diamond, SiC and Related Wide Gap Semiconductors*. Proc. Materials Res. Scic. Symp. **162**, p. 291

[San93] Sangster M J L, Kiflawi I, Woods G S (1993): Diamond and Relat. Mater. **2**, 1247

[San94] Santos L, Pereira E (1994): J. of Luminescence **60/61**, 614

[Sch83] Schmid W, Nieper U, Weber J (1983): Solid State Commun. **45**, 1007

[Sch95] Schneider H., Dischler B., Wild C., Koidl P. (1995): Phys. Rev. B **51**, 16677

[Set94] Setaka N (1994): *In*: Spear H E, Dismukes J P (eds): *Synthetic Diamond*, Wiley, New York, Chap. 4, pp. 57–90

[Shi92] Shigley J E, Fritsch E, Reinitz I, Moon M (1992): Gems Gemmol. **28**, 116

[Shi02] Shirey S (2002): Science **297**, 1683

[Sil94a] Sildos I, Osvet A (1994a): Diamond and Related Materials **3**, 725

[Sil94b] Sildos I, Osvet A (1994b): In *Advances in New Diamond Science and Technology*. Saito S, Fujimory N, Fukunaga O, Kamo M, Kobashi K, Yoshikawa M (eds.). MYU, Tokyo, p. 395

[Sil95] Sildos I, Zavt G, Osvet A (1995): In *Wide Bandgap Electronic Materials*. Prelas M A, Gielisse P, Popovici G, Spitsyn B V, Stacy T (eds.). NATO ASI Series, Kluwer Academic Publishers, p. 89

[Sit95] Sittas G, Kanda H, Kiflawi I, Spear P M (1995): *Growth and Characterization of Si-doped Diamond Single Crystals Grown by the HTHP Method*, Int. Conf Diamond Films-95, Barcelona

[Sit96] Sittas G, Kanda H, Kiflawi I., Spear P M (1996): Diamond and Related Materials **5**, 866

[Smi62] Smith S D, Taylor W (1962): Proc. Phys. Soc. London **79**, 1142

[Sob69c] Sobolev E V, Ilin V E, Yureva O P (1969c): Sov. Phys. Solid State **11**, 938

[Sob68d] Sobolev E V, Ilyin V E., Lenskaya S V, Yureva O P (1968d): J. Appl. Spectroscopy **9**, 1108

[Sob76] Sobolev E V, Yeliseev A P (1976): J. of Struct. Chemistry **17**, 933 (in Russian)

[Sob79a] Sobolev E V, Dubov Yu I (1979a): Synthetic Diamonds, No. 2, 3 (in Russian)

[Sol72] Solin S A (1972): Physics Letters A **38**, 101

[Spe94] Spear H E, Dismukes J P (eds): *Synthetic Diamond*, Wiley, New York

[Ste94] Sternschulte H, Thonke K, Sauer R, Muenzinger P C, Michler P (1994): Phys. Rev. B **50**, 14554

[Ste95] Sternschulte H, Thonke K, Gerster J, Limmer W, Sauer R, Spitzer J, Muenzinger P C (1995): Diamond and Related Materials **4**, 1189

[Ste96a] Stemschulte H, Horseling J, Albrecht T, Thonke K, Sauer R (1996a): Diamond and Related Materials **5**, 585

[Ste96b] Sternschulte H, Albrecht T, Thonke K, Sauer R, Griesser M, Grasserbauer M (1996b): MRS Symp. Proc. **423**, 693

[Ste97a] Sternschulte H, Wahl S, Thonke K, Sauer R, Dalmer M, Ronning C, Hofsäss H, (1997a): Appl. Phys. Letters **71**, 2668

[Ste97b] Sternschulte H, Wahl S, Thonke K, Sauer R, Ruf T, Cardona M, Anthony T R (1997b): Materials Sci. Forum **258–263**, 757

[Ste99a] Sternschulte H, Thonke K, Sauer R (1999a): Phys. Stat. Solidi (a) **172**, 37

[Ste99b] Sternschulte H, Thonke K, Sauer R (1999b): Phys. Rev. B **59**, 12924

[Ste99c] Steeds J W, Davis T J, Charles S J, Hayes J M, Butler J E (1999c): In *Proc. Diamond Conference 1999*, Oxford, UK, p. 26.1

[Ste99d] Steeds J W, Gilmore A, Charles S, Heard P, Howarth B, Butler J E (1999d): Acta mater **47**, 4025

[Sto92] Stoneham A M (1992), *Recent Advances in Theory*, In: Field J E (ed.): *The Properties of Natural and Synthetic Diamond*, Academic Press, London, p. 334

[Sum88] Sumida N, Lang A R (1988): Proc. R. Soc. London A **419**, 235

[Tag87] Taguchi T, Kawaguchi Y, Otera H, Hiraki A (1987): Jap. J. Appl. Phys. **26**, 1923

[Tan01] Tanabe K, Nakazawa K, Susantyo J, Kawarada H, Koizumi S (2001): Diamond and Related Materials **10**, 1652

[Tho72] Thompson F, Nicklin R. (1972): J. Phys. C: Solid State Phys. **5**, L 223

[Tho00] Thonke K, Schliesing R, Teofilov N, Zacharias H, Sauer R, Zaitsev A M, Kanda H, Anthony T R (2000): Diamond and Related Materials **9**, 428

[Tka85b] Tkachev V D, Zaitsev A M, Tkachev V V (1985b): Phys. Stat. Solidi b **129**, 129

[Twi99] Twitchen D J, Hunt D C, Smart V, Newton M E, Baker J M (1999): Diam. Relat. Mater. **8**, 1572

[Twi01] Twitchen D J, Newton M E, Baker J M, Nadolinny V A (2001): In *Properties, Growth and Applications of Diamond*, Nazare M H and Neves A J (eds.), IEE-INSPEC Datareviews **26**, p. 214

[Vac75c] Vachidov Sh A., Nurullaev E, Samoilovich M I (1975b): Dokl. AN Uzbek SSR, No. 6, p. 22 (in Russian)

[Var86a] Varichenko V S, Zaitsev A M, Stelmakh V F (1986a): Phys. Stat. Solidi (a) **95**, K25

[Var88] Varichenko V S, Didyk A Yu, Zaitsev A M, Kuznetsov V I, Kulakov V M, Melnikov A A, Plotnikova S P, Skuratov V A, Stelmakh V F, Shestakov V D (1988): Preprint JINR No. R14–88–44, Dubna (in Russian) 1–12

[Var89] Varichenko V S, Dobrinets I A, Dravin V A, Zaitsev A M (1989): J. of Appl. Spectroscopy (Minsk, Belarus) **51**, 218 (in Russian)

[Ver74] Vermeulen L A, Clark C D, Walker J (1974): *Lattice Defects in Semiconductors*, Inst. Phys. Conf. Ser. (Bristol), **23**, 294

[Ver75] Vermeulen L A, Farrer R G (1975): Diamond Research **18**, 62

[Vin88a] Vins V. G. (1988a): *Spectroscopy of Optically Active Defects in Synthetic Diamonds*. PhD thesis, Novosibirsk (in Russian)

[Vin88b] Vins V G, Yeliseev A P, Malogolovets V G (1988b): Superhard Materials, **4**, 18

[Wag89] Wagner J, Ramsteiner M, Wild Ch, Koidl P (1989): Phys. Rev. B **40**, 1817

[Wal79] Walker J. (1979): *Optical Absorption and Luminescence in Diamond, In* Rep. Prog. Phys. 42, pp 1605–1659

[Wei73] Weigel C, Peak D, Corbett J W, Watkins G D, Messmer P (1973): Phys. Rev. B **8**, 2906

[Wel94] Welbourn C M (1994): In *Advances in New Diamond Science and Technology*. Saito S, Fujimory N, Fukunaga O, Kamo M, Kobashi K, Yoshikawa M (eds.). MYU, Tokyo, p. 327

[Wig67] Wight D R, Dean P J (1967): Phys. Rev. **154**, 689

[Wig71] Wight D R, Dean P J, Lightowlers E C, Mobsby C D (1971): J. Luminscence **4**, 169

[Wil70] Wiley J D, Seman J A (1970): Bell Syst. Tech. J. **50**, 355

[Wil89] Wild C, Herres N, Wagner J, Koidl P, Antony T R (1989): Proc. Electrochem. Soc. **89**, 283

[Wil94] Wild C, Kohl R, Herres N, Müller-Sebert W, Koidl P (1994): Diamond Relat. Mater. **3**, 373

[Wol-10] Wolfer M, Obloh H, Williams O A, Leancu C-C, Kirste L, Gheeraert E, Nebel C E (2010): Phys. Status Solidi A **207**, 2054

[Woo83] Woods G S, Collins A T (1983): J. Phys. Chem. Solids **44**, 471

[Woo86] Woods G S, Collins A T (1986): J. Gemology **20**, 75

[Yac91] Yacobi B G, Badzian A R, Badzian T (1991): J. Appl. Phys. **69**, 1643

[Yel92a] Yelisseyev A P, Nadolinny V A (1992a): Reports Russian Acad. of Sci. **326**, 524 (in Russian)

[Yel92b] Yelisseyev A P, Rylov G M, Fedorova E N, Vins V G, Feigelson B N, Sharapov A P, Ulanov N E, Zhiltsov A N, Kungurov A I. (1992b): Joint Inst. of Geology, Geophysics and Mineralogy, Preprint No. 5, Novosibirsk (in Russian)

[Yel95a] Yelisseyev A P, Nadolinny V A (1995a): Diamond and Related Materials **4**, 177

[Yel96] Yelisseyev A P, Nadilinny V A, Feigelson B, Terentyev S, Nosukhin S (1996): Diamond and Related Materials **5**, 1113

[Yel99] Yelisseyev A P, Lawson S, Osvet A, Nadolinny V, Sildos I, Feigelson B, Baker M, Newton M, Twitchen D (1999): In *Proc. Diamond Conference 1999*, Oxford, UK, p. 35.1

[Yok92] Yokota Y, Kotsuka H, Sogi T, Ma J S, Hiraki A, Kawarada H, Matsuda K, Hatada M (1992): Diamond and Related Materials **1**, 470

[Yok99] Yokota Y, Tachibana T, Miyata K, Hayashi K, Kobashi K, Hatta A, Ito T, Hiraki A, Shintani Y (1999): Diamond and Related Materials **8**, 1587

[Zai79a] Zaitsev A M, Gippius A A (1979a): *Cathodoluminescence of diamond Implanted with Transition Metal Ions*. (unpublished)

[Zai80b] Zaitsev A M, Gippius A A, Vavilov V S (1980b): J. of Experim. and Theor. Phys. **31**, 181 (in Russian)

[Zai82] Zaitsev A M, Gippius A A, Vavilov V S (1982): Sov. Phys. Semicond. **16**, 252

[Zai88a] Zaitsev A M, Tkachev V V (1988a): In *Proc. All-Union Conference on Advanced Technology of Semiconducting Materials*, Odessa University, Odessa, p. 138 (in Russian)

[Zai88c] Zaitsev A M, Gippius A A (1988c): *Piezospectroscopy of nitrogen related centers in ion implanted diamonds*. (unpublished)

[Zai89] Zaitsev D M, Zaitsev A M (1989): *CL Studies of Neutron Irradiated Natural Diamond*. (in Russian, unpublished)

[Zai92a] Zaitsev A M (1992a): *Luminescence of Ion-Implanted Diamond and Cubic Boron Nitride*, Dr. Sc. thesis, University of Minsk, p. 1–367 (in Russian)

[Zai95] Zaitsev A M, Melnikov A A, Denisenko A V, Varichenko V S, Job R, Fahmer W R (1995): In *Proc. Fall MRS Meeting, DD Symp.* Boston, p. DD3.5

[Zai96a] Zaitsev A M, Kanda H, Raiko V (1996a): *Raman Studies of ^{12}C and ^{13}C Diamonds implanted with ^{12}C and ^{13}C High Energy Ions*. (unpublished)

[Zai96b] Zaitsev A M, Filipp A R, Collins A T (1996b): *Cathodoluminescence of CVD diamond films irradiated with high energy ions*. (unpublished)

[Zai97a] Zaitsev A M, Melnikov A A (1997a): *Cathodoluminecsence Investigations of Thermochemically Treated CVD Diamond*. Belarussian State University. (unpublished)

[Zai98] Zaitsev A M (1998): *Optical Properties of Diamond*. In: Prelas M A, Popovici G, Bigelow L K (eds): *Handbook of Industrial Diamonds and Diamond Films*, Marcel Dekker, New York, Chap. 7, pp 227–376

[Zai98b] Zaitsev A M, Denisenko A V., Fahrner W R (1998b): *Diamond Electronics: Where to Go*. Tutorial lecture, Int. Conf on Diamond Technology, La Jolla, CA

[Zai98d] Zaitsev A M, Denisenko A V, Melnikov A A, Varichenko V S, Fahrner W R (1998d): *In SPIE Conf. on Light Emitting Diodes: Research, Manufacturing, and Applications* 11, San Jose, CA.

[Zai00a] Zaitsev A M (2000a): Phys. Rev. B **61**, 12909

[Zai00b] Zaitsev A M, Meijer J, Burchard M (2000b): *Raman and Photoluminescence Spectroscopy of Diamond Heavily Irradiated with Protons*. (to be published)

[Zai01] Zaitsev A M (2001): *Optical Properties of Diamond*, Springer-Verlag, Berlin, Heidelberg

[Zai10] Zaitsev A M (2010) (private communication) Dobrinets I, Zaitsev A M: *HPHT Treated Diamonds*, Springer-Verlag, Berlin, Heidelberg (in preparation)

[Zha96a] Zhang B, Wang X, He J, Liu D, Wang J (1996a): Chin. Sci. Bulletin 41, 1075

Index

B. Dischler, *Handbook of Spectral Lines in Diamond*,
DOI 10.1007/978-3-642-22215-3, © Springer-Verlag Berlin Heidelberg 2012

MIX
Papier aus verantwortungsvollen Quellen
Paper from responsible sources
FSC® C105338

If you have any concerns about our products,
you can contact us on
ProductSafety@springernature.com

In case Publisher is established outside the EU,
the EU authorized representative is:
Springer Nature Customer Service Center GmbH
Europaplatz 3, 69115 Heidelberg, Germany

Printed by Libri Plureos GmbH
in Hamburg, Germany